Derek J. S. Robinson
Abstract Algebra

Also of Interest

Abstract Algebra
Applications to Galois Theory, Algebraic Geometry, Representation
Theory and Cryptography
Celine Carstensen-Opitz, Benjamin Fine, Anja Moldenhauer,
Gerhard Rosenberger, 2019
ISBN 978-3-11-060393-4, e-ISBN (PDF) 978-3-11-060399-6,
e-ISBN (EPUB) 978-3-11-060525-9

The Structure of Compact Groups
A Primer for the Student – A Handbook for the Expert
Karl H. Hofmann, Sidney A. Morris, 2020
ISBN 978-3-11-069595-3, e-ISBN (PDF) 978-3-11-069599-1,
e-ISBN (EPUB) 978-3-11-069601-1

Commutative Algebra
Aron Simis, 2020
ISBN 978-3-11-061697-2, e-ISBN (PDF) 978-3-11-061698-9,
e-ISBN (EPUB) 978-3-11-061707-8

Quantum Mechanics
An Introduction to the Physical Background and Mathematical
Structure
Gregory L. Naber, 2021
ISBN 978-3-11-075161-1, e-ISBN (PDF) 978-3-11-075194-9,
e-ISBN (EPUB) 978-3-11-075204-5

Algebraic Quantum Physics
Volume 1 Quantum Mechanics via Lie Algebras
Arnold Neumaier, Dennis Westra, 2022
ISBN 978-3-11-040610-8, e-ISBN (PDF) 978-3-11-040620-7,
e-ISBN (EPUB) 978-3-11-040624-5

Derek J. S. Robinson

Abstract Algebra

An Introduction with Applications

3rd, extended edition

DE GRUYTER

Mathematics Subject Classification 2020
1101, 1201, 1301, 1501, 1601, 1801, 2001

Author
Prof. Dr. Derek J. S. Robinson
Department of Mathematics,
University of Illinois at Urbana-Champaign,
1409 West Green Street
Urbana, IL 61801,
USA
dsrobins@illinois.edu

NC 01.19.2022 1226

ISBN 978-3-11-068610-4
e-ISBN (PDF) 978-3-11-069116-0
e-ISBN (EPUB) 978-3-11-069121-4

Library of Congress Control Number: 2021947896

Bibliographic information published by the Deutsche Nationalbibliothek
The Deutsche Nationalbibliothek lists this publication in the Deutsche Nationalbibliografie;
detailed bibliographic data are available on the Internet at http://dnb.dnb.de.

© 2022 Walter de Gruyter GmbH, Berlin/Boston
Cover image: thinkstock
Typesetting: VTeX UAB, Lithuania
Printing and Binding: LSC Communications, United States

www.degruyter.com

In Memory of My Parents
Stanley Scott Robinson, Helen Annan Hardie

Preface

The origins of algebra can be traced back to Muhammad ben Musa al-Khwarizmi, who worked at the court of the Caliph al-Ma'mun in Baghdad in the early 9th Century. The word derives from the Arabic al-jabr, which refers to the process of adding the same quantity to both sides of an equation. The work of Arabic scholars was known in Italy by the 13th Century and a lively school of algebraists arose there. Much of their attention was centered on the solution of polynomial equations. This preoccupation of mathematicians lasted through the beginning of the 19th Century, when the possibility of solving the general equation of the fifth degree in terms of radicals was finally disproved by Ruffini and Abel.

This early work led to the introduction of some of the main concepts of abstract algebra: groups, rings and fields. These structures have been studied intensively over the past two hundred years. For an interesting historical account of the origins of algebra the reader may consult the book by van der Waerden [21].

Until recently algebra was very much the domain of the pure mathematician, and its applications were few. But the situation has changed, in part as a result of the rise of information technology, where the precision and power inherent in the language and concepts of algebra have proved to be invaluable. Today many students of computer science and engineering, as well as physics and chemistry, take courses in abstract algebra. The present work represents an attempt to meet the needs of both mathematicians and scientists who seek to acquire a basic knowledge of algebra.

The book should be suitable for students in the final year of undergraduate or initial years of (post)graduate studies at a university in North America or the United Kingdom. What is expected of the reader is a knowledge of matrix algebra and at least the level of mathematical maturity consistent with completion of three semesters of calculus. The objective is to introduce the reader to the principal structures of abstract algebra and to give an account of some of its more convincing applications. In particular there are sections on solution of equations by radicals, ruler and compass constructions, Polya counting theory, Steiner triple systems, orthogonal latin squares and error correcting codes. A less common application is to economics: the final section in the book shows how algebraic concepts may be used to construct abstract models of accounting systems.

The principal change to the book from the second edition is the addition of two new chapters. The first of these is an introduction to the representation theory of finite groups. This is a topic of interest to many working in the fields of chemistry and physics. The second new chapter is an introduction to category theory. Categories and functors play a unifying role throughout mathematics and often point to common features of the different branches. More recently category theory has become an important tool in theoretical computer science, where it provides ways of formalizing data structures and programming language semantics. In addition the section on free groups and generators and relations has been expanded to a new chapter and

https://doi.org/10.1515/9783110691160-201

includes an introduction to free products of groups. Finally, the various applications that were scattered throughout the second edition have been collected together in the final chapter.

Some of the changes have had the effect of raising the level of abstraction in parts of the book. Zorn's Lemma (and its variants) now appears at the end of Chapter 1, which should make plain its central role in algebra, although some readers may prefer to postpone this topic until later in the reading. Nevertheless, the original aim of making algebra accessible to as many readers as possible is maintained in this edition. Naturally, the opportunity has been taken to correct errors and obscurities in the previous edition. The author is grateful to the readers who took the trouble to report corrections: as usual full credit for any remaining errors belongs to the author.

There is more than enough material here for a two semester course in abstract algebra. If just one semester is available, Chapters 1 through 8, Chapter 11 and perhaps some of the applications could be covered. Chapters 1 and 2 contain topics that will be familiar to many readers and can be covered more quickly. In addition, a good deal of the material in Chapter 8 on vector spaces will not be new to a reader who has taken a course in linear algebra. A word about the proofs is in order. As a rule complete proofs are given and the reader is encouraged to read them. It should be kept in mind that the only way to be sure that a statement is correct, or that a computer program will always deliver the correct answer, is to prove it. The first two chapters, which contain much elementary material, are good places for the reader to acquire and polish theorem proving skills. Each section of the book is followed by a selection of problems of varying degrees of difficulty, with hints where appropriate.

This edition of the book, like previous ones, is based on courses given by the author over a period of years at the University of Illinois at Urbana-Champaign, the National University of Singapore and the University of London. I am grateful to colleagues for much good advice, which has led to numerous improvements. The first edition of this book was undertaken with the assistance and encouragement of Otto Kegel and Manfred Karbe. In preparing this third edition I have been aided by Leonardo Milla and Ute Skrambraks at Walter de Gruyter: it was the former who suggested the possibility of a new edition. Finally, I thank my family for their support and encouragement, always essential in such a project. This edition was prepared during the great pandemic of 2020–21, a time of tragic loss and suffering for so many. For the author the project has been a much needed distraction at a sad time in human history.

Urbana, Illinois,   Derek Robinson
December 2021

Contents

List of symbols

$A, B, \ldots$:	sets
$a, b, \ldots$:	elements of a set
$a \in A$:	a is an element of the set A
$|A|$:	the cardinal of the set A
$A \subseteq B, A \subset B$:	A is a subset, proper subset of B
$\emptyset$:	the empty set
$\mathbb{N}, \mathbb{Z}, \mathbb{Q}, \mathbb{R}, \mathbb{C}$:	the sets of natural numbers, integers, rational numbers, real numbers, complex numbers
$\bigcup, \bigcap$:	union and intersection
$A_1 \times \cdots \times A_n$:	a set product
$A - B, \bar{A}$:	complementary sets
$\mathcal{P}(A)$:	the power set of A
$S \circ R$:	the composite of relations or functions
$[x]_E$:	the E-equivalence class of x
$\alpha : A \to B$:	a function from A to B
$\mathrm{Im}(\alpha)$:	the image of the function α
id_A:	the identity function on the set A
α^{-1}:	the inverse of a bijective function α
$\mathrm{Fun}(A, B)$:	the set of all functions from A to B
$\mathrm{Fun}(A)$:	the set of all functions on A
gcd, lcm:	greatest common divisor, least common multiple
$a \equiv b \pmod{m}$:	a congruence
$[x]_m$ or $[x]$:	the congruence class of x modulo m
$\mathbb{Z}_n$:	the integers modulo n
$a \mid b$:	a divides b
ϕ:	Euler's function
$\lambda(n)$:	the number of partitions of n
$\mathrm{sign}(\pi)$:	the sign of a permutation π
$(i_1 i_2 \cdots i_r)$:	a cyclic permutation
$\mathrm{St}_G(x)$:	the stabilizer of x in G
$G \cdot a$:	the G-orbit of a
$\mathrm{Fix}(G)$:	the set of elements fixed by a group G
$\langle X \rangle$:	the subgroup or subspace generated by X
$|x|$:	the order of a group element x
$XY, X + Y$:	product, sum of subsets of a group
$H \leq G, H < G$:	H is a subgroup, proper subgroup of the group G
$\mathrm{Dr}_{\lambda \in \Lambda} G_\lambda, G_1 \times \cdots \times G_n$:	direct products of groups
$\mathrm{Cr}_{\lambda \in \Lambda} G_\lambda$:	the unrestricted direct (or cartesian) product of groups
$\mathrm{Fr}_{\lambda \in \Lambda} G_\lambda, G_1 * \cdots * G_n$:	free products of groups
$\mathrm{Sym}(X)$:	the symmetric group on a set X

https://doi.org/10.1515/9783110691160-202

S_n, A_n:	symmetric and alternating groups of degree n
Dih($2n$):	the dihedral group of order $2n$
$GL_n(F), GL_n(q), GL(V)$:	general linear groups
$SL_n(F), SL_n(q)$:	special linear groups
$\|G : H\|$:	the index of H in G
$N \triangleleft G$:	N is a normal subgroup of the group G
G/N:	the quotient group of N in G
$\simeq$:	an isomorphism
Ker(α):	the kernel of a homomorphism
$Z(G)$:	the center of the group G
$[x, y]$:	the commutator $xyx^{-1}y^{-1}$
G':	the derived subgroup of a group G
$G^{(i)}$:	a term of the derived chain of the group G
$Z_i(G)$:	a term of the upper central chain of the group G
$G^{ab} = G/G'$:	the abelianization of the group G.
$\phi(G)$:	the Frattini subgroup of a group G
$N_G(H), C_G(H)$:	normalizer and centralizer of H in G
Aut(G), Inn(G):	automorphism and inner automorphism groups
Out(G):	an outer automorphism group
$\langle X \mid R \rangle$:	a presentation of a group or module
Rg$\langle X \rangle$:	the ring generated by a subset
$U(R), R^*$:	the group of units of a ring R
R^{op}:	the opposite ring of R
$RX, (x)$:	ideals of a commutative ring R generated by a subset or element
$R[t_1, \ldots, t_n]$:	the ring of polynomials in $t_1, \ldots, t_n$ over a ring R
$F\{t_1, \ldots, t_n\}$:	The field of rational functions in $t_1, \ldots, t_n$ over a field F
$M_{m,n}(R), M_n(R)$:	sets of $m \times n$, $n \times n$ matrices over a ring R
I, I_n:	identity matrices
diag($d_1, d_2, \ldots, d_n$):	the diagonal matrix with $d_1, d_2, \ldots, d_n$ on the principal diagonal
det(A), tr(A):	the determinant and trace of a matrix A
dim(V):	the dimension of a vector space V
$F\langle X \rangle, \langle X \rangle$:	subspace generated by a subset of an F-vector space
$[v]_{\mathcal{B}}$:	the coordinate vector of v with respect to a basis $\mathcal{B}$
$C[a, b]$:	the vector space of continuous functions on the interval $[a, b]$
$L(V, W), L(V)$:	vector spaces of linear mappings
GF(q):	the field with q elements
$(E : F)$:	the degree of a field E over a subfield F
Gal(E/F), Gal(f):	Galois groups
deg(f):	the degree of a polynomial f

f':	the derivative of a polynomial f
$\mathrm{Irr}_F(x)$:	the irreducible polynomial of x over F
Φ_n:	the cyclotomic polynomial of order n
$_RM$ and N_R:	left and right R-modules.
$_RM_S$:	an (R, S)-bimodule.
$R \cdot X, R \cdot a$:	submodules generated by a set or element
$\bigoplus_{\lambda \in \Lambda} M_\lambda, M_1 \oplus \cdots \oplus M_n$:	direct sums of modules
$\mathrm{rank}(F)$:	the rank of a free module F.
M_p:	the p-torsion component of a module M
$\mathrm{Ann}_R(X), \mathrm{Ann}_R(x)$:	annihilators in a ring R
$\mathrm{Hom}_R(M, N)$:	a group of homomorphisms
$\mathrm{End}_R(M)$:	the ring of endomorphisms of a module
α_*, α^*:	induced, coinduced mappings
$a \otimes b$:	a tensor
$M \otimes_R N, M \otimes N$:	tensor products of modules
$\alpha \otimes \beta, A \otimes B$:	tensor products of homomorphisms, matrices
RG, FG:	a group ring, a group algebra
Set, **Mon**, **Gp**, **Ab**, **Rg**, **Mod**:	standard categories
$\mathcal{C}^{op}$:	the opposite of a category $\mathcal{C}$
$\mathrm{obj}(\mathcal{C})$:	the class of objects in category $\mathcal{C}$
ι_A:	an identity morphism
$\mathrm{Mor}_\mathcal{C}(A, B)$:	a set of morphisms in a category $\mathcal{C}$
$\mathrm{Hom}(A, -), \mathrm{Hom}(-, B)$:	Hom functors
$A \otimes -, - \otimes B$:	tensor product functors
$\prod_{i \in I} C_i, \coprod_{i \in I} C_i$:	product, coproduct in a category
$H_n(q)$:	Hamming n-space over a set with q elements
$B_n(v)$:	the n-ball with center v
$d(a, b)$:	the distance between words a and b
$wt(v)$:	the weight of a word v
$\mathrm{Bal}_n(R)$:	balance space
$\mathrm{Trans}_n(R)$:	transaction space
$\tau_\mathbf{v}$:	a transaction
$\mathcal{A} = (A|T|B)$:	an abstract accounting system

1 Sets, Relations and Functions

The concepts introduced in this chapter are truly fundamental and they underlie almost every branch of mathematics. Most of the material is quite elementary and will be familiar to many readers. Nevertheless readers are encouraged to review the material and to check notation and definitions. Because of its nature the pace of this chapter is somewhat faster than in subsequent chapters.

1.1 Sets and subsets

By a *set* we will mean any well-defined collection of objects, which are called the *elements* of the set. Some care must be exercised in using the term "set" because of Bertrand Russell's famous paradox, which shows that not every collection can be regarded as a set. Russell considered the collection C of all sets which are not elements of themselves. If C is allowed to be a set, a contradiction arises when one inquires whether or not C is an element of itself. Now plainly there is something suspicious about the idea of a set being an element of itself and we shall take this as evidence that the qualification "well-defined" needs to be taken seriously. A collection that is not a set is called a *proper class*.

Sets will be denoted by capital letters and their elements by lower case letters. The standard notation

$$a \in A$$

means that a is a element of the set A, or a *belongs* to A. The negation of $a \in A$ is denoted by $a \notin A$. Sets can be defined either by writing their elements out between braces, as in $\{a, b, c, d\}$, or alternatively by giving a formal description of the elements, the general format being

$$A = \{a \mid a \text{ has property } P\},$$

i. e., A is the set of all objects with the property P. If A is a finite set, the number of its elements is written

$$|A|.$$

Subsets
Let A and B be sets. If every element of A is an element of B, we write

$$A \subseteq B$$

https://doi.org/10.1515/9783110691160-001

and say that A is a *subset* of B, or that A *is contained* in B. If $A \subseteq B$ and $B \subseteq A$, so that A and B have exactly the same elements, then A and B are said to be *equal*,

$$A = B.$$

The negation of this is $A \neq B$. The notation $A \subset B$ is used if $A \subseteq B$ and $A \neq B$; then A is called a *proper* subset of B.

Some special sets

A set with no elements at all is called an *empty set*. An empty set E is a subset of any set A; for if this were false, there would be an element of E that is not in A, which is certainly wrong. As a consequence, *there is exactly one empty set*: for if E and E' are two empty sets, then $E \subseteq E'$ and $E' \subseteq E$, so that $E = E'$. The unique empty set is written

$$\emptyset.$$

Some further standard sets with reserved notations are

$$\mathbb{N}, \ \mathbb{Z}, \ \mathbb{Q}, \ \mathbb{R}, \ \mathbb{C},$$

which are respectively the sets of natural numbers $0, 1, 2, \ldots$, integers, rational numbers, real numbers and complex numbers.

Set operations

Next we recall the familiar set operations of union, intersection and complement. Let A and B be sets. The *union* $A \cup B$ is the set of all objects which belong to A or B, or possibly to both; the *intersection* $A \cap B$ consists of all objects that belong to both A and B. Thus

$$A \cup B = \{x \mid x \in A \text{ or } x \in B\},$$

while

$$A \cap B = \{x \mid x \in A \text{ and } x \in B\}.$$

It should be clear how to define the union and intersection of an arbitrary collection of sets $\{A_\lambda \mid \lambda \in \Lambda\}$; these are written

$$\bigcup_{\lambda \in \Lambda} A_\lambda \quad \text{and} \quad \bigcap_{\lambda \in \Lambda} A_\lambda,$$

respectively. The *relative complement* of B in A is

$$A - B = \{x \mid x \in A \text{ and } x \notin B\}.$$

Sometimes one has to deal only with subsets of some fixed set U, called the *universal set*. If $A \subseteq U$, then the *complement* of A in U is

$$\bar{A} = U - A.$$

We list for future reference the fundamental properties of unions, intersections and complements: most of these should be familiar.

(1.1.1). *Let A, B, C, B_λ ($\lambda \in \Lambda$) be sets. Then the following statements are valid:*
(i) $A \cup B = B \cup A$ and $A \cap B = B \cap A$, *(commutative laws).*
(ii) $(A \cup B) \cup C = A \cup (B \cup C)$ and $(A \cap B) \cap C = A \cap (B \cap C)$, *(associative laws).*
(iii) $A \cap (B \cup C) = (A \cap B) \cup (A \cap C)$ and $A \cup (B \cap C) = (A \cup B) \cap (A \cup C)$, *(distributive laws).*
(iv) $A \cup A = A = A \cap A$.
(v) $A \cup \emptyset = A$, $A \cap \emptyset = \emptyset$.
(vi) $A - (\bigcup_{\lambda \in \Lambda} B_\lambda) = \bigcap_{\lambda \in \Lambda} (A - B_\lambda)$ and $A - (\bigcap_{\lambda \in \Lambda} B_\lambda) = \bigcup_{\lambda \in \Lambda} (A - B_\lambda)$, *(De Morgan's Laws).*[1]

The easy proofs of these results are left to the reader as an exercise.

Set products
Let $A_1, A_2, \ldots, A_n$ be sets. By an *n-tuple* of elements from $A_1, A_2, \ldots, A_n$ is to be understood a sequence of elements $a_1, a_2, \ldots, a_n$ with $a_i \in A_i$. The n-tuple is usually written $(a_1, a_2, \ldots, a_n)$ and the set of all n-tuples is denoted by

$$A_1 \times A_2 \times \cdots \times A_n.$$

This is the *set product* (or *cartesian product*) of $A_1, A_2, \ldots, A_n$. For example $\mathbb{R} \times \mathbb{R}$ is the set of coordinates of points in the plane.

The following result is a basic counting tool.

(1.1.2). *If $A_1, A_2, \ldots, A_n$ are finite sets, then*

$$|A_1 \times A_2 \times \cdots \times A_n| = |A_1| \cdot |A_2| \cdots |A_n|.$$

Proof. In forming an n-tuple $(a_1, a_2, \ldots, a_n)$ we have $|A_1|$ choices for a_1, $|A_2|$ choices for $a_2, \ldots, |A_n|$ choices for a_n. Each choice of an a_i yields a different n-tuple. Therefore the total number of n-tuples is $|A_1| \cdot |A_2| \cdots |A_n|$. $\qquad\square$

---

[1] Augustus De Morgan (1806–1871).

Power sets

The *power set* of a set A is the set of all subsets of A, including the empty set and A itself; it is denoted by

$$\mathcal{P}(A).$$

The power set of a finite set is always a larger set, as the next result shows.

(1.1.3). *If A is a finite set, then $|\mathcal{P}(A)| = 2^{|A|}$.*

Proof. Let $A = \{a_1, a_2, \ldots, a_n\}$ with distinct a_i's. Also put $I = \{0, 1\}$. Each subset B of A is to correspond to an n-tuple $(i_1, i_2, \ldots, i_n)$ with $i_j \in I$. Here the rule for forming the n-tuple corresponding to B is this: $i_j = 1$ if $a_j \in B$ and $i_j = 0$ if $a_j \notin B$. Conversely, every n-tuple $(i_1, i_2, \ldots, i_n)$ with $i_j \in I$ determines a subset B of A, defined by $B = \{a_j \mid 1 \leq j \leq n, i_j = 1\}$. It follows that the number of subsets of A equals the number of elements in $I \times I \times \cdots \times I$, where the number of factors is n. By (1.1.2) we obtain $|\mathcal{P}(A)| = 2^n = 2^{|A|}$. $\square$

For an infinite version of the last result see (1.4.5) below. A power set $\mathcal{P}(A)$, together with the operations $\cup$ and $\cap$, constitutes what is known as a *Boolean*[2] *algebra*; such algebras have become very important in logic and computer science.

Exercises (1.1).
(1) Prove as many parts of (1.1.1) as possible.
(2) Let A, B, C be sets such that $A \cap B = A \cap C$ and $A \cup B = A \cup C$. Prove that $B = C$.
(3) If A, B, C are sets, establish the following:
    (i) $(A - B) - C = A - (B \cup C)$.
    (ii) $A - (B - C) = (A - B) \cup (A \cap B \cap C)$.
(4) Let A and B be finite sets. Prove that $|\mathcal{P}(A \times B)| = |\mathcal{P}(A)|^{|B|}$.
(5) Let A and B be finite sets with more than one element in each. Prove that $|\mathcal{P}(A \times B)|$ is larger than both $|\mathcal{P}(A)|$ and $|\mathcal{P}(B)|$.
(6) The *disjoint union* $A \oplus B$ of sets A and B is defined by the rule $A \oplus B = A \cup B - A \cap B$, so its elements are those that belong to exactly one of A and B. Prove the following statements:
    (i) $A \oplus A = \emptyset$, $A \oplus B = B \oplus A$.
    (ii) $(A \oplus B) \oplus C = A \oplus (B \oplus C)$.
    (iii) $(A \oplus B) \cap C = (A \cap C) \oplus (B \cap C)$.
(7) If A and B be finite sets, prove that $|\mathcal{P}(A \cup B)| = \frac{|\mathcal{P}(A)| \cdot |\mathcal{P}(B)|}{|\mathcal{P}(A \cap B)|}$.

---

2 George Boole (1815–1864).

1.2 Relations, equivalence relations, partial orders

In mathematics it is often not sufficient to deal with the individual elements of a set: for it may be critical to understand how elements of the set are related to each other. This leads us to formulate the concept of a relation.

Let A and B be sets. Then a *relation R between A and B* is a subset of the set product $A \times B$. The definition is clarified by use of a suggestive notation: if $(a, b) \in R$, then a and b are said to be *related* by R and we write

$$a \, R \, b.$$

The most important case is of a relation R between A and itself; this is called *a relation on the set A.*

Example (1.2.1).
(i) Let A be a set and define $R = \{(a, a) \mid a \in A\}$. Thus $a_1 \, R \, a_2$ means that $a_1 = a_2$ and R is the relation of equality on A.
(ii) Let P be the set of all points and L the set of all lines in the plane. A relation R from P to L is defined by: $p \, R \, \ell$ if the point p lies on the line ℓ.
(iii) A relation R on the set of integers $\mathbb{Z}$ is defined by: $a \, R \, b$ if $a - b$ is even.

The next result confirms what one might suspect, that a finite set has many relations.

(1.2.1). *If A is a finite set, the number of relations on A equals $2^{|A|^2}$.*

For this is the number of subsets of $A \times A$ by (1.1.2) and (1.1.3).

The concept of a relation on a set is evidently a very broad one. In practice the relations of greatest interest are those which have special properties. The most common of these are listed next. Let R be a relation on a set A.
(i) R is *reflexive* if $a \, R \, a$ for all $a \in A$.
(ii) R is *symmetric* if $a \, R \, b$ always implies that $b \, R \, a$.
(iii) R is *antisymmetric* if $a \, R \, b$ and $b \, R \, a$ imply that $a = b$;
(iv) R is *transitive* if $a \, R \, b$ and $b \, R \, c$ imply that $a \, R \, c$.

Relations which are reflexive, symmetric and transitive are called *equivalence relations*; they are of fundamental importance. Relations which are reflexive, antisymmetric and transitive are also important; they are called *partial orders*. Here are some examples of relations of various types.

Example (1.2.2).
(i) Equality on a set is both an equivalence relation and a partial order.

(ii) A relation R on $\mathbb{Z}$ is defined by: $a\ R\ b$ if and only if $a - b$ is even. This is an equivalence relation, but it is not a partial order.

(iii) If A is any set, the relation of containment $\subseteq$ is a partial order on the power set $P(A)$.

(iv) A relation R on $\mathbb{N}$ is defined by $a\ R\ b$ if a divides b. Here R is a partial order on $\mathbb{N}$.

Equivalence relations and partitions

The structure of an equivalence relation on a set will now be analyzed. The essential conclusion will be that an equivalence relation causes the set to split up into non-overlapping non-empty subsets.

Let E be an equivalence relation on a set A. First of all define the *E-equivalence class* of an element a of A to be the subset

$$[a]_E = \{x \mid x \in A \text{ and } x\ E\ a\}.$$

By the reflexive law $a \in [a]_E$, so

$$A = \bigcup_{a \in A} [a]_E$$

and A is the union of all the equivalence classes.

Next suppose that the equivalence classes $[a]_E$ and $[b]_E$ both contain an element x. Assume that $y \in [a]_E$; then $y\ E\ a$, $a\ E\ x$ and $x\ E\ b$, by the symmetric law. Hence $y\ E\ b$ by two applications of the transitive law. Therefore $y \in [b]_E$ and we have proved that $[a]_E \subseteq [b]_E$. By the same reasoning $[b]_E \subseteq [a]_E$, so that $[a]_E = [b]_E$. It follows that distinct equivalence classes are disjoint, i. e., they have no elements in common.

What has been shown so far is that the set A is the union of the E-equivalence classes and that distinct equivalence classes are disjoint. A decomposition of A into disjoint non-empty subsets is called a *partition* of A. Thus E determines a partition of A.

Conversely, suppose that a partition of A into non-empty disjoint subsets A_λ, $(\lambda \in \Lambda)$, is given. We would like to construct an equivalence relation on A corresponding to the partition. Now each element of A belongs to a unique subset A_λ; thus we may define $a\ E\ b$ to mean that a and b belong to the same subset A_λ. It follows immediately from the definition that the relation E is an equivalence relation; what is more, the equivalence classes are just the subsets A_λ of the original partition. We summarize these conclusions in:

(1.2.2).

(i) *If E is an equivalence relation on a set A, the E-equivalence classes form a partition of A.*

(ii) *Conversely, each partition of A determines an equivalence relation on A for which the equivalence classes are the subsets in the partition.*

Thus the concepts of equivalence relation and partition are in essence the same. For example, in the equivalence relation (ii) above there are two equivalence classes, the sets of even and odd integers; of course these form a partition of $\mathbb{Z}$.

Partial orders

Suppose that R is a partial order on a set A, i. e., R is a reflexive, antisymmetric, transitive relation on A. Instead of writing $a\,R\,b$ it is customary to employ a more suggestive symbol and write

$$a \preceq b.$$

The pair $(A, \preceq)$ then constitutes a *partially ordered set* (or *poset*).

The effect of a partial order is to impose a hierarchy on the set A. When the set is finite, this can be visualized by drawing a picture of the poset called a *Hasse*[3] *diagram*. It consists of vertices and edges drawn in the plane, the vertices representing the elements of A. A sequence of upwardly sloping edges from a to b, as in the diagram below, indicates that $a \preceq b$. Elements a, b not connected by such a sequence of edges do not satisfy $a \preceq b$ or $b \preceq a$. In order to simplify the diagram as far as possible, it is agreed that unnecessary edges are to be omitted.

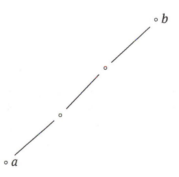

A very familiar poset is the power set of a set A with the partial order $\subseteq$, i. e. $(P(A), \subseteq)$.

Example (1.2.3). Draw the Hasse diagram of the poset $(\mathcal{P}(A), \subseteq)$ where $A = \{1, 2, 3\}$.

---

3 Helmut Hasse (1898–1979).

This poset has $2^3 = 8$ vertices, which can be visualized as the vertices of a cube standing on one corner.

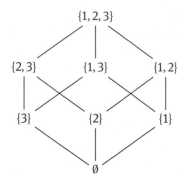

One reason why partially ordered sets are important in algebra is that they provide a useful representation of substructures of standard algebraic structures, for example subsets, subgroups, subrings etc.

A partial order $\preceq$ on a set A is called a *linear order* if, given $a, b \in A$, either $a \preceq b$ or $b \preceq a$ holds. Then $(A, \preceq)$ is called a *linearly ordered set* or *chain*. The Hasse diagram of a chain is a single sequence of edges sloping upwards. Obvious examples of chains are $(\mathbb{Z}, \leq)$ and $(\mathbb{R}, \leq)$ where $\leq$ is the usual "less than or equal to". Finally, a linear order on A is called a *well order* if each non-empty subset X of A contains a *least* element a, i. e., such that $a \preceq x$ for all elements $x \in X$. While it might seem obvious that $\leq$ is a well order on the set of all positive integers, this is actually an axiom, the Well-Ordering Law, which is discussed in Section 2.1.

Lattices

Consider a poset $(A, \preceq)$. If $a, b \in A$, an *upper bound* for a and b in A is an element $x \in A$ such that $a \preceq x$ and $b \preceq x$. An upper bound ℓ is called a *least upper bound* (or lub) of a and b if every upper bound x satisfies $\ell \preceq x$ and $\ell \preceq x$. Similarly a *lower bound* and a *greatest lower bound* (or glb) of a and b are defined by reversing the order of elements around the symbol $\preceq$. If ℓ and g exist, the Hasse diagram of $(A, \preceq)$ will contain the subdiagram

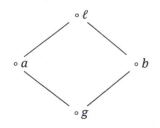

A poset in which each pair of elements has an lub and a glb is called a *lattice*. For example, $(P(S), \subseteq)$ is a lattice since the lub and glb of A and B are just $A \cup B$ and $A \cap B$ respectively.

The composite of relations

Since a relation is a subset, two relations may be combined by forming their union or intersection. However, there is a more useful way of combining relations called *composition*: let R and S be relations between A and B and between B and C respectively. Then the *composite relation*

$$S \circ R$$

is the relation between A and C defined by: a $(S \circ R)$ c if and only if there exists $b \in B$ such that $a\, R\, b$ and $b\, S\, c$.

For example, assume that $A = \mathbb{Z}$, $B = \{a, b, c\}$, $C = \{\alpha, \beta, \gamma\}$. Define relations $R = \{(1, a), (2, b), (4, c)\}$, $S = \{(a, \alpha), (b, \gamma), (c, \beta)\}$. Then $S \circ R = \{(1, \alpha), (2, \gamma), (4, \beta)\}$.

In particular one can form the composite of any two relations R and S on a set A. Notice that the condition for a relation R to be transitive can now be expressed in the form $R \circ R \subseteq R$.

A result of fundamental importance is the associative law for composition of relations.

(1.2.3). *Let R, S, T be relations between A and B, B and C, and C and D respectively. Then $T \circ (S \circ R) = (T \circ S) \circ R$.*

Proof. Let $a \in A$ and $d \in D$. Then a $(T \circ (S \circ R))$ d means that there exists $c \in C$ such that a $(S \circ R)$ c and $c\, T\, d$, i.e., there exists $b \in B$ such that $a\, R\, b$, $b\, S\, c$ and $c\, T\, d$. Therefore b $(T \circ S)$ d and a $((T \circ S) \circ R)$ d. Thus $T \circ (S \circ R) \subseteq (T \circ S) \circ R$, and in a similar way $(T \circ S) \circ R \subseteq T \circ (S \circ R)$. $\square$

Exercises (1.2).

(1) Determine whether the binary relations R defined on the set A below are reflexive, symmetric, antisymmetric or transitive.
  (i) $A = \mathbb{R}$ and $a\, R\, b$ means $a^2 = b^2$.
  (ii) $A = \mathbb{R}$ and $a\, R\, b$ means $a - b \leq 2$.
  (iii) $A = \mathbb{Z} \times \mathbb{Z}$ and $(a, b)\, R\, (c, d)$ means $a + d = b + c$.
  (iv) $A = \mathbb{Z}$ and $a\, R\, b$ means that $b = a + 3c$ for some integer c.

(2) A relation $\sim$ on $\mathbb{R} - \{0\}$ is defined by $a \sim b$ if $ab > 0$. Show that $\sim$ is an equivalence relation and identify the equivalence classes.

(3) Let $A = \{1, 2, \ldots, n\}$ where n is a positive integer. Define $a \preceq b$ to mean that a divides b. Show that $(A, \preceq)$ is a poset. Draw the Hasse diagram for the case $n = 12$.

(4) Let $(A, \preceq)$ be a poset and let $a, b \in A$. Show that a and b have at most one lub and at most one glb.

(5) Given linearly ordered sets $(A_i, \preceq_i)$, $i = 1, 2, \ldots, k$, suggest a way to make the set $A_1 \times A_2 \times \cdots \times A_k$ into a linearly ordered set.

(6) How many equivalence relations are there on sets with 1, 2, 3 or 4 elements?

(7) Suppose that A is a set with n elements. Show that there are exactly 2^{n^2-n} reflexive relations on A and $2^{n(n+1)/2}$ symmetric ones.

(8) Let R be a relation on a set A. Define *powers* of R recursively by $R^1 = R$ and $R^{n+1} = R^n \circ R$ for $n = 1, 2, \ldots$.

    (i) If R is transitive, show that $\cdots R^n \subseteq R^{n-1} \subseteq \cdots \subseteq R^2 \subseteq R$.

    (ii) If in addition R is reflexive, show that $R = R^2 = R^3 =$ etc.

    (iii) If R is a transitive relation on a finite set with n elements, prove that $R^m = R^{m+1} = \cdots$ where $m = n^2 + 1$.

1.3 Functions

A more familiar concept than a relation is a function. While functions are to be found throughout mathematics, they are usually first encountered in calculus as real-valued functions of a real variable. Functions can provide convenient descriptions of complex objects and processes in mathematics and the information sciences.

    Let A and B be (non-empty) sets. A *function* or *mapping* or *map* from A to B, in symbols

$$\alpha : A \rightarrow B,$$

is a rule which assigns to each element a of A a unique element $\alpha(a)$ of B, called the *image of a* under α. The sets A and B are the *domain* and *codomain* of α respectively. The *image of the function α* is

$$\mathrm{Im}(\alpha) = \{\alpha(a) \mid a \in A\},$$

which is a subset of the codomain. The set of all functions from A to B may be written $\mathrm{Fun}(A, B)$ and $\mathrm{Fun}(A)$ is written for $\mathrm{Fun}(A, A)$.

Examples of functions

(i) The functions that appear in calculus are those whose domain and codomain are subsets of $\mathbb{R}$. Such a function can be visualized by drawing its graph in the usual way.

(ii) Given a function $\alpha : A \rightarrow B$, we can define

$$R_\alpha = \{(a, \alpha(a)) \mid a \in A\} \subseteq A \times B.$$

Thus R_α is a relation between A and B. Observe that R_α is a special kind of relation since each a in A is related to a *unique* element of B, namely $\alpha(a)$.

Conversely, suppose that R is a relation between A and B such that each $a \in A$ is related to a unique $b \in B$. We may define a corresponding function $\alpha_R : A \rightarrow B$ by $\alpha_R(a) = b$ where $a\,R\,b$. Thus functions from A to B may be regarded as special types of relation between A and B.

This observation permits us to form the composite of two functions $\alpha : A \rightarrow B$ and $\beta : B \rightarrow C$ by forming the composite of the corresponding relations: thus $\beta \circ \alpha : A \rightarrow C$ is defined by

$$\beta \circ \alpha(a) = \beta(\alpha(a)).$$

Frequently we will write $\beta\alpha$ for $\beta \circ \alpha$.

(iii) *The characteristic function of a subset.* Let A be a fixed set. For each subset X of A define a function $\alpha_X : A \rightarrow \{0,1\}$ by the rule

$$\alpha_X(a) = \begin{cases} 1 & \text{if } a \in X \\ 0 & \text{if } a \notin X. \end{cases}$$

Then α_X is called the *characteristic function* of the subset X. Conversely, a function $\alpha : A \rightarrow \{0,1\}$ is the characteristic function of the subset $\{a \mid \alpha(a) = 1\}$.

(iv) The *identity function* on a set A is the function $\mathrm{id}_A : A \rightarrow A$ defined by $\mathrm{id}_A(a) = a$ for all $a \in A$.

Injectivity and surjectivity

There are two special types of function of critical importance. A function $\alpha : A \rightarrow B$ is called *injective* (or *one-one*) if $\alpha(a) = \alpha(a')$ always implies that $a = a'$, i. e., distinct elements of A have distinct images in B under α. Next $\alpha : A \rightarrow B$ is *surjective* (or *onto*) if each element of B is the image under α of at least one element of A, i. e., $\mathrm{Im}(\alpha) = B$. Finally, $\alpha : A \rightarrow B$ is said to be *bijective* (or a *one-one correspondence*) if it is both injective and surjective.

Here are some examples of various types of functions.

(i) $\alpha : \mathbb{R} \rightarrow \mathbb{R}$ where $\alpha(x) = 2^x$ is injective but not surjective.

(ii) $\alpha : \mathbb{R} \rightarrow \mathbb{R}$ where $\alpha(x) = x^3 - 4x$ is surjective but not injective. Here surjectivity is best seen by drawing the graph of $y = x^3 - 4x$. Note that any line parallel to the x-axis meets the curve at least once. But α is not injective since $\alpha(0) = 0 = \alpha(2)$.

(iii) $\alpha : \mathbb{R} \rightarrow \mathbb{R}$ where $\alpha(x) = x^3$ is bijective.

(iv) $\alpha : \mathbb{R} \rightarrow \mathbb{R}$ where $\alpha(x) = x^2$ is neither injective nor surjective.

Inverse functions

Functions $\alpha : A \rightarrow B$ and $\beta : B \rightarrow A$ are said to be *mutually inverse* if $\alpha \circ \beta = \mathrm{id}_B$ and $\beta \circ \alpha = \mathrm{id}_A$. Also β is an *inverse* of α. Suppose that β' is another inverse of α. Then, with the aid of the associative law, we have

$$\beta = \mathrm{id}_A \circ \beta = (\beta' \circ \alpha) \circ \beta = \beta' \circ (\alpha \circ \beta) = \beta' \circ \mathrm{id}_B = \beta'.$$

Therefore α has a unique inverse, if it has one at all. We will write

$$\alpha^{-1} : B \to A$$

for the unique inverse of α when it exists.

It is important to be able to recognize functions which possess inverses.

(1.3.1). *A function $\alpha : A \to B$ has an inverse if and only if it is bijective.*

Proof. Assume that $\alpha^{-1} : B \to A$ exists. If $\alpha(a_1) = \alpha(a_2)$, then, applying α^{-1} to each side, we arrive at $a_1 = a_2$, which shows that α is injective. Next, to show that α is surjective, let $b \in B$. Then $b = \mathrm{id}_B(b) = \alpha(\alpha^{-1}(b)) \in \mathrm{Im}(\alpha)$, showing that $\mathrm{Im}(\alpha) = B$ and α is surjective. Thus α is bijective.

Conversely, let α be bijective. If $b \in B$, there is precisely one element a in A such that $\alpha(a) = b$ since α is bijective. Define $\beta : B \to A$ by $\beta(b) = a$. Then $\alpha\beta(b) = \alpha(a) = b$ and $\alpha\beta = \mathrm{id}_B$. Also $\beta\alpha(a) = \beta(b) = a$; since every a in A arises in this way, $\beta\alpha = \mathrm{id}_A$ and $\beta = \alpha^{-1}$. $\qquad\square$

The next result records some useful facts about inverses.

(1.3.2). *Let A, B, C be sets.*
(i) *If $\alpha : A \to B$ is bijective, then so is $\alpha^{-1} : B \to A$ and $(\alpha^{-1})^{-1} = \alpha$.*
(ii) *If $\alpha : A \to B$ and $\beta : B \to C$ are bijective functions, then $\beta \circ \alpha : A \to C$ is bijective and $(\beta \circ \alpha)^{-1} = \alpha^{-1} \circ \beta^{-1}$.*

Proof. The equations $\alpha \circ \alpha^{-1} = \mathrm{id}_B$ and $\alpha^{-1} \circ \alpha = \mathrm{id}_A$ tell us that α is the inverse of α^{-1}. Check directly that $\alpha^{-1} \circ \beta^{-1}$ is the inverse of $\beta \circ \alpha$ by using the associative law twice: thus $(\beta \circ \alpha) \circ (\alpha^{-1} \circ \beta^{-1}) = ((\beta \circ \alpha) \circ \alpha^{-1}) \circ \beta^{-1} = (\beta \circ (\alpha \circ \alpha^{-1})) \circ \beta^{-1} = (\beta \circ \mathrm{id}_B) \circ \beta^{-1} = \beta \circ \beta^{-1} = \mathrm{id}_C$. Similarly $(\alpha^{-1} \circ \beta^{-1}) \circ (\beta \circ \alpha) = \mathrm{id}_A$. $\qquad\square$

Application to automata

As an illustration of how the language of sets and functions may be used to describe information systems, we give a brief account of automata. An *automaton* is a theoretical device that is a basic model of a digital computer. It consists of an input tape and an output tape together with two "heads", which are able to read symbols on the input tape and print symbols on the output tape. At any instant the system is in one of a number of states. When the automaton reads a symbol on the input tape, it goes to another state and writes a symbol on the output tape.

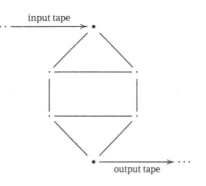

To make this precise we define an automaton A to be a 5-tuple

$$(I, O, S, v, \sigma)$$

where I and O are the respective sets of input and output symbols, S is the set of states, $v : I \times S \rightarrow O$ is the *output function* and $\sigma : I \times S \rightarrow S$ is the *next state function*. The automaton operates in the following manner. If it is in state $s \in S$ and input symbol $i \in I$ is read, the automaton prints the symbol $v(i, s)$ on the output tape and goes to state $\sigma(i, s)$. Thus the mode of operation is determined by the three sets I, O, S and the two functions v, σ.

Exercises (1.3).

(1) Which of the following functions are injective, surjective, bijective?
  (i) $\alpha : \mathbb{R} \rightarrow \mathbb{Z}$ where $\alpha(x) = [x]$, the largest integer $\leq x$.
  (ii) $\alpha : \mathbb{R}^{>0} \rightarrow \mathbb{R}$ where $\alpha(x) = \log_{10}(x)$. (Here $\mathbb{R}^{>0} = \{x \mid x \in \mathbb{R}, x > 0\}$).
  (iii) $\alpha : A \times B \rightarrow B \times A$ where $\alpha((a, b)) = (b, a)$.

(2) Prove that a composite of injective functions is injective and a composite of surjective functions is surjective.

(3) Let $\alpha : A \rightarrow B$ be a function between finite sets. Show that if $|A| > |B|$, then α cannot be injective, and if $|A| < |B|$, then α cannot be surjective.

(4) Define $\alpha : \mathbb{R} \rightarrow \mathbb{R}$ by $\alpha(x) = \frac{x^3}{x^2+1}$. Prove that α is bijective.

(5) Give an example of two functions α, β on a set A such that $\alpha \circ \beta = \mathrm{id}_A$ but $\beta \circ \alpha \neq \mathrm{id}_A$.

(6) Let $\alpha : A \rightarrow B$ be a injective function. Show that there is a surjective function $\beta : B \rightarrow A$ such that $\beta \circ \alpha = \mathrm{id}_A$.

(7) Let $\alpha : A \rightarrow B$ be a surjective function. Show that there is an injective function $\beta : B \rightarrow A$ such that $\alpha \circ \beta = \mathrm{id}_B$.

(8) Describe a simplified version of an automaton with no output tape in which the outputs are states. (This is called a *state output automaton*).

(9) Let $\alpha : A \rightarrow B$ be a function. Define a relation E_α on A by the rule: $a \, E_\alpha \, a'$ means that $\alpha(a) = \alpha(a')$. Prove that E_α is an equivalence relation on A. Then show that,

conversely, if E is any equivalence relation on a set A, then $E = E_\alpha$ for some function α with domain A.

1.4 Cardinality

If we want to compare two sets, a natural basis for comparison is the "size" of each set. If the sets are finite, their sizes are just the numbers of elements in the sets. But how can one measure the size of an infinite set? A reasonable point of view would be to hold that two sets have the same size if their elements can be paired off. Certainly two finite sets have the same number of elements precisely when their elements can be paired. The point to observe is that this idea also applies to infinite sets, making it possible to give a rigorous definition of the size of an infinite set, its *cardinal*.

Let A and B be two sets. Then A and B are said to be *equipollent* if there is a bijection $\alpha : A \to B$: thus the elements of A and B may be paired off as $(a, \alpha(a))$, $a \in A$. It follows from (1.3.2) that equipollence is an equivalence relation on the class of all sets. Thus each set A belongs to a unique equivalence class, which will be written

$$|A|$$

and called the *cardinal* of A. Informally we can think of $|A|$ as the collection of all sets with the same "size" as A. A *cardinal number* is the cardinal of some set.

If A is a finite set with exactly n elements, then A is equipollent to the set $\{0, 1, \ldots, n-1\}$ and $|A| = |\{0, 1, \ldots, n-1\}|$. It is reasonable to identify the finite cardinal $|\{0, 1, \ldots, n-1\}|$ with the non-negative integer n. For then cardinal numbers appear as infinite versions of the non-negative integers.

Let us sum up our very elementary conclusions so far.

(1.4.1).
(i) *Every set A has a unique cardinal number $|A|$.*
(ii) *Two sets are equipollent if and only if they have the same cardinal.*
(iii) *The cardinal of a finite set may be identified with the number of its elements.*

Since we plan to use cardinals to compare the sizes of sets, it makes sense to define a "less than or equal to" relation $\leq$ on cardinals. Define

$$|A| \leq |B|$$

to mean that there is an injective function $\alpha : A \to B$. Of course we will write $|A| < |B|$ if $|A| \leq |B|$ and $|A| \neq |B|$.

It is important to verify that this definition of $\leq$ depends only on the cardinals $|A|$ and $|B|$, not on the choice of sets A and B. Indeed, if $A' \in |A|$ and $B' \in |B|$, then there are bijections $\alpha' : A' \to A$ and $\beta' : B \to B'$; by composing these with the injection $\alpha : A \to B$ we obtain the injection $\beta' \circ \alpha \circ \alpha' : A' \to B'$. Thus $|A'| \leq |B'|$.

Next we prove a famous result about inequality of cardinals.

(1.4.2) (The Cantor-Bernstein[4] Theorem). *If A and B are sets such that $|A| \leq |B|$ and $|B| \leq |A|$, then $|A| = |B|$.*

The proof of (1.4.2) is our most challenging proof so far and some readers may prefer to skip it. However, the basic idea behind it is not difficult to grasp.

Proof. By hypothesis there are injective functions $\alpha : A \to B$ and $\beta : B \to A$. These will be used to construct a bijective function $y : A \to B$, which will show that $|A| = |B|$.

Consider an arbitrary element a in A; either $a = \beta(b)$ for some unique $b \in B$ or else $a \notin \text{Im}(\beta)$: here we use the injectivity of β. Similarly, either $b = \alpha(a')$ for a unique $a' \in A$ or else $b \notin \text{Im}(\alpha)$. Continuing this process, we trace back the "ancestry" of the element a. There are three possible outcomes:

(i)   we reach an element of $A - \text{Im}(\beta)$;
(ii) we reach an element of $B - \text{Im}(\alpha)$;
(iii) the process continues without end.

Partition the set A into three subsets corresponding to possibilities (i), (ii), (iii) and call them AA, AB, $A\infty$ respectively. In a similar fashion the set B is partitioned into three disjoint subsets BA, BB, $B\infty$; for example, if $b \in BA$, we can trace b back to an element of $A - \text{Im}(\beta)$.

Now we are in a position to define the function $y : A \to B$. First observe that the restriction of α to AA is a bijection from AA to BA, the restriction of β to BB is a bijection from BB to AB and the restriction of α to $A\infty$ is a bijection from $A\infty$ to $B\infty$. Also, if $x \in AB$, there is a unique element $x' \in BB$ such that $\beta(x') = x$. Now define

$$y(x) = \begin{cases} \alpha(x) & \text{if } x \in AA \\ \alpha(x) & \text{if } x \in A\infty \\ x' & \text{if } x \in AB. \end{cases}$$

Then y is the desired bijection. □

(1.4.3). *The relation $\leq$ is a partial order on cardinal numbers.*

For we have proved antisymmetry in (1.4.2), while reflexivity and transitivity are clearly true. In fact one can do better since $\leq$ is even a *linear* order. This is because of:

---

4 Georg Cantor (1845–1918), Felix Bernstein (1878–1956).

(1.4.4) (The Law of Trichotomy). *If A and B are sets, then exactly one of the following must hold:*

$$|A| < |B|, \quad |A| = |B|, \quad |B| < |A|.$$

The proof calls for the use of Zorn's Lemma and is given as (1.5.1) below.

The next result establishes the existence of arbitrarily large cardinal numbers.

(1.4.5). *If A is any set, then $|A| < |\mathcal{P}(A)|$.*

Proof. The easy step is to show that $|A| \leq |\mathcal{P}(A)|$. This is because the assignment $a \mapsto \{a\}$ sets up an injection from A to $\mathcal{P}(A)$.

Next assume that $|A| = |\mathcal{P}(A)|$, so that there is a bijection $\alpha : A \to \mathcal{P}(A)$. Of course at this point we are looking for a contradiction. The trick is to consider the subset $B = \{a \mid a \in A, a \notin \alpha(a)\}$ of A. Then $B \in \mathcal{P}(A)$, so $B = \alpha(a)$ for some $a \in A$. Now either $a \in B$ or $a \notin B$. If $a \in B$, then $a \notin \alpha(a) = B$; if $a \notin B = \alpha(a)$, then $a \in B$. This is our contradiction. $\qquad\square$

Countable sets

The cardinal of the set of natural numbers $\mathbb{N} = \{0, 1, 2, \ldots\}$ is denoted by

$$\aleph_0.$$

Here $\aleph$ is the Hebrew letter aleph. A set A is said to be *countable* if $|A| \leq \aleph_0$. Essentially this means that the elements of A can be "labelled" by attaching to each element a natural number as a label. An uncountable set cannot be so labelled.

We need to explain what is meant by an infinite set for the next result to be meaningful. A set A will be called *infinite* if it has a subset that is equipollent with $\mathbb{N}$, i. e., if $\aleph_0 \leq |A|$. An *infinite cardinal* is the cardinal of an infinite set.

(1.4.6). $\aleph_0$ *is the smallest infinite cardinal.*

Proof. If A is an infinite set, then A has a subset B such that $\aleph_0 = |B|$. Hence $\aleph_0 \leq |A|$. $\qquad\square$

It follows that *if A is a countable set, either A is finite or $|A| = \aleph_0$*. As the final topic of the section we consider the cardinals of the sets $\mathbb{Q}$ and $\mathbb{R}$.

(1.4.7).
(i) *The set $\mathbb{Q}$ of rational numbers is countable.*
(ii) *The set $\mathbb{R}$ of real numbers is uncountable.*

Proof. (i) Each positive rational number has the form $\frac{m}{n}$ where m and n are positive integers. Arrange these rationals in a rectangular array, with $\frac{m}{n}$ in the mth row and nth column. Of course each rational will occur infinitely often because of cancellation. Now follow the path indicated by the arrows in the diagram below.

This creates a sequence in which every positive rational number appears infinitely often. Delete repetitions in the sequence. Insert 0 at the beginning of the sequence and insert $-r$ immediately after r for each positive rational r. Now every rational occurs exactly once in the sequence. Hence $\mathbb{Q}$ is countable.

(ii) It is enough to show that the set I of all real numbers r such that $0 \leq r < 1$ is uncountable: this is because $|I| \leq |\mathbb{R}|$. Assume that I is countable, so that it can be written in the form $\{r_1, r_2, r_3, \ldots\}$. Write each r_i as a decimal, say

$$r_i = 0 \cdot r_{i1} r_{i2} \cdots$$

where $0 \leq r_{ij} \leq 9$. We reach a contradiction by producing a number in the set I which does not equal any r_i. Define

$$s_i = \begin{cases} 0 & \text{if } r_{ii} \neq 0 \\ 1 & \text{if } r_{ii} = 0 \end{cases}$$

and let s be the decimal $0 \cdot s_1 s_2 \cdots$; then certainly $s \in I$. Hence $s = r_i$ for some i, so that $s_i = r_{ii}$; but this is impossible by the definition of s_i. $\qquad \square$

Exercises (1.4).
(1) A finite set cannot be equipollent to a proper subset.
(2) A set is infinite if and only if it has the same cardinal as some proper subset.
(3) If there is a surjection from a set A to a set B, then $|B| \leq |A|$.
(4) Show that $|\mathbb{Z}| = \aleph_0$ and $|\mathbb{Z} \times \mathbb{Z}| = \aleph_0$.
(5) Let $A_1, A_2, \ldots$ be countably many, countable sets. Prove that $\bigcup_{i=1,2,\ldots} A_i$ is a countable set. [Hint: write $A_i = \{a_{i0}, a_{i1}, \ldots\}$ and follow the method of proof of (1.4.7)(i).]
(6) Suggest reasonable definitions of the sum and product of two cardinal numbers. [Hint: try using the union and set product]
(7) Let S denote the set of all *restricted sequences* of integers $a_1, a_2, a_3, \ldots$, i.e., $a_i = 0$ for all but a finite number of i. Prove that $|S| = \aleph_0$.

(8) Let A be a countably infinite set and let $\mathcal{P}_f(A)$ denote the set of all finite subsets of A.

(i) Prove that $|\mathcal{P}_f(A)| = |A|$, so that $\mathcal{P}_f(A)$ is countable.

(ii) Prove that on the other hand $\mathcal{P}(A)$ is uncountable.

1.5 Zorn's Lemma and variants

The background to Zorn's Lemma lies in the kind of set theory that is being used.[5] Up to this point we have been functioning – quite naively – in what is called *Gödel–Bernays Theory*. In this the primitive, or undefined, notions are *class*, *membership* and *equality*. On the basis of these concepts and the accompanying axioms, the usual elementary properties of sets can be derived.

However, the set theory just described does not provide an adequate basis for dealing with infinite sets. For many purposes in algebra the most useful additional axiom is what has become known as *Zorn's Lemma*. Despite its name, this is an axiom that must be assumed.

Zorn's Lemma. *Let $(S, \preceq)$ be a non-empty partially ordered set with the property that every chain in S has an upper bound in S. Then S contains at least one maximal element.*

The terminology here calls for some explanation. Recall that a *chain* in the partially ordered set S is a subset C which is linearly ordered by the partial order $\preceq$. Also an *upper bound* for C is an element s of S such that $c \preceq s$ is valid for all c in C. Finally, a *maximal element* of S is an element m such that $m \preceq s \in S$ implies that $m = s$. Note that in general a partially ordered set may contain several maximal elements or none at all.

As will be seen in the sequel, Zorn's Lemma is a vital tool in proving such fundamental theorems in algebra as the existence of a basis in an infinite dimensional vector space and the existence of algebraic closures. For the present as an illustration of the power of Zorn's Lemma, we will establish a result on cardinal numbers which was stated above without proof as (1.4.3).

(1.5.1) (The Law of Trichotomy). *If A and B are sets, then exactly one of the following must hold:*

$$|A| < |B|, \quad |A| = |B|, \quad |B| < |A|.$$

Proof. Because of the Cantor–Bernstein Theorem (1.4.2), it is enough to prove that either $|A| \leq |B|$ or $|B| \leq |A|$ holds. Clearly A and B can be assumed non-empty.

---

[5] Some readers may wish to defer reading the proofs in this section until later.

Consider the set $\mathcal{F}$ of all pairs (X, α) where $X \subseteq A$ and $\alpha : X \to B$ is an injective function. A partial order $\preceq$ on $\mathcal{F}$ is defined by $(X, \alpha) \preceq (X', \alpha')$ if $X \subseteq X'$ and $\alpha'|_X = \alpha$. It is obvious that $\mathcal{F}$ is not empty. Let $\mathcal{C} = \{(X_i, \alpha_i) \mid i \in I\}$ be a chain in $\mathcal{F}$. Put $U = \bigcup_{i \in I} X_i$ and define $\alpha : U \to B$ by extending the α_i, which are consistent functions, to U. Then (U, α) is an upper bound for $\mathcal{C}$ in $\mathcal{F}$.

We can now apply Zorn's Lemma to obtain a maximal element (X, α) of $\mathcal{F}$. We claim that either $X = A$ or $\mathrm{Im}(\alpha) = B$. For suppose that both statements are false, and let $a \in A - X$ and $b \in B - \mathrm{Im}(\alpha)$. Put $Y = X \cup \{a\}$ and define $\beta : Y \to B$ by $\beta(a) = b$ and $\beta|_X = \alpha$. Then β is injective since $b \notin \mathrm{Im}(\alpha)$, and clearly $(\alpha, X) \preceq (\beta, Y)$, which is a contradiction. Therefore, either $X = A$ and hence $|A| \leq |B|$ by definition of the linear ordering of cardinals, or else $\mathrm{Im}(\alpha) = B$. In the latter case for each b in B choose an a_b in A such that $\alpha(a_b) = b$: then the map sending $b \mapsto a_b$ is an injective function from B to A. Therefore $|B| \leq |A|$. □

Axioms equivalent to Zorn's Lemma

We mention in passing three axioms that are logically equivalent to Zorn's Lemma.

(i) *The Axiom of Well-Ordering*. Every non-empty set can be well-ordered.

Recall from Section 1.2 that a *well order* on a set is a linear order such that each non-empty subset has a first element. Compare the Axiom of Well-Ordering with the Well-Ordering Law in Section 2.1, which implies that $\leq$ is a well-order on $\mathbb{N}$.

(ii) *The Principle of Transfinite Induction*. Let S be a well-ordered set and T a non-empty subset of S. Let $t \in S$ and assume that $t \in T$ holds whenever it is true that $x \in T$ for every x in S such that $x < t$. Then $T = S$.

This result, which is the basis for the method of *proof by transfinite induction*, should be compared with the Principal of Mathematical Induction (2.1.1).

(iii) *The Axiom of Choice*. Let $\{S_i \mid i \in I\}$ be a non-empty set whose members are non-empty sets S_i. Then there is at least one function $\alpha : I \to \bigcup_{i \in I} S_i$ such that $\alpha(i) \in S_i$ for each $i \in I$. Such functions are called *choice functions*.

Informally we may express this by saying that it is possible to choose an element simultaneously from every set S_i. The Axiom of Choice is perhaps the most "obvious" of the four axioms. For a very clear proof of the equivalence of the axioms see [7].

2 The Integers

The role of the integers is central in algebra, as it is in all parts of mathematics. One reason for this is that the set of integers $\mathbb{Z}$, together with the standard arithmetic operations of addition and multiplication, serves as a model for several of the fundamental structures of algebra, including groups and rings. In this chapter the most basic properties of the integers are developed.

2.1 Well-ordering and mathematical induction

We begin by listing the properties of the fundamental arithmetic operations on $\mathbb{Z}$, addition and multiplication. In the following a, b, c are arbitrary integers.
(i) $a + b = b + a$, $ab = ba$, (commutative laws);
(ii) $(a + b) + c = a + (b + c)$, $(ab)c = a(bc)$, (associative laws);
(iii) $(a + b)c = ac + bc$, (distributive law);
(iv) $0 + a = a$ and $1 \cdot a = a$, (existence of identities);
(v)   each integer a has a *negative* $-a$ with the property $a + (-a) = 0$;
(vi) if $ab = 0$, then $a = 0$ or $b = 0$.

Next we list properties of the relation $\leq$ on $\mathbb{Z}$.
(vii) $\leq$ is a linear order on $\mathbb{Z}$, i. e., the relation $\leq$ is reflexive, antisymmetric and transitive; in addition, for any pair of integers a, b either $a \leq b$ or $b \leq a$;
(viii) if $a \leq b$ and $c \geq 0$, then $ac \leq bc$;
(ix)   if $a \leq b$, then $-b \leq -a$.

These properties are assumed as axioms. But there is a further property of the linearly ordered set $(\mathbb{Z}, \leq)$ which is independent of the above axioms and is quite vital for the development of the elementary theory of the integers.

The Well-Ordering Law

Let k be a fixed integer and put $U = \{n \mid n \in \mathbb{Z}, n \geq k\}$. Suppose that S is a non-empty subset of U. Then the *Well-Ordering Law* (WO) asserts that S has a smallest element. Thus $\leq$ is a well order on U in the sense of Section 1.2.

While this may seem a harmless assumption, it cannot be deduced from axioms (i)–(ix) and must be adopted as an additional axiom. The importance of WO for us is that it provides a sound basis for the method of proof by mathematical induction. This is embodied in

(2.1.1) (The Principle of Mathematical Induction). *Let k be an integer and let $U = \{n \mid n \in Z, n \geq k\}$. Assume that S is a subset of U with the properties:*
(i) $k \in S$;

https://doi.org/10.1515/9783110691160-002

(ii) *if $n \in S$, then $n + 1 \in S$.*

Then S equals U.

Proof. Once again the assertion sounds fairly obvious, but in order to prove it, we must use WO. To see how WO applies, assume that $S \neq U$, so that $S' = U - S$ is not empty. Then WO guarantees that S' has a smallest element, say s. Notice that $k < s$ since $k \in S$ by hypothesis. Thus $k \leq s - 1$ and $s - 1 \notin S'$ because s is minimal in S'. Hence $s - 1 \in S$, which by (ii) above implies that $s \in S$, a contradiction. Thus (2.1.1) is established. □

The method of proof by induction

Suppose that k is a fixed integer and that for each integer $n \geq k$ there is a proposition $p(n)$, which is either true or false. Assume that the following hold:

(i) $p(k)$ is true;

(ii) if $p(n)$ is true, then $p(n + 1)$ is true.

Then we can conclude that $p(n)$ is true for all $n \geq k$.

For let S be the set of all integers $n \geq k$ for which $p(n)$ is true. Then the hypotheses of PMI (Principle of Mathematical Induction) apply to S. The conclusion is that S equals $\{n \mid n \in \mathbb{Z}, n \geq k\}$, i. e., $p(n)$ is true for all $n \geq k$.

Here is a simple example of proof by mathematical induction.

Example (2.1.1). Use mathematical induction to show that $8^{n+1} + 9^{2n-1}$ is a multiple of 73 for all positive integers n.

Let $p(n)$ denote the statement: $8^{n+1} + 9^{2n-1}$ is a multiple of 73. Then $p(1)$ is certainly true since $8^{n+1} + 9^{2n-1} = 73$ when $n = 1$. Assume that $p(n)$ is true; we have to deduce that $p(n + 1)$ is true. Now we may rewrite $8^{(n+1)+1} + 9^{2(n+1)-1}$ in the form

$$8^{n+2} + 9^{2n+1} = 8(8^{n+1} + 9^{2n-1}) + 9^{2n+1} - 8 \cdot 9^{2n-1}$$
$$= 8(8^{n+1} + 9^{2n-1}) + 73 \cdot 9^{2n-1}.$$

Since both terms in the last expression are multiples of 73, so is $8^{n+2} + 9^{2n+1}$. Thus $p(n + 1)$ is true and by PMI the statement $p(n)$ is true for all $n \geq 1$.

(2.1.2) (Alternate Form of PMI). *Let k be an integer and let $U = \{n \mid n \in \mathbb{Z}, n \geq k\}$. Assume that S is a subset of U with the properties*

(i) $k \in S$;

(ii) *if $m \in S$ for all integers m such that $k \leq m < n$, then $n \in S$.*

Then $S = U$.

This variant of PMI follows from WO just as the original form does. There are situations where proof by induction cannot be easily used but the alternate form is effective. In such a case one has a proposition $p(n)$ for $n \geq k$ such that:

(i) $p(k)$ is true;

(ii) if $p(m)$ is true whenever $k \leq m < n$, then $p(n)$ is true.

The conclusion is that $p(n)$ is true for all $n \geq k$.

A good example of a proposition where this type of induction proof is successful is the Fundamental Theorem of Arithmetic – see (2.2.7).

Our approach to the integers in this section has been quite naive: we have simply stated as axioms all the properties that we need. For a good axiomatic treatment of the construction of the integers, including an account of the axioms of Peano, see [6].

Exercises (2.1).

(1) Use induction to establish the following summation formulas for $n \geq 1$.

   (i) $1 + 2 + 3 + \cdots + n = \frac{1}{2}n(n + 1)$;

   (ii) $1^2 + 2^2 + 3^2 + \cdots + n^2 = \frac{1}{6}n(n + 1)(2n + 1)$;

   (iii) $1^3 + 2^3 + 3^3 + \cdots + n^3 = (\frac{1}{2}n(n + 1))^2$.

(2) Deduce the alternate form of PMI from WO.

(3) Prove that $2^n > n^3$ for all integers $n \geq 10$.

(4) Prove that $2^n > n^4$ for all integers $n \geq 17$.

(5) Prove by mathematical induction that 6 divides $n^3 - n$ for all integers $n \geq 0$.

(6) Use the alternate form of mathematical induction to show that any n cents worth of postage, where $n \geq 12$, can be made up by using only 4-cent and 5-cent stamps. [Hint: first verify the statement for $n \leq 15$.]

2.2 Division in the integers

In this section we establish the basic properties of the integers that relate to division, notably the Division Algorithm, the existence of greatest common divisors and the Fundamental Theorem of Arithmetic.

Recall that if a, b are integers, then a *divides* b, in symbols

$$a \mid b,$$

if there is an integer c such that $b = ac$. The following properties of division are simple consequences of the definition, as the reader should verify.

(2.2.1).

(i) *The relation of division is a partial order on the set of non-negative integers.*

(ii) *If $a \mid b$ and $a \mid c$, then $a \mid bx + cy$ for all integers x, y.*

(iii) $a \mid 0$ *for all* a, *while* $0 \mid a$ *if and only if* $a = 0$.

(iv) $1 \mid a$ *for all* a, *while* $a \mid 1$ *if and only if* $a = \pm 1$.

The division algorithm

The first result about the integers of real significance is the *Division Algorithm*; it codifies the time-honored process of dividing one integer by another to obtain a quotient and remainder. It should be noted that the proof of the result uses WO.

(2.2.2). *Let* a, b *be integers with* $b \neq 0$. *Then there exist unique integers* q *(the quotient) and* r *(the remainder) such that* $a = bq + r$ *and* $0 \leq r < |b|$.

Proof. Let S be the set of all non-negative integers of the form $a - bq$ where $q \in \mathbb{Z}$. In the first place we need to observe that S is not empty. Indeed, if $b > 0$ and we choose an integer $q \leq \frac{a}{b}$, then $a - bq \geq 0$; if $b < 0$, choose an integer $q \geq \frac{a}{b}$, so that again $a - bq \geq 0$. Applying the Well-Ordering Law to the set S, we conclude that it has a smallest element, say $r = a - bq$ for some integer q. Hence $a = bq + r$.

Now suppose that $r \geq |b|$. If $b > 0$, then $a - b(q+1) = r - b < r$, while if $b < 0$, then $a - b(q-1) = r + b < r$. In each case a contradiction is reached since we have found an integer in S which is less than r. Hence $r < |b|$.

Finally, we must show that q and r are unique. Suppose that $a = bq' + r'$ where $q', r' \in \mathbb{Z}$ and $0 \leq r' < |b|$. Then $bq + r = bq' + r'$ and $b(q - q') = r' - r$. Thus $|b| \cdot |q - q'| = |r - r'|$. If $q \neq q'$, then $|r - r'| \geq |b|$, whereas $|r - r'| < |b|$ since $0 \leq r$, $r' < |b|$. Therefore $q = q'$ and it follows at once that $r = r'$. $\qquad\square$

When $a < 0$ or $b < 0$, care must be taken to ensure that a negative remainder is not obtained. For example, if $a = -21$ and $b = -4$, then $-21 = (-4)6 + 3$, so that $q = 6$ and $r = 3$.

Greatest common divisors

Let $a_1, a_2, \ldots, a_n$ be integers. An integer c which divides every a_i is called a *common divisor* of $a_1, a_2, \ldots, a_n$. Our next goal is to establish the existence of a *greatest common divisor*.

(2.2.3). *Let* $a_1, a_2, \ldots, a_n$ *be integers. Then there is a unique integer* $d \geq 0$ *with the properties:*

(i) *d is a common divisor of $a_1, a_2, \ldots, a_n$;*

(ii) *every common divisor of $a_1, a_2, \ldots, a_n$ divides d;*

(iii) *$d = a_1 x_1 + \cdots + a_n x_n$ for some integers x_i.*

Proof. If all of the a_i are 0, we can take $d = 0$ since this fits the description. So assume that at least one a_i is non-zero. Then the set S of all positive integers $a_1x_1+a_2x_2+\cdots+a_nx_n$ with $x_i \in \mathbb{Z}$ is non-empty. By WO there is a least element in S, say

$$d = a_1x_1 + a_2x_2 + \cdots + a_nx_n.$$

If an integer c divides each a_i, then $c \mid d$ by (2.2.1). Thus it only remains to show that $d \mid a_i$ for all i.

By the Division Algorithm we can write $a_i = dq_i + r_i$ where $q_i, r_i \in \mathbb{Z}$ and $0 \le r_i < d$. Then

$$r_i = a_i - dq_i = a_1(-x_1q_i) + \cdots + a_i(1 - x_iq_i) + \cdots + a_n(-x_nq_i).$$

If $r_i \ne 0$, then $r_i \in S$, which contradicts the minimality of d in S. Hence $r_i = 0$ and $d \mid a_i$ for all i.

Finally, we show that d is unique. If d' is another integer satisfying (i) and (ii), then $d \mid d'$ and $d' \mid d$, so that $d = d'$ since $d, d' \ge 0$. $\qquad\square$

The integer d of (2.2.3) is called the *greatest common divisor* of $a_1, a_2, \ldots, a_n$, in symbols

$$d = \gcd\{a_1, a_2, \ldots, a_n\}.$$

If $d = 1$, the integers $a_1, a_2, \ldots, a_n$ are said to be *relatively prime*; of course this means that the integers have no common divisors except ± 1.

The Euclidean[1] Algorithm

The proof of the existence of gcd's which has just been given is not constructive, i. e., it does not provide a method for calculating gcd's. However, there is a well known procedure called the *Euclidean Algorithm* which is effective in this respect.

Assume that a, b are integers with $b \ne 0$. Apply the Division Algorithm to divide a by b to get quotient q_1 and remainder r_1. Next if $r_1 \ne 0$, divide b by r_1 to get quotient q_2 and remainder r_2; then, if $r_2 \ne 0$, divide r_1 by r_2 to get quotient q_3 and remainder r_3. And so on. By WO there is a smallest non-zero remainder, say r_{n-1}. Thus $r_n = 0$ and we

---

1 Euclid of Alexandria (325–265 BC).

have a system of integer equations

$$\begin{cases} a = bq_1 + r_1, \\ b = r_1q_2 + r_2, \\ r_1 = r_2q_3 + r_3, \\ \quad\vdots \\ r_{n-3} = r_{n-2}q_{n-1} + r_{n-1}, \\ r_{n-2} = r_{n-1}q_n + 0. \end{cases}$$

Here $0 \le r_1 < |b|$, $0 \le r_i < r_{i-1}$ and r_{n-1} is the smallest non-zero remainder. With this notation we can state:

(2.2.4) (The Euclidean Algorithm). *The greatest common divisor of a and b equals the last non-zero remainder r_{n-1}.*

Proof. Starting with the second last equation in the system above, we can solve back for r_{n-1}, obtaining eventually an expression of the form $r_{n-1} = ax + by$, where $x, y \in \mathbb{Z}$. This shows that any common divisor of a and b must divide r_{n-1}. We can also use the system of equations above to show successively that $r_{n-1} \mid r_{n-2}$, $r_{n-1} \mid r_{n-3}$, $\ldots$, etc., and finally $r_{n-1} \mid b$, $r_{n-1} \mid a$. It follows that $r_{n-1} = \gcd\{a, b\}$ by uniqueness of gcd's. □

Example (2.2.1). Find $\gcd(76, 60)$. We compute successively: $76 = 60 \cdot 1 + 16$, $60 = 16\cdot3+12$, $16 = 12\cdot1+4$, $12 = 4\cdot3+0$. Hence $\gcd\{76, 60\} = 4$, the last non-zero remainder. By reading back from the third equation we obtain the predicted expression for the gcd, $4 = 76 \cdot 4 + 60 \cdot (-5)$.

The Euclidean algorithm can also be applied to calculate gcd's of more than two integers by using the formula

$$\gcd\{a_1, a_2, \ldots, a_{m+1}\} = \gcd\{\gcd\{a_1, a_2, \ldots a_m\}, a_{m+1}\}$$

and induction on m: see Exercise (2.2.1).

A very useful tool in working with divisibility is:

(2.2.5) (Euclid's Lemma). *Let a, b, m be integers. If m divides ab and m is relatively prime to a, then m divides b.*

Proof. By hypothesis $\gcd\{a, m\} = 1$, so by (2.2.3) there are integers x, y such that $1 = mx + ay$. Multiplying by b, we obtain $b = mbx + aby$. Since m divides ab, it divides the right side of the equation. Hence m divides b. □

Recall that a *prime number* is an integer $p > 1$ such that ±1 and $\pm p$ are its only divisors. If p is a prime and a is any integer, then either $\gcd\{a, p\} = 1$ or $p \mid a$. Thus (2.2.5) has the consequence.

(2.2.6). *If a prime p divides ab where a, b ∈ ℤ, then p divides a or b.*

The Fundamental Theorem of Arithmetic

It is a basic result that every integer greater than 1 can be expressed as a product of primes. The proof of this result is a good example of proof by the alternate form of mathematical induction.

(2.2.7). *Every integer n > 1 can be expressed as a product of primes. Moreover the expression is unique up to the order of the factors.*

Proof. (i) *Existence.* We show that n is a product of primes, which is certainly true if $n = 2$. Assume that every integer m satisfying $2 \le m < n$ is a product of primes. If n itself is a prime, there is nothing to prove. Otherwise $n = n_1 n_2$ where $1 < n_i < n$. Then n_1 and n_2 are both products of primes, whence so is $n = n_1 n_2$. The result now follows by the alternate form of mathematical induction (2.1.2).

(ii) *Uniqueness.* In this part we have to show that n has a unique expression as a product of primes. Again this is clearly correct for $n = 2$. Assume that if $2 \le m < n$, then m is uniquely expressible as a product of primes. Next suppose that $n = p_1 p_2 \cdots p_r = q_1 q_2 \cdots q_s$ where the p_i and q_j are primes. Then $p_1 \mid q_1 q_2 \cdots q_s$ and by (2.2.6) the prime p_1 must divide, and hence equal, one of the q_j's; we can assume $p_1 = q_1$ by relabelling the q_j's if necessary. Now cancel p_1 to get $m = p_2 \cdots p_r = q_2 \cdots q_s$. Since $m = n/p_1 < n$, we deduce that $p_2 = q_2, p_3 = q_3, \ldots, p_r = q_r$, and $r = s$, after further relabelling of the q_j's. Hence the result is proven. □

A convenient expression for an integer $n > 1$ is

$$n = p_1^{e_1} p_2^{e_2} \cdots p_k^{e_k}$$

where the p_i are *distinct* primes and $e_i > 0$. That the p_i and e_i are unique up to order follows from (2.2.7).

Finally in this section we will prove the famous theorem of Euclid on the infinitude of primes.

(2.2.8). *There exist infinitely many prime numbers.*

Proof. Suppose this is false and let $p_1, p_2, \ldots, p_k$ be the list of all the primes. The trick is to produce a prime that is not on the list. To do this put $n = p_1 p_2 \cdots p_k + 1$. Now no p_i can divide n, otherwise $p_i \mid 1$. But n is certainly divisible by at least one prime, so we reach a contradiction. □

Example (2.2.2). If p is a prime, then $\sqrt{p}$ is not a rational number.

For, assume that $\sqrt{p}$ is a rational and $\sqrt{p} = \frac{m}{n}$ where m, n are integers; evidently there is nothing to be lost in assuming that m and n are relatively prime since any common factor can be cancelled. Squaring both sides, we obtain $p = m^2/n^2$ and $m^2 =$

pn^2. Hence $p \mid m^2$ and Euclid's Lemma shows that $p \mid m$. Write $m = pm_1$ for some integer m_1. Then $p^2m_1^2 = pn^2$, so $pm_1^2 = n^2$. Thus $p \mid n^2$ and $p \mid n$: but this means m and n are not relatively prime, a contradiction.

Exercises (2.2).

(1) Let $a_1, a_2, \ldots, a_{n+1}$ be integers. Prove that

$$\gcd\{a_1, a_2, \ldots, a_{n+1}\} = \gcd\{\gcd\{a_1, a_2, \ldots, a_n\}, a_{n+1}\}.$$

(2) Prove that $\gcd\{ka_1, ka_2, \ldots, ka_n\} = k \cdot \gcd\{a_1, a_2, \ldots, a_n\}$ where the a_i and $k \geq 0$ are integers.

(3) Use the Euclidean Algorithm to compute the following: $\gcd\{840, 410\}$, $\gcd\{24, 328, 472\}$. Then express each gcd as a linear combination of the relevant integers.

(4) Consider the equation $ax + by = c$ where a, b, c are given integers.
   (i) Prove that there is a solution in integers x, y if and only if $d = \gcd\{a, b\}$ divides c.
   (ii) Write $d = ua + vb$ where $u, v \in \mathbb{Z}$. Prove that the general solution of the equation is $x = \frac{uc}{d} + \frac{mb}{d}, y = \frac{vc}{d} - \frac{ma}{d}$ where m is an arbitrary integer.

(5) Find all solutions in integers of the equation $6x + 11y = 1$.

(6) If p and q are distinct primes, prove that $\sqrt{pq}$ is irrational.

(7) Let $a_1, a_2, \ldots, a_m$ be positive integers and write $a_i = p_1^{e_{i1}} p_2^{e_{i2}} \cdots p_n^{e_{in}}$ where the e_{ij} are integers ≥ 0 and the primes p_i are all different. Show that $\gcd\{a_1, a_2, \ldots, a_m\} = p_1^{f_1} p_2^{f_2} \cdots p_n^{f_n}$ where $f_j = \min\{e_{1j}, e_{2j}, \ldots, e_{mj}\}$.

(8) A *least common multiple* (or lcm) of integers $a_1, a_2, \ldots, a_m$ is an integer $\ell \geq 0$ such that each a_i divides ℓ and ℓ divides any integer which is divisible by every a_i.
   (i) Let $a_i = p_1^{e_{i1}} p_2^{e_{i2}} \cdots p_n^{e_{in}}$ where the e_{ij} are integers ≥ 0 and the primes p_i are all different. Prove that lcm's exist and are unique by establishing the formula $\mathrm{lcm}\{a_1, a_2, \ldots, a_m\} = p_1^{g_1} p_2^{g_2} \cdots p_n^{g_n}$ with $g_j = \max\{e_{1j}, e_{2j}, \ldots, e_{mj}\}$.
   (ii) Prove that $\gcd\{a, b\} \cdot \mathrm{lcm}\{a, b\} = ab$.

(9) Let r be a rational number and let a and b be relatively prime integers. If ar and br are integers, prove that r is also an integer.

(10) Let a and b be integers with $b > 0$. Prove that there are integers u, v such that $a = bu + v$ and $-\frac{b}{2} \leq v < \frac{b}{2}$. [Hint: start with the Division Algorithm.]

(11) Prove that $\gcd\{4n + 5, 3n + 4\} = 1$ for all integers n.

(12) Prove that $\gcd\{2n + 6, n^2 + 3n + 2\} = 2$ or 4 for any integer n and show that both possibilities can occur.

(13) Show that if $2^n + 1$ is prime, then n must have the form 2^l. (Such primes are called *Fermat*[2] *primes*.)

(14) The only integer n which is expressible as $a^3(3a+1)$ and $b^2(b+1)^3$ with a, b relatively prime and positive is 2000.

---

2 Pierre de Fermat (1601–1665).

2.3 Congruences

The notion of a congruence was introduced by Gauss[3] in 1801, but it had long been implicit in ancient writings concerned with the computation of dates.

Let m be a positive integer. Two integers a, b are said to be *congruent modulo m*, in symbols

$$a \equiv b \ (\text{mod } m),$$

if m divides $a - b$. Thus congruence modulo m is a relation on $\mathbb{Z}$ and an easy check reveals that it is an equivalence relation. Hence the set $\mathbb{Z}$ splits up into equivalence classes, which in this context are called *congruence classes modulo m*: see (1.2.2). The unique congruence class to which an integer a belongs is written

$$[a] \text{ or } [a]_m = \{a + mq \mid q \in \mathbb{Z}\}.$$

By the Division Algorithm any integer a can be written in the form $a = mq + r$ where $q, r \in \mathbb{Z}$ and $0 \le r < m$. Thus $a \equiv r \ (\text{mod } m)$ and $[a] = [r]$. Therefore $[0], [1], \dots, [m-1]$ are all the congruence classes modulo m. Furthermore, if $[r] = [r']$ where $0 \le r, r' < m$, then $m \mid r - r'$, which can only mean that $r = r'$. Thus we have proved:

(2.3.1). *Let m be any positive integer. Then there are exactly m congruence classes modulo m, namely* $[0], [1], \dots, [m - 1]$.

Congruence arithmetic
We will write

$$\mathbb{Z}_m$$

for the set of all congruences classes modulo m. Next we define operations of addition and multiplication for congruence classes, thereby introducing the possibility of arithmetic in $\mathbb{Z}_m$.

The *sum* and *product* of congruence classes modulo m are defined by the rules

$$[a] + [b] = [a + b] \quad \text{and} \quad [a] \cdot [b] = [ab].$$

These definitions are surely the natural ones. However, some care must be exercised in framing definitions of this type. A congruence class can be represented by any one of its elements: we need to ensure that the sum and product specified above depend only on the congruence classes themselves, not on the chosen representatives.

---

3 Carl Friedrich Gauss (1777–1855).

To this end, let $a' \in [a]$ and $b' \in [b]$. It must be shown that $[a + b] = [a' + b']$ and $[ab] = [a'b']$. Now $a' = a + mu$ and $b' = b + mv$ for some $u, v \in \mathbb{Z}$. Therefore $a' + b' = (a + b) + m(u + v)$ and $a'b' = ab + m(av + bu + muv)$; from these equations it follows that $a' + b' \equiv a + b \pmod{m}$ and $a'b' \equiv ab \pmod{m}$. Thus $[a' + b'] = [a + b]$ and $[a'b'] = [ab]$, as required.

Now that we know the sum and product of congruence classes to be well-defined, it is a routine task to establish the basic properties of these operations.

(2.3.2). *Let m be a positive integer and let* $[a], [b], [c]$ *be congruence classes modulo m. Then*

(i) $[a] + [b] = [b] + [a]$ *and* $[a] \cdot [b] = [b] \cdot [a]$;

(ii) $([a] + [b]) + [c] = [a] + ([b] + [c])$ *and* $([a][b])[c] = [a]([b][c])$;

(iii) $([a] + [b])[c] = [a][c] + [b][c]$;

(iv) $[0] + [a] = [a]$ *and* $[1][a] = [a]$;

(v) $[a] + [-a] = [0]$.

Since all of these properties are valid in $\mathbb{Z}$ as well as $\mathbb{Z}_m$ – see Section 2.1 – we recognize some common features of the arithmetics on $\mathbb{Z}$ and $\mathbb{Z}_m$. This commonality can be expressed by saying that $\mathbb{Z}$ and $\mathbb{Z}_m$ are examples of *commutative rings with identity*, as will be explained in Chapter 6.

Fermat's Theorem

Before proceeding to this well-known theorem, we will establish a frequently used property of the binomial coefficients. If n and r are integers satisfying $0 \le r \le n$, the *binomial coefficient* $\binom{n}{r}$ is the number of ways of choosing r objects from a set of n distinct objects. There is the well-known formula

$$\binom{n}{r} = \frac{n!}{r!(n-r)!} = \frac{n(n-1)\cdots(n-r+1)}{r!}.$$

The property needed is:

(2.3.3). *If p is a prime and* $0 < r < p$, *then* $\binom{p}{r} \equiv 0 \pmod{p}$.

Proof. Write $\binom{p}{r} = pm$ where m is the rational number

$$\frac{(p-1)(p-2)\cdots(p-r+1)}{r!}.$$

Notice that each prime appearing as a factor of the numerator or denominator of m is smaller than p. Write $m = \frac{u}{v}$ where u and v are relatively prime integers. Then $v\binom{p}{r} = pmv = pu$ and by Euclid's Lemma v divides p. Now $v \ne p$, so $v = 1$ and $m = u \in \mathbb{Z}$. Hence p divides $\binom{p}{r}$. $\qquad\square$

We are now able to prove what is often called *Fermat's Little Theorem*, to distinguish it from the well known Fermat's Last Theorem.

(2.3.4). *If p is a prime and x is any integer, then $x^p \equiv x \pmod{p}$.*

Proof. Since $(-x)^p \equiv -x^p \pmod{p}$, whether or not p is odd, there is no loss in assuming that $x \geq 0$. We will use induction on x to show that $x^p \equiv x \pmod{p}$, which certainly holds for $x = 0$. Assume it is true for x. Then by the Binomial Theorem

$$(x + 1)^p = \sum_{r=0}^{p} \binom{p}{r} x^r \equiv x^p + 1 \pmod{p}$$

since p divides $\binom{p}{r}$ if $0 < r < p$. Because $x^p \equiv x \pmod{p}$, it follows that $(x + 1)^p \equiv x + 1 \pmod{p}$. The induction is now complete. □

Solving Congruences

Just as we solve equations for unknown real numbers, we can try to solve congruences for unknown integers. The simplest case is that of a *linear congruence* with a single unknown x; this has the form $ax \equiv b \pmod{m}$, where a, b, m are given integers.

(2.3.5). *Let a, b, m be integers with $m > 0$. Then there is a solution x of the congruence $ax \equiv b \pmod{m}$ if and only if $\gcd\{a, m\}$ divides b.*

Proof. Set $d = \gcd\{a, m\}$. If x is a solution of congruence $ax \equiv b \pmod{m}$, then $ax = b + mq$ for some $q \in \mathbb{Z}$, from which it follows that d must divide b. Conversely, assume that $d \mid b$. By (2.2.3) there are integers u, v such that $d = au + mv$. Multiplying this equation by the integer b/d, we obtain $b = a(ub/d) + m(vb/d)$. Put $x = ub/d$, which is an integer; then $ax \equiv b \pmod{m}$ and x is a solution of the congruence. □

The most important case is for $b = 1$.

Corollary (2.3.6). *Let a, m be integers with $m > 0$. Then the congruence $ax \equiv 1 \pmod{m}$ has a solution x if and only if a is relatively prime to m.*

It is worthwhile translating the last result into the language of congruence arithmetic. Given an integer $m > 0$ and a congruence class $[a]$ modulo m, there is a congruence class $[x]$ such that $[a][x] = [1]$ if and only if a is relatively prime to m. Thus we can tell which congruence classes modulo m have "inverses": they are the classes $[x]$ where $0 < x < m$ and x is relatively prime to m. The number of invertible congruence classes modulo m is denoted by

$$\phi(m).$$

This is the number of integers x such that $0 < x < m$ and $\gcd\{x, m\} = 1$: the function ϕ is called *Euler's*[4] *function*. Next we consider systems of linear congruences.

(2.3.7) (The Chinese Remainder Theorem). *Let $a_1, a_2, \ldots, a_k$ and $m_1, m_2, \ldots, m_k$ be integers with $m_i > 0$. Assume that $\gcd\{m_i, m_j\} = 1$ if $i \neq j$. Then there is a common solution x of the system of congruences*

$$\begin{cases} x \equiv a_1 \pmod{m_1} \\ x \equiv a_2 \pmod{m_2} \\ \quad\vdots \\ x \equiv a_k \pmod{m_k}. \end{cases}$$

When $k = 2$, this striking result was discovered by the Chinese mathematician Sun Tse, who lived sometime between the Third and Fifth centuries AD.

Proof of (2.3.7). Put $m = m_1 m_2 \cdots m_k$ and $m_i' = m/m_i$. Then m_i and m_i' are relatively prime, so by (2.3.6) there exists an integer ℓ_i such that $m_i' \ell_i \equiv 1 \pmod{m_i}$. Now let $x = a_1 m_1' \ell_1 + \cdots + a_k m_k' \ell_k$. Then

$$x \equiv a_i m_i' \ell_i \equiv a_i \pmod{m_i}$$

since $m_i \mid m_j'$ if $i \neq j$. $\qquad\square$

As an application of (2.3.7) a well-known formula for Euler's function will be derived.

(2.3.8).
(i) *If m and n are relatively prime positive integers, then $\phi(mn) = \phi(m)\phi(n)$.*
(ii) *If $m = p_1^{l_1} p_2^{l_2} \cdots p_k^{l_k}$ with $l_i > 0$ and distinct primes p_i, then*

$$\phi(m) = \prod_{i=1}^{k} (p_i^{l_i} - p_i^{l_i - 1}).$$

Proof. (i) Let U_m denote the set of invertible congruence classes in $\mathbb{Z}_m$. Thus $|U_m| = \phi(m)$. Define a map $\alpha : U_{mn} \to U_m \times U_n$ by the rule $\alpha([a]_{mn}) = ([a]_m, [a]_n)$. First of all observe that α is well-defined. Next suppose that $\alpha([a]_{mn}) = \alpha([a']_{mn})$. Then $[a]_m = [a']_m$ and $[a]_n = [a']_n$, equations which imply that $a - a'$ is divisible by m and n, and hence by mn. Therefore $[a]_{mn} = [a']_{mn}$ and α is an injective function.

In fact α is also surjective. For, if $[a]_m \in U_m$ and $[b]_n \in U_n$ are given, the Chinese Remainder Theorem assures us that there is an integer x such that $x \equiv a \pmod{m}$ and

---

4 Leonhard Euler (1707–1783).

$x \equiv b \pmod{n}$. Hence $[x]_m = [a]_m$ and $[x]_n = [b]_n$, so that $\alpha([x]_{mn}) = ([a]_m, [b]_n)$. Therefore α is a bijection and consequently $|U_{mn}| = |U_m \times U_n| = |U_m| \cdot |U_n|$, as required.

(ii) Suppose that p is a prime and $n > 0$. There are p^{n-1} multiples of p among the integers $0, 1, \ldots, p^n - 1$; therefore $\phi(p^n) = p^n - p^{n-1}$. Finally apply (2.3.8)(i) to obtain the formula indicated. $\qquad\square$

We end the chapter with several examples which illustrate the utility of congruences.

Example (2.3.1). Show that an integer is divisible by 3 if and only if the sum of its digits is a multiple of 3.

Let $n = a_0 a_1 \ldots a_k$ be the decimal representation of an integer n. Thus $n = a_k + a_{k-1}10 + a_{k-2}10^2 + \cdots + a_0 10^k$ where $0 \le a_i < 10$. The key observation is that $10 \equiv 1 \pmod{3}$, i.e., $[10] = [1]$. Hence $[10^i] = [10]^i = [1]^i = [1]$, i.e., $10^i \equiv 1 \pmod{3}$ for all $i \ge 0$. It therefore follows that $n \equiv a_0 + a_1 + \cdots + a_k \pmod{3}$. The assertion is an immediate consequence of this congruence.

Example (2.3.2) (Days of the week). Congruences have long been used implicitly to compute dates. As an example, let us determine what day of the week September 25 of the year 2020 was.

To keep track of the days assign the integers $0, 1, 2, \ldots, 6$ as labels for the days of the week, say Sunday = 0, Monday = 1, ..., Saturday = 6. Suppose that we reckon from January 5, 2014, which was a Sunday. All we have to do is count the number of days from this date to September 25, 2020. Allowing for leap years, this number is 2455. Now $2455 \equiv 5 \pmod{7}$ and 5 is the label for Friday. Therefore September 25, 2020 was a Friday.

Example (2.3.3) (The Basket of Eggs Problem). What is the smallest number of eggs a basket can contain if, when eggs are removed k at time, there is one egg left when $k = 2, 3, 4, 5$ or 6 and there are no eggs left when $k = 7$? (This ancient problem is mentioned in an Indian manuscript of the 7th Century).

Let x be the number of eggs in the basket. The conditions require that $x \equiv 1 \pmod{k}$ for $k = 2, 3, 4, 5, 6$ and $x \equiv 0 \pmod{k}$ for $k = 7$. Clearly this amounts to x satisfying the four congruences $x \equiv 1 \pmod{3}$, $x \equiv 1 \pmod{4}$, $x \equiv 1 \pmod{5}$ and $x \equiv 0 \pmod{7}$. Furthermore these are equivalent to the congruences

$$x \equiv 1 \pmod{60} \quad \text{and} \quad x \equiv 0 \pmod{7}.$$

By the Chinese Remainder Theorem there is a solution to this pair of congruences: we have to find the smallest positive solution. Applying the method of the proof of (2.3.7), we have $m_1 = 60$, $m_2 = 7$, $m = 420$ and thus $m_1' = 7$, $m_2' = 60$. Also $\ell_1 = 43$, $\ell_2 = 2$. Therefore one solution is given by $x = 1 \cdot 7 \cdot 43 + 0 \cdot 60 \cdot 2 = 301$. If y is any other solution, observe that $y - x$ must be divisible by $60 \times 7 = 420$. Hence the general solution is $x = 301 + 420q$, $q \in \mathbb{Z}$. So the smallest positive solution is 301.

The next example is a refinement of Euclid's Theorem on the infinity of primes – see (2.2.8).

Example (2.3.4). Prove that there are infinitely many primes of the form $3n + 2$ where n is an integer ≥ 0.

In fact the proof is a variant of Euclid's method. Suppose the result is false and let the *odd* primes of the form $3n + 2$ be $p_1, p_2, \ldots, p_k$. Now consider the positive integer $m = 3p_1p_2\cdots p_k + 2$. Notice that m is odd and it is not divisible by any p_i. Therefore m is a product of odd primes different from $p_1, \ldots, p_k$. Hence m must be a product of primes of the form $3n + 1$ since every integer is of the form $3n$, $3n + 1$ or $3n + 2$. It follows that m itself must have the form $3n + 1$ and thus $m \equiv 1 \pmod 3$. On the other hand, $m \equiv 2 \pmod 3$, so we have reached a contradiction.

Actually this example is a special case of a famous theorem of Dirichlet:[5] every arithmetic progression $an + b$, where $n = 0, 1, 2, \ldots$, and the integers a and b are positive and relatively prime, contains infinitely many primes.

Example (2.3.5) (The RSA Cryptosystem). This is a secure system for message encryption which has been widely used for transmitting sensitive data since its invention in 1977 by R. Rivest, A. Shamir and L. Adleman. It has the advantage of being a public key system in which only the deciphering function is not available to the public.

Suppose that a message is to be sent from A to B. The parameters required are two distinct large primes p and q. Put $n = pq$ and $m = \phi(n)$; therefore $m = (p - 1)(q - 1)$ by (2.3.8). Let a be an integer in the range 1 to m which is relatively prime to m. Then by (2.3.6) there is a unique integer b satisfying $0 < b < m$ and $ab \equiv 1 \pmod m$. The sender A is assumed to know the integers a and n, while the receiver B knows b and n.

The message to be sent is first converted to an integer x which is not divisible by p or q and satisfies $0 < x < n$. Then A enciphers x by raising it to the power a and then reducing modulo n. In this form the message is transmitted to B. On receiving the transmitted message, B raises it to the power b and reduces modulo n. The result will be the original message x. What is being claimed here is that $x^{ab} \equiv x \pmod n$, since $0 < x < n$. To see why this holds, first write

$$ab = 1 + lm = 1 + l(p - 1)(q - 1)$$

with l an integer. Then

$$x^{ab} = x^{1+l(p-1)(q-1)} = x\left(x^{p-1}\right)^{l(q-1)} \equiv x \pmod p$$

since $x^{p-1} \equiv 1 \pmod p$ by Fermat's Theorem. Hence p divides $x^{ab} - x$, and in a similar way q also divides this number. Therefore $n = pq$ divides $x^{ab} - x$ as claimed.

---

5 Johann Peter Gustav Lejeune Dirichlet (1805–1859).

Even if n and a become public knowledge, it will be difficult to break the system by finding b. For this would require computation of the inverse of $[a]$ in $\mathbb{Z}_m$. To do this using the Euclidean Algorithm, the result that lies behind (2.3.6), one would need to know the primes p and q. But the problem of factorizing the integer $n = pq$ in order to discover the primes p and q is considered to be computationally very hard. Thus the RSA-system remains secure until much more efficient ways of factorizing large numbers become available.

Exercises (2.3).
(1) Establish the properties of congruences listed in (2.3.2).
(2) In $\mathbb{Z}_{24}$ find the inverses of $[7]$ and $[13]$.
(3) Show that if n is an odd integer, $n^2 \equiv 1 \pmod 8$.
(4) Find the general solution of the congruence $6x \equiv 11 \pmod 5$.
(5) What day of the week will April 1, 2030 be?
(6) Find the smallest positive solution x of the system of congruences $x \equiv 4 \pmod 3$, $x \equiv 5 \pmod 7$, $x \equiv 6 \pmod{11}$.
(7) Prove that there are infinitely many primes of the form $4n + 3$.
(8) Prove that there are infinitely many primes of the form $6n + 5$.
(9) In a certain culture the festivals of the snake, the monkey and the fish occur every 6, 5 and 11 years respectively. The next festivals occur in 3, 4 and 1 years respectively. How many years must pass before all three festivals occur in the same year?
(10) Prove that no integer of the form $4n + 3$ can be written as the sum of two squares of integers.

3 Introduction to Groups

Groups constitute one of the most important and natural structures in algebra. They also feature in other areas of mathematics such as geometry, topology and combinatorics. In addition groups arise in many areas of science, typically in situations where symmetry is important, as in atomic physics and crystallography. More general algebraic structures that have recently come to prominence due to the rise of information science include semigroups and monoids. This chapter serves as an introduction to these types of structure.

There is a continuing debate as to whether it is better to introduce groups or rings first in an introductory course in algebra: here we take the point of view that groups are logically the simpler objects since they involve only one binary operation, whereas rings have two. Accordingly rings are left until Chapter 6.

Historically the first groups to be studied consisted of *permutations*, i. e., bijective functions on a set. Indeed for most of the 19th Century "group" was synonymous with "group of permutations". Since permutation groups have the great advantage that their elements are concrete and easy to compute with, we begin with a discussion of permutations.

3.1 Permutations

If X is any non-empty set, a bijective function $\pi : X \rightarrow X$ is called a *permutation* of X. Thus by (1.3.1) π has a unique inverse function $\pi^{-1} : X \rightarrow X$, which is also a permutation. The set of all permutations of the set X is denoted by

$$\mathrm{Sym}(X),$$

which stands for *the symmetric group on X*.

If π and σ are permutations of X, their composite $\pi \circ \sigma$ is also a permutation; this is because it has an inverse, namely the permutation $\sigma^{-1} \circ \pi^{-1}$ by (1.3.2). In future for the sake of simplicity we will usually write

$$\pi\sigma$$

for $\pi \circ \sigma$. Of course, id $=$ id$_X$, the identity function on X, is a permutation.

At this juncture we pause to note some features of the set Sym(X): this set is "closed" with respect to forming inverses and composites, by which we mean that if $\pi, \sigma \in \mathrm{Sym}(X)$, then π^{-1} and $\pi \circ \sigma$ belong to Sym(X). In addition Sym(X) contains the identity permutation id$_X$, which has the property id$_X \circ \pi = \pi = \pi \circ \mathrm{id}_X$. And finally, the associative law for permutations is valid, $(\pi \circ \sigma) \circ \tau = \pi \circ (\sigma \circ \tau)$. In fact what these properties assert is that the pair $(\mathrm{Sym}(X), \circ)$ is a group, as defined in Section 3.2. Thus the permutations of a set afford a very natural example of a group.

https://doi.org/10.1515/9783110691160-003

Permutations of finite sets

We now begin the study of permutations of a finite set with n elements,

$$X = \{x_1, x_2, \ldots, x_n\}.$$

Let $\pi \in \mathrm{Sym}(X)$. Since π is injective, $\pi(x_1), \pi(x_2), \ldots, \pi(x_n)$ are all different and therefore constitute all n elements of the set X, but possibly in some order different from $x_1, x_2, \ldots, x_n$. Thus we can think of a permutation as a rearrangement of the order $x_1, x_2, \ldots, x_n$. A convenient way to denote the permutation π is

$$\pi = \begin{pmatrix} x_1 & x_2 & \cdots & x_n \\ \pi(x_1) & \pi(x_2) & \cdots & \pi(x_n) \end{pmatrix}$$

where the second row consists of the images under π of the elements of the first row. It should be clear to the reader that nothing essential is lost if we take X to be the set $\{1, 2, \ldots, n\}$. With this choice of X, it is usual to write

$$S_n$$

for $\mathrm{Sym}(X)$; this is called the *symmetric group of degree n*.

Computations with elements of S_n are easily performed by working directly from the definitions. An example will illustrate this.

Example (3.1.1). Let

$$\pi = \begin{pmatrix} 1 & 2 & 3 & 4 & 5 & 6 \\ 6 & 1 & 2 & 5 & 3 & 4 \end{pmatrix} \quad \text{and} \quad \sigma = \begin{pmatrix} 1 & 2 & 3 & 4 & 5 & 6 \\ 6 & 1 & 4 & 3 & 2 & 5 \end{pmatrix}$$

be elements of S_6. Hence

$$\pi\sigma = \begin{pmatrix} 1 & 2 & 3 & 4 & 5 & 6 \\ 4 & 6 & 5 & 2 & 1 & 3 \end{pmatrix}, \quad \sigma\pi = \begin{pmatrix} 1 & 2 & 3 & 4 & 5 & 6 \\ 5 & 6 & 1 & 2 & 4 & 3 \end{pmatrix}$$

and

$$\pi^{-1} = \begin{pmatrix} 1 & 2 & 3 & 4 & 5 & 6 \\ 2 & 3 & 5 & 6 & 4 & 1 \end{pmatrix}.$$

Here $\pi\sigma$ has been computed using the definition $\pi\sigma(i) = \pi(\sigma(i))$, while π^{-1} is readily obtained by reading up from $1, 2, \ldots, 6$ in the second row of π to obtain the second row of π^{-1}. Notice that $\pi\sigma \neq \sigma\pi$, i.e., multiplication of permutations is not commutative in general.

A simple count establishes the number of permutations of a finite set.

(3.1.1). *If X is a set with n elements, then $|\mathrm{Sym}(X)| = n!$.*

Proof. Consider the number of ways of constructing the second row of a permutation

$$\pi = \begin{pmatrix} x_1 & x_2 & \cdots & x_n \\ y_1 & y_2 & \cdots & y_n \end{pmatrix}.$$

There are n choices for y_1, but only $n-1$ choices for y_2 since y_1 cannot be chosen again. Next we cannot choose y_1 or y_2 again, so there are $n-2$ choices for y_3, and so on; finally, there is just one choice for y_n. Each choice of a y_i leads to a different permutation. Therefore the number of different permutations of X is $n(n-1)(n-2)\cdots 1 = n!$. □

Cyclic permutations

Let $\pi \in S_n$, so that π is a permutation of the set $\{1, 2, \ldots, n\}$. The *support* of π is defined to be the set of all i such that $\pi(i) \neq i$, in symbols

$$\operatorname{supp}(\pi).$$

Let r be an integer satisfying $1 \leq r \leq n$. Then π is called an r-cycle if $\operatorname{supp}(\pi) = \{i_1, i_2, \ldots, i_r\}$, with distinct i_j, where $\pi(i_1) = i_2, \pi(i_2) = i_3, \ldots, \pi(i_{r-1}) = i_r$ and $\pi(i_r) = i_1$. To visualize the permutation think of the integers $i_1, i_2, \ldots, i_r$ as being arranged in this order anticlockwise round a circle. Then π has the effect of rotating the circle in the anticlockwise direction. Of course π fixes all the other integers: often π is written in the form

$$\pi = (i_1 i_2 \cdots i_r)(i_{r+1}) \cdots (i_n)$$

where the presence of a 1-cycle (j) means that $\pi(j) = j$. The notation may be abbreviated by omitting all 1-cycles, although if this is done, the integer n may need to be specified.

In particular a 2-cycle has the form (ij): it interchanges i and j and fixes all other integers. 2-cycles are often called *transpositions*.

Example (3.1.2). The permutation $\begin{pmatrix} 1 & 2 & 3 & 4 & 5 \\ 2 & 5 & 3 & 4 & 1 \end{pmatrix}$ is the 3-cycle $(125)(3)(4)$, that is, (125). While

$$\begin{pmatrix} 1 & 2 & 3 & 4 & 5 & 6 & 7 & 8 \\ 6 & 1 & 5 & 8 & 7 & 2 & 3 & 4 \end{pmatrix}$$

is not a cycle, it is the composite of three disjoint cycles of length > 1, namely $(162) \circ (357) \circ (48)$, as one can see by following what happens to each of the integers $1, 2, \ldots, 8$ when the permutation is applied. In fact this is an instance of an important general result, that any permutation is expressible as a composite of cycles: this will be established in (3.1.3).

It should be observed that there are r different ways to write an r-cycle since any element of the cycle can be the initial element: indeed $(i_1 i_2 \ldots i_r) = (i_2 i_3 \ldots i_r i_1) = \cdots = (i_r i_1 i_2 \cdots i_{r-1})$.

Two permutations π, σ in S_n are said to be *disjoint* if their supports are disjoint, i. e., they do not both move the same element. An important fact about disjoint permutations is that they commute, in contrast to permutations in general.

(3.1.2). *If π and σ are disjoint permutations in S_n, then $\pi\sigma = \sigma\pi$.*

Proof. Let $i \in \{1, 2, \ldots, n\}$; we show that $\pi\sigma(i) = \sigma\pi(i)$. If $i \notin \operatorname{supp}(\pi) \cup \operatorname{supp}(\sigma)$, then plainly $\pi\sigma(i) = i = \sigma\pi(i)$. Suppose that $i \in \operatorname{supp}(\pi)$; then $i \notin \operatorname{supp}(\sigma)$ and $\sigma(i) = i$. Thus $\pi\sigma(i) = \pi(i)$. Also $\sigma\pi(i) = \pi(i)$; for otherwise $\pi(i) \in \operatorname{supp}(\sigma)$ and so $\pi(i) \notin \operatorname{supp}(\pi)$, which leads to $\pi(\pi(i)) = \pi(i)$. However, π^{-1} can be applied to both sides of this equation to give $\pi(i) = i$, a contradiction since $i \in \operatorname{supp}(\pi)$. $\qquad\square$

Powers of a permutation

Since we know how to form products of permutations by using composition, it is natural to define powers of a permutation. Let $\pi \in S_n$ and let i be a non-negative integer. Then the *ith power π^i* is defined recursively by the rules:

$$\pi^0 = \mathrm{id}, \quad \pi^{i+1} = \pi^i \pi.$$

The point to note here is that the rule allows us to compute successive powers of the permutation as follows: $\pi^1 = \pi$, $\pi^2 = \pi\pi$, $\pi^3 = \pi^2\pi$, etc. Powers are used in the proof of the following fundamental theorem.

(3.1.3). *Let $\pi \in S_n$. Then π is expressible as a product of disjoint cycles and the cycles appearing in the product are unique.*

Proof. We deal with the existence of the expression first. If π is the identity, then obviously $\pi = (1)(2)\cdots(n)$. Assume that $\pi \neq \mathrm{id}$ and choose an integer i_1 such that $\pi(i_1) \neq i_1$. Now the integers $i_1, \pi(i_1), \pi^2(i_1), \ldots$ belong to the finite set $\{1, 2, \ldots, n\}$ and so they cannot all be different; say $\pi^r(i_1) = \pi^s(i_1)$ where $r > s \geq 0$. Applying $(\pi^{-1})^s$ to both sides of the equation and using associativity, we find that $\pi^{r-s}(i_1) = i_1$. Hence by the Well-Ordering Law there is a least positive integer m_1 such that $\pi^{m_1}(i_1) = i_1$.

Next we argue that the integers $i_1, \pi(i_1), \pi^2(i_1), \ldots, \pi^{m_1-1}(i_1)$ are all different. For if not and $\pi^r(i_1) = \pi^s(i_1)$ where $m_1 > r > s \geq 0$, then, just as above, we can argue that $\pi^{r-s}(i_1) = i_1$; on the other hand, $0 < r - s < m_1$, which contradicts the choice of m_1. It follows that π permutes the m_1 distinct integers $i_1, \pi(i_1), \ldots, \pi^{m_1-1}(i_1)$ in a cycle, so that we have identified the m_1-cycle $(i_1\, \pi(i_1) \ldots \pi^{m_1-1}(i_1))$ as a component of π.

If π fixes all other integers, then $\pi = (i_1\, \pi(i_1)\cdots\pi^{m_1-1}(i_1))$ and π is an m_1-cycle. Otherwise there exists an integer $i_2 \notin \{i_1, \pi(i_1), \ldots, \pi^{m_1-1}(i_1)\}$ such that $\pi(i_2) \neq i_2$. Just as above we identify a second cycle $(i_2\, \pi(i_2) \ldots \pi^{m_2-1}(i_2))$ present in π. This is disjoint from the first cycle. Indeed, if the cycles had a common element, they would have to coincide. It should also be clear that by a finite number of applications of this procedure we can express π as a product of disjoint cycles.

Next we establish uniqueness. Assume that there are two expressions for π as a product of disjoint cycles, say $(i_1 i_2 \cdots)(j_1 j_2 \cdots) \cdots$ and $(i_1' i_2' \cdots)(j_1' j_2' \cdots) \cdots$. By (3.1.2) disjoint cycles commute. Thus without loss of generality we can assume that i_1 occurs in the cycle $(i_1' i_2' \cdots)$. Since any element of a cycle can be moved up to the initial position, it can also be assumed that $i_1 = i_1'$. Then $i_2 = \pi(i_1) = \pi(i_1') = i_2'$; similarly $i_3 = i_3'$, etc. The other cycles are dealt with in the same manner. Therefore the two expressions for π are identical. $\qquad\square$

Corollary (3.1.4). *If $n > 1$, every element of S_n is expressible as a product of transpositions.*

Proof. Because of (3.1.3) it is sufficient to show that each cyclic permutation is a product of transpositions. That this is true follows from the easily verified identity:

$$(i_1 i_2 \cdots i_{r-1} i_r) = (i_1 i_r)(i_1 i_{r-1}) \cdots (i_1 i_3)(i_1 i_2). \qquad\square$$

Example (3.1.3). Express $\pi = \left(\begin{smallmatrix} 1 & 2 & 3 & 4 & 5 & 6 \\ 3 & 6 & 5 & 1 & 4 & 2 \end{smallmatrix}\right)$ as a product of transpositions.

First of all write π as a product of disjoint cycles, following the method of the proof of (3.1.3) to get $\pi = (1354)(26)$. Also $(1354) = (14)(15)(13)$, so that $\pi = (14)(15)(13)(26)$.

On the other hand, not every permutation in S_n is expressible as a product of *disjoint* transpositions: the reader should explain why not.

Even and odd permutations

If π is a permutation in S_n, then π replaces the natural order of integers, 1, 2, ..., n by the new order $\pi(1), \pi(2), \ldots, \pi(n)$. Thus π may cause inversions of the natural order: here an *inversion* occurs if for some $i < j$, we have $\pi(i) > \pi(j)$. To clarify the definition it is convenient to introduce a formal device.

Consider a polynomial f in indeterminates $x_1, x_2, \ldots, x_n$, with integer coefficients. (Here we assume the reader is familiar with the concept of a polynomial). If $\pi \in S_n$, then π determines a new polynomial πf which is obtained by permuting the variables $x_1, x_2, \ldots, x_n$. Thus $\pi f(x_1, \ldots, x_n) = f(x_{\pi(1)}, \ldots, x_{\pi(n)})$. For example, if $f = x_1 - x_2 - 2x_3$ and $\pi = (12)(3)$, then $\pi f = x_2 - x_1 - 2x_3$.

Now consider the polynomial

$$f(x_1, \ldots, x_n) = \prod_{\substack{i,j=1 \\ i<j}}^{n} (x_i - x_j).$$

A typical factor in πf is $x_{\pi(i)} - x_{\pi(j)}$. Now if $\pi(i) < \pi(j)$, this is also a factor of f, while if $\pi(i) > \pi(j)$, then $-(x_{\pi(i)} - x_{\pi(j)})$ is a factor of f. Consequently $\pi f = +f$ if the number of inversions of the natural order in π is even and $\pi f = -f$ if it is odd. This observation permits us to define the *sign* of the permutation π to be

$$\text{sign}(\pi) = \frac{\pi f}{f}.$$

Thus $\text{sign}(\pi) = 1$ or -1 according as the number of inversions in π is even or odd. Call π an *even permutation* if $\text{sign}(\pi) = 1$ and an *odd permutation* if $\text{sign}(\pi) = -1$.

Example (3.1.4). The even permutations in S_3 are (1)(2)(3), (123) and (132), while the odd permutations are (1)(23), (2)(13) and (3)(12).

In deciding if a permutation is even or odd a *crossover diagram* is a useful tool. We illustrate this idea with an example.

Example (3.1.5). Is the permutation

$$\pi = \begin{pmatrix} 1 & 2 & 3 & 4 & 5 & 6 & 7 \\ 3 & 7 & 2 & 5 & 4 & 1 & 6 \end{pmatrix}$$

even or odd?

To construct the crossover diagram simply join equal integers in the top and bottom rows of π and count the intersections or "crossovers", taking care to avoid multiple or unnecessary intersections. A crossover indicates the presence of an inversion of the natural order.

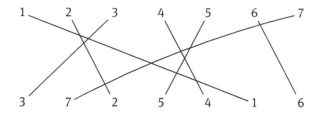

There are 11 crossovers, so $\text{sign}(\pi) = -1$ and π is an odd permutation.

The next result records very significant property of transpositions.

(3.1.5). *Transpositions are always odd.*

Proof. Consider the crossover diagram for the transposition (ij) where $i < j$.

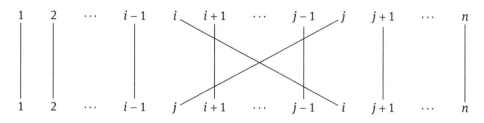

An easy count reveals the presence of $1 + 2(j - i - 1)$ crossovers. Since this integer is certainly odd, (ij) is an odd permutation. □

The basic properties of the sign function are laid out next.

(3.1.6). *Let $\pi, \sigma \in S_n$. Then the following hold:*
(i) $\text{sign}(\pi\sigma) = \text{sign}(\pi)\text{sign}(\sigma)$;
(ii) $\text{sign}(\pi^{-1}) = \text{sign}(\pi)$.

Proof. Let $f = \prod_{i<j=1}^{n}(x_i - x_j)$. Since $\pi f = \text{sign}(\pi)f$, we have

$$\pi\sigma f(x_1, \ldots, x_n) = \pi(\sigma f(x_1, \ldots, x_n))$$
$$= \pi((\text{sign}(\sigma)f(x_1, \ldots, x_n)))$$
$$= \text{sign}(\sigma)\pi f(x_1, \ldots, x_n)$$
$$= \text{sign}(\sigma)\text{sign}(\pi)f(x_1, \ldots, x_n).$$

Since $(\pi\sigma)f = \text{sign}(\pi\sigma)f$, it follows that $\text{sign}(\pi\sigma) = \text{sign}(\pi)\text{sign}(\sigma)$. Finally, by (i) we have $1 = \text{sign}(\text{id}) = \text{sign}(\pi\pi^{-1}) = \text{sign}(\pi)\text{sign}(\pi^{-1})$, so that $\text{sign}(\pi^{-1}) = 1/\text{sign}(\pi) = \text{sign}(\pi)$. □

Corollary (3.1.7). *A permutation π in S_n is even (odd) if and only if it is a product of an even (respectively odd) number of transpositions.*

For, if $\pi = \prod_{i=1}^{k} \pi_i$ with each π_i a transposition, then

$$\text{sign}(\pi) = \prod_{i=1}^{k} \text{sign}(\pi_i) = (-1)^k$$

by (3.1.5) and (3.1.6).

The subset of all even permutations in S_n is denoted by

$$A_n,$$

which is called the *alternating group of degree n*. Obviously $A_1 = S_1$. For $n > 1$ exactly half of the permutations in S_n are even, as the next result shows.

(3.1.8). *If $n > 1$, there are $\frac{1}{2}(n!)$ even permutations and $\frac{1}{2}(n!)$ odd permutations in S_n.*

Proof. Define a function $\alpha : A_n \to S_n$ by the rule $\alpha(\pi) = \pi \circ (12)$, observing that $\alpha(\pi)$ is odd as $\pi \in A_n$. Also α is injective. Every odd permutation σ belongs to $\text{Im}(\alpha)$ since $\alpha(\pi) = \sigma$ where $\pi = \sigma \circ (12) \in A_n$. Thus $\text{Im}(\alpha)$ is precisely the set of all odd permutations and $|\text{Im}(\alpha)| = |A_n|$. Hence $|A_n| = \frac{1}{2}|S_n| = \frac{1}{2}(n!)$. □

(3.1.9) (Cauchy's[1] Formula). *If π in S_n is the product of c disjoint cycles, including 1-cycles, then*

$$\mathrm{sign}(\pi) = (-1)^{n-c}.$$

Proof. Let $\pi = \sigma_1\sigma_2\cdots\sigma_c$ where the σ_i are disjoint cycles and σ_i has length ℓ_i. Now σ_i is expressible as a product of $\ell_i - 1$ transpositions by the proof of (3.1.4). Hence by (3.1.6) we have $\mathrm{sign}(\sigma_i) = (-1)^{\ell_i - 1}$ and thus

$$\mathrm{sign}(\pi) = \prod_{i=1}^{c}\mathrm{sign}(\sigma_i) = \prod_{i=1}^{c}(-1)^{\ell_i-1} = (-1)^{n-c}$$

since $\sum_{i=1}^{c}\ell_i = n$. □

Derangements

We conclude the section with a discussion of a special type of permutation. A permutation of a set is called a *derangement* if it fixes no elements of the set, i. e., its support is the entire set. For example, $(1234)(56)$ is a derangement in S_6. A natural question is: how many derangements does S_n contain? To answer the question we employ a well known combinatorial principle.

(3.1.10) (The Inclusion–Exclusion Principle). *If $A_1, A_2, \ldots, A_r$ are finite sets, then*

$$|A_1 \cup A_2 \cup \cdots \cup A_r| = \sum_{i=1}^{r}|A_i| - \sum_{i<j=1}^{r}|A_i \cap A_j|$$

$$+ \sum_{i<j<k=1}^{r}|A_i \cap A_j \cap A_k| - \cdots + (-1)^{r-1}|A_1 \cap A_2 \cap \cdots \cap A_r|.$$

Proof. We have to count the number of objects that belong to at least one A_i. Our first estimate is $\sum_{i=1}^{r}|A_i|$, but this double counts elements in more than one A_i, so we subtract $\sum_{i<j=1}^{r}|A_i \cap A_j|$. But now elements belonging to three or more A_i's have not been counted at all, so we must add $\sum_{i<j<k=1}^{r}|A_i \cap A_j \cap A_k|$. Now elements in four or more A_i's have been double counted, and so on. After a succession of r such "inclusions" and "exclusions" we arrive at the correct formula. □

It is now relatively easy to count derangements.

(3.1.11). *The number of derangements in S_n is given by the formula*

$$d_n = n!\left(1 - \frac{1}{1!} + \frac{1}{2!} - \frac{1}{3!} + \cdots + (-1)^n\frac{1}{n!}\right).$$

---

1 Augustin Louis Cauchy (1789–1857).

Proof. Let X_i denote the set of all permutations in S_n which fix the integer i, $(1 \le i \le n)$. Then the number of derangements in S_n is

$$d_n = n! - |X_1 \cup \cdots \cup X_n|.$$

Now $|X_i| = (n-1)!$; also $|X_i \cap X_j| = (n-2)!$, $(i < j)$, and $|X_i \cap X_j \cap X_k| = (n-3)!$, $(i < j < k)$, etc. Therefore by the Inclusion–Exclusion Principle

$$d_n = n! - \left\{ \binom{n}{1}(n-1)! - \binom{n}{2}(n-2)! + \binom{n}{3}(n-3)! \right.$$
$$\left. - \cdots + (-1)^{n-1}\binom{n}{n}(n-n)! \right\}.$$

Here the reason is that there are $\binom{n}{r}$ intersections $X_{i_1} \cap X_{i_2} \cap \cdots \cap X_{i_r}$ with $i_1 < i_2 < \cdots < i_r$. The required formula appears after a minor simplification of the terms in the sum. $\square$

Notice that $\lim_{n \to \infty}(\frac{d_n}{n!}) = e^{-1} = 0.36787\ldots$, so roughly 36.8 % of the permutations in S_n are derangements.

Example (3.1.6) (The Hat Problem). There are n people attending a party each of whom wears a different hat. All the hats are checked in on arrival. Afterwards each person is given a hat at random. What is the probability that no one get the correct hat?

A distribution of hats corresponds to a permutation of the original order. The permutations that are derangements give the distributions in which everyone has the wrong hat. So the probability asked for is $\frac{d_n}{n!}$ or roughly e^{-1}.

Exercises (3.1).

(1) Let $\pi = \left(\begin{smallmatrix} 1 & 2 & 3 & 4 & 5 & 6 \\ 2 & 4 & 1 & 5 & 3 & 6 \end{smallmatrix}\right)$ and $\sigma = \left(\begin{smallmatrix} 1 & 2 & 3 & 4 & 5 & 6 \\ 6 & 1 & 5 & 3 & 2 & 4 \end{smallmatrix}\right)$. Compute π^{-1}, $\pi\sigma$ and $\pi\sigma\pi^{-1}$.

(2) Determine which of the permutations in Exercise (3.1.1) are even and which are odd.

(3) Prove that $\text{sign}(\pi\sigma\pi^{-1}) = \text{sign}(\sigma)$ for all $\pi, \sigma \in S_n$.

(4) Prove that if $n > 1$, every non-trivial element of S_n is a product of *adjacent* transpositions, i. e., transpositions of the form $(i\, i+1)$. [Hint: it is enough to prove the statement for a transposition $(i\, j)$ where $i < j$. Now consider $(j\, j+1)(i\, j)(j\, j+1) = (i\, j+1)$.]

(5) Prove that an element π in S_n satisfies $\pi^2 = \text{id}$ if and only if π is a product of disjoint transpositions.

(6) How many elements π in S_n satisfy $\pi^2 = \text{id}$? [Hint: count the permutations which have exactly k disjoint transpositions for $2k \le n$ by first choosing $2k$ integers from $1, 2, \ldots, n$ and then forming k transpositions from them.]

(7) How many permutations in S_n contain at most one 1-cycle? [Hint: count the permutations with exactly one 1-cycle, then the permutations with no 1-cycles.]

(8) In the game of Rencontre there are two players A and B each of whom has a regular pack of 52 cards. The players deal their cards simultaneously. If at some point they

both deal the same card, this is a "rencontre" and player A wins. If no rencontre appears, player B wins. What are the probabilities of each player winning?

3.2 Semigroups, monoids and groups

Many of the structures that occur in algebra consist of a set together with a set of operations that can be applied to elements of the set. To make this precise, let us define a *binary operation on a set S* to be a function

$$\alpha : S \times S \to S.$$

Thus for each ordered pair (a, b) with a, b in S the function α produces a unique element $\alpha((a, b))$ of S. It is better notation if we write

$$a * b$$

instead of $\alpha((a, b))$ and refer to the binary operation as $*$.

Of course binary operations abound: one need think no further than addition or multiplication in sets such as $\mathbb{Z}$, $\mathbb{Q}$, $\mathbb{R}$, or composition on the set of all functions on a given set.

The first algebraic structure of interest to us is a *semigroup*, which is a pair

$$(S, *)$$

consisting of a non-empty set S and a binary operation $*$ on S which satisfies the *associative law*,

(i) $(a * b) * c = a * (b * c)$ for all $a, b, c \in S$.

If the semigroup has an *identity element*, i. e., an element e of S such that
(ii) $a * e = a = e * a$ for all $a \in S$,

then it is called a *monoid*.

Finally, the monoid is called a *group* if each element a has an inverse, i. e., an element a' of S such that
(iii) $a * a' = e = a' * a$.

Also a semigroup $(S, *)$ is said to be *commutative* if
(iv) $a * b = b * a$ for all $a, b \in S$.

A commutative group is called an *abelian*[2] *group*.

---

[2] After Niels Henrik Abel (1802–1829).

Thus semigroups, monoids and groups form successively narrower classes of algebraic structures. The three concepts will now be illustrated by some familiar examples.

Examples of semigroups, monoids and groups

(i) The pairs $(\mathbb{Z}, +)$, $(\mathbb{Q}, +)$, $(\mathbb{R}, +)$ are groups where + is ordinary addition, 0 is an identity element and an inverse of x is its negative $-x$.

(ii) Next consider $(\mathbb{Q}^*, \cdot)$, $(\mathbb{R}^*, \cdot)$ where the dot denotes ordinary multiplication and $\mathbb{Q}^*$ and $\mathbb{R}^*$ are the sets of *non-zero* rational numbers and real numbers respectively. Here $(\mathbb{Q}^*, \cdot)$ and $(\mathbb{R}^*, \cdot)$ are groups, the identity element being 1 and an inverse of x being $\frac{1}{x}$. On the other hand, $(\mathbb{Z}^*, \cdot)$ is only a monoid since the integer 2, for example, has no inverse in $\mathbb{Z}^* = \mathbb{Z} - \{0\}$.

(iii) $(\mathbb{Z}_m, +)$ is a group where m is a positive integer and the usual addition of congruence classes is used.

(iv) $(\mathbb{Z}_m^*, \cdot)$ is a group where m is a positive integer: here $\mathbb{Z}_m^*$ is the set of invertible congruence classes $[a]$ modulo m, i. e., such that $\gcd\{a, m\} = 1$, and multiplication of congruence classes is used. Note that $|\mathbb{Z}_m^*| = \phi(m)$ where ϕ is Euler's function.

(v) Let $M_n(\mathbb{R})$ be the set of all $n \times n$ matrices with real entries. If the usual rule of addition of matrices is used, $(M_n(\mathbb{R}), +)$ is an abelian group.

On the other hand, $M_n(\mathbb{R})$ with matrix multiplication is only a monoid. To obtain a group we must form

$$\mathrm{GL}_n(\mathbb{R}),$$

the subset of all invertible (or *non-singular*) matrices in $M_n(\mathbb{R})$: recall that these are the matrices with non-zero determinant. This group is called the *general linear group of degree n* over $\mathbb{R}$.

(vi) For an example of a semigroup that is not a monoid we need look no further than the set of all even integers with multiplication as the group operation. Clearly there is no identity element here.

(vii) *The monoid of functions on a set.* Let A be any non-empty set, and write $\mathrm{Fun}(A)$ for the set of all mappings or functions α on A. Then

$$(\mathrm{Fun}(A), \circ)$$

is a monoid where $\circ$ is functional composition. Indeed, this binary operation is associative by (1.2.3) and the identity function on A is an identity element.

If we restrict attention to the bijective functions on A, i. e., to those which have inverses, we obtain the *symmetric group* on A

$$(\mathrm{Sym}(A), \circ),$$

consisting of all the permutations of A. This example was the motivation for the definition of a group.

(viii) *Monoids of words*. For a different type of example consider words in an alphabet X. Here X is any non-empty set and a *word* in X is just an n-tuple of elements of X, written for convenience without parentheses in the form $x_1 x_2 \cdots x_n$, $n \geq 0$. The case $n = 0$ is the *empty word* $\emptyset$. Let X^* denote the set of all words in X.

There is a natural binary operation on X, namely juxtaposition. Thus, if $u = x_1 \cdots x_n$ and $v = y_1 \cdots y_m$ are words in X, define uv to be the word $x_1 \cdots x_n y_1 \cdots y_m$. If $u = \emptyset$, then by convention $uz = z = zu$ for all z. It is clear that this binary operation is associative and that $\emptyset$ is an identity element. Thus X^*, with the operation specified, is a monoid: it is known as the *free monoid on X*.

(ix) *Monoids and automata*. There is a somewhat unexpected connection between monoids and automata. Suppose that $A = (I, S, v)$ is a state output automaton with input set I, state set S and next state function $v : I \times S \to S$: see Exercise (1.3.8). Then A determines a monoid M_A in the following way.

Let $i \in I$ and define $\theta_i : S \to S$ by the rule $\theta_i(s) = v(i, s)$ where $s \in S$. Now let M_A consist of the identity function and all composites of finite sequences of θ_i's; thus $M_A \subseteq \mathrm{Fun}(S)$. Clearly $(M_A, \circ)$ is a monoid with respect to functional composition.

In fact one can go in the opposite direction as well. Let $(M, *)$ be a monoid and define an automaton $A_M = (M, M, v)$ where the next state function $v : M \times M \to M$ is given by the rule $v(x_1, x_2) = x_1 * x_2$. Thus a connection between monoids and state output automata has been established.

Symmetry groups

As has been remarked, groups tend to arise wherever symmetry is of importance. The size of the group can be regarded as a measure of the amount of symmetry present. Since symmetry is at heart a geometric notion, it is not surprising that geometry provides many interesting examples of groups.

A bijective function defined on 3-dimensional space or the plane is called an *isometry* if it preserves distances between points. Natural examples of isometries are translations, rotations and reflections. Let X be a non-empty set of points in 3-space or the plane – we will refer to X as a *geometric configuration*. An isometry α which fixes the set X, i. e., such that

$$X = \{\alpha(x) \mid x \in X\},$$

is called a *symmetry of* X. Note that a symmetry can move the individual points of X.

It is easy to see that the symmetries of X form a group with respect to functional composition; this is the *symmetry group* $S(X)$ of X. Thus $S(X)$ is a subset of $\mathrm{Sym}(X)$, usually a proper subset.

The symmetry group of the regular *n*-gon

As an illustration let us analyze the symmetries of the *regular n-gon*: this is a polygon in the plane with n edges of equal length, ($n \geq 3$). It is convenient to label the vertices of the *n*-gon 1, 2, ..., n, so that each symmetry is represented by a permutation of the vertex set $\{1, 2, \ldots, n\}$, i. e., by an element of S_n.

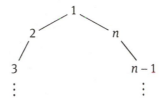

Each symmetry arises from an axis of symmetry of the figure. Of course, in order to obtain a group, we must include the identity symmetry, represented by $(1)(2) \cdots (n)$. There are $n - 1$ anticlockwise rotations about the line perpendicular to the plane of the figure and through the centroid, through angles $i(\frac{2\pi}{n})$, for $i = 1, 2, \ldots, n - 1$. For example, the rotation through $\frac{2\pi}{n}$ is represented by the *n*-cycle $(1\,2\,3\ldots n)$; other rotations correspond to powers of this *n*-cycle. Notice that the inverse of a rotation can be thought of as a rotation through a suitable angle in the clockwise direction.

Then there are n reflections in axes of symmetry in the plane. If n is odd, such axes join a vertex to the midpoint of the opposite edge. For example, $(1)(2\,n)(3\,n-1)\cdots$ corresponds to one such reflection. However, if n is even, there are two types of reflections, in an axis joining a pair of opposite vertices and in an axis joining midpoints of opposite edges: hence there are $\frac{1}{2}n + \frac{1}{2}n = n$ reflections in this case as well.

Since all axes of symmetry of the *n*-gon have now been exhausted, we conclude that the order of the symmetry group is $1 + (n - 1) + n = 2n$. This group is called the *dihedral group of order 2n*,

$$\mathrm{Dih}(2n).$$

Notice that $\mathrm{Dih}(2n)$ is a proper subset of S_n if $2n < n!$, i. e., if $n \geq 4$. Thus not every permutation of the vertices arises from a symmetry when $n \geq 4$.

Simple consequences of the axioms

We end the section by noting three elementary facts that follow quickly from the axioms.

(3.2.1).

(i) *(The Generalized Associative Law) Let $x_1, x_2, \ldots, x_n$ be elements of a semigroup $(S, *)$. If an element u is constructed by combining these elements in the given order,* *using any mode of bracketing, then $u = (\cdots((x_1 * x_2) * x_3) * \cdots) * x_n$, so that u is independent of the positioning of the parentheses.*

(ii) *Every monoid has a unique identity element.*

(iii) *Every element in a group has a unique inverse.*

Proof. (i) We argue by induction on n, which can be assumed to be at least 3. If u is constructed from $x_1, x_2, \ldots, x_n$ in that order, then $u = v * w$ where v is constructed from $x_1, x_2, \ldots, x_i$ and w from $x_{i+1}, \ldots, x_n$; here $1 \le i \le n - 1$. Then $v = (\cdots(x_1 * x_2) * \cdots * x_i)$ by induction on n. If $i = n - 1$, then $w = x_n$ and the result follows at once. Otherwise $i + 1 < n$ and $w = z * x_n$ where z is constructed from $x_{i+1}, \ldots, x_{n-1}$. Then $u = v * w = v * (z * x_n) = (v * z) * x_n$ by the associative law. The result is true for $v * z$ by induction, so it is true for u.

(ii) Suppose that e and e' are two identity elements in a monoid. Then $e = e * e'$ since e' is an identity, and $e * e' = e'$ since e is an identity. Hence $e = e'$.

(iii) Let g be an element of a group and suppose g has two inverses x and x'; we claim that $x = x'$. To see this observe that $(x * g) * x' = e * x' = x'$, while also $(x * g) * x' = x * (g * x') = x * e = x$. Hence $x = x'$. $\square$

Because of (3.2.1)(i) above, we can without ambiguity omit all parentheses from an expression formed from elements $x_1, x_2, \ldots, x_n$ of a semigroup – an enormous gain in simplicity. Also (ii) and (iii) show that it is unambiguous to speak of *the* identity element of a monoid and *the* inverse of an element of a group.

Exercises (3.2).

(1) Let S be the subset of $\mathbb{R} \times \mathbb{R}$ specified below and define $(x, y) * (x', y') = (x+x', y+y')$. Say in each case whether $(S, *)$ is a semigroup, a monoid, a group, or none of these, as is most appropriate.

    (i) $S = \{(x, y) \mid x + y \ge 0\}$;

    (ii) $S = \{(x, y) \mid x + y > 0\}$;

    (iii) $S = \{(x, y) \mid |x + y| \le 1\}$;

    (iv) $S = \{(x, y) \mid 2x + 3y = 0\}$.

(2) Do the sets of even or odd permutations in S_n form a semigroup when functional composition is used as the binary operation?

(3) Show that the set of all 2×2 real matrices with non-negative entries is a monoid, but not a group, when matrix addition used.

(4) Let A be a non-empty set and define a binary operation $*$ on the power set $\mathcal{P}(A)$ by $S * T = (S \cup T) - (S \cap T)$. Prove that $(\mathcal{P}(A), *)$ is an abelian group.

(5) Define powers in a semigroup $(S, *)$ by the rules $x^1 = x$ and $x^{n+1} = x^n * x$ where $x \in S$ and n is a non-negative integer. Prove that $x^m * x^n = x^{m+n}$ and $(x^m)^n = x^{mn}$ where $m, n > 0$.

(6) Let G be a monoid such that for each x in G there is a positive integer n such that $x^n = e$. Prove that G is a group.

(7) Let G be the set consisting of the permutations $(12)(34)$, $(13)(24)$, $(14)(23)$ and the identity permutation $(1)(2)(3)(4)$. Show that G is a group with exactly four elements in which each element is its own inverse. (This group is called the *Klein*[3] 4-group).

(8) Prove that the group S_n is abelian if and only if $n \leq 2$.

(9) Prove that the group $GL_n(\mathbb{R})$ is abelian if and only if $n = 1$.

3.3 Groups and subgroups

From this point on we will concentrate on groups: we begin by improving the notation. In the first place it is customary not to distinguish between a group $(G, *)$ and its underlying set G, provided there is no likelihood of confusion. Then there are two standard ways of writing the group operation. In the *additive notation* we write $x + y$ for $x * y$; the identity is 0_G or 0 and the inverse of an element x is $-x$. The additive notation is most often used for abelian groups, i. e., groups $(G, *)$ such that $x * y = y * x$ for all $x, y \in G$.

For non-abelian groups the *multiplicative* notation is generally employed, with xy being written for $x * y$; the identity element is 1_G or 1 and the inverse of x is x^{-1}. The multiplicative notation will be used here unless the additive notation is clearly preferable, as with a group such as $\mathbb{Z}$.

Isomorphism

It is important to decide when two groups are to be regarded as essentially the same. It is possible that two groups have very different sets of elements, but their elements behave in a similar manner with respect to their respective group operations. This leads us to introduce the concept of isomorphism. Let G and H be (multiplicatively written) groups. An *isomorphism* from G to H is a bijective function $\alpha : G \to H$ such that

$$\alpha(xy) = \alpha(x)\alpha(y)$$

for all $x, y \in G$. Groups G and H are said to be *isomorphic* if there exists an isomorphism from G to H, in symbols

$$G \simeq H.$$

---

3 Felix Klein (1849–1925).

(3.3.1).

(i) *If $\alpha : G \to H$ is an isomorphism of groups, then so is its inverse $\alpha^{-1} : H \to G$.*

(ii) *Isomorphism is an equivalence relation on the class of groups.*

Proof. To establish (i) all we need to do is prove that $\alpha^{-1}(xy) = \alpha^{-1}(x)\alpha^{-1}(y)$. Now $\alpha(\alpha^{-1}(xy)) = xy$, while

$$\alpha(\alpha^{-1}(x)\alpha^{-1}(y)) = \alpha(\alpha^{-1}(x))\alpha(\alpha^{-1}(y)) = xy.$$

Hence $\alpha^{-1}(xy) = \alpha^{-1}(x)\alpha^{-1}(y)$ by injectivity of α.

To prove (ii) note that reflexivity is obvious, while transitivity follows from the observation that a composite of isomorphisms is an isomorphism: of course (i) implies the symmetric property. □

The idea behind isomorphism is that, while the elements in two isomorphic groups may be different, they have the same properties in relation to their respective group operations. Note that isomorphic groups have the same order, where by the *order* of a group G we mean the cardinality of its set of elements $|G|$.

The next result records some very useful techniques for working with group elements.

(3.3.2). *Let x, a, b be elements of a group.*

(i) *If $xa = b$, then $x = ba^{-1}$, and if $ax = b$, then $x = a^{-1}b$.*

(ii) $(ab)^{-1} = b^{-1}a^{-1}$.

Proof. From $xa = b$ we obtain $(xa)a^{-1} = ba^{-1}$ and thus $x(aa^{-1}) = ba^{-1}$. Since $aa^{-1} = 1$ and $x1 = x$, we get $x = ba^{-1}$. The second statement in (i) is dealt with similarly. By (3.2.1) to establish (ii) it is enough to show that $b^{-1}a^{-1}$ is an inverse of ab. This can be checked directly: $(ab)(b^{-1}a^{-1}) = a(bb^{-1})a^{-1} = a1a^{-1} = aa^{-1} = 1$; similarly $(b^{-1}a^{-1})(ab) = 1$. Consequently $(ab)^{-1} = b^{-1}a^{-1}$. □

The group table

Suppose that $(G, *)$ is a group of finite order n whose elements are ordered in some fixed manner, let us say $g_1, g_2, \ldots, g_n$. The rule for combining elements in the group can be displayed in its *group table*. This is the $n \times n$ rectangular array M whose (i, j) entry is $g_i * g_j$. Thus the ith row of M is $g_i * g_1, g_i * g_2, \ldots, g_i * g_n$. From the group table any pair of group elements can be combined. If the group is written multiplicatively, the term *multiplication table* is used.

Notice that all the elements in a row are different: for $g_i * g_j = g_i * g_k$ implies that $g_j = g_k$ by (3.3.2). The same is true of the columns of M. What this means is that each group element appears exactly once in each row and exactly once in each column of M, that is, the group table is a *latin square*. Such configurations are studied in Section 17.3 below.

As an example, consider the group of order 4 whose elements are the identity permutation $1 = (1)(2)(3)(4)$ and the permutations $a = (12)(34)$, $b = (13)(24)$, $c = (14)(23)$. This is the *Klein 4-group*, which was mentioned in Exercise(3.2.7). The multiplication table of this group is the 4×4 array

	1	a	b	c
1	1	a	b	c
a	a	1	c	b
b	b	c	1	a
c	c	b	a	1

Powers of group elements

Let x be an element of a (multiplicative) group G and let n be an integer. The *nth power* x^n of x is defined recursively as follows:

$$x^0 = 1, \quad x^{n+1} = x^n x, \quad x^{-n} = \left(x^n\right)^{-1}$$

where $n \geq 0$. (See also Exercise (3.2.5). Of course, if G were written additively, we would write nx instead of x^n. Fundamental for the manipulation of powers is:

(3.3.3) (The Laws of Exponents). *Let x be an element of a group G and let m, n be integers. Then*

(i) $x^m x^n = x^{m+n} = x^n x^m$;
(ii) $(x^m)^n = x^{mn}$.

Proof. (i) First we assume that $n \geq 0$ and prove that $x^m x^n = x^{m+n}$ for all m by induction on n. This is clear if $n = 0$. Assuming it true for n, we have

$$x^m x^{n+1} = x^m x^n x = x^{m+n} x = x^{m+n+1},$$

thus completing the induction. Now let $n < 0$. Then by the case just dealt with, $x^{m+n} x^{-n} = x^m$ and hence $x^{m+n} = x^m x^n$.

(ii) When $n \geq 0$, use induction on n: clearly it is true when $n = 0$. Assuming the statement true for n, we have $(x^m)^{n+1} = (x^m)^n x^m = x^{mn} x^m = x^{m(n+1)}$ by (i). Next $(x^m)^{-n} = ((x^m)^n)^{-1} = (x^{mn})^{-1} = x^{-mn}$, which covers the case where the second exponent is negative. □

Subgroups

Roughly speaking, a subgroup is a group contained within a larger group that has consistent group operations. To make this concept precise, consider a group $(G, *)$ and a subset S of G. If the group operation $*$ is restricted to S, we obtain a function $*'$

from $S \times S$ to G. If $*'$ is a binary operation on S, i. e., if $x * y \in S$ whenever $x, y \in S$, and if $(S, *')$ is actually a group, then S is called a *subgroup* of G.

The first point to settle is that 1_S, the identity element of $(S, *')$, equals 1_G. Indeed $1_S = 1_S *' 1_S = 1_S * 1_S$, so $1_S * 1_S = 1_S * 1_G$. By (3.3.2) it follows that $1_S = 1_G$. Next let $x \in S$ and denote the inverse of x in $(S, *)$ by x_S^{-1}. We want to be sure that $x_S^{-1} = x^{-1}$. Now $1_G = 1_S = x *' x_S^{-1} = x * x_S^{-1}$. Hence $x * x^{-1} = x * x_S^{-1}$ and so $x_S^{-1} = x^{-1}$. Thus inverses are the same in $(S, *')$ and in $(G, *)$.

On the basis of these observations we are able to formulate a convenient test for a subset of a group to be a subgroup.

(3.3.4). *Let S be a subset of a group G. Then S is a subgroup of G if and only if the following hold:*

(i) $1_G \in S$;

(ii) $xy \in S$ *whenever $x \in S$ and $y \in S$, (closure under products);*

(iii) $x^{-1} \in S$ *whenever $x \in S$, (closure under inverses).*

To indicate that S is a subgroup of a group G we write

$$S \leq G.$$

If in addition $S \neq G$, then S is a *proper* subgroup and we write $S < G$.

Examples of subgroups

(i) $\mathbb{Z} < \mathbb{Q} < \mathbb{R} < \mathbb{C}$. These statements follow at once via (3.3.4). For the same reason $\mathbb{Q}^* < \mathbb{R}^* < \mathbb{C}^*$.

(ii) $A_n < S_n$. Recall that A_n is the set of even permutations in S_n. Here the point to note is that if π and σ are even permutations, then so are $\pi\sigma$ and π^{-1} by (3.1.6): of course the identity permutation is even. However, the odd permutations in S_n do not form a subgroup.

(iii) Two subgroups that are present in every group G are the *trivial* or *identity subgroup* $\{1_G\}$, which is written 1 or 1_G, and the *improper subgroup* G itself. For some groups these are the only subgroups.

(iv) *Cyclic subgroups.* The interesting subgroups of a group are the proper non-trivial ones. An easy way to produce subgroups is to take all the powers of a fixed element. Let G be a group and choose $x \in G$. We denote the set of all powers of the element x by

$$\langle x \rangle.$$

Using (3.3.4) and the Laws of Exponents (3.3.3), we quickly verify that $\langle x \rangle$ is a subgroup. It is called *the cyclic subgroup generated by x*. Since every subgroup of G

which contains x must also contain all powers of x, it follows that $\langle x \rangle$ *is the smallest subgroup of G containing x.*

A group G is said to be *cyclic* if $G = \langle x \rangle$ for some x in G. For example, $\mathbb{Z}$ and $\mathbb{Z}_n$ are cyclic groups since, allowing for the additive notation, $\mathbb{Z} = \langle 1 \rangle$ and $\mathbb{Z}_n = \langle [1]_n \rangle$.

Next we consider intersections of subgroups.

(3.3.5). *If $\{S_\lambda \mid \lambda \in \Lambda\}$ is a non-empty set of subgroups of a group G, then $\bigcap_{\lambda \in \Lambda} S_\lambda$ is also a subgroup of G.*

This follows immediately from (3.3.4). Now suppose that X is a non-empty subset of a group G. There is at least one subgroup that contains X, namely G itself. Thus we may form the intersection of all the subgroups of G that contain X. This is a subgroup by (3.3.5), which is denoted by

$$\langle X \rangle.$$

Obviously $\langle X \rangle$ is the smallest subgroup of G containing X: it is called the *subgroup generated by X*. Note that the cyclic subgroup $\langle x \rangle$ is just the subgroup generated by the singleton set $\{x\}$. More generally a group G is said to be *finitely generated* if $G = \langle X \rangle$ for some finite set X.

It is natural to enquire about the form of elements of $\langle X \rangle$.

(3.3.6). *Let X be a non-empty subset of a group G. Then $\langle X \rangle$ consists of all elements of G of the form*

$$x_1^{e_1} x_2^{e_2} \cdots x_k^{e_k}$$

where $x_i \in X$, $e_i = \pm 1$ and $k \geq 0$, (the case $k = 0$ being interpreted as 1_G).

Proof. Let S denote the set of all elements of the specified form. It is easy to check that S contains 1 and is closed under products and inversion, by using (3.3.2). Thus S is a subgroup. Clearly $X \subseteq S$, so that $\langle X \rangle \subseteq S$ since $\langle X \rangle$ is the smallest subgroup containing X. On the other hand, any element of the form $x_1^{e_1} \cdots x_k^{e_k}$ must belong to $\langle X \rangle$ since $x_i \in \langle X \rangle$. Therefore $S \subseteq \langle X \rangle$ and $\langle X \rangle = S$. $\square$

Notice that if X is the 1-element set $\{x\}$, we recover the fact that the cyclic subgroup $\langle x \rangle$ consists of all powers of x.

The lattice of subgroups

Let G be a group; then set inclusion is a partial order on the set of all subgroups of G

$$S(G),$$

which is therefore a partially ordered set. If H and K are subgroups of G, they have a greatest lower bound in $S(G)$, namely $H \cap K$, and also a least upper bound $\langle H \cup K \rangle$, which is usually written $\langle H, K \rangle$. This last is true because any subgroup containing H and K must also contain $\langle H, K \rangle$. (Notice that $H \cup K$ is not a subgroup). This means that $S(G)$ is a lattice in the sense of Section 1.2. When G is finite, $S(G)$ can be visualized by means of its Hasse diagram; the basic component in the diagram of subgroups of a group is the subdiagram below

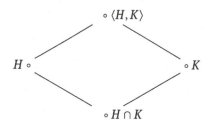

The order of a group element

Let x be an element of a group. If the subgroup $\langle x \rangle$ has a finite number m of elements, x is said to have *finite order m*. If on the other hand $\langle x \rangle$ is infinite, then x is called an element of *infinite order*. We shall write

$$|x|$$

for the order of x. Some basic facts about orders of group elements are contained in the next result.

(3.3.7). *Let x be an element of a group G.*
(i) *If all powers of x are distinct, then x has infinite order.*
(ii) *Assume that two powers of x are equal. Then x has finite order m and $x^\ell = 1$ if and only if ℓ is divisible by m. Thus m is the smallest positive integer such that $x^m = 1$. Furthermore $\langle x \rangle = \{1, x, \ldots, x^{m-1}\}$.*

Proof. (i) This is clearly true.
(ii) Suppose that two powers of x are equal, say $x^i = x^j$ where $i > j$. Then $x^{i-j} = 1$ by (3.3.3). Using Well-Ordering we may choose a smallest positive integer m for which $x^m = 1$. Now let ℓ be any integer and write $\ell = mq + r$ where $q, r \in \mathbb{Z}$ and $0 \leq r < m$, using the Division Algorithm. By (3.3.3) again $x^\ell = (x^m)^q x^r = x^r$. By minimality of m we deduce that $x^\ell = 1$ if and only if $r = 0$, i.e., ℓ is divisible by m. It follows that $\langle x \rangle = \{1, x, x^2, \ldots, x^{m-1}\}$, so that x has finite order m. $\qquad\square$

We will now study cyclic groups with the aim of identifying them up to isomorphism.

(3.3.8). *A cyclic group of order n is isomorphic with $\mathbb{Z}_n$. An infinite cyclic group is isomorphic with $\mathbb{Z}$.*

Proof. Let $G = \langle x \rangle$ be a cyclic group. If $|G| = n$, then $G = \{1, x, \ldots, x^{n-1}\}$. Define $\alpha : \mathbb{Z}_n \to G$ by $\alpha([i]) = x^i$, which is a well-defined function because $x^{i+nq} = x^i(x^n)^q = x^i$. Also

$$\alpha([i] + [j]) = \alpha([i+j]) = x^{i+j} = x^i x^j = \alpha([i])\alpha([j]),$$

while α is clearly bijective. Therefore, allowing for $\mathbb{Z}_n$ being written additively and G multiplicatively, we conclude that α is an isomorphism and $\mathbb{Z}_n \simeq G$. When G is infinite cyclic, the proof is similar, but easier, and is left to the reader. $\qquad\square$

There is a simple way to compute the order of an element of the symmetric group S_n by using least common multiples – see Exercise (2.2.8).

(3.3.9). *Let $\pi \in S_n$ and write $\pi = \pi_1 \pi_2 \cdots \pi_k$ where the π_i are disjoint cycles, with π_i of length ℓ_i. Then the order of π equals the least common multiple of $\ell_1, \ell_2, \ldots, \ell_k$.*

Proof. By (3.1.3) there is a such an expression for π. Also disjoint permutations commute by (3.1.2). Hence $\pi^m = \pi_1^m \pi_2^m \cdots \pi_k^m$ for any $m > 0$. Now the π_i^m affect disjoint sets of integers, so $\pi^m = 1$, (i. e., $\pi^m = $ id), if and only if $\pi_1^m = \pi_2^m = \cdots = \pi_k^m = 1$. By (3.3.7) these conditions are equivalent to m being divisible by the orders of all the π_i. Finally, it is easy to see by forming successive powers that the order of an r-cycle is r. Therefore $|\pi| = \mathrm{lcm}\{\ell_1, \ell_2, \ldots, \ell_k\}$. $\qquad\square$

Example (3.3.1). What is the largest possible order of an element of S_8?

Let $\pi \in S_8$ and write $\pi = \pi_1 \cdots \pi_k$ where the π_i are disjoint cycles. If π_i has length ℓ_i, then $\sum_{i=1}^k \ell_i = 8$ and $|\pi| = \mathrm{lcm}\{\ell_1, \ldots, \ell_k\}$. So the question is: which positive integers $\ell_1, \ldots, \ell_k$ with sum equal to 8 have the largest least common multiple? A little experimentation will convince the reader that the answer is $k = 2$, $\ell_1 = 3$, $\ell_2 = 5$. Hence 15 is the largest order of an element of S_8. For example, the permutation $(123)(45678)$ has order 15.

We conclude with two more examples, including an application to number theory.

Example (3.3.2). Let G be a finite abelian group. Prove that the product of all the elements of G equals the product of all the elements of G of order 2. If there are no elements of order 2, the product is to be interpreted as 1.

The key point to notice here is that if $x \in G$, then $|x| = 2$ if and only if $x = x^{-1} \neq 1$. Since G is abelian, in the product $\prod_{g \in G} g$ we can group together elements of order greater than 2 with their inverses and then cancel each pair xx^{-1}. What is left is the product of the elements of order 2.

Example (3.3.3) (Wilson's[4] Theorem). *If p is a prime, then $(p - 1)! \equiv -1 \pmod{p}$.*

Apply Example (3.3.2) to $\mathbb{Z}_p^*$, the multiplicative group of non-zero congruence classes mod p. Now the only element of order 2 in $\mathbb{Z}_p^*$ is $[-1]$: for $a^2 \equiv 1 \pmod{p}$ implies that $a \equiv \pm 1 \pmod{p}$, i.e., $[a] = [1]$ or $[-1]$. It follows that $[1][2] \cdots [p-1] = [-1]$ and hence $(p - 1)! \equiv -1 \pmod{p}$.

Exercises (3.3).

(1) In each of the following situations say whether or not the subset S is a subgroup of the group G:
   (i) $G = \mathrm{GL}_n(\mathbb{R})$, $S = \{A \in G \mid \det(A) = 1\}$.
   (ii) $G = (\mathbb{R}, +)$, $S = \{x \in R \mid |x| \leq 1\}$.
   (iii) $G = \mathbb{R} \times \mathbb{R}$, $S = \{(x, y) \mid 3x - 2y = 1\}$: here the group operation of G is addition of pairs componentwise.

(2) Let H and K be subgroups of a group G. Prove that $H \cup K$ is a subgroup if and only if $H \subseteq K$ or $K \subseteq H$.

(3) Show that no group can be the union of two proper subgroups. Then exhibit a group which is the union of three proper subgroups.

(4) Find the largest possible order of an element of S_{11}. How many elements of S_{11} have this order?

(5) The same question for S_{12}.

(6) Find the orders of the elements $[3]$ and $[7]$ of $\mathbb{Z}_{11}^*$.

(7) Prove that a group of even order must contain an element of order 2. [Hint: assume this is false and group the non-identity elements in pairs x, x^{-1}.]

(8) Assume that for each pair of elements a, b of a group G there is an integer n such that $(ab)^i = a^i b^i$ holds for $i = n$, $n + 1$ and $n + 2$. Prove that G is abelian.

(9) Let S denote the set product $\mathbb{Z} \times \mathbb{Z}$. Define a relation E on S by $(a, b) \, E \, (a', b') \Leftrightarrow a - b = a' - b'$.
   (i)   Prove that E is an equivalence relation on S and that each E-equivalence class contains a pair (a, b) with $a, b > 0$.
   (ii) Define $(a, b) + (a', b')$ to be $(a + a', b + b')$ and show that this is a well defined binary operation on the set P of all E-equivalence classes.
   (iii) Prove that if $+$ denotes the binary operation in (ii), then $(P, +)$ is an abelian group.
   (iv) By finding a mapping from P to $\mathbb{Z}$, prove that $P \simeq \mathbb{Z}$.

(10) Let $\mathcal{S}$ be a non-empty set of subgroups of a group. Then $\mathcal{S}$ is said to satisfy the *ascending chain condition* (acc) if there does not exist an infinite ascending chain of subgroups $G_1 < G_2 < \cdots$ where $G_i \in \mathcal{S}$. Also $\mathcal{S}$ is said to satisfy the *maximal condition* (max) if each non-empty subset $\mathcal{T}$ of $\mathcal{S}$ has at least one maximal element,

---

4 John Wilson (1741–1793).

i. e., a subgroup in $\mathcal{T}$ which is not properly contained in any other subgroup in $\mathcal{T}$. Prove that the acc and max are the same property.

(11) A group G is said to satisfy the *maximal condition on subgroups* (max) if the set of all its subgroups $S(G)$ satisfies max, or equivalently the acc. Prove that G satisfies max if and only if every subgroup of G is finitely generated. [Hint: use the acc form.]

(12) Prove that $\mathbb{Z}$ satisfies max, but $\mathbb{Q}$ does not.

4 Quotient groups and Homomorphisms

In this chapter we probe more deeply into the nature of the subgroups of a group and we introduce functions between groups that relate their group operations.

4.1 Cosets and Lagrange's Theorem

Consider a group G with a fixed subgroup H. A binary relation $\sim_H$ on G is defined by the following rule: $x \sim_H y$ means that $x = yh$ for some $h \in H$. It is an easy verification that $\sim_H$ is an equivalence relation on G. Therefore by (1.2.2) the group G splits up into disjoint equivalence classes. The equivalence class to which an element x belongs is the subset

$$\{xh \mid h \in H\},$$

which is called a *left coset* of H in G and is written

$$xH.$$

Thus G is the disjoint union of the distinct left cosets of H. Notice that the only coset which is a subgroup is $1H = H$ since no other coset contains the identity element.

Next observe that the assignment $h \mapsto xh$, $(h \in H)$, determines a bijection from H to xH; for $xh_1 = xh_2$ implies that $h_1 = h_2$. From this it follows that

$$|xH| = |H|,$$

so that each left coset of H has the cardinal of H.

Suppose that we label the left cosets of H in some manner, say as $C_\lambda, \lambda \in \Lambda$, and for each λ in Λ we choose an arbitrary element t_λ from C_λ. (If Λ is infinite, we are assuming at this point the set theoretic axiom known as the axiom of choice – see Section 1.5. Then $C_\lambda = t_\lambda H$ and, since every group element belongs to some left coset of H, we have $G = \bigcup_{\lambda \in \Lambda} t_\lambda H$. Furthermore, cosets are equivalence classes and therefore are disjoint, so each element x of G has a unique expression $x = t_\lambda h$, where $h \in H$, $\lambda \in \Lambda$. The set $\{t_\lambda \mid \lambda \in \Lambda\}$ is called a *left transversal* to H in G. Thus we have found a unique way to express elements of G in terms of the transversal and elements of the subgroup H.

In a similar fashion one can define *right cosets* of H in G; these are the equivalence classes of the equivalence relation $_H\sim$, where $x \,_H\!\sim y$ means that $x = hy$ for some h in H. The *right coset* containing x is

$$Hx = \{hx \mid h \in H\}$$

and right transversals are defined analogously.

The next result was the first significant theorem to be discovered in group theory.

https://doi.org/10.1515/9783110691160-004

(4.1.1) (Lagrange's[1] Theorem). *Let H be a subgroup of a finite group G. Then $|H|$ divides $|G|$. Moreover $\frac{|G|}{|H|}$ equals the number of left cosets of H and also the number of right cosets of H.*

Proof. Let ℓ be the number of left cosets of H in G. Since the number of elements in any left coset of H is $|H|$ and distinct left cosets are disjoint, a count of the elements of G yields $|G| = \ell \cdot |H|$; thus $\ell = |G|/|H|$. For right cosets the argument is similar. □

Corollary (4.1.2). *The order of an element of a finite group divides the order of the group.*

For the order of an element equals the order of the cyclic subgroup it generates.

The index of a subgroup

Even in an infinite group G the sets of left and right cosets of a subgroup H have the same cardinal. Indeed the assignment $xH \mapsto Hx^{-1}$ determines a bijection between these sets. To see this note that Hx^{-1} depends only on the coset xH, not on the choice of element x from it. For, if we choose xh, with $h \in H$, we would get $H(xh)^{-1} = Hh^{-1}x^{-1} = Hx^{-1}$.

This allows us to define the *index of H in G* to be simultaneously the cardinal of the set of left cosets and the cardinal of the set of right cosets of H; the index is written

$$|G : H|.$$

When G is finite, we have already seen that

$$|G : H| = |G|/|H|$$

by Lagrange's Theorem.

Example (4.1.1). Let G be the symmetric group S_3 and let $H = \langle (12)(3) \rangle$. Then $|H| = 2$ and $|G : H| = |G|/|H| = 6/2 = 3$, so we expect to find three left cosets and three right ones. The left cosets are

$$H = \{(1)(2)(3), (12)(3)\}, \quad (123)H = \{(123), (13)(2)\}, \quad (132)H = \{(132), (1)(23)\}$$

and the right cosets are

$$H = \{(1)(2)(3), (12)(3)\}, \quad H(123) = \{(123), (1)(23)\}, \quad H(132) = \{(132), (13)(2)\}$$

Notice that the left cosets are disjoint, as are the right ones; but the left and right cosets may intersect.

---

1 Joseph Louis Lagrange (1736–1813).

The next result is useful in calculations with subgroups: it involves the concept of the product of cardinal numbers, for which see Exercise (1.4.6).

(4.1.3). *Let H and K be subgroups of a group G such that $H \subseteq K$. Then*

$$|G : H| = |G : K| \cdot |K : H|.$$

Proof. Let $\{t_\lambda \mid \lambda \in \Lambda\}$ be a left transversal to H in K, and let $\{s_\mu \mid \mu \in M\}$ be a left transversal to K in G. Thus $K = \bigcup_{\lambda \in \Lambda} t_\lambda H$ and $G = \bigcup_{\mu \in M} s_\mu K$. Hence

$$G = \bigcup_{\lambda \in \Lambda, \, \mu \in M} (s_\mu t_\lambda) H.$$

We claim that the elements $s_\mu t_\lambda$ belong to different left cosets of H. Indeed suppose that $s_\mu t_\lambda H = s_{\mu'} t_{\lambda'} H$; then, since $t_\lambda H \subseteq K$, we have $s_\mu K = s_{\mu'} K$, which implies that $\mu = \mu'$. Hence $t_\lambda H = t_{\lambda'} H$, which shows that $\lambda = \lambda'$. It follows that $|G : H|$, which is the cardinal of the set of left cosets of H in G, equals $|M \times \Lambda|$. By definition of the product of cardinals $|M \times \Lambda| = |M| \cdot |\Lambda| = |G : K| \cdot |K : H|$. $\qquad \square$

Groups of prime order

Lagrange's Theorem is sufficiently strong to enable us to describe all groups of prime order. This is our first example of a classification theorem in group theory; it is also a first indication of the importance of arithmetic properties of the group order for the structure of a group.

(4.1.4). *A group G has prime order p if and only if $G \simeq \mathbb{Z}_p$.*

Proof. Assume that $|G| = p$ and let $1 \neq x \in G$. Then $|\langle x \rangle|$ divides $|G| = p$ by (4.1.1). Hence $|\langle x \rangle| = p = |G|$ and $G = \langle x \rangle$, a cyclic group of order p. Thus $G \simeq \mathbb{Z}_p$ by (3.3.8). The converse is obvious. $\qquad \square$

Example (4.1.2). Find all groups of order less than 6.

Let G be a group such that $|G| < 6$. If $|G| = 1$, then G is a trivial group. If $|G| = 2, 3$ or 5, then (4.1.4) tells us that $G \simeq \mathbb{Z}_2$, $\mathbb{Z}_3$ or $\mathbb{Z}_5$ respectively. We are left with the case where $|G| = 4$. If G contains an element x of order 4, then $G = \langle x \rangle$ and $G \simeq \mathbb{Z}_4$ by (3.3.8). Assuming that G has no elements of order 4, we conclude from (4.1.2) that G must consist of 1 and three elements of order 2, say a, b, c.

Now ab cannot equal 1, otherwise $b = a^{-1} = a$. Also it is clear that $ab \neq a$ and $ab \neq b$. Hence ab must equal c; also $ba = c$ by the same argument. Similarly we can prove that $bc = a = cb$ and $ca = b = ac$.

At this point the reader should recognize that G looks very like the Klein 4-group

$$V = \{(1)(2)(3)(4), (12)(34), (13)(24), (14)(23)\}.$$

In fact the assignments $1_G \mapsto 1_V$, $a \mapsto (12)(34)$, $b \mapsto (13)(24)$, $c \mapsto (14)(23)$ determine an isomorphism from G to V. Our conclusion is that *up to isomorphism there are exactly six groups with order less than 6, namely $\mathbb{Z}_1$, $\mathbb{Z}_2$, $\mathbb{Z}_3$, $\mathbb{Z}_4$, V, $\mathbb{Z}_5$.*

The following application of Lagrange's Theorem furnishes another proof of Fermat's Theorem – see (2.3.4).

Example (4.1.3). If p is a prime and n is any integer, then $n^p \equiv n \pmod{p}$.

Apply (4.1.2) to $\mathbb{Z}_p^*$, the multiplicative group of non-zero congruence classes modulo p. If $[n] \neq [0]$, then (4.1.2) implies that the order of $[n]$ divides $|\mathbb{Z}_p^*| = p - 1$. Thus $[n]^{p-1} = [1]$, i.e., $n^{p-1} \equiv 1 \pmod{p}$. Multiply by n to get $n^p \equiv n \pmod{p}$, and observe that this also holds if $[n] = [0]$.

According to Lagrange's Theorem the order of a subgroup of a finite group divides the group order. However, the natural converse of this statement is false: there need not be a subgroup with order equal to a positive divisor of the group order. This is demonstrated by the following example.

Example (4.1.4). The alternating group A_4 has order 12, but it has no subgroups of order 6.

Write $G = A_4$. First note that each non-trivial element of G is either a 3-cycle or the product of two disjoint transpositions. Also all of the latter with the identity form the Klein 4-group V.

Suppose that H is a subgroup of G with order 6. Assume first that $H \cap V = 1$. Then there are $6 \times 4 = 24$ distinct elements of the form hv, $h \in H$, $v \in V$; for if $h_1 v_1 = h_2 v_2$ with $h_i \in H$, $v_i \in V$, then $h_2^{-1} h_1 = v_2 v_1^{-1} \in H \cap V = 1$, so that $h_1 = h_2$ and $v_1 = v_2$. But this is impossible, so $H \cap V \neq 1$.

Let us say $H \cap V$ contains $\pi = (12)(34)$. Now H must also contain a 3-cycle since there are 8 of these in G, say $\sigma = (123) \in H$. Hence H contains $\tau = \sigma\pi\sigma^{-1} = (14)(23)$. Thus H contains $\pi\tau = (13)(24)$ and it follows that $V \subseteq H$. Other choices of elements leads to the same conclusion. However, $|V|$ does not divide $|H|$, a final contradiction.

Subgroups of cyclic groups
Usually a group has many subgroups and it can be a difficult task to find all of them. Thus it is of interest that the subgroups of a cyclic group are easy to describe. The first observation is that such subgroups are themselves cyclic.

(4.1.5). *A subgroup of a cyclic group is cyclic.*

Proof. Let H be a subgroup of a cyclic group $G = \langle x \rangle$. If $H = 1$, then obviously $H = \langle 1 \rangle$; thus we may assume that $H \neq 1$, so that H contains some $x^m \neq 1$; since H must also contain $(x^m)^{-1} = x^{-m}$, we may as well assume that $m > 0$. Now choose m to be the *smallest* positive integer for which $x^m \in H$; of course we have used the Well-Ordering Law here.

Certainly it is true that $\langle x^m \rangle \subseteq H$. We will prove the reverse inclusion, which will show that $H = \langle x^m \rangle$. Let $h \in H$ and write $h = x^i$. By the Division Algorithm $i = mq + r$ where $q, r \in \mathbb{Z}$ and $0 \leq r < m$. By the Laws of Exponents (3.3.3) $x^i = x^{mq} x^r$. Hence $x^r = x^{-mq} x^i$, which belongs to H since $x^m \in H$ and $x^i \in H$. From the minimality of m it follows that $r = 0$ and $i = mq$. Therefore $h = x^i \in \langle x^m \rangle$. □

The next result tells us how to construct the subgroups of a given cyclic group.

(4.1.6). *Let $G = \langle x \rangle$ be a cyclic group.*
(i) *If G is infinite, each subgroup of G has the form $G_i = \langle x^i \rangle$ where $i \geq 0$. Furthermore, the G_i are all distinct and G_i has infinite order if $i > 0$.*
(ii) *If G has finite order n, then it has exactly one subgroup of order d for each positive divisor d of n, namely $\langle x^{n/d} \rangle$.*

Proof. Assume first that G is infinite and let H be a subgroup of G. By (4.1.5) H is cyclic, say $H = \langle x^i \rangle$ where $i \geq 0$. Thus $H = G_i$. If x^i had finite order m, then $x^{im} = 1$, which, since x has infinite order, can only mean that $i = 0$ and $H = 1$. Thus H is certainly infinite cyclic if $i > 0$. Next $G_i = G_j$ implies that $x^i \in \langle x^j \rangle$ and $x^j \in \langle x^i \rangle$, i.e., $j \mid i$ and $i \mid j$, so that $i = j$. Therefore all the G_i's are different.

Next let G have finite order n and suppose that d is a positive divisor of n. Now $(x^{\frac{n}{d}})^d = x^n = 1$, so $\ell = |x^{\frac{n}{d}}|$ must divide d by (3.3.7). But also $x^{\frac{n\ell}{d}} = 1$ and hence n divides $\frac{n\ell}{d}$, i.e., d divides ℓ. It follows that $\ell = d$ and thus $K = \langle x^{n/d} \rangle$ has order exactly d.

To complete the proof, suppose that $H = \langle x^r \rangle$ is another subgroup with order d. Then $x^{rd} = 1$, so n divides rd and $\frac{n}{d}$ divides r. This shows that $H = \langle x^r \rangle \leq \langle x^{n/d} \rangle = K$. But $|H| = |K| = d$, from which it follows that $H = K$. Consequently there is exactly one subgroup of order d. □

Recall from Section 3.3 that the set of all subgroups of a group is a lattice and may be represented by a Hasse diagram. In the case of a finite cyclic group, (4.1.6) shows that the lattice corresponds to the lattice of positive divisors of the group order.

Example (4.1.5). Display the Hasse diagram for the subgroups of a cyclic group of order 12.

Let $G = \langle x \rangle$ have order 12. By (4.1.6) the subgroups of G correspond to the positive divisors of 12, i. e., 1, 2, 3, 4, 6, 12; indeed, if $i \mid 12$, the subgroup $\langle x^{12/i} \rangle$ has order i. It is now easy to draw the Hasse diagram:

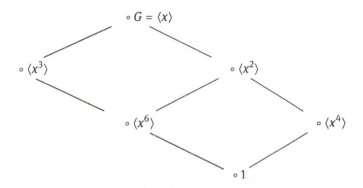

Next comes a useful formula for order of an element in a cyclic group.

(4.1.7). *Let $G = \langle x \rangle$ be a cyclic group with finite order n. Then the order of the element x^i is*

$$\frac{n}{\gcd\{i, n\}}.$$

Proof. In the first place $(x^i)^m = 1$ if and only if $n \mid im$, i.e., $\frac{n}{d} \mid (\frac{i}{d})m$ where $d = \gcd\{i, n\}$. Since $\frac{n}{d}$ and $\frac{i}{d}$ are relatively prime, by Euclid's Lemma this is equivalent to $\frac{n}{d}$ dividing m. Therefore $(x^i)^m = 1$ if and only if $\frac{n}{d}$ divides m, which shows that x^i has order $\frac{n}{d}$, as claimed. □

Corollary (4.1.8). *Let $G = \langle x \rangle$ be a cyclic group of finite order n. Then $G = \langle x^i \rangle$ if and only if $\gcd\{i, n\} = 1$.*

For $G = \langle x \rangle$ if and only if x^i has order n, i.e., $\gcd\{i, n\} = 1$. This means that the number of possible generators of G equals the number of integers i satisfying $1 \leq i < n$ and $\gcd\{i, n\} = 1$. This number is $\phi(n)$ where ϕ is the Eulerian function introduced in Section 2.3.

Every non-trivial group has at least two subgroups, itself and the trivial subgroup: which groups have these two subgroups and no more? The question is easily answered using (4.1.7) and Lagrange's Theorem.

(4.1.9). *A group G has just two subgroups if and only if $G \simeq \mathbb{Z}_p$ for some prime p.*

Proof. Assume that G has only the two subgroups 1 and G. Let $1 \neq x \in G$; then $1 \neq \langle x \rangle \leq G$, so $G = \langle x \rangle$ and G is cyclic. Now G cannot be infinite; for then it would have infinitely many subgroups by (4.1.6). Thus G has finite order n, say. Now if n is not a prime, it has a divisor d such that $1 < d < n$; but then (4.1.6) shows that $\langle x^{n/d} \rangle$ is a subgroup of order d, which is impossible. Therefore G has prime order p and $G \simeq \mathbb{Z}_p$ by (4.1.4).

Conversely, if $G \simeq \mathbb{Z}_p$, then $|G| = p$ and by Lagrange's Theorem G has no non-trivial proper subgroups. $\qquad \square$

Products of subgroups

If H and K are subsets of a group G, the *product* of H and K is defined to be the subset

$$HK = \{hk \mid h \in H, k \in K\}.$$

For example, if $H = \{h\}$ and K is a subgroup, then HK is just the left coset hK. Products of more than two subsets are defined in the obvious way:

$$H_1 H_2 \cdots H_m = \{h_1 h_2 \cdots h_m \mid h_i \in H_i\}.$$

Even if H and K are subgroups, their product HK need not be a subgroup. For example, in S_3 let $H = \langle (12) \rangle$ and $K = \langle (13) \rangle$. Then $HK = \{(1)(2)(3)(4), (12), (13), (132)\}$. But HK cannot be a subgroup since 4 does not divide 6, the order of S_3.

The following result tells us when the product of two subgroups is a subgroup.

(4.1.10). *Let H and K be subgroups of a group G. Then HK is a subgroup if and only if $HK = KH$, and in this event $\langle H, K \rangle = HK$.*

Proof. Assume first that HK is a subgroup of G. Then $H \leq HK$ and $K \leq HK$, so $KH \subseteq HK$. By taking the inverse of each side of this inclusion, we deduce that $HK \subseteq KH$. Hence $HK = KH$. Moreover $\langle H, K \rangle \subseteq HK$ since $H \leq HK$ and $K \leq HK$, while $HK \subseteq \langle H, K \rangle$ is always true. Therefore $\langle H, K \rangle = HK$.

Conversely, assume that $HK = KH$; we will verify that HK is a subgroup by using (3.3.4). Let $h_1, h_2 \in H$ and $k_1, k_2 \in K$. Then $(h_1 k_1)^{-1} = k_1^{-1} h_1^{-1} \in KH = HK$. Also $(h_1 k_1)(h_2 k_2) = h_1(k_1 h_2)k_2$; now $k_1 h_2 \in KH = HK$, so that $k_1 h_2 = h_3 k_3$ where $h_3 \in H$, $k_3 \in K$. Thus $(h_1 k_1)(h_2 k_2) = (h_1 h_3)(k_3 k_2) \in HK$. Obviously $1 \in HK$. Since we have proved that the subset HK is closed under products and inversion, it is a subgroup. $\qquad \square$

It is customary to say that the subgroups H and K *permute* if $HK = KH$. The next result is frequently used in calculations with subgroups.

(4.1.11) (Dedekind's[2] Modular Law). *Let H, K, L be subgroups of a group and assume that $K \subseteq L$. Then*

$$(HK) \cap L = (H \cap L)K.$$

Proof. In the first place $(H \cap L)K \subseteq L$ since $K \subseteq L$; therefore $(H \cap L)K \subseteq (HK) \cap L$. To prove the converse, let $x \in (HK) \cap L$ and write $x = hk$ where $h \in H$, $k \in K$. Hence

---

[2] Richard Dedekind (1831–1916).

$h = xk^{-1} \in LK = L$, from which it follows that $h \in H \cap L$ and $x = hk \in (H \cap L)K$. Thus $(HK) \cap L \subseteq (H \cap L)K$ and the result follows. □

Notice that (4.1.11) is a special case of the distributive law $\langle H, K \rangle \cap L = \langle H \cap L, K \cap L \rangle$. However, this law does not hold in general, (see Exercise (4.1.1) below).

Frequently one wants to count the elements in a product of finite subgroups, which makes the next result useful.

(4.1.12). *If H and K are finite subgroups of a group, then*

$$|HK| = \frac{|H| \cdot |K|}{|H \cap K|}.$$

Proof. Define a function $\alpha : H \times K \to HK$ by the rule $\alpha((h, k)) = hk$ where $h \in H$, $k \in K$; evidently α is surjective. Now $\alpha((h_1, k_1)) = \alpha((h_2, k_2))$ holds if and only if $h_1 k_1 = h_2 k_2$, i. e., $h_2^{-1} h_1 = k_2 k_1^{-1} = d \in H \cap K$. Thus $h_2 = h_1 d^{-1}$ and $k_2 = d k_1$. It follows that the elements of $H \times K$ which have the same image under α as (h_1, k_1) are those of the form $(h_1 d^{-1}, d k_1)$ where $d \in H \cap K$. Now compute the number of the elements of $H \times K$ by counting their images under α and allowing for the number of elements with the same image. This gives $|H \times K| = |HK| \cdot |H \cap K|$. Of course $|H \times K| = |H| \cdot |K|$, so the result is proved. □

The final result of this section provides important information about the index of the intersection of two subgroups.

(4.1.13) (Poincaré[3]). *Let H and K be subgroups of finite index in a group G. Then $H \cap K$ has finite index and*

$$|G : H \cap K| \le |G : H| \cdot |G : K|,$$

with equality if $|G : H|$ and $|G : K|$ are relatively prime.

Proof. To each left coset $x(H \cap K)$ assign the pair of left cosets (xH, xK). This is a well-defined function; for, if we were to replace x by xd with $d \in H \cap K$, then $xH = xdH$ and $xK = xdK$. The function is also injective; for $(xH, xK) = (yH, yK)$ implies that $xH = yH$ and $xK = yK$, i. e., $y^{-1}x \in H \cap K$, so that $x(H \cap K) = y(H \cap K)$. It follows that the number of left cosets of $H \cap K$ in G is at most $|G : H| \cdot |G : K|$.

Now assume that $|G : H|$ and $|G : K|$ are relatively prime. Since

$$|G : H \cap K| = |G : H| \cdot |H : H \cap K|$$

by (4.1.3), we see that $|G : H|$ divides $|G : H \cap K|$, as does $|G : K|$ for a similar reason. But $|G : H|$ and $|G : K|$ are relatively prime, which means that $|G : H \cap K|$ is divisible by $|G : H| \cdot |G : K|$. It follows that $|G : H \cap K|$ must equal $|G : H| \cdot |G : K|$. □

---

3 Henri Poincaré (1854–1912).

Exercises (4.1).

(1) Show that the distributive law for subgroups $\langle H, K \rangle \cap L = \langle H \cap L, K \cap L \rangle$ is false in general.

(2) If H is a subgroup of a finite group, show that there are $|H|^{|G:H|}$ left transversals to H in G and the same number of right transversals.

(3) Let H be a subgroup of a group G such that $G - H$ is finite. Prove that either $H = G$ or G is finite.

(4) Display the Hasse diagram for the subgroup lattices of the following groups: $\mathbb{Z}_{18}$, $\mathbb{Z}_{24}$, V (the Klein 4-group), S_3.

(5) Let G be a group with exactly three subgroups. Show that $G \simeq \mathbb{Z}_{p^2}$ where p is a prime. [Hint: first prove that G is cyclic.]

(6) Let H and K be subgroups of a finite group G with relatively prime indexes in G. Prove that $G = HK$. [Hint: use (4.1.12) and (4.1.13).]

(7) If the product of subsets is used as the binary operation, show that the set of all non-empty subsets of a group is a monoid.

(8) Let H and K be subgroups of a finite group with relatively prime orders. Show that $H \cap K = 1$ and $|HK|$ divides the order of $\langle H, K \rangle$.

(9) Let $G = \langle x \rangle$ be an infinite cyclic group and put $H = \langle x^i \rangle$, $K = \langle x^j \rangle$. Prove that $H \cap K = \langle x^\ell \rangle$ and $\langle H, K \rangle = \langle x^d \rangle$ where $\ell = \mathrm{lcm}\{i, j\}$ and $d = \gcd\{i, j\}$.

(10) Let G be a finite group of order n and let d be the minimum number of generators of G. Prove that $n \geq 2^d$, so that $d \leq [\log_2 n]$.

(11) By applying Lagrange's Theorem to the group $\mathbb{Z}_n^*$, prove that $x^{\phi(n)} \equiv 1 \pmod{n}$ where n is any positive integer and x is an integer relatively prime to n. Here ϕ is Euler's function (cf. (2.3.4)).

(12) Let H be a subgroup with finite index in a finitely generated group G. Use the argument that follows to prove that H is also finitely generated. Let $G = \langle g_1, \ldots, g_n \rangle$ and let $\{t_1, \ldots, t_m\}$ be a left transversal to H in G with $t_1 = 1$. Without loss assume that each g_i^{-1} is also a generator. Write $g_i = t_{\ell_i} h_i$ with $h_i \in H$. Also write $g_i t_j = t_{r(i,j)} h_{ij}$ with $h_{ij} \in H$.

    (i) Prove that $g_i g_j = g_{r(i,\ell_j)} h_{i\ell_j} h_j$.

    (ii) Let $h = g_{i_1} g_{i_2} \cdots g_{i_k} \in H$. By applying (i) repeatedly to segments of h, prove that $h \in \langle h_i, h_{j\ell} \mid i = 1, \ldots, n, j, \ell = 1, \ldots m, \rangle$. Conclude that H is finitely generated.

(13) (*Double cosets*). Let H and K be subgroups of a group G. Define a relation $\sim_{(H,K)}$ on G as follows: $x \sim_{(H,K)} y$ if and only if $x = hyk$ where $h \in H$, $k \in K$. Prove the following.

    (i) $\sim_{(H,K)}$ is an equivalence relation on G and the equivalence class of $x \in G$ is the (H, K)-*double coset* $HxK = \{hxk \mid h \in H, k \in K\}$.

    (ii) If H and K are finite, then $|HxK| = \dfrac{|H| \cdot |K|}{|H \cap xKx^{-1}|} = \dfrac{|H| \cdot |K|}{|K \cap x^{-1}Hx|}$. [Hint: count the number of right cosets of H contained in HxK.]

    (iii) If G is finite, the number of (H, K)-double cosets in G equals $\dfrac{|G| \cdot |H \cap xKx^{-1}|}{|H| \cdot |K|} = \dfrac{|G| \cdot |K \cap x^{-1}Hx|}{|H| \cdot |K|}$.

4.2 Normal subgroups and quotient groups

We focus next on a special type of subgroup called a *normal subgroup*. Such subgroups are important since they can be used to construct new groups, the so-called *quotient groups*. Normal subgroups are characterized in the following result.

(4.2.1). *Let H be a subgroup of a group G. Then the following statements about H are equivalent:*
(i) *$xH = Hx$ for all x in G;*
(ii) *$xhx^{-1} \in H$ whenever $h \in H$ and $x \in G$.*

Proof. First assume that (i) holds and let $x \in G$ and $h \in H$. Then $xh \in xH = Hx$, so $xh = h_1 x$ for some $h_1 \in H$; hence $xhx^{-1} = h_1 \in H$, which establishes (ii).

Conversely, assume that (ii) holds. Again let $x \in G$ and $h \in H$. Then $xhx^{-1} = h_1 \in H$, so $xh = h_1 x \in Hx$ and therefore $xH \subseteq Hx$. Next $x^{-1}hx = x^{-1}h(x^{-1})^{-1} = h_2 \in H$, which shows that $hx = xh_2 \in xH$ and $Hx \subseteq xH$. Thus $xH = Hx$ and (i) is valid. □

A subgroup H with the equivalent properties in (4.2.1) is called a *normal subgroup* of G. The notation

$$H \lhd G$$

is used to denote the fact that H is a normal subgroup of a group G. Also xhx^{-1} is called the *conjugate* of h by x. Thus $H \lhd G$ is valid if and only if H contains all conjugates of its elements by elements of G.

Example (4.2.1).
(i) Obvious examples of normality are: $1 \lhd G$ and $G \lhd G$, and it is possible that these are the only normal subgroups present. If 1 and G are the only normal subgroups of a non-trivial group G, then G is said to be a *simple group*. This is one of the great misnomers of mathematics since simple groups can have extremely complicated structure.

(ii) $A_n \lhd S_n$.
For, if $\pi \in A_n$ and $\sigma \in S_n$, then by (3.1.6) we have

$$\text{sign}(\sigma\pi\sigma^{-1}) = \text{sign}(\sigma)\text{sign}(\pi)(\text{sign}(\sigma))^{-1} = (\text{sign}(\sigma))^2 = 1,$$

so that $\sigma\pi\sigma^{-1} \in A_n$.

(iii) In an abelian group G every subgroup H is normal.
This is because $xhx^{-1} = hxx^{-1} = h$ for all $h \in H$, $x \in G$.

(iv) Recall that $\text{GL}_n(\mathbb{R})$ is the group of all non-singular $n \times n$ real matrices. The subset of matrices in $\text{GL}_n(\mathbb{R})$ with determinant equal to 1 is denoted by

$$\text{SL}_n(\mathbb{R}).$$

First observe that this is a subgroup, the so-called *special linear group* of degree n over $\mathbb{R}$; indeed, if $A, B \in SL_n(\mathbb{R})$, then $\det(AB) = \det(A)\det(B) = 1$ and $\det(A^{-1}) = (\det(A))^{-1} = 1$. In addition

$$SL_n(\mathbb{R}) \lhd GL_n(\mathbb{R}):$$

for if $A \in SL_n(\mathbb{R})$ and $B \in GL_n(\mathbb{R})$,

$$\det(BAB^{-1}) = \det(B)\det(A)(\det(B))^{-1} = \det(B)1\det(B)^{-1} = 1.$$

In these computations two standard results on determinants have been used: $\det(XY) = \det(X)\det(Y)$ and $\det(X^{-1}) = (\det(X))^{-1}$.

(v) A subgroup of S_3 that is *not* normal is $\langle(12)(3)\rangle$.

(vi) *The normal closure.* Let X be a non-empty subset of a group G. The *normal closure*

$$\langle X^G \rangle$$

of X in G is the subgroup generated by all the conjugates gxg^{-1} with $g \in G$ and $x \in X$. Clearly this is the smallest normal subgroup of G which contains X.

(vii) Finally, we introduce two important normal subgroups that can be formed in any group G. The *center* of G,

$$Z(G),$$

consists of all x in G such that $xg = gx$ for every g in G. The reader should check that $Z(G) \lhd G$. Plainly a group G is abelian if and only if $G = Z(G)$.

Next, if x, y are elements of a group G, their *commutator* is the element

$$[x, y] = xyx^{-1}y^{-1}.$$

The significance of commutators arises from the fact that $[x, y] = 1$ if and only if $xy = yx$, i. e., x and y commute. The *derived subgroup* or *commutator subgroup* of G is the subgroup G' generated by all the commutators,

$$G' = \langle [x, y] \mid x, y \in G \rangle.$$

An easy calculation reveals that $z[x, y]z^{-1} = [zxz^{-1}, zyz^{-1}]$, which implies that $G' \lhd G$. Clearly a group G is abelian if and only if $G' = 1$.

Quotient groups

Next we will explain how to form a new group from a normal subgroup N of a group G. This is called the *quotient group* of N in G,

$$G/N.$$

The elements of G/N are the cosets $xN = Nx$, while the group operation is given by the natural rule

$$(xN)(yN) = (xy)N, \ (x, y \in G).$$

Our first concern is to check that this binary operation on G/N is properly defined; it should depend on the two cosets xN and yN, not on the choice of coset representatives x and y. To prove this, let $x_1 \in xN$ and $y_1 \in yN$, so that $x_1 = xa$ and $y_1 = yb$ where $a, b \in N$. Then

$$x_1 y_1 = xayb = xy(y^{-1}ay)b \in (xy)N$$

since $y^{-1}ay = y^{-1}a(y^{-1})^{-1} \in N$ by normality of N. Thus $(xy)N = (x_1 y_1)N$.

It is straightforward to verify that the binary operation just defined is associative. The role of the identity in G/N is played by $1N = N$, while $x^{-1}N$ is the inverse of xN, as is readily checked. It follows that G/N is a group. Note that the elements of G/N are *subsets*, not elements, of G, so that G/N is not a subgroup of G. If G is finite, so is G/N with order

$$|G/N| = |G : N| = \frac{|G|}{|N|}.$$

Example (4.2.2).
(i) $G/1$ is the set of all $x1 = \{x\}$, i. e., one-element subsets of G. Also $\{x\}\{y\} = \{xy\}$. In fact this quotient is not really a new group since $G \simeq G/1$ via the isomorphism in which $x \mapsto \{x\}$. Another trivial example of a quotient group is G/G, which is a group of order 1, with the single element G.

(ii) Let n be a positive integer. Then $\mathbb{Z}/n\mathbb{Z} = \mathbb{Z}_n$. For, allowing for the additive notation, the coset of the subgroup $n\mathbb{Z}$ containing x is $x + n\mathbb{Z} = \{x + nq \mid q \in \mathbb{Z}\}$, which is just the congruence class of x modulo n.

(iii) If G is any group, the quotient group G/G' is an abelian group: indeed $(xG')(yG') = xyG' = (yx)(x^{-1}y^{-1}xy)G' = yxG' = (yG')(xG')$. Also, if G/N is any other abelian quotient group, then

$$(xy)N = (xN)(yN) = (yN)(xN) = (yx)N,$$

which implies that $[x^{-1}, y^{-1}] = x^{-1}y^{-1}xy \in N$ for all $x, y \in N$. Since every commutator has the form $[x^{-1}, y^{-1}]$, it follows that $G' \leq N$. Therefore G/G' is the "largest" abelian quotient group of G.

(iv) *The circle group.* Let r be a real number and suppose that the plane is rotated through angle $2r\pi$ in an anti-clockwise direction about an axis through the origin and perpendicular to the plane. This results in a symmetry of the unit circle, which we will call r'.

Now define $G = \{r' \mid r \in \mathbb{R}\}$, a subset of the symmetry group of the unit circle. Note that $r_1' \circ r_2' = (r_1 + r_2)'$ and $(r')^{-1} = (-r)'$. This shows that G is actually a subgroup of the symmetry group; indeed it is the subgroup of all rotations. Our aim is to identify G as a quotient group.

It is claimed that the assignment $r + \mathbb{Z} \mapsto r'$ determines a function $\alpha : \mathbb{R}/\mathbb{Z} \to G$: first we need to make sure that the function is well-defined. To this end let n be an integer and observe that $(r + n)' = r' \circ n' = r'$ since n' is a rotation through angle $2n\pi$, i. e., it is the identity rotation. Clearly α is surjective; it is also injective because $r_1' = r_2'$ implies that $2r_1\pi = 2r_2\pi + 2n\pi$ for some integer n, i. e., $r_1 = r_2 + n$, and hence $r_1 + \mathbb{Z} = r_2 + \mathbb{Z}$. Thus α is a bijection. Finally $\alpha((r_1 + \mathbb{Z}) + (r_2 + \mathbb{Z})) = \alpha((r_1 + r_2) + \mathbb{Z})$, which equals

$$(r_1 + r_2)' = r_1' \circ r_2' = \alpha(r_1 + \mathbb{Z}) \circ \alpha(r_2 + \mathbb{Z}).$$

Therefore, allowing for the additive and multiplicative notations for the respective groups $\mathbb{R}/\mathbb{Z}$ and G, we conclude that α is an isomorphism from the quotient group $\mathbb{R}/\mathbb{Z}$ to the circle group G. Hence $G \simeq \mathbb{R}/\mathbb{Z}$.

Subgroups of quotient groups

Suppose that N is a normal subgroup of a group G; it is natural to enquire about the subgroups of the quotient group G/N. It is to be expected that they are related to the subgroups of G.

Assume that H is a subgroup of G/N and define a corresponding subset of G,

$$\bar{H} = \{x \in G \mid xN \in H\}.$$

It is easy to verify that $\bar{H}$ is a subgroup of G. Also the definition of $\bar{H}$ shows that $N \subseteq \bar{H}$.

Conversely, suppose we start with a subgroup K of G which contains N. Since $N \lhd G$ implies that $N \lhd K$, we can form the quotient group K/N, which is evidently a subgroup of G/N. Notice that if $N \leq K_1 \leq G$, then $K/N = K_1/N$ implies that $K = K_1$. Thus the assignment $K \mapsto K/N$ determines an injective function from the set of subgroups of G that contain N to the set of subgroups of G/N. The function is also surjective since $\bar{H} \mapsto H$ in the notation of the previous paragraph; therefore it is a bijection.

These arguments establish the following fundamental theorem.

(4.2.2) (The Correspondence Theorem). *Let N be a normal subgroup of a group G. Then the assignment $K \mapsto K/N$ determines a bijection from the set of subgroups of G that contain N to the set of subgroups of G/N. Furthermore, $K/N \lhd G/N$ if and only if $K \lhd G$.*

All of this has been proven except the last statement, which follows from the observation that $(xN)(kN)(xN)^{-1} = (xkx^{-1})N$ for $k \in K$ and $x \in G$.

Example (4.2.3). Let n be a positive integer. The subgroups of $\mathbb{Z}_n = \mathbb{Z}/n\mathbb{Z}$ are of the form $K/n\mathbb{Z}$ where $n\mathbb{Z} \le K \le \mathbb{Z}$. Now by (4.1.5) there is an integer $m > 0$ such that $K = \langle m \rangle = m\mathbb{Z}$, and clearly m divides n since $n\mathbb{Z} \le K$. Thus the Correspondence Theorem tells us that the subgroups of $\mathbb{Z}_n$ correspond to the positive divisors of n, a fact we already know from (4.1.6).

Example (4.2.4). Let N be a normal subgroup of a group G. Call N a *maximal normal subgroup of G* if $N \ne G$ and if $N < L \lhd G$ implies that $L = G$. In short "maximal normal" means "maximal *proper* normal". It follows from the Correspondence Theorem that if N is a proper normal subgroup of G, then G/N is simple if and only if there are no normal subgroups of G lying strictly between N and G, i. e., N is maximal normal in G. Thus maximal normal subgroups lead in a natural way to simple groups.

Direct products
Consider two normal subgroups H and K of a (multiplicatively written) group G such that $H \cap K = 1$. Let $h \in H$ and $k \in K$. Then $[h, k] = (hkh^{-1})k^{-1} \in K$ since $K \lhd G$; also $[h, k] = h(kh^{-1}k^{-1}) \in H$ since $H \lhd G$. But $H \cap K = 1$, so $[h, k] = 1$, i. e., $hk = kh$. Thus elements of H commute with elements of K.

If in addition $G = HK$, then G is said to be the *(internal) direct product* of H and K, in symbols

$$G = H \times K.$$

Each element of G is *uniquely* expressible in the form hk, $(h \in H, k \in K)$. For if $hk = h'k'$ with $h' \in H$, $k' \in K$, then $(h')^{-1}h = k'k^{-1} \in H \cap K = 1$, so that $h = h'$ and $k = k'$. Notice also the form taken by the group operation in G,

$$(h_1 k_1)(h_2 k_2) = (h_1 h_2)(k_1 k_2), \quad (h_i \in H, k_i \in K),$$

since $k_1 h_2 = h_2 k_1$.

For example, consider the Klein 4-group

$$V = \{(1)(2)(3)(4), (12)(34), (13)(24), (14)(23)\} :$$

here $V = A \times B = B \times C = A \times C$ where $A = \langle (12)(34) \rangle$, $B = \langle (13)(24) \rangle$, $C = \langle (14)(23) \rangle$.

The direct product concept may be extended to an arbitrary set of normal subgroups $\{G_\lambda \mid \lambda \in \Lambda\}$ of a group G where
(i) $G_\lambda \cap \langle G_\mu \mid \mu \in \Lambda, \mu \ne \lambda \rangle = 1$ for all $\lambda \in \Lambda$;
(ii) $G = \langle G_\lambda \mid \lambda \in \Lambda \rangle$.

By the argument in the case of two subgroups, elements from different G_λ's commute. Also every element of G has a unique expression of the form $g_1 g_2 \cdots g_m$ where $g_i \in G_{\lambda_i}$

and the $\lambda_i \in \Lambda$ are distinct. (The reader should supply a proof). The direct product is denoted by

$$G = \mathrm{Dr}_{\lambda \in \Lambda} G_\lambda$$

or, in case Λ is a finite set $\{\lambda_1, \lambda_2, \ldots, \lambda_n\}$, by

$$G_{\lambda_1} \times G_{\lambda_2} \times \cdots \times G_{\lambda_n}.$$

For additively written groups the term *direct sum* is used and the notation for direct sums is

$$\bigoplus_{\lambda \in \Lambda} G_\lambda \quad \text{and} \quad G_{\lambda_1} \oplus G_{\lambda_2} \oplus \cdots \oplus G_{\lambda_n}.$$

External direct products

Up to now a direct product can only be formed from subgroups within a given group. We show next how to form the direct product of groups that are not necessarily subgroups of the same group. For simplicity we deal in detail only with the case of a finite set of groups $\{G_1, G_2, \ldots, G_m\}$, but see Exercise (4.2.13) for the infinite case.

First we form the set product

$$G = G_1 \times G_2 \times \cdots \times G_m,$$

consisting of all m-tuples $(g_1, g_2, \ldots, g_m)$ with $g_i \in G_i$. Next a binary operation on G is defined by the rule

$$(g_1, g_2, \ldots, g_m)(g_1', g_2', \ldots, g_m') = (g_1 g_1', g_2 g_2', \ldots, g_m g_m')$$

where $g_i, g_i' \in G_i$. With this operation G becomes a group, with identity element $(1_{G_1}, 1_{G_2}, \ldots, 1_{G_m})$ and inverses given by

$$(g_1, g_2, \ldots, g_m)^{-1} = (g_1^{-1}, g_2^{-1}, \ldots, g_m^{-1}).$$

Call G the *external direct product* of the G_i: it is also written

$$G_1 \times G_2 \times \cdots \times G_m.$$

Although G_i is not a subgroup of G, there is an obvious subgroup of G which is isomorphic with G_i. Let $\bar{G}_i$ consist of all elements of the form $\bar{g}_i = (1_{G_1}, 1_{G_2}, \ldots, g_i, \ldots, 1_{G_m})$ where $g_i \in G_i$ appears as the ith entry of $\bar{g}_i$. Then the assignment $g_i \mapsto \bar{g}_i$ defines an isomorphism $\mu_i : G_i \to \bar{G}_i$: the μ_i are called the *canonical injections*. Also, if $g = (g_1, g_2, \ldots, g_m) \in G$, then $g = \bar{g}_1 \bar{g}_2 \cdots \bar{g}_m$, by the product rule in G. Hence

$G = \bar{G}_1\bar{G}_2 \cdots \bar{G}_m$. It is easy to verify that $\bar{G}_i \lhd G$ and $\bar{G}_i \cap \langle \bar{G}_j \mid j = 1, \ldots, m, j \neq i \rangle = 1$, which shows that G is also the internal direct product

$$G = \bar{G}_1 \times \bar{G}_2 \times \cdots \times \bar{G}_m$$

of subgroups isomorphic with $G_1, G_2, \ldots, G_m$. Thus the external direct product can be regarded as an internal direct product.

One can also define surjective maps $\pi_i : G \to G_i$ by sending $g = (g_1, g_2, \ldots, g_n)$ to its ith component g_i. These are the *canonical projections*.

Example (4.2.5). Let $C_1, C_2, \ldots, C_k$ be finite cyclic groups of orders $n_1, n_2, \ldots, n_k$ where the n_i are pairwise relatively prime. Form the external direct product

$$D = C_1 \times C_2 \times \cdots \times C_k.$$

Therefore $|D| = n_1 n_2 \cdots n_k = n$, say. Now let $C_i = \langle x_i \rangle$ and put $x = (x_1, x_2, \ldots, x_m) \in D$. We claim that x generates D, so that D is a cyclic group of order n.

To prove this statement it is enough to show that an arbitrary element $(x_1^{u_1}, \ldots, x_k^{u_k})$ of D is of the form x^r for some r. This amounts to proving that $x_i^r = x_i^{u_i}$ for each i, so there is a solution r of the system of linear congruences $r \equiv u_i \pmod{n_i}, i = 1, 2, \ldots, k$. This is true by the Chinese Remainder Theorem (2.3.7) since $n_1, n_2, \ldots, n_k$ are pairwise relatively prime.

For example, let n be a positive integer and write $n = p_1^{e_1} p_2^{e_2} \cdots p_k^{e_k}$ where the p_i are distinct primes and $e_i > 0$. Then the preceding discussion shows that $\mathbb{Z}_{p_1^{e_1}} \times \mathbb{Z}_{p_2^{e_2}} \times \cdots \times \mathbb{Z}_{p_k^{e_k}}$ is a cyclic group of order n and hence is isomorphic with $\mathbb{Z}_n$.

External direct products of infinitely many groups

Finally, we briefly explain how to form external direct products of infinite families of groups. Let $\{G_\lambda | \lambda \in \Lambda\}$ be a set of groups. Recall from Section 1.5 the notion of a choice function $f : \Lambda \to \bigcup_{\lambda \in \Lambda} G_\lambda$ where $f(\lambda) \in G_\lambda$. Let G denote the set of all choice functions and define a binary operation $(f, g) \mapsto fg$ on G by $fg(\lambda) = f(\lambda)g(\lambda)$ for $\lambda \in \Lambda$. It is easy to see that this makes G into a group. Call this group the *unrestricted external direct product or cartesian product* of the G_λ,

$$G = \mathrm{Cr}_{\lambda \in \Lambda} G_\lambda.$$

Next we define a particular choice function f_x, where $x \in G_\lambda$, as follows: let $f_x(\lambda) = x$ and $f_x(\mu) = 1_{G_\mu}$ for $\mu \neq \lambda$. It can be verified that $\bar{G}_\lambda = \{f_x \mid x \in G_\lambda\} \lhd G$ and that $G_\lambda \simeq \bar{G}_\lambda$ via the assignment $x \mapsto f_x$.

The functions $\mu_\lambda : G_\lambda \to G$ where $x \mapsto f_x$ are the canonical injections in this general case. In a similar way the assignment $f \mapsto f(\lambda)$ sets up a surjective function $\pi_\lambda : G \to G_\lambda$; these are the canonical projections. The canonical injections and projections are homomorphisms as defined in Section 4.3 below.

A *restricted choice function* for the set $\{G_\lambda, \lambda \in \Lambda\}$ is a choice function $f : \Lambda \to \bigcup_{\lambda \in \Lambda} G_\lambda$ such that $f(\mu) = 1_{G_\mu}$ for all but a finite number of μ. Let G_0 be the set of all restricted choice functions. It is evident that G_0 is a subgroup of G. The group G_0 is called the *restricted external direct product* of the groups G_λ:

$$G_0 = \mathrm{Dr}_{\lambda \in \Lambda} G_\lambda.$$

It is a routine exercise to verify that G_0 is the internal direct product of the subgroups $\bar{G}_\lambda$. Also noteworthy is the fact that when the set Λ is finite, the unrestricted and restricted direct products coincide. Under this circumstance a choice function may be identified with the list of its values. When these are written in an agreed order, we recover the original definition of the direct product of finitely many subgroups.

Exercises (4.2).

(1) Identify all the normal subgroups of the groups S_3, Dih(8) and A_4.

(2) Let H be a subgroup of a group G with index 2. Prove that $H \lhd G$. Is this true if 2 is replaced by 3?

(3) Let $H \lhd K \leq G$ and $L \leq G$. Show that $H \cap L \lhd K \cap L$. Also, if $L \lhd G$, prove that $HL/L \lhd KL/L$.

(4) Let $H \leq G$ and $N \lhd G$. Prove that HN is a subgroup of G.

(5) Assume that $H \leq K \leq G$ and $N \lhd G$. If $H \cap N = K \cap N$ and $HN = KN$, prove that $H = K$.

(6) Show that normality is not a transitive relation in general, i. e., $H \lhd K \lhd G$ does not imply that $H \lhd G$. [Suggestion: consider Dih(8).]

(7) If H, K, L are arbitrary groups, prove that

$$H \times (K \times L) \simeq H \times K \times L \simeq (H \times K) \times L.$$

(8) Let $G = H \times K$ where $H, K \leq G$. Prove that $G/H \simeq K$ and $G/K \simeq H$.

(9) Let $G = \langle x \rangle$ be a cyclic group of order n. If $d \geq 0$, prove that $G/\langle x^d \rangle$ is cyclic with order $\gcd\{n, d\}$.

(10) Prove that $Z(S_n) = 1$ if $n \neq 2$.

(11) Prove that $S_n' \neq S_n$ if $n \neq 1$.

(12) Prove that the center of the group $\mathrm{GL}_n(\mathbb{R})$ of all $n \times n$ non-singular real matrices is the subgroup of all scalar matrices, i. e., scalar multiples of the identity matrix.

(13) This exercise employs the notation used in the discussion of the unrestricted and restricted direct products of a set of groups $\{G_\lambda, \lambda \in \Lambda\}$. Prove that (i) $\bar{G}_\lambda \lhd G$ and $G_0 \lhd G$; (ii) $\bar{G}_\lambda \simeq G_\lambda$; (iii) G_0 is the internal direct product of the subgroups $\bar{G}_\lambda$.

4.3 Homomorphisms

A homomorphism between two groups is a function that links the operations of the groups. More precisely, if G and H are groups, a function $\alpha : G \to H$ is called a *homomorphism* if

$$\alpha(xy) = \alpha(x)\alpha(y)$$

for all $x, y \in G$. The reader will recognize that a bijective homomorphism is what we have been calling an *isomorphism*. A homomorphism from a group to itself is termed an *endomorphism*.

We list next some standard examples of homomorphisms.

Example (4.3.1).

(i) $\alpha : \mathbb{Z} \to \mathbb{Z}_n$ where $\alpha(x) = [x]_n$. Here n is a positive integer. Allowing for the additive notation, what is claimed here is that $\alpha(x + y) = \alpha(x) + \alpha(y)$, i. e. $[x + y]_n = [x]_n + [y_n]$; this is just the definition of the sum of congruence classes.

(ii) The determinant function $\det : \text{GL}_n(\mathbb{R}) \to \mathbb{R}^*$ in which $A \mapsto \det(A)$, is a homomorphism, the reason being the well known identity $\det(AB) = \det(A) \det(B)$.

(iii) The sign function $\text{sign} : S_n \to \{\pm 1\}$ in which $\pi \mapsto \text{sign}(\pi)$, is a homomorphism since $\text{sign}(\pi\sigma) = \text{sign}(\pi)\text{sign}(\sigma)$ by (3.1.6).

(iv) *The canonical homomorphism.* This example provides the first evidence of a link between homomorphisms and normal subgroups. Let N be a normal subgroup of a group G and define a function

$$\alpha : G \to G/N$$

by the rule $\alpha(x) = xN$. Then $\alpha(xy) = \alpha(x)\alpha(y)$, i. e., $(xy)N = (xN)(yN)$, by definition of the group operation in G/N. Thus α is a homomorphism.

(v) For any pair of groups G, H there is always at least one homomorphism from G to H, namely the *trivial homomorphism* in which $x \mapsto 1_H$ for all x in G. Another obvious example is the *identity homomorphism* from G to G, which is the identity function on G.

Next come two very basic properties that all homomorphism possess.

(4.3.1). *Let $\alpha : G \to H$ be a homomorphism of groups. Then:*
(i) $\alpha(1_G) = 1_H$;
(ii) $\alpha(x^n) = (\alpha(x))^n$ *for all $n \in \mathbb{Z}$.*

Proof. Applying α to the equation $1_G 1_G = 1_G$, we obtain $\alpha(1_G)\alpha(1_G) = \alpha(1_G)$, which on cancellation yields $\alpha(1_G) = 1_H$.

If $n > 0$, an easy induction on n shows that $\alpha(x^n) = (\alpha(x))^n$. Next $xx^{-1} = 1_G$, so that $\alpha(x)\alpha(x^{-1}) = \alpha(1_G) = 1_H$; from this it follows that $\alpha(x^{-1}) = (\alpha(x))^{-1}$. Finally, if $n \geq 0$, we have $\alpha(x^{-n}) = \alpha((x^n)^{-1}) = (\alpha(x^n))^{-1} = ((\alpha(x))^n)^{-1} = (\alpha(x))^{-n}$, which completes the proof. $\square$

Image and kernel

Let $\alpha : G \to H$ be a group homomorphism. The *image* of α is the subset $\mathrm{Im}(\alpha) = \{\alpha(x) \mid x \in G\}$. Another significant subset associated with α is the *kernel*, which is defined by

$$\mathrm{Ker}(\alpha) = \{x \in G \mid \alpha(x) = 1_H\}.$$

The fundamental properties of image and kernel appear in the following result.

(4.3.2). *If $\alpha : G \to H$ is a homomorphism of groups, the image $\mathrm{Im}(\alpha)$ is a subgroup of H and the kernel $\mathrm{Ker}(\alpha)$ is a normal subgroup of G.*

Proof. By (4.3.1) $1_H \in \mathrm{Im}(\alpha)$. Let $x, y \in G$; then $\alpha(x)\alpha(y) = \alpha(xy)$ and $(\alpha(x))^{-1} = \alpha(x^{-1})$. These equations show that $\mathrm{Im}(\alpha)$ is a subgroup of H.

Next, if $x, y \in \mathrm{Ker}(\alpha)$, then $\alpha(xy) = \alpha(x)\alpha(y) = 1_H 1_H = 1_H$, and $\alpha(x^{-1}) = (\alpha(x))^{-1} = 1_H^{-1} = 1_H$; thus $\mathrm{Ker}(\alpha)$ is a subgroup of G. Finally, we establish the critical fact that $\mathrm{Ker}(\alpha)$ is normal in G. Let $x \in \mathrm{Ker}(\alpha)$ and $g \in G$; then

$$\alpha(gxg^{-1}) = \alpha(g)\alpha(x)\alpha(g)^{-1} = \alpha(g)1_H\alpha(g)^{-1} = 1_H,$$

so that $gxg^{-1} \in \mathrm{Ker}(\alpha)$, as required. $\square$

Example (4.3.2). Let us compute the image and kernel of some of the homomorphisms in Example (4.3.1).
(i) $\det : \mathrm{GL}_n(\mathbb{R}) \to \mathbb{R}^*$. The kernel is $\mathrm{SL}_n(\mathbb{R})$, the special linear group, and the image is $\mathbb{R}^*$ since each non-zero real number is the determinant of a diagonal matrix in $\mathrm{GL}_n(\mathbb{R})$.
(ii) $\mathrm{sign} : S_n \to \{\pm 1\}$. The kernel is the alternating group A_n and the image is the group $\{\pm 1\}$, unless $n = 1$.
(iii) The kernel of the canonical homomorphism from G to G/N is, as one might expect, the normal subgroup N. The image is G/N.

Clearly one can tell from the image of a homomorphism whether it is surjective. In fact the kernel of a homomorphism shows whether or not it is injective.

(4.3.3). *Let $\alpha : G \to H$ be a group homomorphism. Then:*
(i) *α is surjective if and only if $\mathrm{Im}(\alpha) = H$;*
(ii) *α is injective if and only if $\mathrm{Ker}(\alpha) = 1_G$;*
(iii) *α is an isomorphism if and only if $\mathrm{Im}(\alpha) = H$ and $\mathrm{Ker}(\alpha) = 1_G$.*

Proof. Of course (i) is true by definition. As for (ii), if α is injective and $x \in \mathrm{Ker}(\alpha)$, then $\alpha(x) = 1_H = \alpha(1_G)$, so that $x = 1_G$ by injectivity of α. Conversely, assume that

$\mathrm{Ker}(\alpha) = 1_G$. If $\alpha(x) = \alpha(y)$, then $\alpha(xy^{-1}) = \alpha(x)(\alpha(y))^{-1} = 1_H$, which means that $xy^{-1} \in \mathrm{Ker}(\alpha) = 1_G$ and $x = y$. Thus (ii) is proven, while (iii) follows at once from (i) and (ii). $\qquad\square$

A homomorphism that is injective is sometimes called a *monomorphism* and one that is surjective an *epimorphism*.

The Isomorphism Theorems

We come now to three fundamental results about homomorphisms and quotient groups which are traditionally known as the *Isomorphism Theorems*.

(4.3.4) (First Isomorphism Theorem). *If $\alpha : G \to H$ is a homomorphism of groups, then $G/\mathrm{Ker}(\alpha) \simeq \mathrm{Im}(\alpha)$ via the assignment $x\mathrm{Ker}(\alpha) \mapsto \alpha(x)$.*

Thus the image of a homomorphism may be regarded as a quotient group: conversely, every quotient group is the image of the associated canonical homomorphism. What this means is that up to isomorphism quotient groups and homomorphic images are the same objects.

Proof of (4.3.4). Let $K = \mathrm{Ker}(\alpha)$. We would like to define a function $\theta : G/K \to \mathrm{Im}(\alpha)$ by the natural rule $\theta(xK) = \alpha(x)$, but first we need to check that this makes sense. If $k \in K$, then $\alpha(xk) = \alpha(x)\alpha(k) = \alpha(x)$, showing that $\theta(xK)$ depends only on the coset xK and not on the choice of x from xK. Thus θ is a well-defined function.

Next $\theta((xy)K) = \alpha(xy) = \alpha(x)\alpha(y) = \theta(xK)\theta(yK)$, so θ is a homomorphism. It is obvious that $\mathrm{Im}(\theta) = \mathrm{Im}(\alpha)$. Finally, $\theta(xK) = 1_H$ if and only if $\alpha(x) = 1_H$, i. e., $x \in K$ or equivalently $xK = K = 1_{G/K}$. Therefore $\mathrm{Ker}(\theta)$ is the identity subgroup of G/K and θ is an isomorphism from G/K to $\mathrm{Im}(\alpha)$. $\qquad\square$

(4.3.5) (Second Isomorphism Theorem). *Let G be a group with a subgroup H and a normal subgroup N. Then $HN \leq G$, $H \cap N \lhd H$ and $HN/N \simeq H/H \cap N$.*

Proof. We begin by defining a function $\theta : H \to G/N$ by the rule $\theta(h) = hN$, $(h \in H)$. It is easy to check that θ is a homomorphism. Also $\mathrm{Im}(\theta) = \{hN \mid h \in H\} = HN/N$, which is a subgroup of G/N by (4.3.2); therefore $HN \leq G$. Next $h \in \mathrm{Ker}(\theta)$ if and only if $hN = N$, i. e., $h \in H \cap N$. Therefore $\mathrm{Ker}(\theta) = H \cap N$ and $H \cap N \lhd H$ by (4.3.2). Apply the First Isomorphism Theorem to the homomorphism θ to obtain $H/H \cap N \simeq HN/N$. $\qquad\square$

(4.3.6) (Third Isomorphism Theorem). *Let M and N be normal subgroups of a group G such that $N \subseteq M$. Then $M/N \lhd G/N$ and $(G/N)/(M/N) \simeq G/M$.*

Proof. Define $\theta : G/N \to G/M$ by the rule $\theta(xN) = xM$; the reader should verify that θ is a well-defined homomorphism. Also $\mathrm{Im}(\theta) = G/M$ and $\mathrm{Ker}(\theta) = M/N$; the result now follows via (4.3.4). $\qquad\square$

Thus a quotient group of a quotient group of G is essentially a quotient group of G, which represents a considerable simplification. Next these theorems are illustrated by some examples.

Example (4.3.3). Let m, n be positive integers. We apply (4.3.5) with $H = m\mathbb{Z}$ and $K = n\mathbb{Z}$ Then, allowing for the additive notation, we obtain

$$(m\mathbb{Z} + n\mathbb{Z})/n\mathbb{Z} \simeq m\mathbb{Z}/(m\mathbb{Z} \cap n\mathbb{Z}).$$

What does this say about the integers m, n? Obviously $m\mathbb{Z} \cap n\mathbb{Z} = \ell\mathbb{Z}$ where ℓ is the least common multiple of m and n. Next $m\mathbb{Z} + n\mathbb{Z}$ consists of all $ma + nb$ where $a, b \in \mathbb{Z}$. From (2.2.3) we see that this is just $d\mathbb{Z}$ where $d = \gcd\{m, n\}$. So the assertion is that $d\mathbb{Z}/n\mathbb{Z} \simeq m\mathbb{Z}/\ell\mathbb{Z}$. Now $d\mathbb{Z}/n\mathbb{Z} \simeq \mathbb{Z}/(\frac{n}{d})\mathbb{Z}$ via the mapping $dx + n\mathbb{Z} \mapsto x + \frac{n}{d}\mathbb{Z}$. Similarly $m\mathbb{Z}/\ell\mathbb{Z} \simeq \mathbb{Z}/(\frac{\ell}{m})\mathbb{Z}$. Therefore $\mathbb{Z}/(\frac{n}{d})\mathbb{Z} \simeq \mathbb{Z}/(\frac{\ell}{m})\mathbb{Z}$. Since isomorphic groups have the same order, it follows that $\frac{n}{d} = \frac{\ell}{m}$ or $mn = d\ell$. Hence (4.3.5) implies that

$$\gcd\{m, n\} \cdot \text{lcm}\{m, n\} = mn,$$

(see also Exercise (2.2.8)).

Example (4.3.4). Consider the determinantal homomorphism $\det : GL_n(\mathbb{R}) \rightarrow \mathbb{R}^*$, which has kernel $SL_n(\mathbb{R})$ and image $\mathbb{R}^*$. Then by (4.3.4)

$$GL_n(\mathbb{R})/SL_n(\mathbb{R}) \simeq \mathbb{R}^*.$$

Automorphisms

An *automorphism* of a group G is an isomorphism from G to itself. Thus an automorphism of G is a permutation of the set of group elements which is also a homomorphism. The set of all automorphisms of G,

$$\text{Aut}(G),$$

is therefore a subset of the symmetric group $\text{Sym}(G)$. The first observation to make is:

(4.3.7). *If G is a group, then $\text{Aut}(G)$ is a subgroup of* $\text{Sym}(G)$.

Proof. The identity permutation is certainly an automorphism. Also, if $\alpha \in \text{Aut}(G)$, then $\alpha^{-1} \in \text{Aut}(G)$ by (3.3.1). Finally, if $\alpha, \beta \in \text{Aut}(G)$, then $\alpha\beta$ is certainly a permutation of G, while $\alpha\beta(xy) = \alpha(\beta(x)\beta(y)) = \alpha\beta(x)\alpha\beta(y)$, which leads to $\alpha\beta \in \text{Aut}(G)$. Hence $\text{Aut}(G)$ is a subgroup. □

In fact $\text{Aut}(G)$ is usually quite a small subgroup of $\text{Sym}(G)$, as will be seen in some of the ensuing examples.

Example (4.3.5). Let A be any additively written abelian group and define $\alpha : A \to A$ by $\alpha(x) = -x$. Then α is an automorphism since

$$\alpha(x + y) = -(x + y) = -x - y = \alpha(x) + \alpha(y),$$

while $\alpha^2 = 1$, so $\alpha^{-1} = \alpha$.

Now suppose we choose A to be $\mathbb{Z}$ and let β be any automorphism of A. Thus $\beta(1) = n$ for some integer n. Notice that β is completely determined by n since $\beta(m) = \beta(m1) = m\beta(1) = mn$ by (4.3.1)(ii). Also $\beta(x) = 1$ for some integer x since β is surjective. Furthermore $1 = \beta(x) = \beta(x1) = x\beta(1) = xn$ and it follows that $n = \pm 1$. Hence there are just two possibilities for β, namely the identity and the automorphism α of the last paragraph which forms negatives. Therefore $|\mathrm{Aut}(\mathbb{Z})| = 2$ and $\mathrm{Aut}(\mathbb{Z}) \simeq \mathbb{Z}_2$. On the other hand, it can be shown that the group $\mathrm{Sym}(\mathbb{Z})$ is uncountable.

Inner automorphisms

An easy way to construct automorphisms is to use a fixed element of the group to form conjugates. If g is an element of a group G, define a function $\tau(g)$ on G by the rule

$$\tau(g)(x) = gxg^{-1}, \quad (x \in G).$$

Recall that gxg^{-1} is the conjugate of x by g. Since

$$\tau(g)(xy) = g(xy)g^{-1} = (gxg^{-1})(gyg^{-1}) = \tau(g)(x)\,(\tau(g)(y)),$$

we see that $\tau(g)$ is a homomorphism. Now $\tau(g^{-1})$ is clearly the inverse of $\tau(g)$, therefore $\tau(g)$ is an automorphism of G: it is known as the *inner automorphism induced by g*. Thus we have discovered a function

$$\tau : G \to \mathrm{Aut}(G).$$

The next observation is that τ is a homomorphism, called the *conjugation homomorphism*; for

$$\tau(gh)(x) = (gh)x(gh)^{-1} = g(hxh^{-1})g^{-1},$$

which is also the image of x under the composite $\tau(g)\tau(h)$. Thus $\tau(gh) = \tau(g)\tau(h)$ for all $g, h \in G$.

The image of τ is the set of all inner automorphisms of G, which is denoted by

$$\mathrm{Inn}(G).$$

This is a subgroup of $\mathrm{Aut}(G)$ by (4.3.2). What can be said about $\mathrm{Ker}(\tau)$? It is of course a normal subgroup of G. An element g belongs to $\mathrm{Ker}(\tau)$ if and only if $\tau(g)(x) = x$ for all

x in G, i. e., $gxg^{-1} = x$, or $gx = xg$. Therefore the kernel of τ is exactly $Z(G)$, the center of G, which consists of the elements of G that commute with every element of G.

These conclusions are summed up in the following important result.

(4.3.8). *Let G be a group and let $\tau : G \rightarrow \text{Aut}(G)$ be the conjugation homomorphism. Then $\text{Ker}(\tau) = Z(G)$ and $\text{Im}(\tau) = \text{Inn}(G)$. Hence $\text{Inn}(G) \simeq G/Z(G)$.*

The final statement follows on applying the First Isomorphism Theorem to the homomorphism τ.

Usually a group possesses non-inner automorphisms. For example, if A is an (additively written) abelian group, every inner automorphism is trivial since $\tau(g)(x) = g + x - g = g - g + x = x$. On the other hand, the assignment $x \mapsto -x$ determines an automorphism of A which is not trivial unless $2x = 0$ for all x in A.

(4.3.9). *The relation $\text{Inn}(G) \triangleleft \text{Aut}(G)$ holds for any group G.*

Proof. Let $\alpha \in \text{Aut}(G)$ and $g \in G$; we claim that $\alpha\tau(g)\alpha^{-1} = \tau(\alpha(g))$, which will establish normality. For if $x \in G$, we have

$$\tau(\alpha(g))(x) = \alpha(g)x(\alpha(g))^{-1} = \alpha(g)x\alpha(g^{-1}),$$

which equals

$$\alpha(ga^{-1}(x)g^{-1}) = \alpha(\tau(g)(\alpha^{-1}(x))) = (\alpha\tau(g)\alpha^{-1})(x).$$

Therefore $\alpha\tau(g)\alpha^{-1} = \tau(\alpha(g))$, as required. □

On the basis of (4.3.9) we can form the quotient group

$$\text{Out}(G) = \text{Aut}(G)/\text{Inn}(G),$$

which is termed the *outer automorphism group* of G, (although its elements are not actually automorphisms). Thus all automorphisms of G are inner precisely when $\text{Out}(G) = 1$.

A group G is said to be *complete* if the conjugation homomorphism $\tau : G \rightarrow \text{Aut}(G)$ is an isomorphism: this is equivalent to requiring that $Z(G) = 1$ and $\text{Out}(G) = 1$. It will be shown in Chapter Five that the symmetric group S_n is always complete unless $n = 2$ or 6.

Finally, we point out that the various groups and homomorphisms introduced above fit neatly together in a sequence of groups and homomorphisms

$$1 \rightarrow Z(G) \xrightarrow{\iota} G \xrightarrow{\tau} \text{Aut}(G) \xrightarrow{\nu} \text{Out}(G) \rightarrow 1.$$

Here ι is the inclusion map, τ is the conjugation homomorphism and ν is the canonical homomorphism associated with the normal subgroup $\text{Inn}(G)$. Of course $1 \rightarrow Z(G)$ and $\text{Out}(G) \rightarrow 1$ are trivial homomorphisms.

The sequence above is an example of an *exact sequence*, whose notable feature is that at each group in the interior of the sequence the image of the homomorphism on the left equals the kernel of the homomorphism on the right. For example at $\mathrm{Aut}(G)$ we have $\mathrm{Im}(\tau) = \mathrm{Inn}(G) = \mathrm{Ker}(\nu)$. Exact sequences play a prominent role in algebra, especially in the theory of modules: for more on this see Section 9.1.

In general it is hard to determine the automorphism group of a given group. A useful aid in the process of deciding which permutations of the group are actually automorphisms is the following simple fact.

(4.3.10). *Let G be a group, $g \in G$ and $\alpha \in \mathrm{Aut}(G)$. Then g and $\alpha(g)$ have the same order.*

Proof. By (4.3.1) $\alpha(g^m) = \alpha(g)^m$. Since α is injective, it follows that $\alpha(g)^m = 1$ if and only if $g^m = 1$. Hence $|g| = |\alpha(g)|$. □

The automorphism group of a cyclic group

As a first example consider the automorphism group of a cyclic group $G = \langle x \rangle$. If G is infinite, then $G \simeq \mathbb{Z}$ and we saw in Example (4.3.5) that $\mathrm{Aut}(G) \simeq \mathbb{Z}_2$. Assume from now on that G has finite order m.

First of all notice that α is completely determined by $\alpha(x)$ since $\alpha(x^i) = \alpha(x)^i$. Also $|\alpha(x)| = |x| = m$ by (4.3.10). Thus (4.1.7) shows that $\alpha(x) = x^i$ where $1 \leq i < m$ and i is relatively prime to m. Consequently $|\mathrm{Aut}(G)| \leq \phi(m)$ where ϕ is Euler's function, since $\phi(m)$ is the number of such integers i.

Conversely, suppose that i is an integer satisfying $1 \leq i < m$ and $\gcd\{i, m\} = 1$. Then the assignment $g \mapsto g^i$, $(g \in G)$, determines a homomorphism $\alpha_i : G \rightarrow G$ because $(g_1 g_2)^i = g_1^i g_2^i$, the group G being abelian. Since $|x^i| = m$, the element x^i generates G and so this homomorphism is surjective. But G is finite, so we may conclude that α_i is also injective and thus $\alpha_i \in \mathrm{Aut}(G)$. It follows that $|\mathrm{Aut}(G)| = \phi(m)$, the number of such i's.

It is not hard to identify the group $\mathrm{Aut}(G)$. Recall that $\mathbb{Z}_m^*$ is the multiplicative group of congruence classes $[a]_m$ where a is relatively prime to m. Now there is a natural function $\theta : \mathbb{Z}_m^* \rightarrow \mathrm{Aut}(G)$ given by $\theta([i]_m) = \alpha_i$ where α_i is defined as above; θ is well-defined since $\alpha_{i+\ell m} = \alpha_i$ for all ℓ. In addition θ is a homomorphism because $\alpha_{ij} = \alpha_i \alpha_j$, and the preceding discussion shows that it is surjective and hence bijective. We have therefore established:

(4.3.11). *Let $G = \langle x \rangle$ be a cyclic group of order m. Then $\mathbb{Z}_m^* \simeq \mathrm{Aut}(G)$ via the assignment $[i]_m \mapsto (g \mapsto g^i)$.*

In particular this establishes:

Corollary (4.3.12). *The automorphism group of a cyclic group is abelian.*

The next example is more challenging.

Example (4.3.6). Show that the order of the automorphism group of the dihedral group Dih($2p$) where p is an odd prime is $p(p-1)$.

Recall that Dih($2p$) is the symmetry group of a regular p-gon – see Section 3.2. First we need a good description of the elements of $G = $ Dih($2p$). If the vertices of the p-gon are labelled $1, 2, \ldots, p$, then G contains the p-cycle $\sigma = (1\ 2 \ldots p)$, which corresponds to an anticlockwise rotation through angle $\frac{2\pi}{p}$. It also contains the permutation $\tau = (1)(2\,p)(3\,p{-}1)\ldots(\frac{p+1}{2}\,\frac{p+3}{2})$, which represents a reflection in the line through the vertex 1 and the midpoint of the opposite edge.

The elements σ^r, $\sigma^r\tau$, where $r = 0, 1, \ldots, p-1$, are all different and there are $2p$ of them. Since $|G| = 2p$, we conclude that

$$G = \{\sigma^r, \sigma^r\tau \mid r = 0, 1, \ldots, p-1\}.$$

Notice that $(\sigma^r\tau)^2 = 1$ and in fact $\sigma^r\tau$ is a reflection, while σ^r is a rotation of order 1 or p.

Next let $\alpha \in$ Aut(G). By (4.3.10) $\alpha(\sigma)$ has order p, and hence $\alpha(\sigma) = \sigma^r$ where $1 \le r < p$; also $\alpha(\tau)$ has order 2 and so it must equal $\sigma^s\tau$ where $0 \le s < p$. Observe that α is determined by its effect on σ and τ since $\alpha(\sigma^i) = \alpha(\sigma)^i$ and $\alpha(\sigma^i\tau) = \alpha(\sigma)^i\alpha(\tau)$. It follows that there are at most $p(p-1)$ possibilities for α and hence that $|$Aut(G)$| \le p(p-1)$.

To show that $p(p-1)$ is the order of the automorphism group we need to construct some automorphisms. Now it is easy to see that $Z(G) = 1$; thus by (4.3.8) Inn(G) $\simeq G/Z(G) \simeq G$. Therefore $|$Inn(G)$| = 2p$, and since Inn(G) $\le$ Aut(G), it follows from Lagrange's Theorem that p divides $|$Aut(G)$|$.

Next for $0 < r < p$ we define an automorphism α_r of G by the rules

$$\alpha_r(\sigma) = \sigma^r \quad \text{and} \quad \alpha_r(\tau) = \tau.$$

To verify that α_r is a homomorphism one needs to check that $\alpha_r(xy) = \alpha_r(x)\alpha_r(y)$; this is not difficult, but it does involve some case distinctions, depending on the form of x and y. Now α_r is clearly surjective because σ^r generates $\langle\sigma\rangle$; thus it is an automorphism. Notice also that $\alpha_r\alpha_s = \alpha_{rs}$, so that $[r]_p \mapsto \alpha_r$ determines a homomorphism from $\mathbb{Z}_p^*$ to $H = \{\alpha_r \mid 1 \le r < p\}$. This mapping is surjective, while if $\alpha_r = 1$, then $r \equiv 1$ (mod p), i. e., $[r]_p = [1]_p$. Hence the assignment $[r]_p \mapsto \alpha_r$ determines an isomorphism from $\mathbb{Z}_p^*$ to H. Therefore H has order $p-1$ and $p-1$ divides $|$Aut(G)$|$. Consequently $p(p-1)$ divides the order of Aut(G) and hence $|$Aut(G)$| = p(p-1)$.

Since $|$Inn(G)$| = |G| = 2p$, we see that

$$|\text{Out}(G)| = \frac{p(p-1)}{2p} = \frac{p-1}{2}.$$

Thus $|$Out(G)$| = 1$ if and only if $p = 3$. Since also $Z(G) = 1$, as a consequence Dih($2p$) is a complete group if and only if $p = 3$.

Semidirect products

Suppose that G is a group with a normal subgroup N and a subgroup H such that

$$G = NH \quad \text{and} \quad N \cap H = 1.$$

Then G is said to be the (*internal*) *semidirect product* of N and H. As a simple example, consider the alternating group $G = A_4$; this has a normal subgroup of order 4, namely the Klein 4-group V, and also the subgroup $H = \langle (123)(4) \rangle$ of order 3. Thus $V \cap H = 1$ and $|VH| = |V| \cdot |H| = 12$ by (4.1.12). Hence $G = VH$ and G is the semidirect product of V and H.

Let us analyze the structure of a semidirect product $G = NH$. In the first place each $g \in G$ has a unique expression $g = nh$ with $n \in N$ and $h \in H$. For if $g = n'h'$ is another such expression, $(n')^{-1}n = h'h^{-1} \in N \cap H = 1$, which shows that $n = n'$ and $h = h'$. Secondly, conjugation in N by an element h of H produces an automorphism of N, say $\theta(h)$. Thus $\theta(h)(n) = hnh^{-1}$, $(n \in N)$. Furthermore it is easily verified that $\theta(h_1 h_2) = \theta(h_1)\theta(h_2)$, $(h_i \in H)$. Therefore $\theta : H \to \text{Aut}(N)$ is a homomorphism.

Let us see whether, on the basis of the preceding analysis, we can reconstruct the semidirect product from the groups N and H and a given homomorphism $\theta : H \to \text{Aut}(N)$. This will be the *external semidirect product*. The underlying set of this group is to be the set product $N \times H$, so that

$$G = \{(n,h) \mid n \in N, h \in H\}.$$

A binary operation on G is defined by the rule

$$(n_1, h_1)(n_2, h_2) = (n_1\theta(h_1)(n_2), h_1 h_2).$$

The motivation for this rule is the way that products are formed in an internal semidirect product NH, which is $(n_1 h_1)(n_2 h_2) = n_1(h_1 n_2 h_1^{-1})h_1 h_2$. The identity element of G is $(1_N, 1_H)$ and the inverse of (n, h) is to be $(\theta(h^{-1})(n^{-1}), h^{-1})$: the latter is motivated by the fact that in an internal semidirect product NH inverses are formed according to the rule $(nh)^{-1} = h^{-1}n^{-1} = (h^{-1}n^{-1}h)h^{-1}$. We omit the entirely routine verification of the group axioms for G.

Next we look for subgroups of G which resemble the original groups N and H. There are natural candidates,

$$\bar{N} = \{(n, 1_H) \mid n \in N\} \quad \text{and} \quad \bar{H} = \{(1_N, h) \mid h \in H\}.$$

It is straightforward to show that these are subgroups isomorphic with N and H respectively. The group operation of G shows that

$$(n, 1_H)(1_N, h) = (n\theta(1_H)(1_N), h) = (n, h) \in \bar{N}\bar{H}$$

since $\theta(1_H)$ is the identity automorphism of N. It follows that $G = \bar{N}\bar{H}$, while it is evident that $\bar{N} \cap \bar{H} = 1$.

To show that G is the semidirect product of $\bar{N}$ and $\bar{H}$, it is only necessary to check normality of $\bar{N}$ in G. Let $n, n_1 \in N$ and $h \in H$. Then by definition

$$(n, h)(n_1, 1_H)(n, h)^{-1} = (n, h)(n_1, 1_H)(\theta(h^{-1})(n^{-1}), h^{-1})$$
$$= (n\theta(h)(n_1), h)(\theta(h^{-1})(n^{-1}), h^{-1})$$
$$= (n\theta(h)(n_1)\theta(h)(\theta(h^{-1})(n^{-1})), 1_H)$$
$$= (n\theta(h)(n_1)n^{-1}, 1_H) \in \bar{N}.$$

In particular conjugation in $\bar{N}$ by $(1_N, h)$ sends $(n_1, 1_H)$ to $(\theta(h)(n_1), 1_H)$. Therefore conjugation in $\bar{N}$ by $(1_N, h)$ induces the automorphism $\theta(h)$ in N.

In the special case where θ is chosen to be the trivial homomorphism, elements of $\bar{N}$ and $\bar{H}$ commute, so that G becomes the direct product. Thus the semidirect product is a generalization of the direct product of two groups. Semidirect products provide an important means of constructing new groups.

Example (4.3.7). Let $N = \langle n \rangle$ and $H = \langle h \rangle$ be cyclic groups with respective orders 3 and 4. Suppose we wish to form a semidirect product G of N and H. For this purpose choose a homomorphism $\theta : H \to \mathrm{Aut}(N)$; there is little choice here since N has only one non-identity automorphism, namely $n \mapsto n^{-1}$. Accordingly define $\theta(h)$ to be this automorphism. The resulting group G is known as the *dicyclic group of order* 12. Observe that this group is not isomorphic with A_4 or $\mathrm{Dih}(12)$ since, unlike these groups, G has an element of order 4.

Exercises (4.3).

(1) Let $H \triangleleft K \leq G$ and let $\alpha : G \to G_1$ be a homomorphism. Show that $\alpha(H) \triangleleft \alpha(K) \leq G_1$ where $\alpha(H) = \{\alpha(h) \mid h \in H\}$.

(2) If G and H are groups with relatively prime orders, show that the only homomorphism from G to H is the trivial one.

(3) Let G be a simple group. Show that if $\alpha : G \to H$ is a homomorphism, either α is trivial or H has a subgroup isomorphic with G.

(4) Prove that $\mathrm{Aut}(V) \simeq S_3$ where V is the Klein 4-group.

(5) Prove that $\mathrm{Aut}(\mathbb{Q}) \simeq \mathbb{Q}^*$ where $\mathbb{Q}^*$ is the multiplicative group of non-zero rationals. [Hint: an automorphism is determined by its effect on 1.]

(6) Let G and A be groups, with A abelian and written additively. Let $\mathrm{Hom}(G, A)$ denote the set of all homomorphisms from G to A. Define a binary operation $+$ on $\mathrm{Hom}(G, A)$ by $\alpha + \beta(x) = \alpha(x) + \beta(x)$, $(x \in G)$. Prove that with this operation $\mathrm{Hom}(G, A)$ is an abelian group. Then prove that $\mathrm{Hom}(\mathbb{Z}, A) \simeq A$.

(7) Let $G = \langle x \rangle$ have order 8. Identify all the automorphisms of G and verify that $\mathrm{Aut}(G) \simeq V$: conclude that the automorphism group of a cyclic group need not be cyclic.

(8) If G and H are finite groups of relatively prime orders, prove that $\mathrm{Aut}(G \times H) \simeq \mathrm{Aut}(G) \times \mathrm{Aut}(H)$.

(9) Use Exercise (4.3.8) to prove that $\phi(mn) = \phi(m)\phi(n)$ where ϕ is Euler's function and m, n are relatively prime integers. (A different proof of this fact was given in (2.3.8).)

(10) An $n \times n$ matrix is called a *permutation matrix* if each row and each column contains a single 1 and all other entries are 0. If $\pi \in S_n$, form an $n \times n$ permutation matrix $M(\pi)$ by defining $M(\pi)_{ij}$ to be 1 if $\pi(j) = i$ and 0 otherwise.

   (i)   Prove that the assignment $\pi \mapsto M(\pi)$ determines an injective homomorphism from S_n to $\mathrm{GL}_n(\mathbb{R})$.

   (ii) Deduce that the $n \times n$ permutation matrices form a group which is isomorphic with S_n.

   (iii) How can one tell from $M(\pi)$ whether the permutation π is even or odd?

(11) Express each of the groups $\mathrm{Dih}(2n)$ and S_4 as a semidirect product of groups with smaller orders.

(12) Use the groups $\mathbb{Z}_3 \oplus \mathbb{Z}_3$ and $\mathbb{Z}_2$ to form three non-isomorphic groups of order 18, each with a normal subgroup of order 9.

5 Groups Acting on Sets

Until the end of the 19th Century, a group was usually synonymous with a permutation group, so that the elements acted in a natural way on a set. While group theory has since become more abstract, it remains true that groups are at their most useful when their elements act on a set. In this chapter we develop the basic theory of group actions and illustrate its utility with applications to group theory.

5.1 Group actions

Let G be a group and X a non-empty set. A *left action* of G on X is a function

$$\alpha : G \times X \to X,$$

written for convenience $\alpha((g, x)) = g \cdot x$, with the following properties for $g_i \in G$ and $x \in X$:

(i) $g_1 \cdot (g_2 \cdot x) = (g_1 g_2) \cdot x$;
(ii) $1_G \cdot x = x$.

Here one should think of the group element g as operating or acting on a set element x to produce the set element $g \cdot x$.

There is a corresponding definition of a *right action* of G on X as a function $\beta : X \times G \to X$, with $\beta((x, g))$ written $x \cdot g$, such that $x \cdot 1_G = x$ and $(x \cdot g_1) \cdot g_2 = x \cdot (g_1 g_2)$ for all $x \in X$ and $g_i \in G$.

For example, suppose that G is a subgroup of the symmetric group $\mathrm{Sym}(X)$, in which event G is called *a permutation group on X*. Define $\pi \cdot x$ to be $\pi(x)$ where $\pi \in G$ and $x \in X$; this is a left action of G on X. There may of course be other ways for G to act on X, so we are dealing here with a wide generalization of a permutation group.

Permutation representations
Let G be a group and X a non-empty set. A homomorphism

$$\sigma : G \to \mathrm{Sym}(X)$$

is called a *permutation representation* of G on X. Thus the homomorphism σ represents elements of the abstract group G by concrete objects, namely permutations of X. A permutation representation provides a useful way of visualizing the elements of an abstract group.

What is the connection between group actions and permutation representations? In fact the two concepts are essentially identical. To see why, suppose that a permutation representation $\sigma : G \to \mathrm{Sym}(X)$ is given; then there is a corresponding left action

https://doi.org/10.1515/9783110691160-005

of G on X defined by

$$g \cdot x = \sigma(g)(x),$$

where $g \in G$, $x \in X$; it is easy to check that this is an action.

Conversely, if we start with a left action of G on X, say $(g, x) \mapsto g \cdot x$, there is a corresponding permutation representation $\sigma : G \to \mathrm{Sym}(X)$ defined by

$$\sigma(g)(x) = g \cdot x,$$

where $g \in G$, $x \in X$. Again it is an easy verification that the mapping σ is a homomorphism and hence is a permutation representation of G on X.

The foregoing discussion makes the following result clear.

(5.1.1). *Let G be a group and X a non-empty set. Then there is a bijection from the set of left actions of G on X to the set of permutation representations of G on X.*

If σ is a permutation representation of a group G on a set X, then $G/\mathrm{Ker}(\sigma) \simeq \mathrm{Im}(\sigma)$ by the First Isomorphism Theorem (4.3.4). Thus $G/\mathrm{Ker}(\sigma)$ is isomorphic with a permutation group on X. If $\mathrm{Ker}(\sigma) = 1$, then G itself is isomorphic with a permutation group on X, in which case the representation σ is said to be *faithful*. The term faithful can also be applied to a group action by means of the associated permutation representation.

Next we will describe some natural ways in which a group can act on a set.

Action on a group by multiplication

A group G can act on its underlying set G by left multiplication, that is to say,

$$g \cdot x = gx,$$

where $g, x \in G$; this is an action since $1_G \cdot x = 1_G x = x$ and

$$g_1 \cdot (g_2 \cdot x) = g_1(g_2 x) = (g_1 g_2)x = (g_1 g_2) \cdot x.$$

This action is called the *left regular action* of G and the corresponding permutation representation

$$\lambda : G \to \mathrm{Sym}(G),$$

which is given by $\lambda(g)(x) = gx$, is called the *left regular representation* of G. Observe that $\lambda(g) = 1$ if and only if $gx = x$ for all $x \in G$, i. e., $g = 1$. Thus $\mathrm{Ker}(\lambda) = 1$ and λ is a faithful permutation representation.

It follows at once that G is isomorphic with $\mathrm{Im}(\lambda)$, which is a subgroup of $\mathrm{Sym}(G)$. We have therefore proved the following result, which demonstrates in a striking fashion the significance of permutation groups.

(5.1.2) (Cayley's[1] Theorem). *An arbitrary group G is isomorphic with a subgroup of* Sym(G) *via the left regular representation in which* $g \mapsto (x \mapsto gx)$ *where* $x, g \in G$.

Action on cosets

For the next example of an action take a fixed subgroup H of a group G and let $\mathcal{L}$ be the set of all left cosets of H in G. A left action of G on $\mathcal{L}$ is defined by the rule

$$g \cdot (xH) = (gx)H,$$

where $g, x \in G$. Again it is simple to verify that this is a left action.

Now consider the corresponding permutation representation $\lambda : G \to \text{Sym}(\mathcal{L})$. By definition $g \in \text{Ker}(\lambda)$ if and only if $gxH = xH$ for all x in G, i. e., $x^{-1}gx \in H$ or $g \in xHx^{-1}$. It follows that

$$\text{Ker}(\lambda) = \bigcap_{x \in G} xHx^{-1}.$$

Thus we have:

(5.1.3). *The kernel of the permutation representation of G on the set of left cosets of H by left multiplication is*

$$\bigcap_{x \in G} xHx^{-1},$$

which is the largest normal subgroup of G contained in H.

For the final statement in (5.1.3), note that the intersection is normal in G. Also, if $N \lhd G$ and $N \leq H$, then $N \leq xHx^{-1}$ for all $x \in G$. The normal subgroup $\bigcap_{x \in G} xHx^{-1}$ is called the *normal core* of H in G.

Here is an application of the action on left cosets.

(5.1.4). *Suppose that H is a subgroup of a finite group G such that $|G : H|$ equals the smallest prime dividing $|G|$. Then $H \lhd G$. In particular, a subgroup of index 2 is always normal.*

Proof. Let $|G : H| = p$ and let K be the kernel of the permutation representation of G arising from the left action of G on the set of left cosets of H. Then $K \leq H < G$ and $p = |G : H|$ divides $|G : K|$ by (4.1.3). Now G/K is isomorphic with a subgroup of the symmetric group S_p, so $|G : K|$ divides $|S_p| = p!$ by (4.1.1). But $|G : K|$ divides $|G|$ and thus cannot be divisible by a smaller prime than p. Therefore $|G : K| = p = |G : H|$ and $H = K \lhd G$. $\qquad\square$

---

[1] Arthur Cayley (1821–1895).

Action by conjugation

Another natural way in which a group G can act on its underlying set is by conjugation. Define

$$g \cdot x = gxg^{-1},$$

where $g, x \in G$; by a simple check this is a left action. Again we can ask about the kernel of the action. An element g belongs to the kernel if and only if $gxg^{-1} = x$, i. e., $gx = xg$, for all $x \in G$: this is the condition for g to belong to $Z(G)$, the center of G. It follows that $Z(G)$ is the kernel of the conjugation representation.

A group G can also act on its set of subgroups by conjugation; thus if $H \leq G$, define

$$g \cdot H = gHg^{-1} = \{ghg^{-1} \mid h \in H\}.$$

In this case the kernel consists of all group elements g such that $gHg^{-1} = H$ for all $H \leq G$. This normal subgroup is called the *norm* of G; clearly it contains the center $Z(G)$.

Exercises (5.1).

(1) Complete the proof of (5.1.1).

(2) Let $(x, g) \mapsto x \cdot g$ be a *right* action of a group G on a set X. Define $\rho : G \to \mathrm{Sym}(X)$ by $\rho(g)(x) = x \cdot g^{-1}$. Prove that ρ is a permutation representation of G on X. Why is the inverse necessary here?

(3) Establish a bijection between the set of right actions of a group G on a set X and the set of permutation representations of G on X.

(4) A right action of a group G on its underlying set is defined by $x \cdot g = xg$. Verify that this is an action and describe the corresponding permutation representation of G, (it is called the *right regular representation* of G).

(5) Prove that a permutation representation of a simple group is either faithful or trivial.

(6) The left regular representation of a finite group is surjective if and only if the group has order 1 or 2.

(7) Define a natural right action of a group G on the set of right cosets of a subgroup H and then identify the kernel of the associated representation.

(8) Show that the number of isomorphism types of groups of order n is at most $(n!)^{[\log_2 n]}$. [Hint: a group of order n can be generated by $[\log_2 n]$ elements by Exercise (4.1.10). Now apply (5.1.2).]

5.2 Orbits and stabilizers

In this section we develop the theory of group actions, introducing the fundamental concepts of *orbit* and *stabilizer*.

Let G be a group and X a non-empty set, and suppose that a left action of G on X is given. A binary relation $\underset{G}{\sim}$ on X is defined by the rule:

$$a \underset{G}{\sim} b \quad \text{if and only if} \quad g \cdot a = b$$

for some $g \in G$. A simple verification shows that $\underset{G}{\sim}$ is an equivalence relation on the set X. The $\underset{G}{\sim}$-equivalence class containing a is evidently

$$G \cdot a = \{g \cdot a \mid g \in G\},$$

which is called the *G-orbit* of a. Thus X is the union of the distinct G-orbits and distinct G-orbits are disjoint: these statements follow from general facts about equivalence relations – see (1.2.2).

If X is the only G-orbit, the action of G on X – and the corresponding permutation representation of G – is called *transitive*. Thus the action of G is transitive if for each pair of elements a, b of X, there exists a g in G such that $g \cdot a = b$. For example, the left regular representation is transitive, as is the left action of a group on the set of left cosets of a subgroup.

Another important notion is that of a stabilizer. The *stabilizer* in G of an element $a \in X$ is defined to be

$$\mathrm{St}_G(x) = \{g \in G \mid g \cdot x = x\}.$$

It is easy to verify that $\mathrm{St}_G(a)$ is a subgroup of G. If $\mathrm{St}_G(a) = 1$ for all $a \in X$, the action is called *semiregular*. An action which is both transitive and semiregular is termed *regular*.

We illustrate these concepts by examining the group actions introduced in Section 5.1.

Example (5.2.1). Let G be any group.

(i) The left regular action of G is regular. Indeed $(yx^{-1})x = y$ for any $x, y \in G$, so it is transitive, while $gx = x$ implies that $g = 1$ and the action is semiregular. Regularity now follows.

(ii) In the conjugation action of G on its underlying set the stabilizer of x consists of all g in G such that $gxg^{-1} = x$, i. e., $gx = xg$. This subgroup is called the *centralizer* of x in G: it is denoted by

$$C_G(x) = \{g \in G \mid gx = xg\}.$$

(iii) In the conjugation action of G on its underlying set the G-orbit of x is $\{gxg^{-1} \mid g \in G\}$, i. e., the set of all conjugates of x in G. This is called the *conjugacy class* of x. The number of conjugacy classes in a finite group is called the *class number*.

(iv) In the action of G by conjugation on its set of subgroups, the G-orbit of $H \le G$ is just the set of all conjugates of H in G, i. e., $\{gHg^{-1} \mid g \in G\}$. The stabilizer of H in G is an important subgroup termed the *normalizer* of H in G,

$$N_G(H) = \{g \in G \mid gHg^{-1} = H\}.$$

Centralizers and normalizers feature throughout group theory.

Next we will prove two basic theorems on group actions. The first one counts the number of elements in an orbit.

(5.2.1). *Let G be a group acting on a set X on the left and let $x \in X$. Then the assignment $g\mathrm{St}_G(x) \mapsto g \cdot x$ determines a bijection from the set of left cosets of $\mathrm{St}_G(x)$ in G to the orbit $G \cdot x$. Hence $|G \cdot x| = |G : \mathrm{St}_G(x)|$.*

Proof. In the first place the assignment $g\mathrm{St}_G(x) \mapsto g \cdot x$ determines a well-defined function. For if $s \in \mathrm{St}_G(x)$, then $gs \cdot x = g \cdot (s \cdot x) = g \cdot x$. Next $g_1 \cdot x = g_2 \cdot x$ implies that $(g_2^{-1}g_1) \cdot x = x$, so $g_2^{-1}g_1 \in \mathrm{St}_G(x)$, i. e., $g_1\mathrm{St}_G(x) = g_2\mathrm{St}_G(x)$. Hence the function is injective, while it is obviously surjective. □

Corollary (5.2.2). *Let G be a finite group acting on a finite set X. If the action is transitive, $|X|$ divides $|G|$. If the action is regular, then $|X| = |G|$.*

Proof. If the action is transitive, X is the only G-orbit, so $|X| = |G : \mathrm{St}_G(x)|$ for any $x \in X$ by (5.2.1); hence $|X|$ divides $|G|$. If the action is regular, then in addition $\mathrm{St}_G(x) = 1$ and thus $|X| = |G|$. □

The corollary tells us that if G is a transitive permutation group of degree n, then n divides $|G|$, while $|G| = n$ if G is regular.

The second main theorem on actions counts the number of orbits and has many applications. If a group G acts on a set X on the left and $g \in G$, the *fixed point set* of g is defined to be

$$\mathrm{Fix}(g) = \{x \in X \mid g \cdot x = x\}.$$

(5.2.3) (The Frobenius-Burnside Theorem[2]). *Let G be a finite group acting on a finite set X on the left. Then the number of G-orbits in X equals*

$$\frac{1}{|G|} \sum_{g \in G} |\mathrm{Fix}(g)|,$$

i. e., the average number of fixed points of elements of G.

---

2 Ferdinand Georg Frobenius (1849–1917), William Burnside (1852–1927).

Proof. Consider how often an element x of X is counted in the sum $\sum_{g \in G} |\text{Fix}(g)|$. This happens once for each g in $\text{St}_G(x)$. Thus by (5.2.1) the element x contributes $|\text{St}_G(x)| = |G|/|G \cdot x|$ to the sum. The same is true of each element of the orbit $|G \cdot x|$, so that the total contribution of this orbit to the sum is $(|G|/|G \cdot x|) \cdot |G \cdot x| = |G|$. It follows that $\sum_{g \in G} |\text{Fix}(g)|$ must equal $|G|$ times the number of orbits, so the result is proven. □

We illustrate the Frobenius-Burnside Theorem with a simple example.

Example (5.2.2). The group

$$G = \{(1)(2)(3)(4), (12)(3)(4), (1)(2)(34), (12)(34)\}$$

acts on the set $X = \{1, 2, 3, 4\}$ in the natural way, as a permutation group. There are two G-orbits, namely $\{1, 2\}$ and $\{3, 4\}$. Count the fixed points of the elements of G by looking for 1-cycles. Thus the four elements of the group have respective numbers of fixed points 4, 2, 2, 0. Therefore the number of G-orbits should be

$$\frac{1}{|G|}\left(\sum_{g \in G} |\text{Fix}(g)|\right) = \frac{1}{4}(4 + 2 + 2 + 0) = 2,$$

which is the correct answer.

Example (5.2.3). Show that the average number of fixed points of elements of S_n is 1.

The symmetric group S_n acts on the set $\{1, 2, \ldots, n\}$ in the natural way and the action is clearly transitive. By (5.2.3) the average number of fixed points equals the number of S_n-orbits, which is 1 by transitivity of the action.

Exercises (5.2).

(1) If g is an element of a finite group G, show that the number of conjugates of g divides $|G : \langle g \rangle|$.

(2) If H is a subgroup of a finite group G, show that the number of conjugates of H divides $|G : H|$.

(3) Let $G = \langle (1\,2\ldots p), (1)(2\,p)(3\,p-1)\cdots\rangle$ be the dihedral group $\text{Dih}(2p)$ where p is an odd prime. Check the validity of (5.2.3) for the group G acting on the set $\{1, 2, \ldots, p\}$ as a permutation group.

(4) Let G be a finite group acting as a finite set X. If the action is semiregular, prove that $|G|$ divides $|X|$.

(5) Prove that the class number of a finite group G is given by the formula

$$\frac{1}{|G|}\left(\sum_{x \in G} |C_G(x)|\right).$$

(6) Prove that the class number of a direct product $H \times K$ equals the product of the class numbers of H and K.

(7) Let G be a finite group acting transitively on a finite set X where $|X| > 1$. Prove that G contains at least $|X| - 1$ *fixed-point-free elements*, i. e., elements g such that $\mathrm{Fix}(g)$ is empty. [Hint: assume this is false and apply (5.2.3).]

(8) Let H be a proper subgroup of a finite group G. Prove that $G \neq \bigcup_{x \in G} xHx^{-1}$. [Hint: consider the action of G on the set of left cosets of H by multiplication. The action is transitive, so Exercise (5.2.7) is applicable.]

(9) Let X be a subset of a group G. Define the *centralizer* $C_G(X)$ of X in G to be the set of elements of G that commute with every element of X. Prove that $C_G(X)$ is a subgroup and then show that $C_G(C_G(C_G(X))) = C_G(X)$.

(10) Let G be a finite group with class number h. An element is chosen at random from G and replaced. Then another group element is chosen. Prove that the probability of the two elements commuting is $\frac{h}{|G|}$. What will the answer be if the first group element is not replaced? [Hint: use Exercise (5.2.5).]

5.3 Applications to the structure of groups

The aim of this section is to demonstrate that group actions can be a highly effective tool for investigating the structure of groups. The first result provides important arithmetic information about the conjugacy classes of a finite group.

(5.3.1) (The Class Equation). *Let G be a finite group with distinct conjugacy classes $C_1, C_2, \ldots, C_h$. Then*
(i) $|C_i| = |G : C_G(x_i)|$ *for any x_i in C_i; thus $|C_i|$ divides $|G|$.*
(ii) $|G| = |C_1| + |C_2| + \cdots + |C_h|$.

Here (i) follows on applying (5.2.1) to the conjugation action of G on its underlying set. For in this action the G-orbit of x is its conjugacy class, while the stabilizer of x is the centralizer $C_G(x)$; thus $|G \cdot x| = |G : \mathrm{St}_G(x)| = |G : C_G(x)|$. Finally, (ii) holds because the C_i are disjoint.

There are other ways to express the class equation. Choose any $x_i \in C_i$ and put $n_i = |C_G(x_i)|$. Then $|C_i| = |G|/n_i$. On division by $|G|$, the class equation becomes

$$\frac{1}{n_1} + \frac{1}{n_2} + \cdots + \frac{1}{n_h} = 1,$$

an interesting diophantine equation for the orders of the centralizers.

It is clear that a one element set $\{x\}$ is a conjugacy class of G if and only if x is its only conjugate, i. e., x belongs to the center of the group G. Now suppose we order the conjugacy classes in such a way that $|C_i| = 1$ for $i = 1, 2, \ldots, r$ and $|C_i| > 1$ if $r < i \leq h$. With this notation the class equation takes the form:

(5.3.2). $|G| = |Z(G)| + |C_{r+1}| + \cdots + C_h|$.

A natural question is: what are the conjugacy classes of the symmetric group S_n? First note that any two r-cycles in S_n are conjugate. For

$$\pi(i_1 i_2 \cdots i_r)\pi^{-1} = (j_1 j_2 \cdots j_r)$$

where π is any permutation in S_n such that $\pi(i_1) = j_1$, $\pi(i_2) = j_2$, ..., $\pi(i_r) = j_r$. From this remark and (3.1.3) it follows that *any two permutations which have the same cycle type are conjugate in S_n*. Here "cycle type" refers to the numbers of 1-cycles, 2-cycles, etc. which are present in the disjoint cycle decomposition. Conversely, it is easy to see that conjugate permutations have the same cycle type. Thus we have the answer to our question.

(5.3.3). *The conjugacy classes of the symmetric group S_n are the sets of permutations with the same cycle type.*

It follows that the class number of S_n is the number of different cycle types, which equals

$$\lambda(n),$$

the number of partitions of n, i. e., the number of ways of writing the positive integer n as a sum of positive integers when order of summands is not significant. This is a well-known number theoretic function which has been studied intensively.

Example (5.3.1). The symmetric group S_6 has 11 conjugacy classes. For $\lambda(6) = 11$, as is seen by enumerating the partitions of 6.

As a deeper application of our knowledge of the conjugacy classes of S_n we will prove next:

(5.3.4). *The symmetric group S_n has no non-inner automorphisms if $n \neq 6$.*

Proof. Since S_2 has only the trivial automorphism, we can assume that $n > 2$ as well as $n \neq 6$. First a general remark: in any group G the automorphism group $\mathrm{Aut}(G)$ permutes the conjugacy classes of G. Indeed, if $\alpha \in \mathrm{Aut}(G)$, then $\alpha(xgx^{-1}) = \alpha(x)\alpha(g)(\alpha(x))^{-1}$, so α maps the conjugacy class of g to that of $\alpha(g)$.

Now let C_1 denote the conjugacy class consisting of all the 2-cycles in S_n. If π is a 2-cycle, $\alpha(\pi)$ also has order 2 and so is a product of, say, k disjoint 2-cycles. Hence $\alpha(C_1) = C_k$ where C_k is the conjugacy class of all (disjoint) products of k 2-cycles. The first step in the proof is to show by a counting argument that $k = 1$, i. e., α maps 2-cycles to 2-cycles. Assume to the contrary that $k \geq 2$.

Clearly $|C_1| = \binom{n}{2}$, and more generally

$$|C_k| = \binom{n}{2k}\frac{(2k)!}{(2!)^k k!}.$$

For, in order to form a product of k disjoint 2-cycles, first choose the $2k$ integers from $1, 2, \ldots, n$ in $\binom{n}{2k}$ ways. Then divide these $2k$ elements into k pairs, with order of pairs unimportant; this can be done in $\frac{(2k)!}{(2!)^k k!}$ ways. Forming the product, we obtain the formula for $|C_k|$.

Since $\alpha(C_1) = C_k$, it must be the case that $|C_1| = |C_k|$ and hence

$$\binom{n}{2} = \binom{n}{2k} \frac{(2k)!}{(2!)^k k!}.$$

After cancellation this becomes

$$(n-2)(n-3)\cdots(n-2k+1) = 2^{k-1}(k!).$$

This is impossible if $k = 2$, while if $k = 3$, it can only hold if $n = 6$, which is forbidden. Therefore $k > 3$. Clearly $n \geq 2k$, so $(n-2)(n-3)\cdots(n-2k+1) \geq (2k-2)!$. This leads to $(2k-2)! \leq 2^{k-1}(k!)$, which implies that $k = 3$, a contradiction.

The argument so far has established that $k = 1$ and $\alpha(C_1) = C_1$. Write

$$\alpha((ab)) = (b'b'') \quad \text{and} \quad \alpha((ac)) = (c'c'').$$

Since $(ac)(ab) = (abc)$, which has order 3, also $\alpha((ac)(ab)) = (c'c'')(b'b'')$ has order 3. Therefore b', b'', c', c'' cannot all be different and we can write

$$\alpha((ab)) = (a'b') \quad \text{and} \quad \alpha((ac)) = (a'c').$$

Next suppose there is a d such that $\alpha((ad)) = (b'c')$ with $a' \neq b', c'$. Then $(ac)(ad)(ab) = (abdc)$, an element of order 4, whereas its image $(a'c')(b'c')(a'b') = (a')(b'c')$ has order 2, another contradiction.

This argument shows that for each a there is a unique a' such that $\alpha((ab)) = (a'b')$ for all b and some b'. Therefore α determines a permutation $\pi \in S_n$ such that $\pi(a) = a'$. Thus $\alpha((ab)) = (a'b') = (\pi(a)\,\pi(b))$, which equals the conjugate $\pi(ab)\pi^{-1}$ because the latter interchanges a' and b' and fixes all other integers. Since S_n is generated by 2-cycles by (3.1.4), it follows that α is conjugation by π, so it is an inner automorphism. $\qquad\square$

Recall that a group G is said to be complete if the conjugation homomorphism $\tau : G \to \mathrm{Aut}(G)$ is an isomorphism, i.e., $\mathrm{Ker}(\tau) = Z(G) = 1$ and $\mathrm{Aut}(G) = \mathrm{Inn}(G)$ by (4.3.8). Now $Z(S_n) = 1$ if $n \neq 2$ – see Exercise (4.2.10). Hence we obtain:

Corollary (5.3.5). *The symmetric group S_n is complete if $n \neq 2$ or 6.*

Of course S_2 is not complete since it is abelian. It is known that the group S_6 has a non-inner automorphism, so it too is not complete.

Finite *p*-groups

If p is a prime number, a finite group is called a *p-group* if its order is a power of p. Finite p-groups form an important and highly complex class of groups. A first indication that these groups have special features is provided by the following result.

(5.3.6). *If G is a non-trivial finite p-group, then $Z(G) \neq 1$.*

Proof. Consider the class equation of G in the form

$$|G| = |Z(G)| + |C_{r+1}| + \cdots + |C_h|,$$

– see (5.3.1) and (5.3.2). Here $|C_i|$ divides $|G|$ and hence is a power of p; also $|C_i| > 1$. If $Z(G) = 1$, then it would follow that $|G| \equiv 1 \pmod{p}$, which is impossible because $|G|$ is a power of p. Therefore $Z(G) \neq 1$. □

This behavior of finite p-groups stands in contrast to finite groups in general, which can easily have trivial center.

Corollary (5.3.7). *If p is a prime, every group of order p^2 is abelian.*

Proof. Let G be a group of order p^2. Then $|Z(G)| = p$ or p^2 by (5.3.6) and (4.1.1). If $|Z(G)| = p^2$, then $G = Z(G)$ is abelian. Thus we can assume that $|Z(G)| = p$, so that $|G/Z(G)| = p$. By (4.1.4) both $G/Z(G)$ and $Z(G)$ are cyclic, say $G/Z(G) = \langle aZ(G) \rangle$ and $Z(G) = \langle b \rangle$. It follows that each element of G has the form $a^i b^j$ where i, j are integers. However,

$$(a^i b^j)(a^{i'} b^{j'}) = a^{i+i'} b^{j+j'} = (a^{i'} b^{j'})(a^i b^j)$$

since $b \in Z(G)$, which shows that G is abelian and $Z(G) = G$, a contradiction. □

On the other hand, there are non-abelian groups of order $2^3 = 8$, for example Dih(8), so (5.3.7) does not generalize to groups of order p^3.

Sylow's[3] Theorem

Group actions will now be used to give a proof of *Sylow's Theorem*, which is probably the most celebrated and frequently used result in elementary group theory.

Let G be a finite group and p a prime, and write $|G| = p^a m$ where p does not divide the integer m. Thus p^a is the highest power of p dividing $|G|$. Lagrange's Theorem guarantees that the order of a p-subgroup of G is at most p^a. That p-subgroups of this order actually occur is the first part of Sylow's Theorem. A subgroup of G with the order p^a is called a *Sylow p-subgroup*.

(5.3.8) (Sylow's Theorem). *Let G be a finite group and let p^a denote largest power of the prime p that divides $|G|$. Then the following are true.*

---

3 Peter Ludwig Mejdell Sylow (1832–1918).

(i)   *Every p-subgroup of G is contained in some subgroup of order p^a: in particular, Sylow p-subgroups exist.*

(ii) *If n_p is the number of Sylow p-subgroups, $n_p \equiv 1 \pmod{p}$.*

(iii) *Any two Sylow p-subgroups are conjugate in G.*

Proof. Write $|G| = p^a m$ where p does not divide the integer m. Three group actions will be used during the course of the proof.

(a) Let S be the set of all *subsets* of G with exactly p^a elements. Then S has s elements where

$$s = \binom{p^a m}{p^a} = \frac{m(p^a m - 1) \cdots (p^a m - p^a + 1)}{1 \cdot 2 \cdots (p^a - 1)}.$$

First we prove that p does not divide s. To this end consider the rational number $\frac{p^a m - i}{i}$ where $1 \le i < p^a$. If $p^j \mid i$, then $j < a$ and hence $p^j \mid p^a m - i$. On the other hand, if $p^j \mid p^a m - i$, then $j < a$ since otherwise $p^a \mid i$. Therefore $p^j \mid i$. It follows that the integers $p^a m - i$ and i involve the same highest power of p, which can of course be cancelled in the fraction $\frac{p^a m - i}{i}$; thus no p's occur in this rational number. It follows that p does not divide s, as claimed.

Now we introduce the first group action. The group G acts on the set S via left multiplication, i.e., $g \cdot X = gX$ where $X \subseteq G$ and $|X| = p^a$. Thus S splits up into disjoint G-orbits. Since $|S| = s$ is not divisible by p, there must be at least one G-orbit S_1 such that $|S_1|$ is not divisible by p. Choose $X \in S_1$ and put $P = \mathrm{St}_G(X)$, which is, of course, a subgroup. Then $|G : P| = |S_1|$, from which it follows that p does not divide $|G : P|$. However p^a divides $|G| = |G : P| \cdot |P|$, which implies that p^a divides $|P|$.

Now fix x in X; then the number of elements gx with $g \in P$ equals $|P|$. Also $gx \in X$; hence $|P| \le |X| = p^a$ and consequently $|P| = p^a$. Therefore P is a Sylow p-subgroup of G and we have shown that a Sylow p-subgroup exists.

(b) Let T denote the set of all conjugates of the Sylow p-subgroup P constructed in (a). We argue next that $|T| \equiv 1 \pmod{p}$.

The group P acts on the set T *by conjugation*, i.e., $g \cdot Q = gQg^{-1}$ where $g \in P$ and $Q \in T$; clearly $|gQg^{-1}| = |Q| = |P| = p^a$. In this action $\{P\}$ is a P-orbit since $gPg^{-1} = P$ if $g \in P$. Suppose that $\{P_1\}$ is another one element P-orbit. Then $P_1 \triangleleft \langle P, P_1 \rangle$; for $xP_1x^{-1} = P_1$ if $x \in P \cup P_1$, so $N_{\langle P,P_1 \rangle}(P_1) = \langle P, P_1 \rangle$. (Here $N_{\langle P,P_1 \rangle}(P_1)$ denotes the normalizer of P_1.) By (4.3.5) PP_1 is a subgroup and its order is

$$|PP_1| = \frac{|P| \cdot |P_1|}{|P \cap P_1|},$$

which is certainly a power of p. But $P \subseteq PP_1$ and P already has the maximum order possible for a p-subgroup. Therefore $P = PP_1$, so $P_1 \subseteq P$ and hence $P_1 = P$ since $|P_1| = |P|$.

Consequently there is only one P-orbit of $\mathcal{T}$ with a single element. Every other P-orbit has order a power of p greater than 1. Therefore $|\mathcal{T}| \equiv 1 \pmod{p}$.

(c) Finally, let P_2 be an arbitrary p-subgroup of G. We aim to show that P_2 is contained in some conjugate of the Sylow p-subgroup P found in (a); this will complete the proof of Sylow's Theorem.

Let P_2 act on $\mathcal{T}$ by conjugation, where as before $\mathcal{T}$ is the set of all conjugates of P. Assume that P_2 is not contained in any member of $\mathcal{T}$. If $\{P_3\}$ is a one element P_2-orbit of $\mathcal{T}$, then, arguing as in (b), we see that P_2P_3 is a p-subgroup containing P_3, so $P_3 = P_2P_3$ because $|P_3| = p^a$. Thus $P_2 \subseteq P_3 \in \mathcal{T}$, contrary to assumption. It follows that there are no single element P_2-orbits in $\mathcal{T}$; this means that $|\mathcal{T}| \equiv 0 \pmod{p}$, which contradicts the conclusion of (b). □

An important special case of Sylow's Theorem is:

(5.3.9) (Cauchy's Theorem). *If the order of a finite group G is divisible by a prime p, then G has an element of order p.*

Proof. Let P be a Sylow p-subgroup of G. Then $P \neq 1$ since p divides $|G|$. Choose $1 \neq g \in P$; then $|g|$ divides $|P|$, and hence $|g| = p^m$ where $m > 0$. Thus $g^{p^{m-1}}$ has order p, as required. □

While (5.3.8) does not tell us the exact number of Sylow p-subgroups, it provides valuable information which may be sufficient to determine how many such subgroups are present. Let us review what is known. Suppose P is a Sylow p-subgroup of a finite group G. Then, since every Sylow p-subgroup is a conjugate of P, the number of Sylow p-subgroups of G equals the number of conjugates of P, which by (5.2.1) is

$$n_p = |G : N_G(P)|,$$

where $N_G(P)$ is the normalizer of P in G. Hence n_p divides $|G : P|$ since $P \leq N_G(P)$. Also of course

$$n_p \equiv 1 \pmod{p}.$$

Example (5.3.2). Find the numbers of Sylow p-subgroups of the alternating group A_5.

Let $G = A_5$. We can assume that p divides $|G|$, so that $p = 2, 3$ or 5. Note that a non-trivial element of G has one of three cycle types,

$$(**)(**)(*), \quad (***)(*)(*), \quad (*****)$$

If $p = 2$, then $n_2 \mid \frac{60}{4} = 15$ and $n_2 \equiv 1 \pmod 2$, so $n_2 = 1, 3, 5$ or 15. There are $5 \times 3 = 15$ elements of order 2 in G, with three of them in each Sylow 2-subgroup. Hence $n_2 \geq 5$. If $n_2 = 15$, then $P = N_G(P)$ where P is a Sylow 2-subgroup, since $P \leq N_G(P) \leq G$ and $|G : N_G(P)| = 15 = |G : P|$. But this is wrong since P is normalized by a 3-cycle – note that the Klein 4-group is normal in A_4. Consequently $n_2 = 5$.

Next $n_3 \mid \frac{60}{3} = 20$ and $n_3 \equiv 1 \pmod{3}$. Thus $n_3 = 1$, 4 or 10. Now G has $\binom{5}{3} \times 2 = 20$ elements of order 3, which shows that $n_3 > 4$. Hence $n_3 = 10$. Finally, $n_5 \mid 12$ and $n_5 \equiv 1 \pmod{5}$, so $n_5 = 6$ since $n_5 = 1$ would give only four elements of order 5.

The next result provides some very important information about the group A_5.

(5.3.10). *The alternating group A_5 is simple.*

Proof. Let $G = A_5$ and suppose N is a proper non-trivial normal subgroup of G. The possible orders of elements of G are 1, 2, 3, or 5, (note that 4-cycles are odd). If N contains an element of order 3, it contains a Sylow 3-subgroup of G, and by normality it contains all such. Hence N contains all 3-cycles. Now the easily verified equations $(ab)(ac) = (acb)$ and $(ac)(bd) = (abc)(abd)$, together with the fact that every permutation in G is a product of an *even* number of transpositions, shows that G is generated by 3-cycles. Therefore $N = G$, which is a contradiction.

Next suppose N has an element of order 5; then N contains a Sylow 5-subgroup and hence all 5-cycles. But $(12345)(12543) = (132)$, which gives the contradiction that N contains a 3-cycle.

The argument thus far tells us that each element of N has order a power of 2, which implies that $|N|$ is a power of 2 by Cauchy's Theorem. Since $|N|$ divides $|G| = 60$, this order must be 2 or 4. We leave it to the reader to disprove these possibilities. This final contradiction shows that G is a simple group. □

More generally, A_n is simple for all $n \geq 5$: this is proved in (10.1.7) below. We will see in Section 12.4 that the simplicity of A_5 is intimately connected with the insolvability of polynomial equations of degree 5 by radicals.

Example (5.3.3). Find all groups of order 21.

Let G be a group of order 21. Then G contains elements a and b with orders 7 and 3 respectively by (5.3.9). Now the order of $\langle a \rangle \cap \langle b \rangle$ divides both 7 and 3, i. e., $\langle a \rangle \cap \langle b \rangle = 1$, and thus $|\langle a \rangle \langle b \rangle| = |a| \cdot |b| = 21$, which means that $G = \langle a \rangle \langle b \rangle$. Next $\langle a \rangle$ is a Sylow 7-subgroup of G, and $n_7 \equiv 1 \pmod{7}$ and $n_7 \mid 3$. Hence $n_7 = 1$, so that $\langle a \rangle \triangleleft G$ and $bab^{-1} = a^i$ where $1 \leq i < 7$. If $i = 1$, then G is abelian and $|ab| = 21$. In this case $G = \langle ab \rangle \simeq \mathbb{Z}_{21}$.

Next assume $i \neq 1$. Now $b^3 = 1$ and $bab^{-1} = a^i$, with $2 \leq i < 7$, imply that $a = b^3 ab^{-3} = a^{i^3}$. Hence $7 \mid i^3 - 1$, which shows that $i = 2$ or 4. Now $[2]_7 = [4]_7^{-1}$ since $8 \equiv 1 \pmod{7}$. Since we can replace b by b^{-1} if necessary, there is nothing to be lost in assuming that $i = 2$.

Thus far we have discovered that $G = \{a^u b^v \mid 0 \leq u < 7, 0 \leq v < 3\}$ and that the relations $a^7 = 1 = b^3$, $bab^{-1} = a^2$ hold. But is there really such a group? An example can be found by using permutations. Put $\pi = (1234567)$ and $\sigma = (235)(476)$: thus $\langle \pi, \sigma \rangle$ is a subgroup of S_7. One easily verifies that $\pi^7 = 1 = \sigma^3$ and $\sigma \pi \sigma^{-1} = \pi^2$. A brief computation reveals that the assignments $a \mapsto \pi$, $b \mapsto \sigma$ determine an isomorphism from G to the group $\langle \pi, \sigma \rangle$. It follows that *up to isomorphism there are exactly two groups of order* 21.

Example (5.3.4). Show that there are no simple groups of order 300.

Suppose that G is a simple group of order 300. Since $300 = 2^2 \cdot 3 \cdot 5^2$, a Sylow 5-subgroup P has order 25. Now $n_5 \equiv 1 \pmod 5$ and n_5 divides $300/25 = 12$. Thus $n_5 = 1$ or 6. But $n_5 = 1$ implies that $P \lhd G$, which is impossible. Hence $n_5 = 6$ and $|G : N_G(P)| = 6$. The left action of G on the set of left cosets of $N_G(P)$, – see Section 5.1 – leads to a homomorphism θ from G to S_6. Also $\mathrm{Ker}(\theta) = 1$ since G is simple. Thus θ is injective and $G \simeq \mathrm{Im}(\theta) \leq S_6$. However, $|G| = 300$, which does not divide $|S_6| = 6!$, so we have a contradiction.

Exercises (5.3).

(1) A finite p-group cannot be simple unless its order is p.

(2) Let G be a group of order pq where p and q are primes such that $p \not\equiv 1 \pmod q$ and $q \not\equiv 1 \pmod p$. Prove that G is cyclic.

(3) Show that if p is a prime, a group of order p^2 is isomorphic with Z_{p^2} or $Z_p \times Z_p$.

(4) Let P be a Sylow p-subgroup of a finite group G and let $N \lhd G$. Prove that $P \cap N$ and PN/N are Sylow p-subgroups of N and G/N respectively.

(5) Show that there are no simple groups of order 312.

(6) Let G be a finite simple group which has a subgroup of index n. Prove that G is isomorphic with a subgroup of A_n. [Hint: let H have index n in G. Consider the action of G on left cosets of H and the normal core of H in G.]

(7) Prove that there are no simple groups of order 1960. [Hint: assume there is one and find n_7; then apply Exercise (5.3.6).]

(8) Prove that there are no simple groups of order 616. [Hint: if there is one, prove that for this group one must have $n_{11} = 56$ and $n_7 \geq 8$; then count the elements of orders 7 and 11.]

(9) Prove that every group of order 561 is cyclic. [Hint: show that there is a cyclic normal subgroup $\langle x \rangle$ of order $11 \times 17 = 187$; then use the fact that 3 does not divide $|\mathrm{Aut}(\langle x \rangle)|$.]

(10) Let G be a group of order $2m$ where m is odd. Prove that G has a normal subgroup of order m. [Hint: let λ be the left regular representation of G. By (5.3.9) there is an element g of order 2 in G. Now argue that $\lambda(g)$ must be an odd permutation.]

(11) Find all finite groups with class number at most 2.

(12) Show that every group of order 10 is isomorphic with $\mathbb{Z}_{10}$ or $\mathrm{Dih}(10)$. [Hint: follow the method of Example (5.3.3).]

(13) Show that up to isomorphism there are exactly two groups of order 55.

(14) If H is a proper subgroup of a finite p-group G, prove that $H < N_G(H)$. [Hint: use induction on $|G| > 1$, noting that $H \lhd HZ(G)$.]

(15) Let P be a Sylow p-subgroup of a finite group G and let H be a subgroup of G containing $N_G(P)$. Prove that $H = N_G(H)$. [Hint: if $g \in N_G(H)$, then P and gPg^{-1} are conjugate in H.]

(16) Let G be a finite group and suppose it is possible to choose one element from each conjugacy class in such a way that all the selected elements commute. Prove that G is abelian. [Hint: use (5.3.2).]

6 Introduction to rings

A *ring* is a set equipped with two binary operations called *addition* and *multiplication* which are subject to a number of natural requirements. Thus, from the logical point of view, a ring is a more complex object than a group, which is a set with a single binary operation. Yet some of the most familiar mathematical objects are rings – for example, the sets of integers, real polynomials, continuous functions – and for this reason some readers may feel more comfortable with rings than with groups. One motivation for the study of rings is to see how far properties of the ring of integers extend to rings in general.

6.1 Elementary properties of rings

A *ring* is a triple

$$(R, +, \times)$$

where R is a set and $+$ and $\times$ are binary operations on R called *addition* and *multiplication* such that the following properties hold: here $a \times b$ is written ab:
(i) $(R, +)$ is an abelian group;
(ii) $(R, \times)$ is a semigroup;
(iii) the left and right distributive laws hold, i. e.,

$$a(b + c) = ab + ac, \quad (a + b)c = ac + bc, \quad (a, b, c \in R).$$

If in addition the commutative law for multiplication holds, i. e.
(iv) $ab = ba$ for all $a, b \in R$,

then R is called a *commutative ring.*
   If R contains an element $1_R \neq 0_R$ such that $1_R a = a = a 1_R$ for all $a \in R$, then R is called a *ring with identity* and 1_R is the (clearly unique) *identity element* of R. Care must be taken to distinguish between the additive identity (or zero element) 0_R, which exists in any ring R, and the multiplicative identity 1_R in a ring R with identity. These will often be written simply 0 and 1. As with groups, we usually prefer to speak of "the ring R", rather than the triple $(R, +, \times)$.
   There are many familiar examples of rings at hand.

Example (6.1.1).
(i) $\mathbb{Z}$, $\mathbb{Q}$, $\mathbb{R}$, $\mathbb{C}$ are commutative rings with identity where the ring operations are the usual addition and multiplication of arithmetic.

https://doi.org/10.1515/9783110691160-006

(ii) Let m be a positive integer. Then $\mathbb{Z}_m$, the set of congruence classes modulo m, is a commutative ring with identity where the ring operations are addition and multiplication of congruence classes.

(iii) The set of all continuous real-valued functions defined on the interval $[0,1]$ is a ring when addition and multiplication are given by $f + g(x) = f(x) + g(x)$ and $fg(x) = f(x)g(x)$. This is a commutative ring in which the identity element is the constant function 1.

(iv) Let R be any ring with identity and define $M_n(R)$ to be the set of all $n \times n$ matrices with entries in R. The usual rules for adding and multiplying matrices are to be used. By the elementary properties of matrices $M_n(R)$ is a ring with identity. It is not hard to see that $M_n(R)$ is commutative if and only if R is commutative and $n = 1$.

Of course the ring axioms must be verified in these examples, but this is a routine exercise.

Rings of polynomials

Next we introduce rings of polynomials, which are one of the most fruitful sources of rings.

First we must give a clear definition of a polynomial, not involving vague terms like "indeterminate". In essence a polynomial is just the list of its coefficients, of which only finitely many can be non-zero. We proceed to refine this idea. Let R be a ring with identity. A *polynomial over R* is a sequence of elements $a_i \in R$, one for each natural number i,

$$f = (a_0, a_1, a_2, \ldots)$$

such that $a_i = 0$ for all but a finite number of i; the a_i are called the *coefficients* of f. The *zero polynomial* is $(0_R, 0_R, 0_R, \ldots)$. If $f = (a_0, a_1, \ldots)$ is not zero, there is a largest integer i such that $a_i \neq 0$; thus $f = (a_0, a_1, \ldots, a_i, 0, 0, \ldots)$. The integer i is called the *degree* of f, in symbols

$$\deg(f).$$

It turns out to be convenient to assign to the zero polynomial the degree $-\infty$. A polynomial whose degree is less than 1, i. e., one of the form $(a_0, 0, 0, \ldots)$, is called a *constant polynomial*.

The definitions of addition and multiplication of polynomials are just the familiar rules from elementary algebra, but adapted to the current notation. Let $f = (a_0, a_1, \ldots)$ and $g = (b_0, b_1, \ldots)$ be polynomials over R. Their sum and product are defined by

$$f + g = (a_0 + b_0, a_1 + b_1, \ldots, a_i + b_i, \ldots)$$

and

$$fg = \left(a_0 b_0, a_0 b_1 + a_1 b_0, a_0 b_2 + a_1 b_1 + a_2 b_0, \ldots, \sum_{j=0}^{n} a_j b_{n-j}, \ldots \right).$$

Notice that these really are polynomials; for all but a finite number of the coefficients are 0. Negatives are defined by $-f = (-a_0, -a_1, -a_2, \ldots)$.

(6.1.1). *If f and g are polynomials over a ring with identity, then $f + g$ and fg are polynomials. Also*

(i) $\deg(f + g) \leq \max\{\deg(f), \deg(g)\}$;
(ii) $\deg(fg) \leq \deg(f) + \deg(g)$.

This follows quickly from the definitions of sum and product. It is also quite routine to verify that the ring axioms hold for polynomials with these binary operations. Thus we have:

(6.1.2). *If R is a ring with identity, the set of all polynomials over R forms a ring with identity.*

Of course, the multiplicative identity in the polynomial ring over R is the constant polynomial $(1_R, 0_R, 0_R, \ldots)$.

Now we would like to recover the traditional notation for polynomials, involving an "indeterminate" t. This is accomplished as follows. Let t denote the polynomial $(0, 1, 0, 0, \ldots)$; then the product rule shows that $t^2 = (0, 0, 1, 0, \ldots)$, $t^3 = (0, 0, 0, 1, 0, \ldots)$ etc. If we define the *multiple* of a polynomial by a ring element r by the rule

$$r(a_0, a_1, \ldots) = (ra_0, ra_1, \ldots),$$

then it follows that

$$(a_0, a_1, \ldots, a_n, 0, 0, \ldots) = a_0 + a_1 t + \cdots + a_n t^n,$$

which is called a *polynomial in t*. Thus we can return with confidence to the traditional notation for polynomials knowing that it is soundly based. The ring of polynomials in t over R will be written

$$R[t].$$

Polynomial rings in more than one indeterminate are defined recursively by the equation

$$R[t_1, \ldots, t_n] = (R[t_1, \ldots, t_{n-1}])[t_n],$$

where $n > 1$. A typical element of $R[t_1, \ldots, t_n]$ is a multinomial expression

$$\sum_{\ell_i = 0,1,\ldots} r_{\ell_1 \cdots \ell_n} t_1^{\ell_1} \cdots t_n^{\ell_n},$$

where the ℓ_i are non-negative integers and $r_{\ell_1 \cdots \ell_n} \in R$ equals zero for all but a finite number of $(\ell_1, \ell_2, \ldots, \ell_n)$.

We list next some elementary and frequently used consequences of the ring axioms.

(6.1.3). *Let R be ring, let a, b be elements of R and let n be an integer. Then:*
(i) $a0 = 0 = 0a$;
(ii) $a(-b) = (-a)b = -(ab)$;
(iii) $(na)b = n(ab) = a(nb)$.

Proof. By the distributive law $a(0 + 0) = a0 + a0$. Hence $a0 = a0 + a0$ and so $a0 = 0$ after cancellation. Similarly $0a = 0$. This proves (i). As for (ii) we have $a(-b) + ab = a(-b + b) = a0 = 0$. Thus $a(-b) = -(ab)$. Similarly $(-a)b = -(ab)$. To prove (iii) assume that $n \geq 0$; then $(na)b = n(ab)$ by an easy induction on n. Next $(-na)b + nab = (-na+na)b = 0b = 0$, so $(-na)b = -(nab)$. Similarly $a(-nb) = -(nab)$, which completes the proof. $\qquad\square$

Group rings

A group ring unites the concepts of group and ring. Let G be a group and R a ring with identity. The *group ring RG* is the set of all formal finite sums $r_1 g_1 + r_2 g_2 + \cdots + r_n g_n$ where $r_i \in R$ and $g_i \in G$. Frequently such elements are written

$$\sum_{g \in G} r_g g$$

with the understanding that $r_g = 0$ for all but a finite number of g. The ring operations for RG are given by

$$\sum_{g \in G} r_g g + \sum_{g \in G} s_g g = \sum_{g \in G} (r_g + s_g) g$$

and

$$\left(\sum_{g \in G} r_g g \right) \left(\sum_{g \in G} s_g g \right) = \sum_{g \in G} \left(\sum_{x,y \in G, xy = g} r_x s_y \right) g.$$

Thus to form the sum one simply adds components of corresponding group elements: to form the product think of multiplying out the sums in the product term by term and collecting like terms. It is straight forward to verify that RG is a ring with identity. The zero element 0_{RG} has all its coefficients equal to 0_R, while the identity is $1_{RG} = 1_R 1_G$, which is usually written 1.

If the ring is a field F, then FG is also a vector space over F. Then FG is called the *group algebra* of G over F. Group algebras play a critical role in the theory of group representations, which is the topic of Chapter 14.

Units in rings

Suppose that R is a ring with identity. An element $r \in R$ is called a *unit* if it has a *multiplicative inverse*, i. e., an element $s \in R$ such that $rs = 1 = sr$. Notice that 0 cannot be a unit since $0s = 0 \neq 1$ for all $s \in S$ by (6.1.3). Also, if r is a unit, it has a *unique* inverse, written r^{-1}: this is proved in the same way as (3.2.1)(iii).

Now suppose that r_1 and r_2 are two units of R. Then $r_1 r_2$ is also a unit since $(r_1 r_2)^{-1} = r_2^{-1} r_1^{-1}$, as is seen by forming products with $r_1 r_2$. Also of course $(r^{-1})^{-1} = r$, so that r^{-1} is a unit. Since 1 is its own inverse, we can state:

(6.1.4). *If R is a ring with identity, the set of units of R is a multiplicative group in which the group operation is the ring multiplication.*

The group of units of R is written

$$U(R)$$

or sometimes R^*. Here are some examples of groups of units.

Example (6.1.2).
(i) $U(\mathbb{Z}) = \{\pm 1\}$, a group of order 2.
(ii) $U(\mathbb{Q}) = \mathbb{Q} - 0$, the multiplicative group of non-zero rational numbers.
(iii) If $m > 0$, then $U(\mathbb{Z}_m)$ is the multiplicative group $\mathbb{Z}_m^*$ of all congruence classes $[i]_m$ where $\gcd(i, m) = 1$. This is an abelian group of order $\phi(m)$.
(iv) $U(\mathbb{R}[t])$ is the group of non-zero constant polynomials. For if $fg = 1$, the polynomials f and g must be constant.

Exercises (6.1).
(1) Which of the following are rings?
    (i)   The sets of even and odd integers, with the usual arithmetic operations;
    (ii) the set of all differentiable functions on $[0, 1]$ where $f + g(x) = f(x) + g(x)$ and $fg(x) = f(x)g(x)$;
    (iii) the set of all singular 2×2 real matrices, with the usual matrix operations.
(2) Let S be a non-empty set. Define two binary operations on the power set $\mathcal{P}(S)$ by $X + Y = (X \cup Y) - (X \cap Y)$ and $X \cdot Y = X \cap Y$. Prove that $(\mathcal{P}(S), +, \cdot)$ is a commutative ring with identity. Show also that $X^2 = X$ and $2X = 0_{\mathcal{P}(S)}$.

(3) A ring R is called *Boolean* if $r^2 = r$ for all $r \in R$, (cf. Exercise (6.1.2)). If R is a Boolean ring, prove that $2r = 0$ and that R is commutative.

(4) Let A be an arbitrary (additively written) abelian group. Prove that A is the underlying additive group of some commutative ring.

(5) Find the unit groups of the following rings:
   (i) $\{\frac{m}{2^n} \mid m, n \in \mathbb{Z}\}$, with the usual addition and multiplication;
   (ii) $M_n(\mathbb{R})$ with the standard matrix operations;
   (iii) the ring of continuous functions on $[0, 1]$.

(6) Prove that the Binomial Theorem is valid in any commutative ring with identity R, i. e., $(a + b)^n = \sum_{i=0}^{n} \binom{n}{i} a^{n-i} b^i$ where $a, b \in R$ and n is a non-negative integer. [Hint: use induction on n.]

(7) Let R be a ring with identity. Suppose that a is an element of R with a *unique left inverse* b, i. e., b is the unique element in R such that $ba = 1$. Prove that $ab = 1$, so that a is a unit. [Hint: consider the element $ab - 1 + b$.]

(8) Let R be a ring with identity. Explain how to define a *formal power series* over R of the form $\sum_{n=0}^{\infty} a_n t^n$ with $a_n \in R$. Then verify that these form a ring with identity with respect to appropriate sum and product operations. (This is called the *ring of formal power series* in t over R, in symbols $R[[t]]$.)

(9) Let R be a ring with identity. Prove that $M_n(R)$ is a commutative ring if and only if R is commutative and $n = 1$.

(10) Let R be a ring with identity. The *center* of R is defined to be the set $C(R)$ of $r \in R$ such that $rs = sr$ for all $s \in R$. Prove that $C(R)$ is commutative subring of R.

(11) Let R be a ring with identity. Prove that the center of the ring $M_n(\mathbb{R})$ consists of all scalar $n \times n$ matrices.

6.2 Subrings and ideals

In Chapter 3 the concept of a subgroup of a group was introduced and already this has proved to be valuable in the study of groups. We aim to pursue a similar course for rings by introducing subrings.

Let $(R, +, \times)$ be a ring and S a subset of the underlying set R. Then S is called a *subring* of R if $(S, +_S, \times_S)$ is a ring where $+_S$ and $\times_S$ denote the binary operations $+$ and $\times$ when restricted to S. In particular S is a subgroup of the additive group $(R, +)$. With the aid of (3.3.4), we obtain a more useful description of a subring.

(6.2.1). *Let S be a subset of a ring R. Then S is a subring of R if and only if S contains 0_R and it is closed with respect to addition, multiplication and the formation of negatives, i. e., if $a, b \in S$, then $a + b \in S$, $ab \in S$ and $-a \in S$.*

Example (6.2.1).

(i) $\mathbb{Z}$, $\mathbb{Q}$, $\mathbb{R}$ are successively larger subrings of the ring of complex numbers $\mathbb{C}$.

(ii) The set of even integers $2\mathbb{Z}$ is a subring of $\mathbb{Z}$. Notice that it does not contain the identity element, which is not a requirement for a subring.

(iii) In any ring R there are at least two subrings, the *zero subring* $0 = \{0_R\}$ and the *improper* subring R itself.

(iv) Let $S = \frac{1}{2}\mathbb{Z}$, i. e., $S = \{\frac{m}{2} \mid m \in \mathbb{Z}\}$. Then S is an additive subgroup of the ring $\mathbb{Q}$, but it is not a subring since $\frac{1}{2} \times \frac{1}{2} = \frac{1}{4} \notin S$. Thus the concept of a subring is more special than that of an additive subgroup.

Ideals

It is reasonable to expect there to be an analogy between groups and rings in which subgroups correspond to subrings. The question then arises: what is to correspond to normal subgroups? This is where ideals enter the picture.

Let R be an arbitrary ring. A *left ideal* of R is an additive subgroup L such that $ra \in L$ whenever $r \in R$ and $a \in L$. Similarly a *right ideal* of R is an additive subgroup S such that $ar \in S$ whenever $r \in R$ and $a \in S$. If I is both a left and a right ideal of R, it is called a *2-sided ideal*, or simply an *ideal* of R. Thus an ideal is an additive subgroup which is closed with respect to multiplication of its elements by *arbitrary ring elements* on the left and the right. Notice that *left ideals and right ideals are subrings*.

Example (6.2.2).

(i) Let R be a ring and let $x \in R$. Define subsets of R

$$Rx = \{rx \mid r \in R\} \quad \text{and} \quad xR = \{xr \mid r \in R\}.$$

Then Rx and xR are respectively a left ideal and a right ideal of R. For the first statement Rx is a subgroup since $r_1x + r_2x = (r_1 + r_2)x$ and $-(rx) = (-r)x$; also $s(rx) = (sr)x$ for all $r \in R$, so Rx is a left ideal. Similarly xR is a right ideal. If R is a commutative ring, $Rx = xR$ is an ideal. An ideal of this type is called a *principal ideal*.

(ii) Every subgroup of $\mathbb{Z}$ has the form $n\mathbb{Z}$ where $n \geq 0$ by (4.1.5). Hence every subgroup of $\mathbb{Z}$ is a principal ideal.

(iii) On the other hand, $\mathbb{Z}$ is a subring, but not an ideal, of $\mathbb{Q}$ since $\frac{1}{2}(1) \notin \mathbb{Z}$. Thus *subrings are not always ideals*.

Thus we have a hierarchy of five distinct substructures of rings:

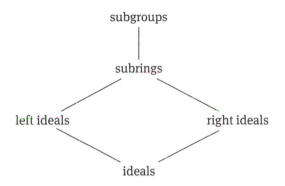

(6.2.2). *The intersection of a non-empty set of subrings (ideals, left ideals, right ideals) of a ring R is a subring (respectively ideal, left ideal, right ideal) of R.*

The easy proofs are left to the reader. Let R be any ring and let X be a non-empty subset of R. On the basis of (6.2.2) we can assert that the intersection of all the subrings of R which contain X is a subring, clearly the smallest subring containing X. This is called the *subring generated* by X and it will be denoted by

$$\mathrm{Rg}\langle X\rangle.$$

If $X = \{x_1, x_2, \ldots, x_n\}$, this subring is denoted by $\mathrm{Rg}\langle x_1, x_2, \ldots, x_n\rangle$. When R has an identity element, the general form of elements of $\mathrm{Rg}\langle X\rangle$ is not hard to determine.

(6.2.3). *Let R be a ring with identity and let X be a non-empty subset of R. Then $\mathrm{Rg}\langle X\rangle$ consists of all elements of the form*

$$\sum_{\ell_1, \ell_2, \ldots, \ell_n} m_{\ell_1, \ell_2, \ldots, \ell_n} x_1^{\ell_1} \cdots x_n^{\ell_n}$$

where $x_i \in X$, $n \geq 0$, $m_{\ell_1, \ell_2, \ldots, \ell_n} \in \mathbb{Z}$ and the ℓ_i are non-negative integers.

Again the easy proof is left to the reader. A ring R is said to be *finitely generated* if $R = \mathrm{Rg}\langle x_1, x_2, \ldots, x_n\rangle$ for some finite set of elements $\{x_1, \ldots, x_n\}$. In a similar vein we define the left, right or two-sided *ideal generated* by a non-empty subset X of a ring R to be the intersection of all the respective types of ideals that contain X.

(6.2.4). *Let R be a ring with identity and let X be a non-empty subset of R. Then the left ideal generated by X consists of all elements of the form*

$$\sum_{i=1}^{n} r_i x_i$$

where $x_i \in X$, $r_i \in R$, $n \geq 0$.

There are similar statements for right and two-sided ideals. The simple proofs are left as an exercise. The left ideal of R generated by X is denoted by

$$RX.$$

A left ideal I of a ring R is said to be *finitely generated as a left ideal* if it can be generated by finitely many elements $x_1, x_2, \ldots, x_n$. If R has an identity, the left ideal I has the form $I = R\{x_1, x_2, \ldots x_n\} = Rx_1 + Rx_2 + \cdots + Rx_n$.

If R is a commutative ring with identity, the ideal $R\{x_1, x_2, \ldots, x_n\}$ is often written $(x_1, x_2, \ldots, x_n)$. In particular

$$(x)$$

is the *principal ideal* $R\{x\}$, consisting of all elements of the form rx where $r \in R$.

Maximal ideals in rings

A *maximal ideal* I of a ring R is defined to be a largest proper ideal. Maximal left ideals and maximal right ideals are defined in a similar manner. For example, if p is a prime, $p\mathbb{Z}$ is a maximal ideal of $\mathbb{Z}$: for $|\mathbb{Z}/p\mathbb{Z}| = p$ and no ideal can occur strictly between $p\mathbb{Z}$ and $\mathbb{Z}$. Zorn's Lemma can be used to establish the existence of maximal ideals.

(6.2.5). *An arbitrary ring R with identity has at least one maximal ideal.*

Proof. Let S denote the set of all proper ideals of R. Now the zero ideal is proper since it does not contain 1_R, so S is not empty. Of course, S is partially ordered by inclusion. Let C be a chain in S and define U to be $\bigcup_{I \in C} I$. It is easily seen that U is an ideal. If $U = R$, then 1_R belongs to some I in C, from which it follows that $R = RI \subseteq I$ and $I = R$. From this contradiction we infer that $U \neq R$, so that $U \in S$. Now Zorn's Lemma can be applied to produce a maximal element of S, i. e., a maximal ideal of R. $\qquad\square$

On the other hand, not every ring has a maximal ideal.

Example (6.2.3). There exist non-zero commutative rings without maximal ideals.

An easy way to get an example is to take the additive abelian group $\mathbb{Q}$ and turn it into a ring by declaring all products of elements to be 0. Then $\mathbb{Q}$ becomes a commutative ring in which subgroups and ideals are the same. But $\mathbb{Q}$ cannot have a maximal subgroup: for if S were one, $\mathbb{Q}/S$ would be a group without proper non-trivial subgroups and so $|\mathbb{Q}/S| = p$, a prime. But this is impossible since $\mathbb{Q} = p\mathbb{Q}$. It follows that this ring has no maximal ideals.

Homomorphisms of rings

It is still not apparent why ideals as defined above should be the analogs of normal subgroups. The decisive test of the appropriateness of the definition will come when ring homomorphisms are defined. If we are right, the kernel of a homomorphism will be an ideal.

It is fairly clear how one should define a *homomorphism from a ring R to a ring S*: this is a function $\theta : R \to S$ which relates the ring operations in the sense that

$$\theta(a + b) = \theta(a) + \theta(b) \quad \text{and} \quad \theta(ab) = \theta(a)\theta(b)$$

for all $a, b \in R$. Thus in particular θ is a homomorphism of groups.

If in addition θ is bijective, θ is called an *isomorphism of rings*. If there is an isomorphism from ring R to ring S, then R and S are said to be *isomorphic rings*, in symbols

$$R \simeq S.$$

Example (6.2.4).

(i) Let m be a positive integer. The function $\theta_m : \mathbb{Z} \to \mathbb{Z}_m$ defined by $\theta_m(x) = [x]_m$ is a ring homomorphism. This is a consequence of the way in which sums and products of congruence classes were defined.

(ii) The *zero homomorphism* $0 : R \to S$ sends every $r \in R$ to 0_S. Also the *identity isomorphism* from R to R is just the identity function.

Complex numbers

For a more interesting example of a ring isomorphism, consider the set R of matrices of the form

$$\begin{bmatrix} a & b \\ -b & a \end{bmatrix}, \quad (a, b \in \mathbb{R}).$$

These are quickly seen to form a subring of the matrix ring $M_2(\mathbb{R})$. Now define a function $\theta : R \to \mathbb{C}$ by the rule

$$\theta\left(\begin{bmatrix} a & b \\ -b & a \end{bmatrix}\right) = a + ib$$

where $i = \sqrt{-1}$. Then θ is a ring homomorphism: for

$$\begin{bmatrix} a_1 & b_1 \\ -b_1 & a_1 \end{bmatrix}\begin{bmatrix} a_2 & b_2 \\ -b_2 & a_2 \end{bmatrix} = \begin{bmatrix} a_1a_2 - b_1b_2 & a_1b_2 + a_2b_1 \\ -a_1b_2 - a_2b_1 & a_1a_2 - b_1b_2 \end{bmatrix},$$

which is mapped by θ to $(a_1a_2 - b_1b_2) + i(a_1b_2 + a_2b_1)$, i.e., to the product $(a_1 + ib_1)(a_2 + ib_2)$. An easier calculation shows that θ sends

$$\begin{bmatrix} a_1 & b_1 \\ -b_1 & a_1 \end{bmatrix} + \begin{bmatrix} a_2 & b_2 \\ -b_2 & a_2 \end{bmatrix}$$

to $(a_1 + ib_1) + (a_2 + ib_2)$.

Certainly θ is surjective; it is also injective since $a + ib = 0$ implies that $a = 0 = b$. Therefore θ is an isomorphism and we obtain the interesting fact that $R \simeq \mathbb{C}$. Thus complex numbers can be represented by real 2×2 matrices. In fact this provides a way to define complex numbers without resorting to the square root of -1.

Next we consider the nature of the kernel and image of a ring homomorphism. The following result should be compared with (4.3.2).

(6.2.6). *If $\theta : R \to S$ is a homomorphism of rings, then $\mathrm{Ker}(\theta)$ is an ideal of R and $\mathrm{Im}(\theta)$ is a subring of S.*

Proof. We know already from (4.3.2) that $\mathrm{Ker}(\theta)$ and $\mathrm{Im}(\theta)$ are subgroups. Let $k \in \mathrm{Ker}(\theta)$ and $r \in R$. Then $\theta(kr) = \theta(k)\theta(r) = 0_S$ and $\theta(rk) = \theta(r)\theta(k) = 0_S$ since $\theta(k) = 0_S$. Therefore $\mathrm{Ker}(\theta)$ is an ideal of R. Furthermore $\theta(r_1)\theta(r_2) = \theta(r_1 r_2)$, so that $\mathrm{Im}(\theta)$ is a subring of S. $\square$

(6.2.7). *If $\theta : R \to S$ is an isomorphism of rings, then so is $\theta^{-1} : S \to R$.*

Proof. Recall from (3.3.1) that θ^{-1} is an isomorphism of groups. It must still be shown that $\theta^{-1}(s_1 s_2) = \theta^{-1}(s_1)\theta^{-1}(s_2)$, $(s_i \in S)$. Observe that the image of each side under θ is $s_1 s_2$. Since θ is injective, it follows that $\theta^{-1}(s_1 s_2) = \theta^{-1}(s_1)\theta^{-1}(s_2)$. $\square$

Quotient rings

Since ideals appear to be the natural ring theoretic analog of normal subgroups, we expect to define a quotient of a ring by means of an ideal. Let I be an ideal of a ring R. Certainly I is a normal subgroup of the additive abelian group R, so we can form the quotient group R/I. This is an additive abelian group whose elements are the cosets of I. To make R/I into a ring, a rule for multiplying cosets must be specified: the natural one to try is

$$(r_1 + I)(r_2 + I) = r_1 r_2 + I, \quad (r_i \in R).$$

To prove that this is well-defined, let $i_1, i_2 \in I$ and note that

$$(r_1 + i_1)(r_2 + i_2) = r_1 r_2 + (r_1 i_2 + i_1 r_2 + i_1 i_2) \in r_1 r_2 + I$$

since I is an ideal. Thus the product is independent of the choice of coset representatives r_1 and r_2.

A further easy check shows that the ring axioms hold; therefore R/I is a ring, the *quotient ring* of I in R. Note also that the assignment $r \mapsto r + I$ leads to a surjective ring homomorphism from R to R/I with kernel I; this is the *canonical homomorphism*, (cf. Section 4.3).

As one might anticipate, there are isomorphism theorems for rings similar to those for groups.

(6.2.8) (First Isomorphism Theorem). *If $\alpha : R \to S$ is a homomorphism of rings, then $R/\mathrm{Ker}(\alpha) \simeq \mathrm{Im}(\alpha)$.*

(6.2.9) (Second Isomorphism Theorem). *If I is an ideal and S is a subring of a ring R, then $S + I$ is a subring of R and $S \cap I$ is an ideal of S. Also $S + I/I \simeq S/S \cap I$.*

(6.2.10) (Third Isomorphism Theorem). *Let I and J be ideals of a ring R such that $I \subseteq J$. Then J/I is an ideal of R/I and $(R/I)/(J/I) \simeq R/J$.*

Fortunately we can apply the isomorphism theorems for groups – see (4.3.4), (4.3.5), (4.3.6). The isomorphisms constructed in the proofs of these theorems still stand, if we allow for the additive notation. Thus we have only to check that they are homomorphisms of rings.

For example, take the case of (6.2.8). From (4.3.4) we know that the assignment $r + \mathrm{Ker}(\alpha) \mapsto \alpha(r)$ yields a group isomorphism $\theta : R/\mathrm{Ker}(\alpha) \mapsto \mathrm{Im}(\alpha)$. Next

$$\theta((r_1 + \mathrm{Ker}(\alpha))(r_2 + \mathrm{Ker}(\alpha))) = \theta(r_1 r_2 + \mathrm{Ker}(\alpha)) = \alpha(r_1 r_2) = \alpha(r_1)\alpha(r_2),$$

which is equal to $\theta(r_1 + \mathrm{Ker}(\alpha))\theta(r_2 + \mathrm{Ker}(\alpha))$. Therefore θ is an isomorphism of rings: this proves (6.2.8). It is left to the reader to complete the proofs of the other two isomorphism theorems.

(6.2.11) (The Correspondence Theorem). *Let I be an ideal of a ring R. Then the assignment $S \mapsto S/I$ determines a bijection from the set of subrings of R that contain I to the set of subrings of R/I. Furthermore S/I is an ideal of R/I if and only if S is an ideal of R.*

Proof. The correspondence between subgroups described in (4.2.2) applies here. It remains only to verify that S is a subring or ideal if and only if S/I is. It is left to the reader to fill in the details. □

Exercises (6.2).
(1) Classify the following subsets of a ring R as an additive subgroup, subring, left or right ideal, or ideal, *as is most appropriate*:
   (i) $\{f \in \mathbb{R}[t] \mid f(a) = 0\}$ where $R = \mathbb{R}[t]$ and $a \in \mathbb{R}$ is fixed;
   (ii) the set of twice differentiable functions on $[0, 1]$ which satisfy the differential equation $f'' + f' = 0$: here R is the ring of continuous functions on $[0, 1]$;
   (iii) $n\mathbb{Z}$ where $R = \mathbb{Z}$;
   (iv) $\frac{1}{2}\mathbb{Z}$ where $R = \mathbb{Q}$.
   (v) the set of real $n \times n$ matrices with zero first row where $R = M_n(\mathbb{R})$.
(2) Prove (6.2.2).
(3) Prove (6.2.3) and (6.2.4).
(4) Which of the following rings are finitely generated? $\mathbb{Z}$; $\mathbb{Q}$; $\mathbb{Z}[t_1, t_2, \ldots, t_n]$.
(5) Let R be a ring with identity. If I is a left ideal containing a unit, show that $I = R$.
(6) Let I and J be ideals of a ring R such that $I \cap J = 0$. Prove that $ab = 0$ for all $a \in I$, $b \in J$.

(7) Let $a \in \mathbb{R}$ and define $\theta_a : \mathbb{R}[t] \to \mathbb{R}$ by $\theta_a(f) = f(a)$. Prove that θ_a is a ring homomorphism. Identify $\mathrm{Im}(\theta_a)$ and $\mathrm{Ker}(\theta_a)$.

(8) Let $\alpha : R \to S$ be a surjective ring homomorphism and assume that R has an identity element and S is not the zero ring. Prove that S has an identity element.

(9) Give examples of a left ideal that is not a right ideal and a right ideal that is not a left ideal.

(10) Give an example of an ideal of a commutative ring with identity that is not principal.

(11) What is the form of elements of the left ideal generated by a subset X in a ring R that does not have an identity element?

(12) Prove that the subring of $\mathbb{Q}$ consisting of all $\frac{m}{2^n}$ is a finitely generated ring.

(13) Let R be a ring with identity. Prove that R has a maximal left ideal and a maximal right ideal. [Use Zorn's Lemma.]

6.3 Integral domains, division rings and fields

The purpose of this section is to introduce some special types of ring with desirable properties. Specifically we are interested in rings having a satisfactory theory of division. For this reason it is necessary to exclude the phenomenon in which the product of two non-zero ring elements equals zero.

If R is a ring, a *left zero divisor* is a non-zero element a such that $ab = 0$ for some $b \neq 0$ in R. Of course b is called a *right zero divisor*. Clearly the presence of zero divisors will make it difficult to construct a reasonable theory of division.

Example (6.3.1). Let n be a positive integer. The zero divisors in $\mathbb{Z}_n$ are the congruence classes $[m]$ where m and n are *not* relatively prime and $1 < m < n$. Thus $\mathbb{Z}_n$ has zero divisors if and only if n is not a prime.

For, if m and n are not relatively prime and $d > 1$ is a common divisor of m and n, then $[m][\frac{n}{d}] = [\frac{m}{d}][n] = [0]$ since $[n] = [0]$, while $[m] \neq 0$ and $[\frac{n}{d}] \neq [0]$; thus $[m]$ is a zero divisor.

Conversely, suppose that $[m]$ is a zero divisor and $[m][\ell] = [0]$ where $[\ell] \neq [0]$. Then $n \mid m\ell$; thus, if m and n are relatively prime, $n \mid \ell$ and $[\ell] = [0]$ by Euclid's Lemma. This contradiction shows that m and n cannot be relatively prime.

Next we introduce an important class of rings with no zero divisors. An *integral domain* (or more briefly a *domain*) is a commutative ring with identity which has no zero divisors. For example, $\mathbb{Z}$ is a domain, while $\mathbb{Z}_n$ is a domain if and only if n is a prime, by Example (6.3.1). Domains can also be characterized by a cancellation property.

(6.3.1). *Let R be a commutative ring with identity. Then R is a domain if and only if the cancellation law is valid in R, that is, $ab = ac$ and $a \neq 0$ always imply that $b = c$.*

Proof. If $ab = ac$ and $b \neq c$, $a \neq 0$, then $a(b - c) = 0$, so that a is a zero divisor and R is not a domain. Conversely, if R is not a domain and $ab = 0$ with $a, b \neq 0$, then $ab = a0$, so the cancellation law fails. $\qquad\square$

The next result shows that it is much simpler to work with polynomials if the co-efficient ring is a domain.

(6.3.2). *Let R be an integral domain and let $f, g \in R[t]$. Then*

$$\deg(fg) = \deg(f) + \deg(g).$$

Hence $fg \neq 0$ if $f \neq 0$ and $g \neq 0$, so that $R[t]$ is an integral domain.

Proof. If $f = 0$, then $fg = 0$ and $\deg(f) = -\infty = \deg(fg)$; hence formally the formula is valid in this case. Assume that $f \neq 0$ and $g \neq 0$, and let at^m and bt^n be the terms of highest degree in f and g respectively; thus $a \neq 0$ and $b \neq 0$. Then $fg = abt^{m+n}$ + terms of lower degree, and $ab \neq 0$ since R is a domain. Therefore $\deg(fg) = m + n = \deg(f) + \deg(g)$. $\qquad\square$

Recall that a unit in a ring with identity is an element with a multiplicative inverse. A ring with identity in which every non-zero element is a unit is termed a *division ring*. Commutative division rings are called *fields*. Clearly $\mathbb{Q}$, $\mathbb{R}$ and $\mathbb{C}$ are examples of fields, while $\mathbb{Z}$ is not a field. Fields are one of the most frequently used types of rings since the ordinary operations of arithmetic can be performed in a field.

Notice that *a division ring cannot have zero divisors*: for if $ab = 0$ and $a \neq 0$, then $b = a^{-1}ab = a^{-1}0 = 0$. Thus the rings without zero divisors include domains and division rings.

The ring of quaternions

The examples of division rings given so far are commutative, i. e., they are fields. We will now describe a famous example of a non-commutative division ring, the *ring of Hamilton's[1] quaternions*. Consider the following 2×2 matrices over $\mathbb{C}$,

$$I = \begin{bmatrix} i & 0 \\ 0 & -i \end{bmatrix}, \quad J = \begin{bmatrix} 0 & 1 \\ -1 & 0 \end{bmatrix}, \quad K = \begin{bmatrix} 0 & i \\ i & 0 \end{bmatrix}$$

where $i = \sqrt{-1}$. These are known in physics as the *Pauli[2] spin matrices*. Simple matrix computations show that the following relations hold:

$$I^2 = J^2 = K^2 = -1,$$

---

1 William Rowan Hamilton (1805–1865).

2 Wolfgang Ernst Pauli (1900–1958).

$$IJ = K = -JI, \quad JK = I = -KJ, \quad KI = J = -IK.$$

Here 1 is being used to denote the identity 2×2 matrix, as it must be distinguished from the quaternion matrix I.

If a, b, c, d are rational numbers, we can form the matrix

$$a1 + bI + cJ + dK = \begin{bmatrix} a + bi & c + di \\ -c + di & a - bi \end{bmatrix},$$

which is called a *rational quaternion*. Let R be the set of all rational quaternions. Then R is a subring of the matrix ring $M_2(\mathbb{C})$ containing the identity: for

$$(a1 + bI + cJ + dK) + (a'1 + b'I + c'J + d'K)$$
$$= (a + a')1 + (b + b')I + (c + c')J + (d + d')K,$$

while $(a1 + bI + cJ + dK)(a'1 + b'I + c'J + d'K)$ equals

$$(aa' - bb' - cc' - dd')1 + (ab' + a'b + cd' - c'd)I$$
$$+ (ac' + a'c + b'd - bd')J + (ad' + a'd + bc' - b'c)K,$$

as is seen by multiplying out and using the properties of I, J, K above.

The significant property of the ring R is that each non-zero element is a unit. For, if $0 \neq Q = a1 + bI + cJ + dK$, then

$$\det(Q) = \begin{vmatrix} a + bi & c + di \\ -c + di & a - bi \end{vmatrix} = a^2 + b^2 + c^2 + d^2 \neq 0,$$

and by elementary matrix algebra

$$Q^{-1} = \frac{1}{\det(Q)} \begin{bmatrix} a - bi & -c - di \\ c - di & a + bi \end{bmatrix} \in R.$$

This allows us to state:

(6.3.3). *The ring of rational quaternions is a non-commutative division ring.*

Notice that the ring of quaternions is infinite. This is no accident since, by a famous theorem of Wedderburn,[3] *a finite division ring is a field*. This will not be proved here; however, we will prove the corresponding statement for domains, which is much easier.

(6.3.4). *A finite integral domain is a field.*

---

3 Joseph Henry Maclagan Wedderburn (1881–1948).

Proof. Let R be a finite domain and let $0 \neq r \in R$; we need to show that r has an inverse. Consider the function $\alpha : R \to R$ defined by $\alpha(x) = rx$. Now α is injective since $rx = ry$ implies that $x = y$ by (6.3.1). However, R is a *finite* set, so it follows that α must also be surjective. Therefore $1 = rx$ for some $x \in R$ and $x = r^{-1}$. $\qquad\qquad\square$

Next we consider the role of ideals in commutative ring theory. A first observation is that the presence of proper non-zero ideals is counter-indicative for the existence of units.

(6.3.5). *Let R be a commutative ring with identity. Then the set of non-units of R is equal to the union of all the proper ideals of R.*

Proof. Suppose that r is not a unit of R; then $Rx = \{rx \mid x \in R\}$ is a proper ideal containing r since $1 \notin Rx$. Conversely, if a unit r belongs to an ideal I, then for any x in R we have $x = (xr^{-1})r \in I$, showing that $I = R$. Thus a unit cannot belong to a proper ideal. $\qquad\qquad\square$

Recalling that fields are exactly the commutative rings with identity in which each non-zero element is a unit, we deduce:

Corollary (6.3.6). *A commutative ring with identity is a field if and only if it has no proper non-zero ideals.*

Prime ideals

Let R be a commutative ring with identity. Recall that a *maximal ideal* of R is a proper ideal I such that the only ideals containing I are I itself and R. Thus a maximal ideal is a *maximal proper ideal*. A related concept is that of a prime ideal. If R is a commutative ring with identity, a *prime ideal* of R is a proper ideal with the property: $ab \in I$ implies that $a \in I$ or $b \in I$, where $a, b \in R$.

There are enlightening characterizations of prime and maximal ideals in terms of quotient rings.

(6.3.7). *Let I be a proper ideal of a commutative ring R with identity.*
(i) *I is a prime ideal of R if and only if R/I is an integral domain;*
(ii) *I is a maximal ideal of R if and only if R/I is a field.*

Proof. Let $a, b \in R$; then $ab \in I$ if and only if $(a + I)(b + I) = I = 0_{R/I}$. Thus I is prime precisely when R/I has no zero divisors, i. e., it is a domain, so (i) is established. By (6.2.11) I is maximal in R if and only if R/I has no proper non-zero ideals and by (6.3.6) this is equivalent to R/I being a field. $\qquad\qquad\square$

Since every field is a domain, there follows at once:

Corollary (6.3.8). *Every maximal ideal of a commutative ring with identity is a prime ideal.*

On the other hand, prime ideals need not be maximal. Indeed, if R is any domain, the zero ideal is certainly prime, but it is not maximal unless R is a field. More interesting examples of non-maximal prime ideals can be constructed in polynomial rings.

Example (6.3.2). Let $R = \mathbb{Q}[t_1, t_2]$, the ring of polynomials in t_1, t_2 with rational coefficients. Let $I = Rt_1$ be the subset of all polynomials over R which are multiples of t_1. Then I is a prime ideal of R, but it is not maximal.

For consider the function $\alpha : R \to \mathbb{Q}[t_2]$ which carries a polynomial $f(t_1, t_2)$ to $f(0, t_2)$. This is a surjective ring homomorphism. Now if $f(0, t_2) = 0$, then f is a multiple of t_1, which shows that the kernel of α is I. From (6.2.8) we deduce that $R/I \simeq \mathbb{Q}[t_2]$. Since $\mathbb{Q}[t_2]$ is a domain, but not a field, it follows from (6.3.7) that I is a prime ideal of R which is not maximal.

The characteristic of an integral domain

Let R be a domain and let $S = \langle 1 \rangle$, the additive subgroup of R generated by 1. Suppose for the moment that S is finite, with order n say; we claim that n must be a prime. For suppose that $n = n_1 n_2$ where $n_i \in \mathbb{Z}$ and $1 < n_i < n$. Then $0 = n1 = (n_1 n_2)1 = (n_1 1)(n_2 1)$ by (6.1.3). However, R is a domain, so $n_1 1 = 0$ or $n_2 1 = 0$, which shows that n divides n_1 or n_2, a contradiction. Therefore n is a prime.

This observation is the basis for:

(6.3.9). *Let R be an integral domain and put $S = \langle 1 \rangle$. Then either S is infinite or else it has prime order p. In the latter event $pa = 0$ for all $a \in R$.*

To prove the final statement, simply note that $pa = (p1_R)a = 0a = 0$.

If R is an integral domain and $\langle 1_R \rangle$ has prime order p, then R is said to have *characteristic p*. The other possibility is that $\langle 1_R \rangle$ is infinite, in which event R is said to have *characteristic 0*. Thus the characteristic of R,

$$\mathrm{char}(R),$$

is either 0 or a prime. For example, $\mathbb{Z}_p$ and $\mathbb{Z}_p[t]$ are domains with characteristic p, while $\mathbb{Q}$, $\mathbb{R}$ and $\mathbb{R}[t]$ all have characteristic 0.

The field of fractions of an integral domain

Suppose that F is a field and R is a subring of F containing 1_F. Then R is a domain since there cannot be zero divisors in F. Conversely, one can ask if every domain arises in this way as a subring of a field. We will answer the question positively by showing how to construct the *field of fractions of a domain*. It will be helpful for the reader to keep in mind that the procedure to be described is a generalization of the way in which the rational numbers are constructed from the integers.

Let R be any integral domain. First we have to decide how to define a fraction over R. Consider the set

$$S = \{(a, b) \mid a, b \in R,\ b \neq 0\}.$$

Here a will correspond to the numerator and b to the denominator of the fraction. A binary relation $\sim$ on S will now be introduced which allows for cancellation between numerator and denominator:

$$(a_1, b_1) \sim (a_2, b_2) \quad \Leftrightarrow \quad a_1 b_2 = a_2 b_1.$$

Of course this relation is motivated by a familiar arithmetic rule: $\frac{m_1}{n_1} = \frac{m_2}{n_2}$ if and only if $m_1 n_2 = m_2 n_1$.

We verify that $\sim$ is an equivalence relation on S. Only transitivity requires a comment: suppose that $(a_1, b_1) \sim (a_2, b_2)$ and $(a_2, b_2) \sim (a_3, b_3)$; then $a_1 b_2 = a_2 b_1$ and $a_2 b_3 = a_3 b_2$. Multiply the first equation by b_3 and use the second equation to derive $a_1 b_3 b_2 = a_2 b_3 b_1 = a_3 b_2 b_1$. Cancel b_2 to obtain $a_1 b_3 = a_3 b_1$; thus $(a_1, b_1) \sim (a_3, b_3)$. Now define a *fraction* over R to be a $\sim$-equivalence class

$$\frac{a}{b} = [(a, b)]$$

where $a, b \in R$, $b \neq 0$. Note that $\frac{ac}{bc} = \frac{a}{b}$ since $(a, b) \sim (ac, bc)$; thus cancellation can be performed within a fraction.

Let F denote the set of all fractions over R: we wish to make F into a ring. To this end define addition and multiplication in R by the rules

$$\frac{a}{b} + \frac{a'}{b'} = \frac{ab' + a'b}{bb'} \quad \text{and} \quad \left(\frac{a}{b}\right)\left(\frac{a'}{b'}\right) = \frac{aa'}{bb'}.$$

Here we have been guided by familiar arithmetic rules for adding and multiplying fractions. However, it is necessary to show that these operations are well-defined, i. e., there is no dependence on the chosen representative (a, b) from the equivalence class $\frac{a}{b}$. For example, take the case of addition. Let $(a, b) \sim (c, d)$ and $(a', b') \sim (c', d')$: then in fact $(ab' + a'b, bb') \sim (cd' + c'd, dd')$ because

$$(ab' + a'b)dd' = ab'dd' + a'bdd' = bcb'd' + b'c'bd = (cd' + c'd)bb'.$$

The next step is to verify the ring axioms: as an example we will check the validity of the distributive law

$$\left(\frac{a}{b} + \frac{c}{d}\right)\left(\frac{e}{f}\right) = \left(\frac{a}{b}\right)\left(\frac{e}{f}\right) + \left(\frac{c}{d}\right)\left(\frac{e}{f}\right),$$

leaving the reader to verify the other axioms. By definition

$$\left(\frac{a}{b}\right)\left(\frac{e}{f}\right) + \left(\frac{c}{d}\right)\left(\frac{e}{f}\right) = \frac{ae}{bf} + \frac{ce}{df} = \frac{aedf + cebf}{bdf^2} = \frac{ade + bce}{bdf},$$

which equals

$$\left(\frac{ad + bc}{bd}\right)\left(\frac{e}{f}\right) = \left(\frac{a}{b} + \frac{c}{d}\right)\left(\frac{e}{f}\right),$$

as claimed.

Once all the axioms have been checked, we can be sure that F is a ring; note that the zero element of F is $0_F = \frac{0_R}{1_R}$. Clearly F is commutative and it has identity element $1_F = \frac{1_R}{1_R}$. Furthermore, if $a, b \neq 0$,

$$\left(\frac{a}{b}\right)\left(\frac{b}{a}\right) = \frac{ab}{ab} = \frac{1_R}{1_R} = 1_F,$$

so that, as expected, the inverse of $\frac{a}{b}$ is $\frac{b}{a}$. Therefore F is a field, the *field of fractions of the domain R*.

In order to relate F to R we introduce the natural function

$$\theta : R \to F$$

defined by $\theta(a) = \frac{a}{1}$. It is straightforward to check that θ is an injective ring homomorphism. Therefore $R \simeq \mathrm{Im}(\theta)$ and of course $\mathrm{Im}(\theta)$ is a subring of F containing 1_F. Thus the original domain R is isomorphic with a subring of the field F. Our conclusions are summed up in the following result.

(6.3.10). *Let R be an integral domain and let F denote the set of all fractions over R, with the addition and multiplication specified above. Then F is a field and the assignment $a \mapsto \frac{a}{1}$ determines is an injective ring homomorphism from R to F.*

Example (6.3.3).
(i) When $R = \mathbb{Z}$, the field of fractions is, up to isomorphism, the field of rational numbers $\mathbb{Q}$. This example motivated the general construction.
(ii) Let K be any field and put $R = K[t]$; this is a domain by (6.3.2). The field of fractions F of R is the *field of rational functions* in t over K; these are formal quotients of polynomials in t over K

$$\frac{f}{g}$$

where $f, g \in R$, $g \neq 0$. The notation $K\{t\}$ is often used denote the field of rational functions in t over K.

Exercises (6.3).

(1) Find all zero divisors in the following rings: $\mathbb{Z}_6$, $\mathbb{Z}_{15}$, $\mathbb{Z}_2[t]$, $\mathbb{Z}_4[t]$, $M_n(\mathbb{R})$.

(2) Let R be a commutative ring with identity such that the degree formula $\deg(fg) = \deg(f) + \deg(g)$ is valid in $R[t]$. Prove that R is a domain.

(3) If R is a division ring, prove that the only left ideals and right ideals are 0 and R.

(4) Let R be a ring with identity. If R has no left or right ideals except 0 and R, prove that R is a division ring.

(5) Let $\theta : D \rightarrow R$ be a non-zero ring homomorphism. If D is a division ring, show that it is isomorphic with a subring of R.

(6) Let $I_1, I_2, \ldots, I_k$ be non-zero ideals of a domain. Prove that $I_1 \cap I_2 \cap \cdots \cap I_k \neq 0$. Is this necessarily true for an infinite set of non-zero ideals?

(7) Let I denote the principal ideal $(\mathbb{Z}[t])t$ of $\mathbb{Z}[t]$. Prove that I is prime but not maximal.

(8) The same problem for $I = (\mathbb{Z}[t])(t^2 - 2)$.

(9) Let F be a field. If $a, b \in F$ and $a \neq 0$, define a function $\theta_{a,b} : F \rightarrow F$ by the rule $\theta_{a,b}(x) = ax + b$. Prove that the set of all $\theta_{a,b}$'s is a group with respect to functional composition.

(10) Let F be the field of fractions of a domain R and let $\alpha : R \rightarrow F$ be the canonical injective homomorphism $r \mapsto \frac{r}{1}$. Suppose that $\beta : R \rightarrow K$ is an injective ring homomorphism from R to some other field K. Prove that there is an injective homomorphism $\theta : F \rightarrow K$ such that $\theta\alpha = \beta$. (Thus in a sense F is the smallest field containing an isomorphic copy of R.)

(11) Let R be a commutative ring and let $0 \neq r \in R$. Prove that there is an ideal I which is maximal subject to not containing r. Then prove that I is an *irreducible ideal*, i. e., it is not the intersection of two larger ideals.

(12) Deduce from Exercise (6.2.11) that every proper ideal of a commutative ring R is an intersection of irreducible ideals.

(13) Prove that quaternions $\pm I$, $\pm J$, $\pm K$ generate a group of order 8 with respect to matrix multiplication: this is the *quaternion group*, denoted by Q_8. Then show that every subgroup of Q_8 is normal, although the group is not abelian.

6.4 Finiteness conditions on ideals

In this section we introduce certain finiteness properties of ideals that are possessed by some important classes of rings.

(6.4.1). *Let $\mathcal{I}$ be a non-empty set of left ideals of a ring R. Then the following statements about $\mathcal{I}$ are equivalent.*

(i) *The set $\mathcal{I}$ satisfies the ascending chain condition, i. e., there does not exist an infinite ascending chain of left ideals $I_1 \subset I_2 \subset \cdots$ with $I_i \in \mathcal{I}$.*

(ii) *The set $\mathcal{I}$ satisfies the* maximal condition, *i. e., every non-empty subset of $\mathcal{I}$ has a maximal element, that is to say, an element which is not properly contained in any other element of $\mathcal{I}$.*

Proof. Assume that $\mathcal{I}$ satisfies condition (i) and suppose that $\mathcal{S}$ is a non-empty subset of $\mathcal{I}$ that does not contain a maximal element. Let $I_1 \in \mathcal{S}$; then there exists $I_2 \in \mathcal{S}$ which is strictly larger than I_1 since I_1 is not maximal in $\mathcal{I}$. Similarly there exists $I_3 \in \mathcal{S}$ which is strictly larger that I_2, and so on. But clearly this leads to an infinite ascending chain $I_1 \subset I_2 \subset \cdots$ in $\mathcal{I}$, a contradiction.

Conversely, assume that $\mathcal{I}$ satisfies condition (ii). If there is an infinite ascending chain $I_1 \subset I_2 \subset \cdots$ in $\mathcal{I}$, the maximal condition can be applied to the set $\{I_1, I_2, \ldots, \}$ to give a maximal element. This is obviously impossible. □

We remark that similar properties for subgroups of a group were introduced in Exercise (3.3.10).

There is of course a corresponding result for right ideals. The case of greatest interest to us is when $\mathcal{I}$ is the set of all left ideals of the ring R. If this set satisfies one of the two equivalent conditions of (6.4.1), then R is called a *left noetherian*[4] *ring*. There is a corresponding definition of a *right noetherian ring*. In case of a commutative ring, the ring is simply said to be *noetherian*. The following result sheds some light on the nature of the noetherian condition.

(6.4.2). *Let R be a ring with identity. Then R is left noetherian if and only if every left ideal of R is finitely generated as a left ideal of R.*

Proof. First suppose that I is a left ideal of R which is not finitely generated. Certainly $I \neq 0$, so there exists $r_1 \in I - 0$. Then $r_1 \in Rr_1$ since R has an identity element. Also $Rr_1 \subset I$, since I is not finitely generated. Let $r_2 \in I - Rr_1$. Then $R\{r_1, r_2\} = Rr_1 + Rr_2 \subset I$. Let $r_3 \in I - (Rr_1 + Rr_2)$ and note that $I \neq R\{r_1, r_2, r_3\} = Rr_1 + Rr_2 + Rr_3 \subset I$, and so on. But this leads to an infinite ascending chain of left ideals $Rr_1 \subset Rr_1 + Rr_2 \subset Rr_1 + Rr_2 + Rr_3 \subset \cdots$ contained in I. Hence R is not left noetherian.

Conversely, assume R is not left noetherian, so that there exists an infinite ascending chain of left ideals $I_1 \subset I_2 \subset \cdots$. Set $I = \bigcup_{i=1,2,\ldots} I_i$, which is clearly a left ideal of R. Then I cannot be generated by finitely many elements $r_1, r_2, \ldots, r_k$, since all the r_i would have to belong to some I_j, which leads to the contradiction $I_j = I_{j+1} = I$. □

Obvious examples of noetherian rings include the ring of integers and any field. More interesting examples are provided by (6.4.3) below, which is one of the most celebrated results in the theory of noetherian rings.

---

4 Emmy Noether (1882–1935).

(6.4.3) (Hilbert's[5] Basis Theorem). *Let R be a commutative noetherian ring with identity. Then the polynomial ring $R[t_1, t_2, \ldots, t_n]$ is also noetherian.*

Proof. In the first place is enough to prove the theorem for $n = 1$. For assume that this case has been dealt with and that $n > 1$. Now $R[t_1, t_2, \ldots, t_n] = S[t_n]$ where $S = R[t_1, t_2, \ldots, t_{n-1}]$ and S is noetherian by induction on n. Therefore the result is true by the case $n = 1$. From now on we will work with the ring $T = R[t]$.

By (6.4.2) it suffices to prove that an arbitrary ideal J of T is finitely generated as an ideal. Suppose that J is not finitely generated; then $J \neq 0$ and there is a polynomial $f_1 \in J - 0$ of smallest degree d_1. Since J is not finitely generated, $J \neq J_1 = (f_1)$ and $J - J_1$ contains a polynomial f_2 of smallest degree d_2. Furthermore $J \neq J_2 = (f_1) + (f_2)$ and $J - J_2$ contains a polynomial f_3 of smallest degree d_3, and so on. This gives rise to infinite sequences of ideals $J_1 \subset J_2 \subset \cdots$ where $J_i = (f_1) + (f_2) + \cdots + (f_i)$, and non-zero polynomials $f_1, f_2, \ldots$ with $\deg(f_i) = d_i$ and $d_1 \leq d_2 \leq \cdots$; moreover $f_{i+1} \notin J_i$. Let us write $f_i = a_i t^{d_i} +$ terms of lower degree, where $0 \neq a_i \in R$.

Set $I_i = (a_1) + (a_2) + \cdots + (a_i)$, so that $I_1 \subseteq I_2 \subseteq \cdots$ is an ascending sequence of ideals of R. This sequence must have finite length since R is noetherian, so $I_m = I_{m+1}$ for some integer m. Hence $a_{m+1} \in I_m$ and consequently there is an expression $a_{m+1} = r_1 a_1 + r_2 a_2 + \cdots + r_m a_m$ with $r_i \in R$. Now define a new polynomial $g \in R[t]$ by

$$g = f_{m+1} - \sum_{i=1}^{m} (r_i f_i) t^{d_{m+1} - d_i}.$$

Thus $g \in J_{m+1}$. Observe that $g \notin J_m$ since $f_{m+1} \notin J_m$. Now the highest power of t that could occur in g is certainly $t^{d_{m+1}}$, but by inspection we see that its coefficient is

$$a_{m+1} - r_1 a_1 - r_2 a_2 - \cdots - r_m a_m = 0.$$

Therefore $\deg(g) < d_{m+1} = \deg(f_{m+1})$, which is contrary to the choice of f_{m+1} as a polynomial of smallest degree in $J - J_m$. This contradiction establishes the theorem. $\square$

Corollary (6.4.4). *The rings $\mathbb{Z}[t_1, t_2, \ldots, t_n]$ and $F[t_1, t_2, \ldots, t_n]$ are noetherian, where F is any field.*

Using this result we can find a large class of noetherian rings.

(6.4.5). *Every finitely generated commutative ring with identity is noetherian.*

Proof. Let R be the ring in question and suppose that it has generators $x_1, x_2, \ldots, x_n$. By (6.2.3) every element of R has the form

$$\sum_{\ell_i \geq 0} m_{\ell_1, \ell_2, \ldots, \ell_n} x_1^{\ell_1} x_2^{\ell_2} \cdots x_n^{\ell_n}$$

---

5 David Hilbert (1862–1943).

where $m_{\ell_1,\ell_2,\ldots,\ell_n} \in \mathbb{Z}$ and the (finite) sum is over all non-negative integers $\ell_1, \ell_2, \ldots, \ell_n$. Let $S = \mathbb{Z}[t_1, t_2, \ldots, t_n]$ and define a map $\theta : S \to R$ by the assignment $t_i \mapsto x_i$: thus

$$\theta\left(\sum_{\ell_i \geq 0} m_{\ell_1,\ell_2,\ldots,\ell_n} t_1^{\ell_1} t_2^{\ell_2} \cdots t_n^{\ell_n}\right) = \sum_{\ell_i \geq 0} m_{\ell_1,\ell_2,\ldots,\ell_n} x_1^{\ell_1} x_2^{\ell_2} \cdots x_n^{\ell_n}.$$

Then θ is a ring homomorphism since sums and products of elements in R and in S are formed by the same rules, and clearly θ is also surjective. Hence $S/\text{Ker}(\theta) \simeq R$ by (6.2.8). By (6.4.4) the ring S is noetherian and thus every quotient of S is also noetherian, which establishes the result. $\qquad\square$

Exercises (6.4).

(1) Prove that every non-zero commutative noetherian ring has at least one maximal (proper) ideal.

(2) If R is a non-zero commutative noetherian ring, prove that R has a quotient ring which is a field.

(3) Let R be a commutative noetherian ring and I an ideal of R. Prove that R/I is also noetherian. [Hint: use the Correspondence Theorem.]

(4) Let R be the ring of all rational numbers of the form $\frac{m}{2^n}$ where $m, n \in \mathbb{Z}$. Show that R is a noetherian ring. [Hint: use (6.4.5).]

(5) Prove that the ring $\mathbb{Z}[t_1, t_2, \ldots] = \bigcup_{n=1,2,\ldots} \mathbb{Z}[t_1, t_2, \ldots, t_n]$ of polynomials in *infinitely* many indeterminates t_i cannot be noetherian.

(6) Prove that if R is a commutative ring with identity which can be generated by n elements, then $R \simeq \mathbb{Z}[t_1, t_2, \ldots, t_n]/(f_1, f_2, \ldots, f_k)$ for certain polynomials f_i. Conclude that R is determined up to isomorphism by finitely many polynomials in $t_1, t_2, \ldots, t_n$.

(7) Establish the following analogue of (6.4.1). Let $\mathcal{I}$ be a non-empty set of left ideals of a ring R. Then the following statements about $\mathcal{I}$ are equivalent.

    (i) The set $\mathcal{I}$ satisfies the descending chain condition, i. e., there does not exist an infinite descending chain of left ideals $I_1 \supset I_2 \supset \cdots$ with $I_i \in \mathcal{I}$.

    (ii) The set $\mathcal{I}$ satisfies the minimal condition, i. e., every non-empty subset of $\mathcal{I}$ has a minimal element, that is, an element which does not properly contain any other element of $\mathcal{I}$.

    (A ring for which the set of all left ideals satisfies the minimal condition is said to be a left artinian[6] ring.)

---

6 Emil Artin (1898–1962).

7 Division in Commutative Rings

The aim of this chapter is to construct a theory of division in rings that mirrors, as closely as possible, the familiar theory of division in the ring of integers. To simplify matters let us agree to restrict attention to commutative rings – in non-commutative rings questions of left and right divisibility arise. Also, remembering from Section 6.3 the phenomenon of zero divisors, we will further restrict ourselves to integral domains. In fact even this class of rings is too wide, although it provides a reasonable target for our theory. For this reason we will introduce some well-behaved types of domains.

7.1 Euclidean domains

Let R be a commutative ring with identity and let $a, b \in R$. Then a is said to *divide b*, in symbols

$$a \mid b,$$

if $ac = b$ for some $c \in R$. From the definition there quickly follow some elementary facts about division.

(7.1.1). *Let R be a commutative ring with identity and let a, b, c, x, y be elements of R. Then:*
(i) $a \mid a$ *and* $a \mid 0$ *for all* $a \in R$;
(ii) $0 \mid a$ *if and only if* $a = 0$;
(iii) *if* $a \mid b$ *and* $b \mid c$, *then* $a \mid c$, *so division is a transitive relation;*
(iv) *if* $a \mid b$ *and* $a \mid c$, *then* $a \mid bx + cy$ *for all* $x, y \in R$;
(v) *if u is a unit,* $u \mid a$ *for all* $a \in R$, *while* $a \mid u$ *if and only if a is a unit.*

For example, taking the case of (iv), we have $b = ad$ and $c = ae$ for some $d, e \in R$. Then $bx + cy = a(dx + ey)$, so that a divides $bx + cy$. The other proofs are equally simple exercises which are left to the reader.

One situation we may encounter in a ring is a pair of elements each of which divides the other: such elements are called *associates*.

(7.1.2). *Let R be an integral domain and let a, b $\in$ R. Then a $\mid$ b and b $\mid$ a if and only if b = au where u is a unit of R.*

Proof. Let u be a unit; then $a|au$. Also $(au)u^{-1} = a$, so $au|a$. Conversely, assume that $a \mid b$ and $b \mid a$. If $a = 0$, then $b = 0$ and the statement is certainly true, so let $a \neq 0$. Now $a = bc$ and $b = ad$ for some $c, d \in R$. Therefore $a = bc = adc$ and by (6.3.1) we obtain $dc = 1$, so that d is a unit. $\square$

For example, two integers a and b are associates if and only if $b = \pm a$.

https://doi.org/10.1515/9783110691160-007

Irreducible elements

Let R be a commutative ring with identity. An element a of R is called *irreducible* if it is neither 0 nor a unit and if its only divisors are units and associates, i. e., the elements that we know must divide a. Thus irreducible elements have as few divisors as possible.

Example (7.1.1).
(i)   The irreducible elements of $\mathbb{Z}$ are the prime numbers and their negatives.
(ii) A field has no irreducible elements since every non-zero element is a unit.
(iii) If F is a field, the irreducible elements of the polynomial ring $F[t]$ are the so-called *irreducible polynomials*, i. e., the non-constant polynomials which are not expressible as a product of polynomials of lower degree.

Almost every significant property of division in $\mathbb{Z}$ depends ultimately on the Division Algorithm. Thus it is natural to focus on rings in which some version of this property is valid. This motivates us to introduce a special class of domains, the so-called Euclidean domains.

A domain R is called *Euclidean* if there is a function

$$\delta : R - \{0_R\} \to \mathbb{N}$$

with the following properties:
(i) $\delta(a) \le \delta(ab)$ if $0 \ne a, b \in R$;
(ii) if $a, b \in R$ and $b \ne 0$, there exist $q, r \in R$ such that $a = bq + r$ and either $r = 0$ or $\delta(r) < \delta(b)$.

The standard example of a Euclidean domain is $\mathbb{Z}$ where δ is the absolute value function, i. e., $\delta(a) = |a|$. Note that property (i) holds since $|ab| = |a| \cdot |b| \ge |a|$ if $b \ne 0$. Of course (ii) is the usual statement of the Division Algorithm for $\mathbb{Z}$.

New and important examples of Euclidean domains are given by the next result.

(7.1.3). *If F is a field, the polynomial ring $F[t]$ is a Euclidean domain with associated function δ given by $\delta(f) = \deg(f)$.*

Proof. We already know from (6.3.2) that $R = F[t]$ is a domain. Also, by the same result, if $f, g \ne 0$, then $\delta(fg) = \deg(fg) = \deg(f) + \deg(g) \ge \deg(f) = \delta(f)$. Hence property (i) is valid. To establish the validity of (ii), put

$$S = \{f - gq \mid q \in R\}.$$

If $0 \in S$, then $f = gq$ for some $q \in R$ and we may take r to be 0. Assuming that $0 \notin S$, note that every element of S has degree ≥ 0, so by the Well-Ordering Principle there is an element r in S with smallest degree, say $r = f - gq$ where $q \in R$. Thus $f = gq + r$.

Suppose that $\deg(r) \geq \deg(g)$. Write $g = at^m + \cdots$ and $r = bt^n + \cdots$ where $m = \deg(g)$, $n = \deg(r)$, $0 \neq a, b \in F$ and the dots represent terms of lower degree in t. Since $m \leq n$, we can form the polynomial

$$s = r - (a^{-1}bt^{n-m})g \in R.$$

Now the term in t^n cancels in s, so either $s = 0$ or $\deg(s) < n$. But $s = f - (q + a^{-1}bt^{n-m})g \in S$ and hence $s \neq 0$. Thus $\deg(s) < n$, which contradicts the minimality of $\deg(r)$. Therefore $\deg(r) < \deg(g)$, as required. ☐

A less familiar example of a Euclidean domain is the ring of Gaussian integers. A *Gaussian integer* is a complex number of the form

$$u + iv$$

where $u, v \in \mathbb{Z}$ and of course $i = \sqrt{-1}$. It is easily seen that the Gaussian integers form a subring of $\mathbb{C}$ containing 1 and hence constitute a domain.

(7.1.4). *The ring R of Gaussian integers is a Euclidean domain.*

Proof. In this case an associated function $\delta : R - \{0\} \to \mathbb{N}$ is defined by the rule

$$\delta(u + iv) = |u + iv|^2 = u^2 + v^2.$$

We must show that δ satisfies the two requirements for a Euclidean domain. In the first place, if $0 \neq a, b \in R$, then $\delta(ab) = |ab|^2 = |a|^2|b|^2 \geq |a|^2$ since $|b| \geq 1$.

Verification of the second requirement is harder. First write $ab^{-1} = u' + iv'$ where u', v' are rational numbers. Now choose integers u and v that are as close as possible to u' and v' respectively; thus $|u - u'| \leq \frac{1}{2}$ and $|v - v'| \leq \frac{1}{2}$. Next

$$a = b(u' + iv') = b(u + iv) + b(u'' + iv'')$$

where $u'' = u' - u$ and $v'' = v' - v$. Finally, let $q = u + iv$ and $r = b(u'' + iv'')$. Then $a = bq + r$; also $q \in R$ and hence $r = a - bq \in R$. If $r \neq 0$, then, since $|u''| \leq \frac{1}{2}$ and $|v''| \leq \frac{1}{2}$,

$$\delta(r) = |b|^2|u'' + iv''|^2 = |b|^2(u''^2 + v''^2) \leq |b|^2\left(\frac{1}{4} + \frac{1}{4}\right) = \frac{1}{2}|b|^2,$$

so that $\delta(r) < |b|^2 = \delta(b)$. Therefore $\delta(r) < \delta(b)$ as required. ☐

Exercises (7.1).

(1) Complete the proof of (7.1.1).

(2) Identify the irreducible elements in the following rings:
   (i) the ring of rational numbers with odd denominators;
   (ii) $\mathbb{Z}[t]$.

(3) Let R be a commutative ring with identity. If R has no irreducible elements, show that either R is a field or there exists an infinite strictly increasing chain of principal ideals $I_1 \subset I_2 \subset \cdots$ in R. Deduce that if R is noetherian, it is a field.

(4) Let $R = F[[t]]$ be the ring of formal power series in t over a field F, (see Exercise (6.1.8)). Prove that the irreducible elements of R are those of the form tf where $f \in R$ and $f(0) \neq 0$.

(5) Let $f = t^5 - 3t^2 + t + 1$ and $g = t^2 + t + 1$ be polynomials in $\mathbb{Q}[t]$. Find $q, r \in \mathbb{Q}[t]$ such that $f = gq + r$ and $\deg(r) \leq 1$.

(6) Let R be a Euclidean domain with associated function $\delta : R - \{0\} \to \mathbb{N}$.
   (i) Show that $\delta(a) \geq \delta(1)$ for all $a \neq 0$ in R.
   (ii) If a is a unit of R, prove that $\delta(a) = \delta(1)$.
   (iii) Conversely, show that if $\delta(a) = \delta(1)$, then a is a unit of R.

(7) Prove that $t^3 + t + 1$ is irreducible in $\mathbb{Z}_2[t]$, but $t^3 + t^2 + t + 1$ is reducible.

7.2 Principal ideal domains

Let R be a commutative ring with identity. If $r \in R$, recall from Section 6.2 that the subset $Rr = \{rx \mid x \in R\} = (r)$ is an ideal of R containing r called a principal ideal. If every ideal of R is principal, then R is a *principal ideal ring*. A domain in which every ideal is principal is called a *principal ideal domain* or PID: these rings form an extremely important class of domains. For example, $\mathbb{Z}$ is a PID; for an ideal of $\mathbb{Z}$ is a cyclic subgroup and thus has the form $\mathbb{Z}n$ where $n \geq 0$.

A good source of PID's is indicated by the next result.

(7.2.1). *Every Euclidean domain is a principal ideal domain.*

Proof. Let R be a Euclidean domain with associated function $\delta : R - 0 \to \mathbb{N}$ and let I be an ideal of R; we need to show that I is principal. If I is the zero ideal, $I = (0)$ and I is principal. So we assume that $I \neq 0$ and apply the Well-Ordering Law to pick an x in $I - 0$ such that $\delta(x)$ is minimal. Now certainly $(x) \subseteq I$; the claim is that $I \subseteq (x)$. To substantiate this, let $y \in I$ and write $y = xq + r$ with $q, r \in R$ where either $r = 0$ or $\delta(r) < \delta(x)$. This is possible since δ is an associated function for the Euclidean domain R. If $r = 0$, then $y = xq \in (x)$. Otherwise $\delta(r) < \delta(x)$; but this is impossible since $r = y - xq \in I$, which contradicts the minimality of $\delta(x)$ for $x \in I - 0$. Therefore $I = (x)$. □

The following important result is a consequence of (7.1.3) and (7.2.1).

Corollary (7.2.2). *If F is a field, then F[t] is a principal ideal domain.*

Another example of a PID is the ring of Gaussian integers by (7.2.1) and (7.1.4). Our next objective is to show that PID's have good division properties, despite the lack of a division algorithm.

Greatest common divisors

Let a, b be elements in a domain R. A *greatest common divisor* (or gcd) of a and b is a ring element d such that the following hold:
(i) $d \mid a$ and $d \mid b$;
(ii) if $c \mid a$ and $c \mid b$ for some $c \in R$, then $c \mid d$.

The definition here has been carried over directly from the integers – see Section 2.2.

Notice that if d and d' are two gcd's of a, b, then $d \mid d'$ and $d' \mid d$, so that d and d' are associates. Thus by (7.1.2) $d' = du$ with u a unit of R. It follows that gcd's are unique only up to a unit. Of course in the case of $\mathbb{Z}$, where the units are ± 1, we were able to make gcd's unique by insisting that they be positive. This course of action is not possible in arbitrary domains since there is no concept of positivity.

There is no reason why gcd's should exist in an arbitrary domain. However, the situation is very satisfactory for PID's.

(7.2.3). *Let a and b be elements of a principal ideal domain R. Then a and b have a greatest common divisor d which has the form $d = ax + by$ with $x, y \in R$.*

Proof. Define $I = \{ax + by \mid x, y \in R\}$ and observe that I is an ideal of R. Hence $I = (d)$ for some $d \in I$, with $d = ax + by$ say. If $c \mid a$ and $c \mid b$, then $c \mid ax + by = d$ by (7.1.1). Also $a \in I = (d)$, so $d \mid a$, and similarly $d \mid b$. Hence d is a gcd of a and b. □

Elements a and b of a domain R are said to be *relatively prime* if 1 is a gcd of a and b, which means that $ax + by = 1$ for some $x, y \in R$.

(7.2.4) (Euclid's Lemma). *Let a, b, c be elements of a principal ideal domain and assume that $a \mid bc$ where a and b are relatively prime. Then $a \mid c$.*

Corollary (7.2.5). *If R is a principal ideal domain and $p \mid bc$ where $p, b, c \in R$ and p is irreducible, then $p \mid b$ or $p \mid c$.*

The proofs of these results are exactly the same as those given in Section 2.2 for $\mathbb{Z}$.

Maximal ideals in principal ideal domains

In a PID the maximal ideals and the prime ideals coincide and admit a nice description in terms of irreducible elements.

(7.2.6). *Let I be a non-zero ideal of a principal ideal domain R. Then the following statements about I are equivalent:*

(i)   *I is maximal;*

(ii) *I is prime;*

(iii) *I = (p) where p is an irreducible element of R.*

Proof. (i) $\Rightarrow$ (ii). This was proved in (6.3.8).

(ii) $\Rightarrow$ (iii). Assume that I is prime. Since R is a PID, we have $I = (p)$ for some $p \in R$. Note that p cannot be a unit since $I \neq R$. Suppose that $p = ab$ where neither a nor b is associate to p. Then $ab \in I$ and I is prime, so $a \in I$ or $b \in I$, i. e., $p \mid a$ or $p \mid b$. Since we also have $a \mid p$ and $b \mid p$, we obtain the contradiction that a or b is associate to p. This shows that p is irreducible.

(iii) $\Rightarrow$ (i). Assume that $I = (p)$ with p irreducible, and let $I \subseteq J \subseteq R$ where J is an ideal of R. Then $J = (x)$ for some $x \in R$, and $p \in (x)$, so that $x \mid p$. Hence either x is a unit or it is associate to p, so that $J = R$ or $J = I$. Therefore I is maximal as claimed. □

Corollary (7.2.7). *Let F be a field. Then the maximal ideals of the polynomial ring F[t] are exactly those of the form (f) where f is an irreducible polynomial which is monic, (i. e., its leading coefficient is 1).*

This is because $F[t]$ is a PID by (7.2.2) and the irreducible elements of $F[t]$ are just the irreducible polynomials. The corollary provides us with an important method for constructing a field from an irreducible polynomial $f \in F[t]$: indeed $F[t]/(f)$ is a field. This will be exploited in Section 7.4 below.

We conclude the section by noting a property of PID's which will be crucial when we address the issue of unique factorization in Section 7.4.

(7.2.8). *Every principal ideal domain is noetherian.*

Proof. Let R be a PID. By definition every ideal of R is principal and hence can be generated by a single element. Therefore R is noetherian by (6.4.2). □

Exercises (7.2).

(1) Prove (7.2.4) and (7.2.5).

(2) Show that $\mathbb{Z}[t]$ is not a PID.

(3) Show that $F[t_1, t_2]$ is not a PID for any field F.

(4) Let R be a commutative ring with identity. If $R[t]$ is a PID, prove that R must be a field.

(5) Let $f = t^3 + t + 1 \in \mathbb{Z}_2[t]$. Show that $\mathbb{Z}_2[t]/(f)$ is finite field and find its order.

(6) Prove that the ring of rational numbers with odd denominators is a PID.

(7) Prove that $F[[t]]$, the ring of formal power series in t over a field F, is a PID by describing its ideals.

(8) Let R be a commutative noetherian ring with identity. Assume that R has the property that each pair of elements a, b has a greatest common divisor which is a linear

combination of a and b. Prove that R is a PID. [Hint: let I be an ideal of R. Note that I is a finitely generated ideal and reduce to the case where it is generated by two elements].

(9) Prove that the Chinese Remainder Theorem holds in a Euclidean domain, (see (2.3.7)).

(10) State and prove the Euclidean algorithm for a Euclidean domain, (see (2.2.4)).

7.3 Unique factorization in integral domains

The present section is concerned with domains in which there is unique factorization in terms of irreducible elements. Our model here is the Fundamental Theorem of Arithmetic (2.2.7), which asserts that such factorizations exist in $\mathbb{Z}$. First it is necessary to clarify what is meant by a unique factorization.

Let R be a domain and let S denote the set of all irreducible elements in R, which might of course be empty. Observe that "being associate to" is an equivalence relation on S, so that S splits up into equivalence classes. Choosing one element from each equivalence class, we form a subset C of S. (Strictly speaking this procedure involves the Axiom of Choice – see Section 1.5.) Now observe that the set C has the following properties:

(i) every irreducible element of R is associate to some element of C;

(ii) distinct elements of C are not associate.

A subset C with these properties is called a *complete set of irreducibles for R*. We have just established the following simple fact.

(7.3.1). *Every integral domain has a (possibly empty) complete set of irreducible elements.*

Our interest in complete sets of irreducibles stems from the observation that if there is to be *unique* factorization in terms of irreducibles, then only irreducibles from a complete set can be used: otherwise there will be different factorizations of the type $ab = (ua)(u^{-1}b)$ where a, b are irreducible and u is a unit.

An integral domain R is called a *unique factorization domain*, or UFD, if there exists a complete set of irreducibles C for R such that each non-zero element a of R has an expression of the form

$$a = up_1 p_2 \cdots p_k$$

where u is a unit and $p_i \in C$, and furthermore this expression is unique up to order of the factors.

At present the only example of a UFD we know is $\mathbb{Z}$, where C can be taken to be the set of prime numbers. The next theorem provides us with more examples.

(7.3.2). *Every principal ideal domain is a unique factorization domain.*

Proof. Let R be a PID and let C be any complete set of irreducibles for R. It will be shown that there is unique factorization for elements of R in terms of units and elements of C. This is accomplished in four steps, the first of which establishes the existence of irreducibles when R contains a non-zero, non-unit element, i.e., R is not a field.

(i) *If a is a non-zero, non-unit element of R, it is divisible by at least one irreducible element of R.*

Suppose this is false. Then a itself must be reducible, so $a = a_1 a_1'$ where a_1 and a_1' are non-units and $(a) \subseteq (a_1)$. Also $(a) \neq (a_1)$. For otherwise $a_1 \in (a)$, so that $a \mid a_1$, as well as $a_1 \mid a$; by (7.1.2) this implies that a_1' is a unit. Therefore $(a) \subset (a_1)$.

Next a_1 cannot be irreducible since $a_1 \mid a$. Thus $a_1 = a_2 a_2'$ where a_2, a_2' are non-units and it follows that $(a_1) \subset (a_2)$ by the argument just given. Continuing in this way, we recognize that the procedure cannot terminate: for otherwise an irreducible divisor of a will appear. Hence there is an infinite strictly ascending chain of ideals $(a) \subset (a_1) \subset (a_2) \subset \cdots$; but this is impossible since R is noetherian by (7.2.8).

(ii) *If a is a non-zero, non-unit element of R, then a is a product of irreducibles.*

Again suppose this is false. By (i) there is an irreducible p_1 dividing a, with $a = p_1 a_1$ say. Now a_1 cannot be a unit, so there is an irreducible element p_2 dividing a_1, with say $a_1 = p_2 a_2$ and $a = p_1 p_2 a_2$, and so on indefinitely. However, this leads to $(a) \subset (a_1) \subset (a_2) \subset \cdots$, a strictly ascending infinite chain of ideals, which again contradicts (7.2.8).

(iii) *If a is a non-zero element of R, then a is the product of a unit and irreducible elements in C.*

This is clear if a is a unit – no irreducibles are needed. Otherwise by (ii) the element a is a product of irreducibles, each of which is associate to an element of C. The result now follows on replacing each irreducible factor of a by an irreducible in C multiplied by a unit.

(iv) The final step in the proof establishes uniqueness. Suppose that

$$a = u p_1 p_2 \cdots p_k = v q_1 q_2 \cdots q_\ell$$

where u, v are units of R and $p_i, q_j \in C$. Argue by induction on k: if $k = 0$, then $a = u$, a unit, so $\ell = 0$ and $u = v$. Now assume that $k > 0$.

Since $p_1 \mid a = v q_1 q_2 \cdots q_\ell$, Euclid's Lemma shows that p_1 must divide one of $q_1, \ldots, q_\ell$. By relabelling the q_j's, we may assume that $p_1 \mid q_1$. Thus p_1 and q_1 are associate members of C, which can only mean that $p_1 = q_1$. Hence, on cancelling p_1, we obtain $a' = u p_2 \cdots p_k = v q_2 \cdots q_\ell$. By the induction hypothesis $k - 1 = \ell - 1$, so $k = \ell$ and, after further relabelling, $p_i = q_i$ for $i = 2, 3, \ldots, k$, and $u = v$. Therefore uniqueness has been established. □

Corollary (7.3.3). *If F is a field, the polynomial ring $F[t]$ is a unique factorization domain.*

This is because $F[t]$ is a PID by (7.2.2). The natural choice for a complete set of irreducibles in $F[t]$ is the set of all monic irreducible polynomials. Thus we have unique factorization in $F[t]$ in terms of constants and monic irreducible polynomials. Another example of a UFD is the ring of Gaussian integers $\{a + b\sqrt{-1} \mid a, b \in \mathbb{Z}\}$, which by (7.1.4) is a Euclidean domain and hence a PID. However, some domains of similar appearance are not UFD's.

Example (7.3.1). Let R be the subring of $\mathbb{C}$ consisting of all $a + b\sqrt{-3}$ where $a, b \in \mathbb{Z}$. Then R is not a unique factorization domain.

First observe that ± 1 are the only units of R. For, let $0 \neq r = a + b\sqrt{-3} \in R$. Then

$$r^{-1} = \frac{1}{a^2 + 3b^2}(a - b\sqrt{-3}),$$

which is in R if and only if $\frac{a}{a^2+3b^2}$ and $\frac{b}{a^2+3b^2}$ are integers. This happens only when $b = 0$ and $\frac{1}{a} \in \mathbb{Z}$, i.e., $r = a = \pm 1$. It follows that no two of the elements $2, 1 + \sqrt{-3}, 1 - \sqrt{-3}$ are associate.

Next we claim that $2, 1 + \sqrt{-3}, 1 - \sqrt{-3}$ are irreducible elements of R. Fortunately all three elements can be handled simultaneously. Suppose that

$$(a + \sqrt{-3}b)(c + \sqrt{-3}d) = 1 \pm \sqrt{-3} \quad \text{or} \quad 2$$

where $a, b, c, d \in \mathbb{Z}$. Taking the modulus squared of both sides, we obtain $(a^2 + 3b^2)(c^2 + 3d^2) = 4$ in every case. But this implies that $a^2 = 1$ and $b = 0$ or $c^2 = 1$ and $d = 0$, i.e., either $a + \sqrt{-3}b$ or $c + \sqrt{-3}d$ is a unit.

Finally, unique factorization fails because

$$4 = 2 \cdot 2 = (1 + \sqrt{-3})(1 - \sqrt{-3})$$

and $2, 1 + \sqrt{-3}, 1 - \sqrt{-3}$ are non-associate irreducibles. It follows that R is not a UFD.

Two useful properties of UFD's are recorded in the next result.

(7.3.4). *Let R be a unique factorization domain. Then:*
(i) *gcd's exist in R;*
(ii) *Euclid's Lemma holds in R.*

Proof. To prove (i) let $a = up_1^{e_1} p_2^{e_2} \cdots p_k^{e_k}$ and $b = vp_1^{f_1} p_2^{f_2} \cdots p_k^{f_k}$ where u, v are units of R, the p_i belong to a complete set of irreducibles for R, and $e_i, f_i \geq 0$. Define $d = p_1^{g_1} p_2^{g_2} \cdots p_k^{g_k}$ where g_i is the minimum of e_i and f_i. Then d is a gcd of a and b. For clearly $d \mid a$ and $d \mid b$, and, on the other hand, if $c \mid a$ and $c \mid b$, the unique factorization property shows that c must have the form $wp_1^{h_1} p_2^{h_2} \cdots p_k^{h_k}$ where w is a unit and $0 \leq h_i \leq g_i$. Hence $c \mid d$. The proof of (ii) is left to the reader as an exercise. $\square$

Although polynomial rings in more than one variable over a field are not PID's – see Exercise (7.2.3) – they are in fact UFD's. It is our aim in the remainder of the section to prove this important result.

Primitive polynomials

Let R be a UFD and let $0 \neq f \in R[t]$. Since gcd's exist in R by (7.3.4), we can form the gcd of the coefficients of f; this is called the *content* of f,

$$c(f).$$

Keep in mind that content is unique only up to a unit of R, and equations involving content have to be interpreted in this light. If $c(f) = 1$, i.e., $c(f)$ is a unit, the polynomial f is said to be *primitive*. For example $2 + 4t - 3t^3 \in \mathbb{Z}[t]$ is a primitive polynomial. Next two useful results about the content of polynomials will be established.

(7.3.5). *Let $0 \neq f \in R[t]$ where R is a unique factorization domain. Then $f = cf_0$ where $c = c(f)$ and $f_0 \in R[t]$ is primitive.*

Proof. Write $f = a_0 + a_1 t + \cdots + a_n t^n$; then $c(f) = \gcd\{a_0, a_1, \ldots, a_n\} = c$, say. Write $a_i = cb_i$ with $b_i \in R$ and put $f_0 = b_0 + b_1 t + \cdots + b_n t^n \in R[t]$. Thus $f = cf_0$. If $d = \gcd\{b_0, b_1, \ldots, b_n\}$, then $d \mid b_i$ and so $cd \mid cb_i = a_i$. Since c is the gcd of the a_i, it follows that cd divides c, which shows that d is a unit and f_0 is primitive. $\qquad\square$

(7.3.6). *Let R be a unique factorization domain and let f, g be non-zero polynomials over R. Then $c(fg) = c(f)c(g)$. In particular, if f and g are primitive, then so is fg.*

Proof. Consider first the special case where f and g are primitive. If fg is not primitive, $c(fg)$ is not a unit, so it must be divisible by an irreducible element p of R. Write $f = \sum_{i=0}^{m} a_i t^i$ and $g = \sum_{j=0}^{n} b_j t^j$, so that

$$fg = \sum_{k=0}^{m+n} c_k t^k$$

where $c_k = \sum_{i=0}^{k} a_i b_{k-i}$. (Here $a_i = 0$ if $i > m$ and $b_j = 0$ if $j > n$.) Since f is primitive, p cannot divide all its coefficients and there is an integer $r \geq 0$ such that $p \mid a_0, a_1, \ldots, a_{r-1}$, but $p \nmid a_r$. Similarly there is an $s \geq 0$ such that p divides each of b_0, $b_1, \ldots, b_{s-1}$, but not b_s. Now consider c_{r+s}, which can be written

$$(a_0 b_{r+s} + a_1 b_{r+s-1} + \cdots + a_{r-1} b_{s+1}) + a_r b_s + (a_{r+1} b_{s-1} + \cdots + a_{r+s} b_0).$$

We know that $p \mid c_{r+s}$; also p divides both the expressions in parentheses in the expression above. It follows that $p \mid a_r b_s$. By Euclid's Lemma for UFD's (see (7.3.4)), it follows that $p \mid a_r$ or $p \mid b_s$, both of which are impossible. By this contradiction fg is primitive.

Now we are ready for the general case. Using (7.3.5), we write $f = cf_0$ and $g = dg_0$ where $c = c(f)$, $d = c(g)$ and the polynomials f_0, g_0 are primitive in $R[t]$. Then $fg = cd(f_0g_0)$ and, as has just been shown, f_0g_0 is primitive. In consequence $c(fg) = cd = c(f)c(g)$. □

The next result is frequently helpful in deciding whether a polynomial is irreducible.

(7.3.7) (Gauss's Lemma). *Let R be a unique factorization domain and let F denote its field of fractions. If $f \in R[t]$, then f is irreducible over R if and only if it is irreducible over F.*

Proof. We can assume that $R \subseteq F$. Of course irreducibility over F implies irreducibility over R. It is the converse implication that requires proof. Assume that f is irreducible over R but reducible over F. We can assume that f is primitive on the basis of (7.3.5). Then $f = gh$ where $g, h \in F[t]$ are non-constant. Since F is the field of fractions of R, there exist elements $a, b \neq 0$ in R such that $g_1 = ag \in R[t]$ and $h_1 = bh \in R[t]$. Write $g_1 = c(g_1)g_2$ where $g_2 \in R[t]$ is primitive. Then $ag = c(g_1)g_2$, so we can divide both sides by $\gcd\{a, c(g_1)\}$. On these grounds it is permissible to assume that $c(g_1)$ and a are relatively prime, and for similar reasons the same can be assumed of $c(h_1)$ and b.

Next $(ab)f = (ag)(bh) = g_1h_1$. Taking the content of each side and using (7.3.6), we obtain $ab = c(g_1)c(h_1)$ since f is primitive. But $c(g_1)$ and a are relatively prime, so $a \mid c(h_1)$, and for a similar reason $b \mid c(g_1)$. Therefore we have the factorization $f = (b^{-1}g_1)(a^{-1}h_1)$ in which both factors are polynomials over R. But this contradicts the irreducibility of f over R, so the proof is complete. □

For example, to show that a polynomial in $\mathbb{Z}[t]$ is $\mathbb{Q}$-irreducible, it is enough to show that it is $\mathbb{Z}$-irreducible, usually an easier task.

Polynomial rings in several variables

Let us now use the theory of content to show that unique factorization occurs in polynomial rings with more than one variable. Here one should keep in mind that such rings are not PID's and so are not covered by (7.3.2). The main result is:

(7.3.8). *If R is a unique factorization domain, then so is the polynomial ring $R[t_1, \ldots, t_k]$.*

Proof. In the first place we need only prove the theorem for $k = 1$. Indeed, if $k > 1$, we have

$$R[t_1, \ldots, t_k] = (R[t_1, \ldots, t_{k-1}])[t_k],$$

so that induction on k will succeed once the case $k = 1$ is settled. From now on consider the ring $S = R[t]$. The first step in the proof is to establish:

(i) *Any non-constant polynomial f in S is expressible as a product of irreducible elements of R and primitive irreducible polynomials over R.*

The key idea in the proof is to introduce the field of fractions F of R, and exploit the fact that $F[t]$ is known to be a PID and hence a UFD. First of all write $f = c(f)f_0$ where $f_0 \in S$ is primitive, using (7.3.5). Here $c(f)$ is either a unit or a product of irreducibles of R. Thus we can assume that f is primitive. Regarding f as an element of the UFD $F[t]$, we write $f = p_1 p_2 \cdots p_m$ where $p_i \in F[t]$ is irreducible over F. Now find $a_i \neq 0$ in R such that $f_i = a_i p_i \in S$. Writing $c(f_i) = c_i$, we have $f_i = c_i q_i$ where $q_i \in R[t]$ is primitive. Hence $p_i = a_i^{-1} f_i = a_i^{-1} c_i q_i$ and q_i is F-irreducible since p_i is F-irreducible. Thus q_i is certainly R-irreducible.

Combining these expressions for p_i, we find that

$$f = (a_1^{-1} a_2^{-1} \cdots a_m^{-1} c_1 c_2 \cdots c_m) q_1 q_2 \cdots q_m,$$

and hence $(a_1 a_2 \cdots a_m)f = (c_1 c_2 \cdots c_m) q_1 q_2 \cdots q_m$. Now take the content of both sides of this equation to get $a_1 a_2 \cdots a_m = c_1 c_2 \cdots c_m$ up to a unit, since f and the q_i are primitive. Consequently $f = u q_1 q_2 \cdots q_m$ for some unit u of R. This is what was to be proved.

(ii) The next step is to assemble a complete set of irreducibles for S. First take a complete set of irreducibles C_1 for R. Then consider the set of all primitive irreducible polynomials in S. Now being associate is an equivalence relation on this set, so we can choose an element from each equivalence class. This yields a set of non-associate primitive irreducible polynomials C_2 with the property that every primitive irreducible polynomial in $R[t]$ is associate to an element of C_2. Now put

$$C = C_1 \cup C_2.$$

Since distinct elements of C cannot be associate, it is a complete set of irreducibles for S. If $0 \neq f \in S$, it follows from step (i) that f is expressible as a product of elements of C and a unit of R.

(iii) There remains the question of uniqueness. Suppose that

$$f = u a_1 a_2 \cdots a_k f_1 f_2 \cdots f_r = v b_1 b_2 \cdots b_\ell g_1 g_2 \cdots g_s$$

where u, v are units, $a_k, b_\ell \in C_1$ and $f_i, g_j \in C_2$. By Gauss's Lemma (7.3.7) the f_i and g_j are F-irreducible. Since $F[t]$ is a UFD and C_2 is a complete set of irreducibles for $F[t]$, we conclude that $r = s$ and $f_i = w_i g_i$, (after possible relabelling), where $w_i \in F$. Write $w_i = c_i d_i^{-1}$ where $c_i, d_i \in R$. Then $d_i f_i = c_i g_i$, which, on taking contents, yields $d_i = c_i$ up to a unit. This implies that w_i is a unit of R. Therefore f_i and g_i are associate and thus $f_i = g_i$.

Cancelling the f_i and g_i, we are left with $u a_1 a_2 \cdots a_k = v b_1 b_2 \cdots b_\ell$. Since R is a UFD with a complete set of irreducibles C_1, it follows that $k = \ell$, $u = v$ and $a_i = b_i$ after further relabelling. This completes the proof. $\qquad\square$

This theorem provides us with some important new examples of UFD's.

Corollary (7.3.9). *The following rings are unique factorization domains:*

$$\mathbb{Z}[t_1, \ldots, t_k] \quad \text{and} \quad F[t_1, \ldots, t_k]$$

where F is any field.

Exercises (7.3).
(1) Prove that a UFD satisfies the ascending chain condition on *principal* ideals, i. e., there does not exist an infinite strictly ascending chain of principal ideals.
(2) If R is a UFD and C is *any* complete set of irreducible elements for R, show that there is unique factorization in terms of C.
(3) If C_1 and C_2 are two complete sets of irreducibles for a domain R, prove that $|C_1| = |C_2|$.
(4) Show that the domain $\{a + b\sqrt{-5} \mid a, b \in \mathbb{Z}\}$ is *not* a UFD. [Hint: First show that ± 1 are the only units.]
(5) Prove that $t^3 + at + 1 \in \mathbb{Z}[t]$ is reducible over $\mathbb{Q}$ if and only if $a = 0$ or -2.
(6) Explain why the ring of rational numbers with odd denominators is a UFD and find a complete set of irreducibles for it.
(7) The same question for the power series ring $F[[t]]$ where F is a field.
(8) Prove that Euclid's Lemma is valid in any UFD.

7.4 Roots of polynomials and splitting fields

Let R be a commutative ring with identity, let $f = b_0 + b_1 t + \cdots + b_n t^n \in R[t]$ and let $a \in R$. Then the *value* of the polynomial f at a is defined to be

$$f(a) = b_0 + b_1 a + \cdots + b_n a^n \in R.$$

Thus we have a function $\theta_a : R[t] \rightarrow R$ which evaluates polynomials at a, i. e., $\theta_a(f) = f(a)$. Now $f + g(a) = f(a) + g(a)$ and $(fg)(a) = f(a)g(a)$, because the ring elements $f(a)$ and $g(a)$ are added and multiplied by the same rules as the polynomials f and g. It follows that $\theta_a : R[t] \rightarrow R$ is a ring homomorphism. Its kernel consists of all $f \in R[t]$ such that $f(a) = 0$, that is, all polynomials that have a as a *root*.

The following criterion for an element to be a root of a polynomial should be familiar from elementary algebra.

(7.4.1) (The Remainder Theorem). *Let R be an integral domain, let $f \in R[t]$ and let $a \in R$. Then a is a root of f if and only if $t - a$ divides f in the ring $R[t]$.*

Proof. If $t - a$ divides f, then $f = (t-a)g$ where $g \in R[t]$. Then $f(a) = \theta_a(f) = \theta_a((t-a)g) = \theta_a(t - a)\theta_a(g) = 0$. Hence a is a root of f.

Conversely, assume that $f(a) = 0$ and let F denote the field of fractions of R. Since $F[t]$ is a Euclidean domain, we can divide f by $t - a$ to get a quotient and remainder in $F[t]$, say $f = (t - a)q + r$ where $q, r \in F[t]$ and $\deg(r) < 1$, i. e., r is constant. However, notice that by the usual long division process q and r actually belong to $R[t]$. Finally, apply the evaluation homomorphism θ_a to $f = (t - a)q + r$ to obtain $0 = r$ since r is constant. Therefore $t - a$ divides f. $\qquad\square$

Corollary (7.4.2). *The kernel of the evaluation homomorphism θ_a is the principal ideal $(t - a)$.*

This is simply a restatement of (7.4.1).

The multiplicity of a root

Let R be a domain. Suppose that $f \in R[t]$ is not constant and it has a root a in R; thus $t - a | f$. There is a largest positive integer n such that $(t - a)^n \mid f$, since the degree of a divisor of f cannot exceed $\deg(f)$. In this situation a is said to be a *root of f with multiplicity n*. If $n > 1$, then a is called a *multiple root* of f.

(7.4.3). *Let R be a domain and let $0 \neq f \in R[t]$ have degree n. Then the sum of the multiplicities of all the roots of f that lie in R is at most n.*

Proof. Let a be a root of f. Then $t - a$ divides f by (7.4.2), so that $f = (t - a)g$ where $g \in R[t]$ has degree $n - 1$. By induction on n the sum of the multiplicities of the roots of g is at most $n - 1$. Now a root of f either equals a or else is a root of g. Consequently the sum of the multiplicities of the roots of f is at most $1 + (n - 1) = n$. $\qquad\square$

Example (7.4.1).
(i) The polynomial $t^2 + 1 \in \mathbb{Q}[t]$ has no roots in $\mathbb{Q}$, which shows that the sum of the multiplicities of the roots of a polynomial can be less than the degree.
(ii) Consider the polynomial $t^4 - 1 \in R[t]$ where R is the ring of rational quaternions (see Section 6.3). Then f has 8 roots in R, namely $\pm 1, \pm I, \pm J, \pm K$. Therefore (7.4.3) is not valid for non-commutative rings, which is another reason to keep our rings commutative.

Next comes another well-known theorem.

(7.4.4) (The Fundamental Theorem of Algebra). *Let f be a non-zero polynomial of degree n over the field of complex numbers $\mathbb{C}$. Then the sum of the multiplicities of the roots of f in $\mathbb{C}$ equals n, i. e., f is a product of n linear factors over $\mathbb{C}$.*

The proof of this theorem will be postponed until Chapter 12 – see (12.3.6). Despite its name, all the known proofs of the theorem employ some analysis.

Derivatives

Derivatives are useful in detecting multiple roots of polynomials. Since we are not dealing with polynomials over $\mathbb{R}$ here, limits cannot be used. For this reason we adopt the following formal definition of the *derivative f'* of the polynomial $f \in R[t]$ where R is a commutative ring with identity. If $f = a_0 + a_1 t + \cdots + a_n t^n$, then

$$f' = a_1 + 2a_2 t + \cdots + na_n t^{n-1} \in R[t].$$

On the basis of this definition the usual rules of differentiation can be established.

(7.4.5). *Let $f, g \in R[t]$ and $c \in R$ where R is a commutative ring with identity. Then*
(i) $(f + g)' = f' + g'$;
(ii) $(cf)' = cf'$;
(iii) $(fg)' = f'g + fg'$.

Proof. Only the statement (iii) will be proved. Write $f = \sum_{i=0}^{m} a_i t^i$ and $g = \sum_{j=0}^{n} b_j t^j$; then

$$fg = \sum_{i=0}^{m+n} \left(\sum_{k=0}^{i} a_k b_{i-k} \right) t^i.$$

The coefficient of t^{i-1} in $(fg)'$ is therefore equal to $i(\sum_{k=0}^{i} a_k b_{i-k})$.
On the other hand, the coefficient of t^{i-1} in $f'g + fg'$ is

$$\sum_{k=0}^{i-1} (k+1)a_{k+1}b_{i-k-1} + \sum_{k=0}^{i-1} (i-k)a_k b_{i-k},$$

which equals

$$ia_i b_0 + \sum_{k=0}^{i-2} (k+1)a_{k+1}b_{i-k-1} + ia_0 b_i + \sum_{k=1}^{i-1} (i-k)a_k b_{i-k}.$$

On adjusting the summation in the second sum, this becomes

$$ia_i b_0 + \sum_{k=0}^{i-2} (k+1)a_{k+1}b_{i-k-1} + \sum_{k=0}^{i-2} (i-k-1)a_{k+1}b_{i-k-1} + ia_0 b_i.$$

On combining the two sums, this reduces to

$$i\left(a_0 b_i + \sum_{k=0}^{i-2} a_{k+1}b_{i-k-1} + a_i b_0 \right) = i\left(\sum_{k=0}^{i} a_k b_{i-k} \right).$$

It follows that $(fg)' = f'g + fg'$. □

Corollary (7.4.6). $(f^m)' = mf^{m-1}f'$ *where m is a positive integer.*

This is proved by induction on m using (7.4.5). A criterion for a polynomial to have multiple roots can now be given.

(7.4.7). *Let R be a domain and let $a \in R$ be a root of a polynomial $f \in R[t]$. Then a is a multiple root if and only if $t - a$ divides both f and f'.*

Proof. Let ℓ be the multiplicity of the root a. Then $\ell \geq 1$ and $f = (t - a)^\ell g$ where $t - a \nmid g \in R[t]$. Hence $f' = \ell(t-a)^{\ell-1}g + (t-a)^\ell g'$ by (7.4.5) and (7.4.6). If a is a multiple root of f, then $\ell \geq 2$ and $f'(a) = 0$; by (7.4.1) it follows that $t - a$ divides f', as well as f.

Conversely, suppose that $t - a \mid f' = \ell(t-a)^{\ell-1}g + (t-a)^\ell g'$. If $\ell = 1$, then $t - a$ divides ℓg, which implies that $t - a$ divides g, a contradiction. Therefore $\ell > 1$ and a is a multiple root. $\square$

Example (7.4.2). Let F be a field whose characteristic does not divide the positive integer n. Then $t^n - 1 \in F[t]$ has no multiple roots in F.

For, with $f = t^n - 1$, we have $f' = nt^{n-1} \neq 0$ since char(F) does not divide n. Hence $t^n - 1$ and nt^{n-1} are relatively prime and thus f and f' have no common roots. Therefore f has no multiple roots by (7.4.7).

Splitting fields

We will now consider roots of polynomials over a field F. If $f \in F[t]$ is not constant, we know that f has at most deg(f) roots in F, including multiplicities, by (7.4.3). However, f need not have any roots in F, as the example $t^2 + 1 \in \mathbb{R}[t]$ shows. On the other hand, $t^2 + 1$ has two roots in the larger field $\mathbb{C}$.

The question to be addressed is this: can we construct a field K, larger than F in some sense, in which f has exactly deg(f) roots up to multiplicity, i. e., over which f splits into a product of linear factors? A smallest such field is called a *splitting field* of f. In the case of the polynomial $t^2 + 1 \in \mathbb{R}[t]$, the situation is quite clear; its splitting field is $\mathbb{C}$ since $t^2 + 1 = (t + i)(t - i)$ where $i = \sqrt{-1}$. However, for a general field F we do not have a convenient larger field like $\mathbb{C}$ at hand. Thus splitting fields will have to be constructed from scratch.

We begin by formulating precisely the definition of a splitting field. If F is a field, by a *subfield* of F is meant a subring containing the identity element which is closed under forming inverses of non-zero elements. Let f be a non-constant polynomial over F. A *splitting field* for f over F is a field K containing an isomorphic copy F_1 of F as a subfield such that the polynomial in $F_1[t]$ corresponding to f can be expressed in the form

$$a(t - c_1)(t - c_2) \cdots (t - c_n)$$

where K is a smallest field containing F_1 and the elements $a, c_1, c_2, \dots c_n$. There is nothing to be lost in assuming that $F \subseteq K$ since F can be replaced by the isomorphic field F_1. Thus F is a subfield of K.

Our first objective is to demonstrate that splitting fields actually exist.

(7.4.8). *If f is a non-constant polynomial over a field F, then f has a splitting field over F.*

Proof. We argue by induction on $n = \deg(f)$; note that we may assume $n > 1$ since otherwise F itself is a splitting field for f. Assume the result is true for all polynomials of degree less than n. Consider first the case where f is reducible, so $f = gh$ where g, h in $F[t]$ both have degree less than n. By induction hypothesis g has a splitting field over F, say K_1, which we may suppose contains F as a subfield. For the same reason h has a splitting field over K_1, say K, with $K_1 \subseteq K$. Clearly f is a product of linear factors over K. If f were such a product over some subfield S of K containing F, then we would obtain first that $K_1 \subseteq S$ and then $K \subseteq S$. Hence $K = S$ and K is a splitting field of f.

Therefore we can assume f is irreducible. By (7.2.6) the ideal (f) is maximal in $F[t]$ and consequently the quotient ring

$$K_1 = F[t]/(f)$$

is a field. Next the assignment $a \mapsto a + (f)$, where $a \in F$, determines an injective ring homomorphism from F to K_1. The image is a subfield F_1 of K_1 and $F \simeq F_1$. Thus we may regard f as a polynomial over F_1.

The critical observation to make is that f has a root in K_1, namely $a_1 = t + (f)$; for $f(a_1) = f(t) + (f) = (f) = 0_{K_1}$. By (7.4.1) we have $f = (t - a_1)g$ where $g \in K_1[t]$, and of course $\deg(g) = n - 1$. By induction hypothesis g has a splitting field K containing K_1. Since $a_1 \in K_1 \subseteq K$, we see that K is a splitting field for f: for any subfield of K containing F and the roots of f must contain K_1 since each element of K_1 has the form $h + (f) = h(a_1)$ for some $h \in F[t]$. This completes the proof. □

Example (7.4.3). Let $f = t^3 - 2 \in \mathbb{Q}[t]$. The roots of f are $2^{1/3}, c2^{1/3}, c^2 2^{1/3}$ where $c = e^{2\pi i/3}$, a complex cube root of unity. Then f has as its splitting field the smallest subfield of $\mathbb{C}$ containing $\mathbb{Q}$, $2^{1/3}$ and c.

The next example shows how finite fields can be constructed from irreducible polynomials.

Example (7.4.4). Show that $f = t^3 + 2t + 1 \in \mathbb{Z}_3[t]$ is irreducible and use it to construct a field of order 27. Prove that this is a splitting field of f.

First of all notice that the only way a cubic polynomial can be reducible is if it has a linear factor, i. e., it has a root in the field. But we easily verify that f has no roots in $\mathbb{Z}_3 = \{0, 1, 2\}$ since $f(0) = f(1) = f(2) = 1$. (For conciseness we have written i for the congruence class $[i]$). It follows that f is irreducible and

$$K = \mathbb{Z}_3[t]/(f)$$

is a field.

If $g \in \mathbb{Z}_3[t]$, then by the Division Algorithm $g = fq + r$ where $q, r \in \mathbb{Z}_3[t]$ and $0 \le \deg r < 3$. Hence $g + (f) = r + (f)$. This shows that every element of K has the form $a_0 + a_1 t + a_2 t^2 + (f)$ where $a_i \in \mathbb{Z}_3$. Thus $|K| \le 3^3 = 27$. On the other hand, all such elements are distinct. Indeed, if $r + (f) = s + (f)$ with r and s both of degree < 3, then $f \mid r - s$, so that $r = s$. Therefore $|K| = 27$ and we have constructed a field of order 27.

As in the proof of (7.4.8), we see that f has the root $a = t + (f)$ in K. To prove that K is actually a splitting field, note that f has two further roots in K, namely $a + 1$ and $a - 1$. Thus $f = (t - a)(t - a - 1)(t - a + 1)$.

Further discussion of fields is postponed until Chapter 11. However, we have seen enough to realize that irreducible polynomials play a vital role in the theory of fields. Thus a practical criterion for irreducibility is sure to be useful. Probably the best known test for irreducibility is:

(7.4.9) (Eisenstein's[1] Criterion). *Let R be a unique factorization domain and let $f = a_0 + a_1 t + \cdots + a_n t^n$ be a non-constant polynomial over R. Suppose that there is an irreducible element p of R such that $p \mid a_0, p \mid a_1, \ldots, p \mid a_{n-1}$, but $p \nmid a_n$ and $p^2 \nmid a_0$. Then f is irreducible over R.*

Proof. Assume that f is reducible and

$$f = (b_0 + b_1 t + \cdots + b_r t^r)(c_0 + c_1 t + \cdots + c_s t^s)$$

where $b_i, c_j \in R$, $r, s < n$, and $r + s = n$. By hypothesis $p \mid a_0 = b_0 c_0$, but $p^2 \nmid a_0$; thus p must divide exactly one of b_0 and c_0, say $p \mid b_0$ and $p \nmid c_0$. Also p does not divide $a_n = b_r c_s$, so it cannot divide b_r. Therefore there is a smallest positive integer $k \le r$ such that $p \nmid b_k$. Now p divides each of $b_0, b_1, \ldots, b_{k-1}$, and also $p \mid a_k$ because $k \le r < n$. Since $a_k = (b_0 c_k + b_1 c_{k-1} + \cdots + b_{k-1} c_1) + b_k c_0$, (where $c_i = 0$ if $i > s$), it follows that $p \mid b_k c_0$. By Euclid's Lemma – which by (7.3.4) is valid in a UFD – either $p \mid b_k$ or $p \mid c_0$, both of which are forbidden. $\qquad\square$

Eisenstein's Criterion is often applied in conjunction with Gauss's Lemma (7.3.7) to give a test for irreducibility over the field of fractions of a domain.

Example (7.4.5). Prove that $t^5 - 9t + 3$ is irreducible over $\mathbb{Q}$.

First of all $f = t^5 - 9t + 3$ is irreducible over $\mathbb{Z}$ by Eisenstein's Criterion with $p = 3$. Then Gauss's Lemma shows that f is irreducible over $\mathbb{Q}$.

---

1 Ferdinand Gotthold Max Eisenstein (1823–1852).

Example (7.4.6). Show that if p is a prime, the polynomial $f = 1 + t + t^2 + \cdots + t^{p-1}$ is irreducible over $\mathbb{Q}$.

By Gauss's Lemma it suffices to prove that f is $\mathbb{Z}$-irreducible. Since (7.4.9) is not immediately applicable to f, we resort to a trick. Consider the polynomial $g = f(t + 1)$; then

$$g = 1 + (t + 1) + \cdots + (t + 1)^{p-1} = \frac{(t + 1)^p - 1}{t},$$

by the formula for the sum of a geometric series. On expanding $(t+1)^p$ by the Binomial Theorem – see Exercise (6.1.6) – we arrive at the formula

$$g = t^{p-1} + \binom{p}{p-1} t^{p-2} + \cdots + \binom{p}{2} t + \binom{p}{1}.$$

Now $p \mid \binom{p}{i}$ if $0 < i < p$ by (2.3.3). Therefore g is irreducible over $\mathbb{Z}$ by Eisenstein's Criterion. Clearly this implies that f is irreducible over $\mathbb{Z}$. (The polynomial f is called the *cyclotomic polynomial of order p*.)

Exercises (7.4).

(1) Let $f \in F[t]$ have degree ≤ 3 where F is a field. Show that f is reducible over F if and only if it has a root in F.

(2) Find the multiplicity of the root 2 of the polynomial $t^3 + 2t^2 + t + 2 \in \mathbb{Z}_5[t]$.

(3) List all irreducible polynomials of degree at most 3 in $\mathbb{Z}_2[t]$.

(4) Use $t^3 + t + 1 \in \mathbb{Z}_5[t]$ to construct a field of order 125.

(5) Let $f = 1 + t + t^2 + t^3 + t^4 \in \mathbb{Q}[t]$.
  (i) Prove that $K = \mathbb{Q}[t]/(f)$ is a field.
  (ii) Show that every element of K can be uniquely written in the form $a_0 + a_1 x + a_2 x^2 + a_3 x^3$ where $x = t + (f)$ and $a_i \in \mathbb{Q}$.
  (iii) Prove that K is a splitting field of f. [Hint: note that $x^5 = 1$ and check that x^2, x^3, x^4 are roots of f.]
  (iv) Compute $(1 + x^2)^3$ and $(1 + x)^{-1}$ in K.

(6) Show that $t^6 + 6t^5 + 4t^4 + 2t + 2$ is irreducible over $\mathbb{Q}$.

(7) Show that $t^6 + 12t^5 + 49t^4 + 96t^3 + 99t^2 + 54t + 15$ is irreducible over $\mathbb{Q}$. [Hint: use a suitable change of variable.]

(8) Let $F = \mathbb{Z}_p\{t_1\}$, the field of rational functions in t_1, and $R = F[t]$ where t and t_1 are distinct indeterminates. Prove that $t^n - t_1^2 t + t_1 \in R$ is irreducible over F for all $n \geq 1$.

(9) Find a polynomial of degree 4 in $\mathbb{Z}[t]$ which has $\sqrt{3} - \sqrt{2}$ as a root and is irreducible over $\mathbb{Q}$.

(10) Let n be a positive integer that is not a prime. Prove that $1 + t + t^2 + \cdots + t^{n-1}$ is reducible over any field.

(11) Show that $\mathbb{Q}[t]$ contains an irreducible polynomial of every degree $n \geq 1$.

(12) Let R be a commutative ring with identity containing a zero divisor. Find a linear polynomial in $R[t]$ which has at least two roots in R, so that (7.4.3) fails for R.

8 Vector Spaces

We have already encountered groups and rings, two of the most commonly used algebraic structures. A third structure of great importance is a vector space. Vector spaces
appear throughout mathematics and they also turn up in many applied areas, for example, in quantum theory and coding theory.

8.1 Vector spaces and subspaces

Let F be a field. A *vector space over F* is an additively written abelian group V with an
action of F on V called *scalar multiplication*, that is, a function from $F \times V$ to V written
$(a, v) \mapsto av$, $(a \in F, v \in V)$, such that the following axioms hold for all $u, v \in V$ and
$a, b \in F$.

(i) $a(u + v) = au + av$;
(ii) $(a + b)v = av + bv$;
(iii) $(ab)v = a(bv)$;
(iv) $1_F v = v$.

Notice that (iii) and (iv) assert that the multiplicative group of F *acts* on the set V in the
sense of Section 5.1. Elements of V are called *vectors* and elements of F *scalars*. When
there is no chance of confusion, it is usual to refer to the set V as the vector space.

First of all we record two elementary consequences of the axioms.

(8.1.1). *Let v be a vector in a vector space V over a field F and let $a \in F$. Then:*
(i) $0_F v = 0_V$ *and* $a0_V = 0_V$;
(ii) $(-1_F)v = -v$.

Proof. Put $a = 0_F = b$ in vector space axiom (ii) to get $0_F v = 0_F v + 0_F v$. Hence $0_F v = 0_V$
by the cancellation law for the group $(V, +)$. Similarly, setting $u = 0_V = v$ in (i) yields
$a0_V = 0_V$. This establishes (i).

Using axioms (ii) and (iv) and property (i), we obtain

$$v + (-1_F)v = 1_F v + (-1_F)v = (1_F + (-1_F))v = 0_F v = 0_V.$$

Therefore $(-1_F)v$ equals $-v$, which completes the proof.   □

Examples of vector spaces
Before proceeding further we review some standard sources of vector spaces.
(i) *Vector spaces of matrices.* Let F be a field and define

$$M_{m,n}(F)$$

https://doi.org/10.1515/9783110691160-008

to be the set of all $m \times n$ matrices over F. This is already an abelian group with respect to ordinary matrix addition. There is also a natural scalar multiplication here: if $A = [a_{ij}] \in M_{m,n}(F)$ and $f \in F$, then fA is the matrix which has fa_{ij} as its (i,j) entry. That the vector space axioms hold is guaranteed by elementary results from matrix algebra.

Two special cases of interest are the vector spaces

$$F^m = M_{m,1}(F) \quad \text{and} \quad F_n = M_{1,n}(F).$$

Thus F^m is the vector space of m-column vectors over F, while F_n is the vector space of n-row vectors over F. The space $\mathbb{R}^n$ is called *Euclidean n-space*. For $n \leq 3$ there is a well-known geometric interpretation of $\mathbb{R}^n$. Consider for example $\mathbb{R}^3$. A vector in $\mathbb{R}^3$

$$v = \begin{bmatrix} a \\ b \\ c \end{bmatrix}$$

is represented by a line segment $\vec{v}$ in 3-dimensional space drawn from an arbitrary initial point (p, q, r) to the point $(p + a, q + b, r + c)$.

With this interpretation of vectors, the rule of addition of vectors u and v in $\mathbb{R}^3$ is equivalent to the well-known *triangle rule* for addition of line segments $\vec{u}$ and $\vec{v}$; this is illustrated in the diagram below.

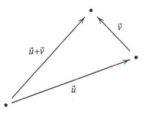

A detailed account of the geometric interpretations of euclidean 2-space and 3-space may be found in any text on linear algebra – see for example [16].

(ii) *Vector spaces of polynomials.* The set $F[t]$ of all polynomials in t over a field F is a vector space over F with the usual addition and scalar multiplication of polynomials.

(iii) *Fields as vector spaces.* Suppose that F is a *subfield* of a field K, i. e., F is a subring of K containing 1 which is closed with respect to taking inverses of non-zero elements. We can regard K as a vector space over F, using the field operations as vector space operations. At first sight this example may seem confusing since elements of F are simultaneously vectors and scalars. However, this point of view will be particularly valuable when we come to investigate the structure of fields in Chapter 11.

Subspaces

By analogy with subgroups of groups and subrings of rings, it is natural to introduce the concept of a subspace of a vector space. Let V be a vector space over a field F and let S be a subset of V. Then S is called a *subspace* of V if, when we restrict the vector space operations of V to S, we obtain a vector space over F. Taking note of the analysis of the subgroup concept in Section 3.3 – see especially (3.3.4) – we conclude that *a subspace is a subset of V containing 0_V which is closed with respect to addition and multiplication by scalars.*

Obvious examples of subspaces of V are $0 = 0_V$, the *zero subspace* which contains just the zero vector, and V itself, the *improper subspace*. A more interesting source of examples is given in:

Example (8.1.1). Let A be an $m \times n$ matrix over a field F and define S to be the subset of all X in F^n such that $AX = 0$. Then S is a subspace of F^n, verification of the closure properties being very easy. The subspace S is called the *null space* of the matrix A.

Linear combinations of vectors

Suppose that V is a vector space over a field F and $v_1, v_2, \ldots, v_k$ are vectors in V. A *linear combination* of these vectors has the form

$$a_1 v_1 + a_2 v_2 + \cdots + a_k v_k$$

where $a_1, a_2, \ldots, a_k \in F$. If X is any non-empty set of vectors in V, we will write either $F\langle X \rangle$ or, if we do not wish to emphasize the field, $\langle X \rangle$ for the set of all linear combinations of vectors in the set X. It is a fundamental fact that this is always a subspace.

(8.1.2). *Let X be a non-empty subset of a vector space V over a field F. Then $F\langle X \rangle$ is the smallest subspace of V that contains X.*

Proof. In the first place it is easy to verify that $F\langle X \rangle$ is closed with respect to addition and scalar multiplication; of course it also contains the zero vector 0_V. Therefore $F\langle X \rangle$ is a subspace. Also it contains X since $x = 1_F x \in F\langle X \rangle$ for all $x \in X$. Finally, any subspace that contains X automatically contains every linear combination of vectors in X, i. e., it must contain $F\langle X \rangle$ as a subset. $\square$

The subspace $\langle X \rangle$ is called the subspace *generated* (or *spanned*) by X. If $V = \langle X \rangle$, then X is said to *generate* the vector space V. If V can be generated by some finite set of vectors, we say that V is a *finitely generated vector space*. What this means is that every vector in V can be expressed as a linear combination of the vectors in some finite set.

Example (8.1.2). F^n is a finitely generated vector space. To see why, consider the so-called *elementary vectors* in F^n,

$$E_1 = \begin{bmatrix} 1 \\ 0 \\ 0 \\ \vdots \\ 0 \end{bmatrix}, \quad E_2 = \begin{bmatrix} 0 \\ 1 \\ 0 \\ \vdots \\ 0 \end{bmatrix}, \quad \ldots, \quad E_n = \begin{bmatrix} 0 \\ 0 \\ \vdots \\ 0 \\ 1 \end{bmatrix}.$$

A general vector in F^n,

$$\begin{bmatrix} a_1 \\ a_2 \\ \vdots \\ a_n \end{bmatrix},$$

can be written as $a_1 E_1 + a_2 E_2 + \cdots + a_n E_n$. Hence $F^n = \langle E_1, E_2, \ldots, E_n \rangle$ and F^n is finitely generated.

On the other hand, infinitely generated, i. e., non-finitely generated, vector spaces are not hard to find.

Example (8.1.3). If F is an arbitrary field, the vector space $F[t]$ is infinitely generated.

Indeed suppose that $F[t]$ could be generated by finitely many polynomials p_1, p_2, ..., p_k and let m be the maximum degree of the p_i. Since the degree of a linear combination of the p_i cannot exceed m, it follows that t^{m+1} cannot be expressed as a linear combination of $p_1, \ldots, p_k$, a contradiction.

Exercises (8.1).

(1) Which of the following are vector spaces? The operations of addition and scalar multiplication are the natural ones.

   (i) The set of all real 2×2 matrices with determinant 0.

   (ii) The set of all solutions $y(x)$ of the homogeneous linear differential equation $a_n(x)y^{(n)} + a_{n-1}(x)y^{(n-1)} + \cdots + a_1(x)y' + a_0(x)y = 0$, where the $a_i(x)$ are real-valued functions of the real variable x.

   (iii) The set of all solutions X of the matrix equation $AX = B$.

(2) In the following cases say whether S is a subspace of the vector space V.

   (i) $V = \mathbb{R}^2$, $S = $ all $\begin{bmatrix} a^2 \\ a \end{bmatrix}$, $a \in \mathbb{R}$;

   (ii) V is the vector space of all real valued continuous functions defined on the interval $[0, 1]$ and S is the set of all infinitely differentiable functions in V.

   (iii) $V = F[t]$, $S = \{f \in V \mid f(a) = 0\}$ where a is a fixed element of F.

(3) Verify that the rule for adding the vectors in $\mathbb{R}^3$ corresponds to the usual triangle rule for the addition of line segments.

(4) Does $\begin{bmatrix} 4 & 3 \\ 1 & -2 \end{bmatrix}$ belong to the subspace of $M_2(\mathbb{R})$ generated by the matrices $\begin{bmatrix} 3 & 4 \\ 1 & 2 \end{bmatrix}$, $\begin{bmatrix} 0 & 2 \\ -\frac{1}{3} & 4 \end{bmatrix}$, $\begin{bmatrix} 0 & 2 \\ 0 & 1 \end{bmatrix}$?

(5) Let V be a vector space over a finite field. Prove that V is finitely generated if and only if it is finite.

8.2 Linear independence, basis and dimension

A concept of critical importance in vector space theory is linear independence. For an understanding of this topic some knowledge of systems of linear equations, and in particular row and column operations on matrices, is essential and will be assumed. See for example [16] for an account.

Let V be a vector space over a field F and let X be a non-empty subset of V. Then X is called *linearly dependent* if there exist distinct vectors $x_1, x_2, \ldots, x_k$ in X and scalars $a_1, a_2, \ldots, a_k \in F$, not all of them zero, such that

$$a_1 x_1 + a_2 x_2 + \cdots + a_k x_k = 0.$$

This amounts to saying that some x_i can be expressed as a linear combination of the others. For if, say, $a_i \neq 0$, we can solve for x_i, obtaining

$$x_i = \sum_{\substack{j=1 \\ j \neq i}}^{k} (-a_i^{-1}) a_j v_j.$$

A subset which is not linearly dependent is called *linearly independent*. For example, the elementary vectors $E_1, E_2, \ldots, E_n$ form a linearly independent subset of F^n for any field F.

Homogeneous linear systems

To make significant progress with linear independence, some knowledge of systems of linear equations is needed. Let F be a field and consider a system of m homogeneous linear equations over F

$$\begin{cases} a_{11}x_1 + \cdots + a_{1n}x_n = 0 \\ a_{21}x_1 + \cdots + a_{2n}x_n = 0 \\ \vdots \\ a_{m1}x_1 + \cdots + a_{mn}x_n = 0 \end{cases}$$

Here $a_{ij} \in F$ and $x_1, x_2, \ldots, x_n$ are the unknowns.

Clearly the system has the *trivial solution* $x_1 = x_2 = \cdots = x_n = 0$. The interesting question is whether there are any non-trivial solutions. A detailed account of the theory of systems of linear equations can be found in any book on linear algebra, for example [16].

The linear system can be written in the matrix form

$$AX = 0,$$

where $A = [a_{ij}]_{m,n}$ is the coefficient matrix and X is the n-column vector formed by the unknowns $x_1, x_2, \ldots, x_n$. The following result is sufficient for our present purposes.

(8.2.1). *The homogenous linear system $AX = 0$ has a non-trivial solution X if and only if the rank of the coefficient matrix A is less than the number of unknowns.*

Proof. Write $A = [a_{ij}]$, which is an $m \times n$ matrix. We adopt the method of systematic elimination known as *Gaussian elimination*. It may be assumed that $a_{11} \neq 0$; for, if this is not true, replace equation 1 by the first equation in which x_1 appears. Since equation 1 can be multiplied by a_{11}^{-1}, we may also assume that $a_{11} = 1$. Then, by subtracting multiples of equation 1 from equations 2 through m, the unknown x_1 can be eliminated from these equations.

Next find the first of equations 2 through m which contains an unknown with smallest subscript > 1, say x_{i_2}. Move this equation up to second position. Now make the coefficient of x_{i_2} equal to 1 and subtract multiples of equation 2 from equations 3 through m so as to eliminate x_{i_2}. Repeat this procedure until the remaining equations involve no further unknowns, i. e., they are of the trivial form $0 = 0$. Let us say this happens after r steps. At this point the matrix of coefficients is in *row echelon form* with r linearly independent rows. The integer r is the *rank* of A.

Unknowns other than $x_1 = x_{i_1}, x_{i_2}, \ldots, x_{i_r}$ can be given arbitrary values. The non-trivial equations may then be used to solve back for $x_{i_r}, x_{i_{r-1}}, \ldots, x_{i_1}$ successively. Therefore there is a non-trivial solution if and only if $r < n$; for then at least one unknown can be given an arbitrary value. $\qquad\square$

Corollary (8.2.2). *A homogeneous linear system $AX = 0$ of n equations in n unknowns has a non-trivial solution if and only if $\det(A) = 0$.*

For $\det(A) = 0$ if and only if the rank of A is less than n. This result is used to establish the fundamental theorem on linear dependence in vector spaces.

(8.2.3). *Let $v_1, v_2, \ldots, v_k$ be vectors in a vector space V over a field F. Then any set of $k + 1$ or more vectors in the subspace $\langle v_1, v_2, \ldots, v_k \rangle$ is linearly dependent.*

Proof. Let $u_1, u_2, \ldots, u_{k+1} \in S = \langle v_1, \ldots, v_k \rangle$. It is enough to show that $\{u_1, u_2, \ldots, u_{k+1}\}$ is a linearly dependent set. This amounts to finding field elements $a_1, a_2, \ldots, a_{k+1}$, not all of them zero, such that $a_1 u_1 + a_2 u_2 + \cdots + a_{k+1} u_{k+1} = 0$.

Since $u_i \in S$, there is an expression $u_i = d_{1i}v_1 + d_{2i}v_2 + \cdots + d_{ki}v_k$ where $d_{ji} \in F$. On substituting for the u_i, we obtain

$$a_1 u_1 + a_2 u_2 + \cdots + a_{k+1} u_{k+1} = \sum_{i=1}^{k+1} a_i \left(\sum_{j=1}^{k} d_{ji} v_j \right) = \sum_{j=1}^{k} \left(\sum_{i=1}^{k+1} d_{ji} a_i \right) v_j.$$

Therefore $a_1 u_1 + a_2 u_2 + \cdots + a_{k+1} u_{k+1} = 0$ if the a_i satisfy the equations

$$\sum_{i=1}^{k+1} d_{ji} a_i = 0, \quad j = 1, \ldots, k.$$

But this is a system of k linear homogeneous equations in the $k+1$ unknowns a_i, so the rank of the coefficient matrix $[d_{ij}]$ is at most k. By (8.2.1) there is a non-trivial solution $a_1, a_2, \ldots, a_{k+1}$. Therefore $\{u_1, u_2, \ldots, u_{k+1}\}$ is linearly dependent, as claimed. $\square$

Corollary (8.2.4). *If a vector space V can be generated by k elements, then every subset of V with $k + 1$ or more elements is linearly dependent.*

Bases

A *basis* of a vector space V is a non-empty subset X such that:
(i) X is linearly independent;
(ii) X generates V.

These are contrasting properties in the sense that (i) means that X is not too large and (ii) that X is not too small.

For example, the elementary vectors E_1, E_2, ..., E_n form a basis of the vector space F^n called the *standard basis*. More generally a basis of $M_{m,n}(F)$ is obtained by taking all the $m \times n$ matrices over F with a single non-zero entry which is equal to 1_F.

A important property of a basis is unique expressibility.

(8.2.5). *If $\{v_1, v_2, \ldots, v_n\}$ is a basis of a vector space V over a field F, then every vector v in V is uniquely expressible in the form $v = a_1 v_1 + \cdots + a_n v_n$ with $a_i \in F$.*

Proof. In the first place such expressions for v exist by definition. If v in V had two such expressions $v = \sum_{i=1}^{n} a_i v_i = \sum_{i=1}^{n} b_i v_i$, we would have $\sum_{i=1}^{n} (a_i - b_i) v_i = 0$, from which it follows that $a_i = b_i$ by linear independence of the v_i. $\square$

This result shows that a basis may be used to introduce coordinates in a vector space. Suppose that V is a vector space over field F and that $\mathcal{B} = \{v_1, v_2, \ldots, v_n\}$ is a basis of V with its elements written in a specific order, i. e., an *ordered basis*. Then by (8.2.5) each $v \in V$ has a unique expression $v = \sum_{i=1}^{n} c_i v_i$ with $c_i \in F$. Thus v is determined by the column vector in F^n whose entries are $c_1, c_2, \ldots, c_n$; this is called the *coordinate column vector* of v with respect to $\mathcal{B}$ and is written

$$[v]_{\mathcal{B}}.$$

Coordinate vectors provide a concrete representation of vectors in an abstract vector space.

The existence of bases

There is nothing in the definition to make us certain that bases exist. Our first task will be to show that this is true for any non-zero vector space. Notice that the zero space does not have a basis since it has no linearly independent subsets.

(8.2.6). *Let V be a vector space and suppose that B_0 is a linearly independent subset of V. Then B_0 is contained in some basis of V.*

Proof. Define S to be the set of all linearly independent subsets of V which contain B_0. Then S is non-empty since it contains B_0. Furthermore, inclusion is a partial order on S, so $(S, \subseteq)$ is a partially ordered set. We wish to apply Zorn's Lemma to S; to do so we need to verify that every chain in S has an upper bound.

Let C be a chain in S. There is an obvious candidate for an upper bound, namely the union $U = \bigcup_{X \in C} X$. Certainly U is linearly independent: for any relation of linear dependence in U will involve a *finite* number of elements of S and so the relation will hold in some $X \in C$. Here it is vital that C be linearly ordered. Also $B_0 \subseteq U$ and $U \in S$; moreover U is an upper bound for C.

It is now possible to apply Zorn's Lemma to obtain a maximal element in S, say B. By definition B is linearly independent: to show that it is a basis we must prove that it generates V. Assume this is false and let v be a vector in V that is not expressible as a linear combination of vectors in B; then certainly $v \notin B$ and hence B is a proper subset of $\{v\} \cup B = B'$. By maximality of B, the set B' does not belong to S and hence is linearly dependent. Therefore there is a linear relation $a_1 u_1 + a_2 u_2 + \cdots + a_m u_m + cv = 0$ where $u_i \in B$ and $c, a_i \in F$, with not all the coefficients being zero. If $c = 0$, then $a_1 u_1 + a_2 u_2 + \cdots + a_m u_m = 0$, so that $a_1 = a_2 = \cdots = a_m = 0$ since $u_1, u_2, \ldots, u_m$ are linearly independent. Therefore $c \neq 0$ and we can solve the equation for v, obtaining

$$v = (-c^{-1} a_1) u_1 + (-c^{-1} a_2) v_2 + \cdots + (-c^{-1} a_m) u_m,$$

which contradicts the choice of v. Hence B generates V. □

Corollary (8.2.7). *Every non-zero vector space V has a basis.*

Proof. By hypothesis there is a non-zero vector v in V. Apply (8.2.6) with $B_0 = \{v\}$. □

Next it will be shown any two bases of a vector space have the same cardinal.

(8.2.8). *Let X and Y be two bases of a vector space V over a field F. Then $|X| = |Y|$.*

Proof. First of all assume that X is finite. Then, since Y is linearly independent, it is finite and $|Y| \leq |X|$ by (8.2.4). Similarly $|X| \leq |Y|$. Thus $|X| = |Y|$ and we can assume that X and Y are both infinite. In this case the proof is harder.

Let $\mathcal{P}_f(Y)$ denote the set of all *finite* subsets of Y. Define a function $\alpha : X \to \mathcal{P}_f(Y)$ as follows. If $x \in X$ and $x = k_1 y_1 + \cdots + k_n y_n$ with distinct $y_i \in Y$ and $0 \neq k_i \in F$, define $\alpha(x) = \{k_1, \ldots, k_n\}$. The function α is not injective, so we will modify it.

Let $T = \{y_1, y_2, \ldots, y_n\} \subseteq Y$, so that $T \in \mathcal{P}_f(Y)$. We claim that $\bar{T} = \{x \in X \mid \alpha(x) \in T\}$ is finite. For if not, there are infinitely many elements of X that are linear combinations of $y_1, y_2, \ldots, y_n$, contradicting (8.2.3). Let the elements of each $\bar{T}$ be linearly ordered in some way. Next define a new function $\beta : X \to \operatorname{Im}(\alpha) \times \mathbb{N}$ by the rule $\beta(x) = (T, m)$ where $\alpha(x) = T$, (so that $x \in \bar{T}$), and x is the mth element of $\bar{T}$. It is clear that β is injective. Therefore

$$|X| \leq |\operatorname{Im}(\alpha) \times \mathbb{N}| \leq |\mathcal{P}_f(Y)| \cdot \aleph_0 = |Y| \cdot \aleph_0 = |Y|.$$

In a similar way $|Y| \leq |X|$, so by the Cantor-Bernstein Theorem (1.4.2) we arrive at $|X| = |Y|$. □

In the foregoing proof we used two facts about cardinals: (i) $|\mathcal{P}_f(Y)| = |Y|$ – cf. Exercise (1.4.8); (ii) $|Y| \cdot \aleph_0 = |Y|$ if Y is infinite. For the latter statement see for example [8].

Dimension

A vector space usually has many bases, but by (8.2.8) all bases have the same cardinal. This fact enables us to define the *dimension* of a vector space V,

$$\dim(V).$$

If $V = 0$, define $\dim(V)$ to be 0, and if $V \neq 0$, let $\dim(V)$ be the cardinal of a basis of V. The definition is unambiguous by (8.2.8).

By (8.2.4) a finitely generated vector space cannot have an infinite linearly independent subset and thus must have finite dimension. In future we will refer to finite dimensional vector spaces instead of finitely generated ones.

The dimensions of vector spaces of row or column vectors can be computed from the ranks of matrices.

(8.2.9). *Let $X_1, X_2, \ldots, X_k$ be vectors in F^n where F is a field. Let $A = [X_1, X_2, \ldots, X_k]$ be the $n \times k$ matrix which has the X_i as columns. Then $\dim(\langle X_1, \ldots, X_k \rangle) = r$ where r is the rank of the matrix A.*

Proof. We will use some elementary facts about matrices here. In the first place, $S = \langle X_1, \ldots, X_k \rangle$ is the *column space* of the matrix A, and it is unaffected when column operations are applied to A. By applying column operations to A, just as we did for row operations during Gaussian elimination in the proof of (8.2.1), we can replace A by a matrix with the same column space S which has the so-called *column echelon form* with r non-zero columns. Here r is the rank of A. Since the r columns are linearly independent, they form a basis of S (if $r > 0$). Hence $\dim(S) = r$. □

Next we consider the relation between the dimension of a vector space and that of a subspace.

(8.2.10). *If V is a vector space with finite dimension n and U is a subspace of V, then $\dim(U) \leq \dim(V)$. Furthermore $\dim(U) = \dim(V)$ if and only if $U = V$.*

Proof. If $U = 0$, then $\dim(U) = 0 \leq \dim(V)$. Assume that $U \neq 0$ and let X be a basis of U. By (8.2.6) the subset X is contained in a basis Y of V. Hence $\dim(U) = |X| \leq |Y| = \dim(V)$. Finally, suppose that $\dim(U) = \dim(V)$, but $U \neq V$. Then $U \neq 0$. As before, a basis X of U is contained in a basis Y of V. Since $|X| = |Y|$, it follows that $X = Y$. Therefore $U = V$, a contradiction. □

The next result can simplify the task of showing that a subset of a finite dimensional vector space is a basis.

(8.2.11). *Let V be a finite dimensional vector space with dimension n and let X be a subset of V with n elements. Then the following statements about X are equivalent:*
(i) *X is a basis of V;*
(ii) *X is linearly independent;*
(iii) *X generates V.*

Proof. Of course (i) implies (ii). Assume that (ii) holds. Then X is a basis of $\langle X \rangle$, the subspace it generates; hence $\dim(\langle X \rangle) = n = \dim(V)$ and (8.2.10) shows that $\langle X \rangle = V$. Thus (ii) implies (iii).

Finally, assume that (iii) holds. If X is not a basis of V, it must be linearly dependent, so one of its elements can be written as a linear combination of the others. Hence V can be generated by fewer than n elements, which is a contradiction by (8.2.4). □

Change of basis

As previously remarked, vector spaces usually have many bases and a vector is represented with respect to each basis by a coordinate column vector. A natural question is: how are these coordinate vectors related?

Let $\mathcal{B} = \{v_1, v_2, \ldots, v_n\}$ and $\mathcal{B}' = \{v'_1, v'_2, \ldots, v'_n\}$ be two ordered bases of a finite dimensional vector space V over a field F. Then each v'_i can be expressed as a linear combination of $v_1, v_2, \ldots, v_n$, say

$$v'_i = \sum_{j=1}^{n} s_{ji} v_j,$$

where $s_{ji} \in F$. The change of basis $\mathcal{B}' \to \mathcal{B}$ is described by the *transition matrix* $S = [s_{ij}]$. Observe that S is $n \times n$ and its ith column is the coordinate vector $[v'_i]_{\mathcal{B}}$.

To understand how S determines the change of basis $\mathcal{B}' \to \mathcal{B}$, choose an arbitrary vector v from V and write $v = \sum_{i=1}^n c'_i v'_i$ where $c'_1, c'_2, \ldots, c'_n$ are the entries of the coordinate vector $[v]_{\mathcal{B}'}$. Replace v'_i by $\sum_{j=1}^n s_{ji} v_j$ to get

$$v = \sum_{i=1}^n c'_i \left(\sum_{j=1}^n s_{ji} v_j \right) = \sum_{j=1}^n \left(\sum_{i=1}^n s_{ji} c'_i \right) v_j.$$

Therefore the entries of the coordinate vector $[v]_{\mathcal{B}}$ are $\sum_{i=1}^n s_{ji} c'_i$ for $j = 1, 2, \ldots, n$. This shows that

$$[v]_{\mathcal{B}} = S[v]_{\mathcal{B}'},$$

i. e., left multiplication by the transition matrix S transforms coordinate vectors with respect to $\mathcal{B}'$ into vectors with respect to $\mathcal{B}$.

Notice that the transition matrix S must be non-singular. For otherwise by standard matrix theory there would exist a non-zero $X \in F^n$ such that $SX = 0$; however, if $u \in V$ is defined by $[u]_{\mathcal{B}'} = X$, then $[u]_{\mathcal{B}} = SX = 0$, which can only mean that $u = 0$ and $X = 0$. From $[v]_{\mathcal{B}} = S[v]_{\mathcal{B}'}$ we deduce that $S^{-1}[v]_{\mathcal{B}} = [v]_{\mathcal{B}'}$. Thus S^{-1} is the transition matrix for the change of basis $\mathcal{B} \to \mathcal{B}'$. These conclusions are summed up in the next result.

(8.2.12). *Let $\mathcal{B}$ and $\mathcal{B}'$ be ordered bases of an n-dimensional vector space V. Define S to be the $n \times n$ matrix whose ith column is the coordinate vector of the ith vector of $\mathcal{B}'$ with respect to $\mathcal{B}$. Then S is non-singular and for all v in V*

$$[v]_{\mathcal{B}} = S[v]_{\mathcal{B}'} \quad \text{and} \quad [v]_{\mathcal{B}'} = S^{-1}[v]_{\mathcal{B}}.$$

Example (8.2.1). Let V be the vector space of all real polynomials in t with degree at most 2. Then $\mathcal{B} = \{1, t, t^2\}$ is clearly a basis for V and so is $\mathcal{B}' = \{1 + t, 2t, 4t^2 - 2\}$, as it is quickly seen that this set is linearly independent. Write the coordinate vectors of $1 + t$, $2t$, $4t^2 - 2$ with respect to $\mathcal{B}$ as columns of the matrix

$$S = \begin{bmatrix} 1 & 0 & -2 \\ 1 & 2 & 0 \\ 0 & 0 & 4 \end{bmatrix}.$$

This is the transition matrix for the change of basis $\mathcal{B}' \to \mathcal{B}$. The transition matrix for $\mathcal{B} \to \mathcal{B}'$ is

$$S^{-1} = \begin{bmatrix} 1 & 0 & \frac{1}{2} \\ -\frac{1}{2} & \frac{1}{2} & -\frac{1}{4} \\ 0 & 0 & \frac{1}{4} \end{bmatrix}.$$

For example, to express $f = a + bt + ct^2$ in terms of the basis $\mathcal{B}'$, we compute

$$[f]_{\mathcal{B}'} = S^{-1}[f]_{\mathcal{B}} = \begin{bmatrix} 1 & 0 & \frac{1}{2} \\ -\frac{1}{2} & \frac{1}{2} & -\frac{1}{4} \\ 0 & 0 & \frac{1}{4} \end{bmatrix} \begin{bmatrix} a \\ b \\ c \end{bmatrix} = \begin{bmatrix} a + \frac{c}{2} \\ -\frac{1}{2}a + \frac{1}{2}b - \frac{1}{4}c \\ \frac{1}{4}c \end{bmatrix}.$$

Thus $f = (a + \frac{c}{2})(1 + t) + (-\frac{1}{2}a + \frac{1}{2}b - \frac{1}{4}c)(2t) + \frac{1}{4}c(4t^2 - 2)$, which is clearly correct.

Dimension of the sum and intersection of subspaces

Since a vector space V is an additively written abelian group, one can form the sum of two subspaces U and W; thus

$$U + W = \{u + w \mid u \in U, w \in W\}.$$

It is easily verified that $U + W$ is a subspace of V. Also $U \cap W$ is a subspace. There is a useful formula connecting the dimensions of $U + W$ and $U \cap W$.

(8.2.13). *If U and W are subspaces of a finite dimensional vector space V, then*

$$\dim(U + W) + \dim(U \cap W) = \dim(U) + \dim(W).$$

Proof. If $U = 0$, then $U + W = W$ and $U \cap W = 0$; in this case the formula is certainly true. Thus we can assume that $U \neq 0$ and $W \neq 0$.

Choose a basis for $U \cap W$, say $z_1, \ldots, z_r$, if $U \cap W \neq 0$; should $U \cap W$ be 0, just ignore the z_i. By (8.2.6) we can extend $\{z_1, \ldots, z_r\}$ to bases of U and of W, say

$$\{z_1, \ldots, z_r, u_{r+1}, \ldots, u_m\} \quad \text{and} \quad \{z_1, \ldots, z_r, w_{r+1}, \ldots, w_n\}.$$

Now the vectors $z_1, z_2, \ldots z_r, u_{r+1}, \ldots, u_m, w_{r+1}, \ldots, w_n$ surely generate $U + W$: for any vector in $U + W$ is expressible as a linear combination of them. In fact these elements are also linearly independent, so they form a basis of $U + W$. To establish this claim, assume there is a linear relation

$$\sum_{i=1}^{r} e_i z_i + \sum_{j=r+1}^{m} c_j u_j + \sum_{k=r+1}^{n} d_k w_k = 0$$

where e_i, c_j, d_k are scalars. Then

$$\sum_{k=r+1}^{n} d_k w_k = \sum_{i=1}^{r}(-e_i)z_i + \sum_{j=r+1}^{m}(-c_j)u_j,$$

which belongs to U and to W and so to $U \cap W$. Hence $\sum_{k=r+1}^{n} d_k w_k$ is a linear combination of the z_i. But $z_1, \ldots, z_r, w_{r+1}, \ldots, w_n$ are linearly independent, which implies that

$d_k = 0$ for all k. The linear relation now reduces to

$$\sum_{i=1}^{r} e_i z_i + \sum_{j=r+1}^{m} c_j u_j = 0.$$

But $z_1, \ldots, z_r, u_{r+1}, \ldots, u_m$ are linearly independent. Therefore all the c_j and e_i equal zero, which establishes the claim of linear independence.

Finally, $\dim(U + W)$ equals the number of the vectors $z_1, \ldots, z_r, u_{r+1}, \ldots, u_m$, $v_{r+1}, \ldots, v_n$: this is, $r + (m - r) + (n - r) = m + n - r$, which equals $\dim(U) + \dim(W) - \dim(U \cap W)$, so the required formula follows. $\square$

Direct sums of vector spaces

Since a vector space V is an additive abelian group, we can form the direct sum of subspaces $U_1, U_2, \ldots, U_k$ – see Section 4.2. This is an additive abelian group which is written

$$U = U_1 \oplus U_2 \cdots \oplus U_k.$$

Thus $U = \{u_1 + u_2 + \cdots + u_k \mid u_i \in U_i\}$ and $U_i \cap \sum_{j \neq i} U_j = 0$. Clearly U is a subspace of V. Note that by (8.2.13) and induction on k

$$\dim(U_1 \oplus U_2 \oplus \cdots \oplus U_k) = \dim(U_1) + \dim(U_2) + \cdots + \dim(U_k).$$

Next if $\{v_1, v_2, \ldots, v_n\}$ is a basis of V, then $V = \langle v_1 \rangle \oplus \langle v_2 \rangle \oplus \cdots \oplus \langle v_n \rangle$, so that we have established:

(8.2.14). *An n-dimensional vector space is the direct sum of n 1-dimensional subspaces.*

This result is also true when $n = 0$ if the direct sum is interpreted as 0.

Quotient spaces

Suppose that V is a vector space over a field F and U is a subspace of V. Since V is an abelian group and U is a subgroup, the quotient

$$V/U = \{v + U \mid v \in V\}$$

already exists as an abelian group. Now make V/U into a vector space over F by defining scalar multiplication in the natural way,

$$a(v + U) = av + U, \quad (a \in F).$$

This is evidently a well-defined operation. After an easy check of the axioms, we conclude that V/U is a vector space over F, the *quotient space* of U in V. The dimension of a quotient space is easily computed.

(8.2.15). *Let U be a subspace of a finite dimensional space V. Then* $\dim(V/U) = \dim(V) - \dim(U)$.

Proof. If $U = 0$, the statement is obviously true. Assuming $U \neq 0$, we choose a basis $\{v_1, v_2, \ldots, v_m\}$ of U and extend it to a basis of V, say $\{v_1, v_2, \ldots, v_m, v_{m+1}, \ldots, v_n\}$. We will argue that $\{v_{m+1} + U, \ldots, v_n + U\}$ is a basis of V/U.

Assume that $\sum_{i=m+1}^{n} a_i(v_i + U) = 0_{V/U} = U$ where $a_i \in F$. Then $\sum_{i=m+1}^{n} a_i v_i \in U$, so this element is a linear combination of $v_1, \ldots, v_m$. It follows by linear independence that each $a_i = 0$, which shows that $\{v_{m+1} + U, \ldots, v_n + U\}$ is linearly independent. Next, if $v \in V$, write $v = \sum_{i=1}^{n} a_i v_i$, with scalars a_i, and observe that $v + U = \sum_{i=m+1}^{n} a_i(v_i + U)$ since $v_1, \ldots, v_m \in U$. It follows that $v_{m+1} + U, \ldots, v_n + U$ form a basis of V/U and $\dim(V/U) = n - m = \dim(V) - \dim(U)$, as required. □

To conclude this section let us show that the mere existence of a basis in a finite dimensional vector space is enough to prove two important results about abelian groups and finite fields.

Let p be a prime. An additively written abelian group A is called an *elementary abelian p-group* if $pa = 0$ for all a in A, i.e., each element of A has order 1 or p. For example, the Klein 4-group is an elementary abelian 2-group. The structure of finite elementary abelian p-groups is given by the next result.

(8.2.16). *Let A be a finite abelian group. Then A is an elementary abelian p-group if and only if A is a direct sum of copies of $\mathbb{Z}_p$.*

Proof. The idea behind the proof is to view A as a vector space over the field $\mathbb{Z}_p$. Here the scalar multiplication is the natural one, namely $(i + p\mathbb{Z})a = ia$ where $i \in \mathbb{Z}, a \in A$. One has to verify that this operation is well-defined, which is true since $(i + pm)a = ia + mpa = ia$ for all $a \in A$. Since A is finite, it is a finite dimensional vector space over $\mathbb{Z}_p$. By (8.2.14) $A = A_1 \oplus A_2 \oplus \cdots \oplus A_n$ where each A_i is a 1-dimensional subspace; thus $|A_i| = p$ and $A_i \simeq \mathbb{Z}_p$. Conversely, any direct sum of copies of $\mathbb{Z}_p$ certainly satisfies $pa = 0$ for every element a and so is an elementary abelian p-group. □

The second application is to prove that the number of elements in a finite field is always a prime power. This is in marked contrast to the behavior of groups and rings, examples of which exist with any finite order.

(8.2.17). *Let F be a finite field. Then $|F|$ is a power of a prime.*

Proof. By (6.3.9) the field F has characteristic a prime p and $pa = 0$ for all $a \in F$. Thus, as an additive group, F is elementary abelian p. It now follows from (8.2.16) that $|F|$ is a power of p. □

Exercises (8.2).

(1) Show that

$$
X_1 = \begin{bmatrix} 4 \\ 2 \\ 1 \end{bmatrix}, \quad X_2 = \begin{bmatrix} -5 \\ 2 \\ -3 \end{bmatrix}, \quad X_3 = \begin{bmatrix} 1 \\ 3 \\ 0 \end{bmatrix}
$$

form a basis of $\mathbb{R}^3$, and express the elementary vectors E_1, E_2, E_3 in terms of X_1, X_2, X_3.

(2) Find a basis for the null space of the matrix

$$
\begin{bmatrix} 2 & 3 & 1 & 1 \\ -3 & 1 & 4 & -7 \\ 1 & 2 & 1 & 0 \end{bmatrix}.
$$

(3) Find the dimension of the vector space $M_{m,n}(F)$ where F is an arbitrary field.

(4) Let $v_1, v_2, \ldots, v_n$ be vectors in a vector space V. Assume that each element of V is uniquely expressible as a linear combination of $v_1, v_2, \ldots, v_n$. Prove that the v_i's form a basis of V.

(5) Let $\mathcal{B} = \{E_1, E_2, E_3\}$ be the standard ordered basis of $\mathbb{R}^3$ and let

$$
\mathcal{B}' = \left\{ \begin{bmatrix} 2 \\ 0 \\ 0 \end{bmatrix}, \begin{bmatrix} -1 \\ 2 \\ 0 \end{bmatrix}, \begin{bmatrix} 1 \\ 1 \\ 1 \end{bmatrix} \right\}.
$$

Show that $\mathcal{B}'$ is a basis of $\mathbb{R}^3$ and find the transition matrices for the changes of bases $\mathcal{B}' \to \mathcal{B}$ and $\mathcal{B} \to \mathcal{B}'$.

(6) Let V be a vector space of dimension n and let i be an integer such that $0 \le i \le n$. Prove that V has at least one subspace of dimension i.

(7) The same as Exercise (8.2.6) with "subspace" replaced by "quotient space".

(8) Let U be a subspace of a finite dimensional vector space V. Prove that there is a subspace W such that $V = U \oplus W$.

(9) Let V be a vector space of dimension $2n$ and assume that U and W are subspaces of dimensions n and $n+1$ respectively. Prove that $U \cap W \ne 0$.

(10) Let the vectors $v_1, v_2, \ldots, v_m$ generate a vector space V. Prove that some subset of $\{v_1, v_2, \ldots, v_m\}$ is a basis of V.

8.3 Linear mappings

Just as there are homomorphisms of groups and rings, there are homomorphisms of vector spaces. Traditionally these are called *linear mappings* or *linear transformations*.

Let V and W be vector spaces over the same field F. Then a function

$$\alpha : V \to W$$

is called a *linear mapping from V to W* if the following rules are valid for all $v_1, v_2 \in V$ and $a \in F$:

(i) $\alpha(v_1 + v_2) = \alpha(v_1) + \alpha(v_2)$;
(ii) $\alpha(av_1) = a\alpha(v_1)$.

If α is also bijective, it is called an *isomorphism of vector spaces*. Should there exist an isomorphism between vector spaces V and W over a field F, then V and W are said to be *isomorphic* and we write

$$V \overset{F}{\simeq} W \quad \text{or} \quad V \simeq W.$$

Notice that a linear mapping is automatically a homomorphism of additive groups by (i) above, so all results established for group homomorphisms can be carried over to linear mappings. A linear mapping $\alpha : V \to V$ is called a *linear operator* on V.

Example (8.3.1). Let A be an $m \times n$ matrix over a field F and define a function $\alpha : F^n \to F^m$ by the rule $\alpha(X) = AX$ where $X \in F^n$. Simple properties of matrices show that α is a linear mapping.

Example (8.3.2). Let V be an n-dimensional vector space over a field F and let $\mathcal{B} = \{v_1, v_2, \ldots, v_n\}$ be an ordered basis of V. Recall that to each vector v in V there corresponds a unique coordinate vector $[v]_{\mathcal{B}}$ with respect to $\mathcal{B}$.

Use this correspondence to define a function $\alpha : V \to F^n$ by $\alpha(v) = [v]_{\mathcal{B}}$. By simple calculations we see that $[u + v]_{\mathcal{B}} = [u]_{\mathcal{B}} + [v]_{\mathcal{B}}$ and $[av]_{\mathcal{B}} = a[v_{\mathcal{B}}]$ where $u, v \in V, a \in F$. Hence α is a linear mapping. Clearly $[v]_{\mathcal{B}} = 0$ implies that $v = 0$; thus α is injective and it is obviously surjective. The conclusion is that α is an isomorphism and $V \overset{F}{\simeq} F^n$.

We state this conclusion as:

(8.3.1). *If V is a vector space with dimension n over a field F, then $V \overset{F}{\simeq} F^n$. Thus two finite dimensional vector spaces over F are isomorphic if and only if they have the same dimension.*

Here the converse statement follows from the evident fact that isomorphic vector spaces have the same dimension.

An important way of defining a linear mapping is by specifying its effect on a basis.

(8.3.2). *Let $\{v_1, \ldots, v_n\}$ be a basis of a vector space V over a field F and let $w_1, \ldots, w_n$ be any n vectors in another F-vector space W. Then there is a unique linear mapping $\alpha : V \to W$ such that $\alpha(v_i) = w_i$ for $i = 1, 2, \ldots, n$.*

Proof. Let $v \in V$ and write $v = \sum_{i=1}^{n} a_i v_i$ where $a_i \in F$. Define a function $\alpha : V \to W$ by the rule

$$\alpha(v) = \sum_{i=1}^{n} a_i w_i.$$

Then an easy check shows that α is a linear mapping, and of course $\alpha(v_i) = w_i$. If $\alpha' : V \to W$ is another such linear mapping, then $\alpha'(v) = \sum_{i=1}^{n} a_i \alpha'(v_i) = \sum_{i=1}^{n} a_i w_i = \alpha(v)$. Hence $\alpha = \alpha'$. □

Our experience with groups and rings suggests that it may be worthwhile to examine the kernel and image of a linear mapping.

(8.3.3). *Let $\alpha : V \to W$ be a linear mapping. Then $\mathrm{Ker}(\alpha)$ and $\mathrm{Im}(\alpha)$ are subspaces of V and W respectively.*

Proof. Since α is a group homomorphism, it follows from (4.3.2) that $\mathrm{Ker}(\alpha)$ and $\mathrm{Im}(\alpha)$ are additive subgroups. We leave the reader to complete the proof by showing that these subgroups are also closed under scalar multiplication. □

Just as for groups and rings, there are isomorphism theorems for vector spaces.

(8.3.4) (First Isomorphism Theorem). *If $\alpha : V \to W$ is a linear mapping between vector spaces over a field F, then $V/\mathrm{Ker}(\alpha) \overset{F}{\simeq} \mathrm{Im}(\alpha)$.*

(8.3.5) (Second Isomorphism Theorem). *Let U and W be subspaces of a vector space over a field F. Then $(U + W)/W \overset{F}{\simeq} U/(U \cap W)$.*

(8.3.6) (Third Isomorphism Theorem). *Let U and W be subspaces of a vector space over a field F such that $U \subseteq W$. Then $(V/U)/(W/U) \overset{F}{\simeq} V/W$.*

Since the isomorphism theorems for groups are applicable, all one has to prove here is that the functions introduced in the proofs of (4.3.4), (4.3.5) and (4.3.6) are linear mappings, i. e., they act appropriately on scalar multiples.

For example, in (8.3.4) the function in question is $\theta : V/\mathrm{Ker}(\alpha) \to \mathrm{Im}(\alpha)$ where $\theta(v + \mathrm{Ker}(\alpha)) = \alpha(v)$. Then

$$\theta(a(v + \mathrm{Ker}(\alpha))) = \theta(av + \mathrm{Ker}(\alpha)) = \alpha(av) = a\alpha(v) = a\theta(v + \mathrm{Ker}(\alpha)).$$

It follows that θ is a linear mapping.

There is an important formula connecting the dimensions of kernel and image.

(8.3.7). *If $\alpha : V \to W$ is a linear mapping between finite dimensional vector spaces, then $\dim(\mathrm{Ker}(\alpha)) + \dim(\mathrm{Im}(\alpha)) = \dim(V)$.*

This follows directly from (8.3.4) and (8.2.15). There is an immediate application to the null space of a matrix.

Corollary (8.3.8). *Let A be an m×n matrix with rank r over a field F. Then the dimension of the null space of A is n − r.*

Proof. Let α be the linear mapping from F^n to F^m defined by $\alpha(X) = AX$. Then Ker(α) is the null space of A and it is readily seen that Im(α) is just the column space. By (8.2.9) dim(Im(α)) = r, the rank of A, and by (8.3.7) dim(Ker(α)) = $n - r$. □

As another application of (8.3.7) we give a different proof of the dimension formula for sum and intersection of subspaces – see (8.2.13).

(8.3.9). *If U and W are subspaces of a finite dimensional vector space, then*

$$\dim(U + W) + \dim(U \cap W) = \dim(U) + \dim(W).$$

Proof. By (8.3.5) $(U + W)/W \simeq U/(U \cap W)$. Hence, taking dimensions and applying (8.2.15), we find that $\dim(U + W) - \dim(W) = \dim(U) - \dim(U \cap W)$, and the result follows. □

Vector spaces of linear mappings

It is useful to endow sets of linear mappings with the structure of a vector space. Suppose that V and W are vector spaces over the same field F. We will write

$$L(V, W)$$

for the set of all linear mappings from V to W. Define addition and scalar multiplication in $L(V, W)$ by the natural rules

$$\alpha + \beta(v) = \alpha(v) + \beta(v), \quad (a \cdot \alpha)(v) = a(\alpha(v)),$$

where $\alpha, \beta \in L(V, W)$, $v \in V$, $a \in F$. It is simple to verify that $\alpha + \beta$ and $a \cdot \alpha$ are linear mappings. The basic result about $L(V, W)$ is:

(8.3.10). *Let V and W be vector spaces over a field F. Then:*
(i) *$L(V, W)$ is a vector space over F;*
(ii) *if V and W are finite dimensional, then so is $L(V, W)$ and in addition*

$$\dim(L(V, W)) = \dim(V) \cdot \dim(W).$$

Proof. We omit the routine proof of (i) and concentrate on (ii). Let $\{v_1, \ldots, v_m\}$ and $\{w_1, \ldots, w_n\}$ be bases of V and W respectively. By (8.3.2), for $i = 1, 2, \ldots, m$ and $j = 1, 2, \ldots, n$, there is a unique linear mapping $\alpha_{ij} : V \to W$ such that

$$\alpha_{ij}(v_k) = \begin{cases} w_j & \text{if } k = i \\ 0 & \text{if } k \neq i. \end{cases}$$

Thus α_{ij} sends basis element v_i to basis element w_j and all other v_k's to 0. First we show that the α_{ij} are linearly independent in the vector space $L(V, W)$.

Let $a_{ij} \in F$; then by definition of α_{ij} we have for each k

$$\left(\sum_{i=1}^{m} \sum_{j=1}^{n} a_{ij}\alpha_{ij} \right)(v_k) = \sum_{j=1}^{n} \sum_{i=1}^{m} a_{ij}(\alpha_{ij}(v_k)) = \sum_{j=1}^{n} a_{kj}w_j. \tag{8.1}$$

Therefore $\sum_{i=1}^{m} \sum_{j=1}^{n} a_{ij}\alpha_{ij} = 0$ if and only if $a_{kj} = 0$ for all j, k. It follows that the α_{ij} are linearly independent.

Finally, we claim that the α_{ij} actually generate $L(V, W)$. To prove this let $\alpha \in L(V, W)$ and write $\alpha(v_k) = \sum_{j=1}^{n} a_{kj}w_j$ where $a_{kj} \in F$. Then from equation (8.1) above we see that $\alpha = \sum_{i=1}^{m} \sum_{j=1}^{n} a_{ij}\alpha_{ij}$. Therefore the α_{ij}'s form a basis of $L(V, W)$ and $\dim(L(V, W)) = mn = \dim(V) \cdot \dim(W)$. □

The dual space

If V is a vector space over a field F, the vector space

$$V^* = L(V, F)$$

is called the *dual space* of V; here F is regarded as a 1-dimensional vector space over F. The elements of V^* are linear mappings from V to F which are called *linear functionals* on V.

Example (8.3.3). Let $Y \in F^n$ be fixed and define $\alpha : F^n \to F$ by the rule $\alpha(X) = Y^T X$ where Y^T is the transpose of Y. Then α is a linear functional on F^n.

If V is an n-dimensional vector space over F,

$$\dim(V^*) = \dim(L(V, F)) = \dim(V)$$

by (8.3.10). Thus V, V^* and the *double dual* $V^{**} = (V^*)^*$ all have the same dimension, so these vector spaces are isomorphic by (8.3.1).

In fact there is a *canonical* linear mapping $\theta : V \to V^{**}$. Let $v \in V$ and define $\theta(v) \in V^{**}$ by the rule

$$\theta(v)(\alpha) = \alpha(v)$$

where $\alpha \in V^*$. Thus $\theta(v)$ evaluates each linear functional on V at v. Regarding the function θ, we prove:

(8.3.11). *If V is a finite dimensional vector space, then $\theta : V \to V^{**}$ is an isomorphism.*

Proof. In the first place $\theta(v) \in V^{**}$ for all $v \in V$: indeed, if $\alpha, \beta \in V^*$,

$$\theta(v)(\alpha + \beta) = (\alpha + \beta)(v) = \alpha(v) + \beta(v) = \theta(v)(\alpha) + \theta(v)(\beta).$$

Also $\theta(v)(a \cdot \alpha) = (a \cdot \alpha)(v) = a(\alpha(v)) = a(\theta(v)(\alpha))$ where a is a scalar.

Next for any $\alpha \in V^*$ and $v_i \in V$, we have

$$
\begin{aligned}
\theta(v_1 + v_2)(\alpha) &= \alpha(v_1 + v_2) \\
&= \alpha(v_1) + \alpha(v_2) \\
&= \theta(v_1)(\alpha) + \theta(v_2)(\alpha) \\
&= (\theta(v_1) + \theta(v_2))(\alpha),
\end{aligned}
$$

which shows that $\theta(v_1 + v_2) = \theta(v_1) + \theta(v_2)$. We leave the reader to verify that $\theta(a \cdot v) = a(\theta(v))$ where a is a scalar and $v \in V$. Hence θ is a linear mapping from V to V^{**}.

Next suppose that $\theta(v) = 0$. Then $0 = \theta(v)(\alpha) = \alpha(v)$ for *all* $\alpha \in V^*$. This can only mean that $v = 0$: for if $v \neq 0$, then v can be included in a basis of V. Then by (8.3.2) we can construct a linear functional α such that $\alpha(v) = 1_F$ and other basis elements are mapped by α to 0. It follows that θ is injective.

Finally, $\dim(V) = \dim(V^*) = \dim(V^{**})$ and also $\dim(V) = \dim(\operatorname{Im}(\theta))$ since θ is injective. By (8.2.10) we have $\operatorname{Im}(\theta) = V^{**}$, so that θ^* is an isomorphism. $\qquad\square$

Representing linear mappings by matrices

A linear mapping between finite dimensional vector spaces can be described by matrix multiplication, which provides us with a concrete way of representing linear mappings.

Let V and W be vector spaces over a field F with respective finite dimensions $m > 0$ and $n > 0$. Choose ordered bases for V and W, say $\mathcal{B} = \{v_1, v_2, \ldots, v_m\}$ and $\mathcal{C} = \{w_1, w_2, \ldots, w_n\}$ respectively. Now let $\alpha \in L(V, W)$; then

$$\alpha(v_i) = \sum_{j=1}^{n} a_{ji} w_j, \quad i = 1, 2, \ldots, m,$$

where $a_{ji} \in F$. This enables us to form the $n \times m$ matrix over F

$$A = [a_{ji}],$$

which is to represent α. Notice that the ith column of A is precisely the coordinate column vector of $\alpha(v_i)$ with respect to the basis $\mathcal{C}$. Thus we have a function

$$\theta : L(V, W) \to M_{n,m}(F)$$

defined by the rule that column i of $\theta(\alpha)$ is $[\alpha(v_i)]_{\mathcal{C}}$.

To understand how the matrix $A = \theta(\alpha)$ reproduces the effect of α on an arbitrary vector $v = \sum_{i=1}^{m} b_i v_i$ of V, we compute

$$\alpha(v) = \sum_{i=1}^{m} b_i(\alpha(v_i)) = \sum_{i=1}^{m} b_i\left(\sum_{j=1}^{n} a_{ji} w_j\right) = \sum_{j=1}^{n}\left(\sum_{i=1}^{m} a_{ji} b_i\right) w_j.$$

Hence the coordinate column vector of $\alpha(v)$ with respect to C has entries $\sum_{i=1}^{m} a_{ji} b_i$, for $j = 1, \ldots, n$ and we have

$$A\begin{bmatrix} b_1 \\ \vdots \\ b_m \end{bmatrix} = A[v]_{\mathcal{B}}.$$

Thus we arrive at the basic formula

$$[\alpha(v)]_C = A[v]_{\mathcal{B}} = \theta(\alpha)[v]_{\mathcal{B}}.$$

Concerning the function θ we prove:

(8.3.12). *If V and W are finite dimensional vector spaces over a field F, the function $\theta : L(V, W) \to M_{n,m}(F)$ is an isomorphism of vector spaces.*

Proof. In the first place θ is a linear mapping. For, let $\alpha, \beta \in L(V, W)$ and $v \in V$; then the formula above shows that

$$\theta(\alpha + \beta)[v]_{\mathcal{B}} = [(\alpha + \beta)(v)]_C = [\alpha(v) + \beta(v)]_C = [\alpha(v)]_C + [\beta(v)]_C,$$

which equals

$$\theta(\alpha)[v]_{\mathcal{B}} + \theta(\beta)[v]_{\mathcal{B}} = (\theta(\alpha) + \theta(\beta))[v]_{\mathcal{B}}.$$

Hence $\theta(\alpha + \beta) = \theta(\alpha) + \theta(\beta)$, and in a similar fashion it may be shown that $\theta(a \cdot \alpha) = a(\theta(\alpha))$ where $a \in F$.

Next if $\theta(\alpha) = 0$, then $[\alpha(v)]_C = 0$ for all $v \in V$, so $\alpha(v) = 0$ and $\alpha = 0$. Hence θ is injective. If V and W have respective dimensions m and n, then $L(V, W) \simeq \text{Im}(\theta) \subseteq M_{n,m}(F)$. But the vector spaces $L(V, W)$ and $M_{n,m}(F)$ both have dimension mn – see (8.3.10). Therefore $\text{Im}(\theta) = M_{n,m}(F)$ by (8.2.10) and θ is an isomorphism. $\qquad\square$

Example (8.3.4). Consider the dual space $V^* = L(V, F)$, where V is an n-dimensional vector space over a field F. Choose an ordered basis $\mathcal{B}$ of V and use the basis $\{1_F\}$ for F. Then a linear functional $\alpha \in V^*$ is represented by an n-row vector, i. e., by X^T where $X \in F^n$, according to the rule $\alpha(v) = X^T[v]_{\mathcal{B}}$. Thus the effect of a linear functional is produced by left multiplication of coordinate vectors by a row vector, (cf. Example (8.3.3)).

The effect of a change of basis

We have seen that any linear mapping between finite dimensional vector spaces can be represented by multiplication by a matrix. However, the matrix depends on the choice of ordered bases of the vector spaces. The precise nature of this dependence will now be investigated.

Let $\mathcal{B}$ and $\mathcal{C}$ be ordered bases of respective finite dimensional vector spaces V and W over a field F, and let $\alpha : V \rightarrow W$ be a linear mapping. Then α is represented by a matrix A over F where $[\alpha(v)]_{\mathcal{C}} = A[v]_{\mathcal{B}}$. Now suppose now that two different ordered bases $\mathcal{B}'$ and $\mathcal{C}'$ are chosen for V and W respectively. Then α will be represented by another matrix A'. The question is: how are A and A' related?

To answer the question we introduce the transition matrices S and T for the respective changes of bases $\mathcal{B} \rightarrow \mathcal{B}'$ and $\mathcal{C} \rightarrow \mathcal{C}'$ (see (8.2.12)). Thus for any $v \in V$ and $w \in W$ we have

$$[v]_{\mathcal{B}'} = S[v]_{\mathcal{B}} \quad \text{and} \quad [w]_{\mathcal{C}'} = T[w]_{\mathcal{C}}.$$

Therefore

$$[\alpha(v)]_{\mathcal{C}'} = T[\alpha(v)]_{\mathcal{C}} = TA[v]_{\mathcal{B}} = TAS^{-1}[v]_{\mathcal{B}'},$$

and it follows that $A' = TAS^{-1}$. We record this conclusion in:

(8.3.13). *Let V and W be non-zero finite dimensional vector spaces over the same field. Let $\mathcal{B}$, $\mathcal{B}'$ be ordered bases of V and $\mathcal{C}$, $\mathcal{C}'$ ordered bases of W. Suppose further that S and T are the transition matrices for the changes of bases $\mathcal{B} \rightarrow \mathcal{B}'$ and $\mathcal{C} \rightarrow \mathcal{C}'$ respectively. If the linear mapping $\alpha : V \rightarrow W$ is represented by matrices A and A' with respect to the respective pairs of bases $(\mathcal{B}, \mathcal{C})$ and $(\mathcal{B}', \mathcal{C}')$, then $A' = TAS^{-1}$.*

The case where α is a linear operator on V is especially important. Here $V = W$ and we can take $\mathcal{B} = \mathcal{C}$ and $\mathcal{B}' = \mathcal{C}'$. Thus $S = T$ and $A' = SAS^{-1}$, i. e., A and A' are similar matrices. Consequently, *the matrices which represent a given linear operator are all similar.*

The algebra of linear operators

Let V be a vector space over a field F and suppose also that V is a ring with respect to some product operation. Then V is said to be an *F-algebra* if, in addition to the vector space and ring axioms, the following law is valid:

$$a(uv) = (au)v = u(av)$$

for all $a \in F$, $u, v \in V$. For example, the set of all $n \times n$ matrices $M_n(F)$ is an F-algebra with respect to the usual matrix operations.

Now let V be any vector space over a field F; we will write

$$L(V)$$

for the vector space $L(V, V)$ of all linear operators on V. Our aim is to make $L(V)$ into an F-algebra: it is already an F-vector space. There is a natural product operation on $L(V)$, namely functional composition. Indeed, if $\alpha_1, \alpha_2 \in L(V)$, then $\alpha_1\alpha_2 \in L(V)$ by an easy check. We claim that with this product operation $L(V)$ becomes an F-algebra.

The first step is to verify that $L(V)$ is a ring. This is fairly routine; for example, if $\alpha_i \in L(V)$ and $v \in V$,

$$\alpha_1(\alpha_2 + \alpha_3)(v) = \alpha_1(\alpha_2(v) + \alpha_3(v)) = \alpha_1\alpha_2(v) + \alpha_1\alpha_3(v),$$

which equals $(\alpha_1\alpha_2 + \alpha_1\alpha_3)(v)$. Hence $\alpha_1(\alpha_2 + \alpha_3) = \alpha_1\alpha_2 + \alpha_1\alpha_3$.

Once the ring axioms have been verified, we have to check that $a(\alpha_1\alpha_2) = (a\alpha_1)\alpha_2 = \alpha_1(a\alpha_2)$ for $a \in F$. This is not hard to see; indeed all three mappings send v to $a(\alpha_1(\alpha_2(v)))$. Therefore $L(V)$ is an F-algebra.

The group of units of $L(V)$ consists of the invertible linear operators on V; this is written

$$GL(V).$$

Here "GL" stands for general linear group.

A function $\alpha : A_1 \rightarrow A_2$ between two F-algebras is called an *algebra isomorphism* if it is bijective and it is both a linear mapping of vector spaces and a homomorphism of rings.

(8.3.14). *Let V be a vector space with finite dimension n over a field F. Then $L(V)$ and $M_n(F)$ are isomorphic as F-algebras.*

Proof. Choose an ordered basis $\mathcal{B}$ of V and let $\Phi : L(V) \rightarrow M_n(F)$ be the function which associates to a linear operator α the $n \times n$ matrix that represents α with respect to $\mathcal{B}$. Thus $[\alpha(v)]_{\mathcal{B}} = \Phi(\alpha)[v]_{\mathcal{B}}$ for all $v \in V$. Clearly Φ is bijective, so to prove that it is an F-algebra isomorphism we need to establish that $\Phi(\alpha + \beta) = \Phi(\alpha) + \Phi(\beta)$, $\Phi(a \cdot \alpha) = a \cdot \Phi(\alpha)$ and $\Phi(\alpha\beta) = \Phi(\alpha)\Phi(\beta)$.

For example, take the third statement. If $v \in V$, then

$$[\alpha\beta(v)]_{\mathcal{B}} = \Phi(\alpha)[\beta(v)]_{\mathcal{B}} = \Phi(\alpha)(\Phi(\beta)[v]_{\mathcal{B}}) = (\Phi(\alpha)\Phi(\beta))[v]_{\mathcal{B}}.$$

Therefore $\Phi(\alpha\beta) = \Phi(\alpha)\Phi(\beta)$. The other statements are dealt with in a similar fashion. $\square$

Thus (8.3.14) tells us in a precise way that linear operators on an n-dimensional vector space over F behave in very much the same manner as $n \times n$ matrices over F.

Corollary (8.3.15). *If V is an n-dimensional vector space over a field F, then $\mathrm{GL}(V) \simeq \mathrm{GL}_n(F)$.*

Exercises (8.3).

(1) Which of the following functions are linear mappings?

    (i) $\alpha : \mathbb{R}_3 \to \mathbb{R}$ where $\alpha([x_1 \; x_2 \; x_3]) = \sqrt{x_1^2 + x_2^2 + x_3^2}$;

    (ii) $\alpha : M_{m,n}(F) \to M_{n,m}(F)$ where $\alpha(A) = A^T$, the transpose of A;

    (iii) $\alpha : M_n(F) \to F$ where $\alpha(A) = \det(A)$.

(2) A linear mapping $\alpha : \mathbb{R}^4 \to \mathbb{R}^3$ sends $[x_1 \; x_2 \; x_3 \; x_4]^T$ to $[x_1 - x_2 + x_3 - x_4 \; 2x_1 + x_2 - x_3 \; x_2 - x_3 + x_4]^T$. Find the matrix which represents α when the standard bases of $\mathbb{R}^4$ and $\mathbb{R}^3$ are used.

(3) Answer Exercise (8.3.2) when the ordered basis $\{[1 \; 1 \; 1]^T, [0 \; 1 \; 1]^T, [0 \; 0 \; 1]^T\}$ of $\mathbb{R}^3$ is used, together with the standard basis of $\mathbb{R}^4$.

(4) Find bases for the kernel and image of the following linear mappings:

    (i) $\alpha : F^4 \to F$ where α maps a column vector to the sum of its entries;

    (ii) $\alpha : \mathbb{R}[t] \to \mathbb{R}[t]$ where $\alpha(f) = f'$, the derivative of f;

    (iii) $\alpha : \mathbb{R}^2 \to \mathbb{R}^2$ where $\alpha([x \; y]^T) = [2x + 3y \; 4x + 6y]^T$.

(5) Prove that a linear mapping $\alpha : V \to W$ is injective if and only if α maps linearly independent subsets of V to linearly independent subsets of W.

(6) Prove that a linear mapping $\alpha : V \to W$ is surjective if and only if α maps generating subsets of V to generating subsets of W.

(7) Let U and W be subspaces of a finite dimensional vector space V. Prove that there is a linear operator α on V such that $\mathrm{Ker}(\alpha) = U$ and $\mathrm{Im}(\alpha) = W$ if and only if $\dim(U) + \dim(W) = \dim(V)$.

(8) Suppose that $\alpha : V \to W$ is a linear mapping. Explain how to define a corresponding "induced" linear mapping $\alpha^* : W^* \to V^*$ of dual spaces. Then prove that $(\alpha\beta)^* = \beta^*\alpha^*$.

(9) Let $U \xrightarrow{\alpha} V \xrightarrow{\beta} W \to 0$ be an *exact sequence* of vector spaces and linear mappings. (This means that $\mathrm{Im}(\alpha) = \mathrm{Ker}(\beta)$ and $\mathrm{Im}(\beta) = \mathrm{Ker}(W \to 0) = W$, i.e., β is surjective). Prove that the corresponding sequence of dual spaces and induced linear mappings $0 \to W^* \xrightarrow{\beta^*} V^* \xrightarrow{\alpha^*} U^*$ is exact, i.e., β^* is injective and $\mathrm{Im}(\beta^*) = \mathrm{Ker}(\alpha^*)$. (For more general results of this kind see (9.1.25)).

8.4 Eigenvalues and eigenvectors

Let α be a linear operator on a vector space V over a field F. An *eigenvector* of α is a non-zero vector v of V such that $\alpha(v) = cv$ for some $c \in F$ called an *eigenvalue*. For example, if α is a rotation in $\mathbb{R}^3$, the eigenvectors of α are the non-zero vectors parallel to the axis of rotation and all the eigenvalues are equal to 1. A large amount of information about a linear operator is carried by its eigenvectors and eigenvalues. In addition the theory of eigenvectors and eigenvalues has many applications, for example to systems of linear recurrence relations and linear differential equations.

Let A be an $n \times n$ matrix over a field F. Define α to be the linear operator on F^n which sends X to AX. Then an eigenvector of α is a non-zero vector $X \in F^n$ such that $AX = cX$ for some $c \in F$. We will also call X an eigenvector and c an eigenvalue of the matrix A.

Conversely, suppose we start with a linear operator α on a finite dimensional vector space V over a field F. Choose an ordered basis $\mathcal{B}$ for V, so that α is represented by an $n \times n$ matrix A with respect to $\mathcal{B}$ and $[\alpha(v)]_\mathcal{B} = A[v]_\mathcal{B}$. Let v be an eigenvector for α with corresponding eigenvalue $c \in F$. Then $\alpha(v) = cv$, which translates into $A[v]_\mathcal{B} = c[v]_\mathcal{B}$. Thus $[v]_\mathcal{B}$ is an eigenvector and c is an eigenvalue of A.

These considerations show that the theory of eigenvalues and eigenvectors can be developed for either matrices or linear operators on a finite dimensional vector space. We will follow both approaches here, as is convenient.

Example (8.4.1). Let D denote the vector space of infinitely differentiable real valued functions on the interval $[a, b]$. Consider the linear operator α on D defined by $\alpha(f) = f'$, the derivative of the function f. The condition for $f \neq 0$ to be an eigenvector of α is that $f' = cf$ for some constant c. The general solution of this simple differential equation is $f = de^{cx}$ where d is a constant. Thus the eigenvectors of α are the exponential functions de^{cx} with $d \neq 0$, while the eigenvalues are arbitrary real numbers c.

Example (8.4.2). A linear operator α on the vector space $\mathbb{C}^2$ is defined by $\alpha(X) = AX$ where

$$A = \begin{bmatrix} 2 & -1 \\ 2 & 4 \end{bmatrix}.$$

Thus α is represented with respect to the standard basis by the matrix A. The condition for a vector $X = \begin{bmatrix} x_1 \\ x_2 \end{bmatrix}$ to be an eigenvector of A (or α) is that $AX = cX$ for some scalar c. This is equivalent to $(cI_2 - A)X = 0$, which asserts that X is a solution of the linear system

$$\begin{bmatrix} c - 2 & 1 \\ -2 & c - 4 \end{bmatrix} \begin{bmatrix} x_1 \\ x_2 \end{bmatrix} = \begin{bmatrix} 0 \\ 0 \end{bmatrix}.$$

By (8.2.2) this linear system has a non-trivial solution $[x_1, x_2]^T$ if and only if the determinant of the coefficient matrix vanishes, i. e.,

$$\begin{vmatrix} c - 2 & 1 \\ -2 & c - 4 \end{vmatrix} = 0.$$

On expansion this becomes $c^2 - 6c + 10 = 0$. The roots of this quadratic equation are $c_1 = 3 + i$ and $c_2 = 3 - i$ where $i = \sqrt{-1}$, so these are the eigenvalues of A.

The eigenvectors for each eigenvalue are found by solving the linear systems $(c_1 I_2 - A)X = 0$ and $(c_2 I_2 - A)X = 0$. For example, in the case of c_1 we have to solve

$$\begin{cases} (1 + i)x_1 + x_2 = 0 \\ -2x_1 + (-1 + i)x_2 = 0 \end{cases}$$

The general solution of this system is $x_1 = \frac{d}{2}(-1 + i)$, $x_2 = d$ where d is an arbitrary scalar. Thus the eigenvectors of A associated with the eigenvalue c_1 are the non-zero vectors of the form

$$d \begin{bmatrix} \frac{-1+i}{2} \\ 1 \end{bmatrix}.$$

Notice that these, together with the zero vector, form a 1-dimensional subspace of $\mathbb{C}^2$. In a similar manner the eigenvectors for the eigenvalue $c_2 = 3 - i$ are found to be the vectors of the form

$$d \begin{bmatrix} -\left(\frac{1+i}{2}\right) \\ 1 \end{bmatrix}$$

where $d \neq 0$. Again these form with the zero vector a subspace of $\mathbb{C}^2$.

This example is an illustration of the general procedure for finding eigenvectors and eigenvalues of a matrix.

The characteristic equation

Let A be an $n \times n$ matrix over a field F and let X be a non-zero n-column vector over F. The condition for X to be an eigenvector of A is $AX = cX$ or

$$(cI_n - A)X = 0,$$

where c is the corresponding eigenvalue. Thus the eigenvectors associated with c, together with the zero vector, form the null space of the matrix $cI_n - A$. This subspace is called the *eigenspace* of the eigenvalue c.

Next $(cI_n - A)X = 0$ is a homogeneous linear system of n equations in n unknowns, namely the entries of X. By (8.2.2) the condition for there to be a non-trivial solution of the system is

$$\det(cI_n - A) = 0.$$

Conversely, if $c \in$ satisfies this equation, there is a non-zero solution of the system and c is an eigenvalue. These considerations show that the determinant

$$
\det(tI_n - A) = \begin{vmatrix} t - a_{11} & -a_{12} & \cdots & -a_{1n} \\ -a_{21} & t - a_{22} & \cdots & -a_{2n} \\ \cdot & \cdot & \cdots & \cdot \\ -a_{n1} & -a_{n2} & \cdots & t - a_{nn} \end{vmatrix}
$$

plays a critical role. This is a polynomial of degree n in t with coefficients in F called the *characteristic polynomial* of A. The equation obtained by setting the characteristic polynomial equal to zero is the *characteristic equation*. The eigenvalues of A are the roots of the characteristic polynomial that lie in the field F.

One should keep in mind that A may well have no eigenvalues in F. For example, the characteristic polynomial of the real matrix

$$
\begin{bmatrix} 0 & 1 \\ -1 & 0 \end{bmatrix}
$$

is $t^2 + 1$, which has no real roots, so the matrix has no eigenvalues in $\mathbb{R}$.

In general the eigenvalues of a linear operator or a matrix lie in the splitting field of the characteristic polynomial – see Section 7.4. If $F = \mathbb{C}$, all roots of the characteristic equation lie in $\mathbb{C}$ by the Fundamental Theorem of Algebra – see (12.3.6). Because of this we can be sure that a complex matrix has all its eigenvalues in $\mathbb{C}$.

Let us sum up our conclusions about the eigenvalues of matrices so far.

(8.4.1). *Let A be an $n \times n$ matrix over a field F.*
(i) *The eigenvalues of A in F are precisely the roots of the characteristic polynomial $\det(tI_n - A)$ which lie in F.*
(ii) *The eigenvectors of A associated with the eigenvalue c are the non-zero vectors in the null space of the matrix $cI_n - A$.*

Example (8.4.3). Find the eigenvalues of the upper triangular matrix

$$
A = \begin{bmatrix} a_{11} & a_{12} & a_{13} & \cdots & a_{1n} \\ 0 & a_{22} & a_{23} & \cdots & a_{2n} \\ \cdot & \cdot & \cdot & \cdots & \cdot \\ 0 & 0 & 0 & \cdots & a_{nn} \end{bmatrix}.
$$

The characteristic polynomial of A is

$$
\begin{vmatrix} t - a_{11} & -a_{12} & -a_{13} & \cdots & -a_{1n} \\ 0 & t - a_{22} & -a_{23} & \cdots & -a_{2n} \\ \cdot & \cdot & \cdot & \cdots & \cdot \\ 0 & 0 & 0 & \cdots & t - a_{nn} \end{vmatrix},
$$

which equals $(t - a_{11})(t - a_{22}) \cdots (t - a_{nn})$. The eigenvalues of the matrix are therefore just the diagonal entries $a_{11}, a_{22}, \ldots, a_{nn}$.

Example (8.4.4). Consider the 3×3 matrix

$$A = \begin{bmatrix} 2 & -1 & -1 \\ -1 & 2 & -1 \\ -1 & -1 & 0 \end{bmatrix}.$$

The characteristic polynomial of A is

$$\begin{vmatrix} t-2 & 1 & 1 \\ 1 & t-2 & 1 \\ 1 & 1 & t \end{vmatrix} = t^3 - 4t^2 + t + 6.$$

By inspection one root of this cubic polynomial is -1. Dividing the polynomial by $t + 1$ using long division, we obtain the quotient $t^2 - 5t + 6 = (t - 2)(t - 3)$. Hence the characteristic polynomial factorizes completely as $(t+1)(t-2)(t-3)$ and the eigenvalues of A are $-1, 2$ and 3.

To find the corresponding eigenvectors, solve the three linear systems $(-I_3 - A)X = 0$, $(2I_3 - A)X = 0$ and $(3I_3 - A)X = 0$. On solving these, we find that the respective eigenvectors are the non-zero scalar multiples of the vectors

$$\begin{bmatrix} 1 \\ 1 \\ 2 \end{bmatrix}, \begin{bmatrix} 1 \\ 1 \\ -1 \end{bmatrix}, \begin{bmatrix} 1 \\ -1 \\ 0 \end{bmatrix},$$

so that the eigenspaces all have dimension 1.

Properties of the characteristic polynomial

Let us see what can be said about the characteristic polynomial of an arbitrary $n \times n$ matrix $A = [a_{ij}]$ over a field F. This is

$$p(t) = \begin{vmatrix} t - a_{11} & -a_{12} & \cdots & -a_{1n} \\ -a_{21} & t - a_{22} & \cdots & -a_{2n} \\ \cdot & \cdot & \cdots & \cdot \\ -a_{n1} & -a_{n2} & \cdots & t - a_{nn} \end{vmatrix}.$$

At this point recall the definition of a determinant as an alternating sum of $n!$ terms, each term being a product of n entries, one from each row and column. The term of $p(t)$ with highest degree in t arises from the product

$$(t - a_{11})(t - a_{22}) \cdots (t - a_{nn})$$

and is clearly t^n. The terms of degree $n - 1$ are easily identified as they arise from the same product. Thus the coefficient of t^{n-1} is $-(a_{11} + a_{22} + \cdots + a_{nn})$. The sum of the diagonal entries of A is called the *trace* of A,

$$\mathrm{tr}(A) = a_{11} + a_{22} + \cdots + a_{nn},$$

so the term in $p(t)$ of degree $n - 1$ is $-\mathrm{tr}(A)t^{n-1}$.

The constant term in $p(t)$ is $p(0) = \det(-A) = (-1)^n \det(A)$. Our knowledge of $p(t)$ so far is summarized by the formula

$$p(t) = t^n - \mathrm{tr}(A)t^{n-1} + \cdots + (-1)^n \det(A).$$

The other coefficients in the characteristic polynomial are not so easy to describe, but they are expressible in terms of subdeterminants of $\det(A)$. For example, take the case of t^{n-2}. A term in t^{n-2} arises in two ways: from the product $(t - a_{11})(t - a_{22}) \cdots (t - a_{nn})$ or from products like $-a_{12}a_{21}(t - a_{33}) \cdots (t - a_{nn})$. So a typical contribution to the coefficient of t^{n-2} is

$$(a_{11}a_{22} - a_{12}a_{21}) = \begin{vmatrix} a_{11} & a_{12} \\ a_{21} & a_{22} \end{vmatrix}.$$

From this one can see that the term of degree $n - 2$ in $p(t)$ is t^{n-2} times the sum of all the 2×2 sub-determinants of the form

$$\begin{vmatrix} a_{ii} & a_{ij} \\ a_{ji} & a_{jj} \end{vmatrix}$$

where $i < j$.

In general it can be shown by similar considerations that the following is true.

(8.4.2). *The characteristic polynomial of the $n \times n$ matrix A is*

$$t^n + \sum_{i=1}^{n} (-1)^i d_i t^{n-i}$$

where d_i is the sum of all the $i \times i$ subdeterminants of $\det(A)$ whose principal diagonals are part of the principal diagonal of A.

Next let $c_1, c_2, \ldots, c_n$ be the eigenvalues of A in the splitting field of its characteristic polynomial $p(t)$. Since $p(t)$ is monic, we have

$$p(t) = (t - c_1)(t - c_2) \cdots (t - c_n).$$

The constant term in this product is evidently $(-1)^n c_1 c_2 \ldots c_n$, while the term in t^{n-1} has coefficient $-(c_1 + \cdots + c_n)$. On the other hand, we found these coefficients to be

$(-1)^n \det(A)$ and $-\text{tr}(A)$ respectively. Thus we have discovered two important relations between the eigenvalues and the entries of A.

Corollary (8.4.3). *If A is a square matrix, the product of the eigenvalues equals the determinant $\det(A)$ and the sum of the eigenvalues equals $\text{tr}(A)$, the trace of A.*

Let A and B be $n \times n$ matrices over a field F. Recall that A and B are similar over F if there is an invertible $n \times n$ matrix S over F such that $B = SAS^{-1}$. The next result indicates that similar matrices have much in common.

(8.4.4). *Similar matrices have the same characteristic polynomial. Hence they have the same eigenvalues, trace and determinant.*

Proof. Let A and S be $n \times n$ matrices over a field with S invertible. Then the characteristic polynomial of the matrix SAS^{-1} is

$$\det(tI - SAS^{-1}) = \det(S(tI - A)S^{-1}) = \det(S)\det(tI - A)\det(S)^{-1}$$
$$= \det(tI - A).$$

Here we have used the property of determinants: $\det(PQ) = \det(P)\det(Q)$. The statements about trace and determinant follow from (8.4.3). $\qquad\square$

On the other hand, similar matrices need not have the same eigenvectors. Indeed the condition for X to be an eigenvector of SAS^{-1} with eigenvalue c is $(SAS^{-1})X = cX$, which is equivalent to $A(S^{-1}X) = c(S^{-1}X)$. Thus X is an eigenvector of SAS^{-1} if and only if $S^{-1}X$ is an eigenvector of A.

Diagonalizable matrices

Next we consider when a square matrix is similar to a diagonal matrix. This is an important question since diagonal matrices have much simpler properties than arbitrary matrices. For example, when a diagonal matrix is raised to the mth power, the effect is merely to raise each element on the diagonal to the mth power, whereas there is no simple expression for the mth power of an arbitrary matrix. Suppose we want to compute A^m where A is similar to a diagonal matrix D, with say $A = SDS^{-1}$. Then $A^m = (SDS^{-1})^m = SD^mS^{-1}$ after cancellation. Thus it is possible to calculate A^m quite simply if we have explicit knowledge of S and D.

Let A be a square matrix over a field F. Then A is said to be *diagonalizable* over F if it is similar to a diagonal matrix D over F, that is, there is an invertible matrix S over F such that $A = SDS^{-1}$ or equivalently $D = S^{-1}AS$. We also say that S *diagonalizes* A.

The terminology extends naturally to linear operators on a finite dimensional vector space V. A linear operator α on V is said to be *diagonalizable* if there is a basis $\{v_1, \ldots, v_n\}$ such that $\alpha(v_i) = c_i v_i$ where $c_i \in F$, for $i = 1, \ldots, n$. Thus α is represented by the diagonal matrix $\text{diag}(c_1, c_2, \ldots, c_n)$ with respect to this basis.

It is an important observation that if a matrix A is diagonalizable and its eigenvalues are $c_1, \ldots, c_n$, then A must be similar to the diagonal matrix with $c_1, \ldots, c_n$ on the principal diagonal. This is true because similar matrices have the same eigenvalues and the eigenvalues of a diagonal matrix are just the entries on the principal diagonal.

We aim to find a criterion for a square matrix to be diagonalizable. A key step in the search is next.

(8.4.5). *Let A be an $n \times n$ matrix over a field F and let $c_1, \ldots, c_r$ be distinct eigenvalues of A with associated eigenvectors $X_1, \ldots, X_r$. Then $\{X_1, \ldots, X_r\}$ is a linearly independent subset of F^n.*

Proof. Assume the theorem is false; then there is a positive integer i such that $\{X_1, \ldots, X_i\}$ is linearly independent, but adjunction of the vector X_{i+1} produces a linearly dependent set $\{X_1, \ldots, X_i, X_{i+1}\}$. Hence there are scalars $d_1, \ldots, d_{i+1}$, not all of them zero, such that

$$d_1 X_1 + \cdots + d_i X_i + d_{i+1} X_{i+1} = 0.$$

Premultiply both sides of this equation by A and use the equations $AX_j = c_j X_j$ to get

$$c_1 d_1 X_1 + \cdots + c_i d_i X_i + c_{i+1} d_{i+1} X_{i+1} = 0.$$

On subtracting c_{i+1} times the first equation from the second, we arrive at the equation

$$(c_1 - c_{i+1}) d_1 X_1 + \cdots + (c_i - c_{i+1}) d_i X_i = 0.$$

Since $X_1, \ldots, X_i$ are linearly independent, the coefficients $(c_j - c_{i+1}) d_j$ must vanish. But $c_1, \ldots, c_{i+1}$ are all different, so it follows that $d_j = 0$ for $j = 1, \ldots, i$. Hence $d_{i+1} X_{i+1} = 0$ and $d_{i+1} = 0$, contrary to assumption, so the theorem is proved. $\qquad\square$

A criterion for diagonalizability can now be established.

(8.4.6). *Let A be an $n \times n$ matrix over a field F. Then A is diagonalizable over F if and only if A has n linearly independent eigenvectors in F^n.*

Proof. First of all assume that A has n linearly independent eigenvectors in F^n, say $X_1, X_2, \ldots, X_n$, and let the associated eigenvalues be $c_1, c_2, \ldots, c_n$. Define S to be the $n \times n$ matrix whose columns are the eigenvectors; thus

$$S = [X_1 \, X_2 \ldots X_n].$$

The first thing to note is that S is invertible since its columns are linearly independent. Forming the product of A and S in partitioned form, we find that

$$AS = [AX_1 \, AX_2 \ldots AX_n] = [c_1 X_1 \, c_2 X_2 \, \cdots \, c_n X_n],$$

so that

$$AS = [X_1 \, X_2 \, \ldots \, X_n] \begin{bmatrix} c_1 & 0 & 0 & \cdots & 0 \\ 0 & c_2 & 0 & \cdots & 0 \\ . & & . & \cdots & . \\ 0 & 0 & . & \cdots & c_n \end{bmatrix} = SD,$$

where $D = \text{diag}(c_1, c_2, \ldots, c_n)$ is the diagonal matrix with diagonal entries $c_1, \ldots, c_n$. Therefore $A = SDS^{-1}$ and A is diagonalizable.

Conversely, assume that A is diagonalizable and $S^{-1}AS = D = \text{diag}(c_1, c_2, \ldots, c_n)$. Here the c_i must be the eigenvalues of A. Then $AS = SD$, which implies that $AX_i = c_i X_i$ where X_i is the ith column of S. Therefore $X_1, X_2, \ldots, X_n$ are eigenvectors of A with associated eigenvalues $c_1, c_2, \ldots, c_n$. Since $X_1, X_2, \ldots, X_n$ are columns of the invertible matrix S, they are linearly independent. Consequently A has n linearly independent eigenvectors. $\square$

Corollary (8.4.7). *An $n \times n$ complex matrix with n distinct eigenvalues is diagonalizable.*

This follows at once from (8.4.5) and (8.4.6). On the other hand, it is easy to find matrices that are not diagonalizable: for example, the matrix

$$A = \begin{bmatrix} 1 & 1 \\ 0 & 1 \end{bmatrix}.$$

Indeed, if A were diagonalizable, it would be similar to the identity matrix I_2, since both eigenvalues of A equal to 1. But then $A = SI_2S^{-1} = I_2$ for some S, a contradiction.

A feature of the proof of (8.4.6) is that it provides a method for finding a matrix S which diagonalizes A. It suffices to find a largest set of linearly independent eigenvectors of A; if there are enough of them, they can be taken to form the columns of the diagonalizing matrix S.

Example (8.4.5). Find a matrix which diagonalizes

$$A = \begin{bmatrix} 2 & -1 \\ 2 & 4 \end{bmatrix}.$$

From Example (8.4.2) we know that the eigenvalues of A are $3 \pm i$, so A is diagonalizable over $\mathbb{C}$ by (8.4.7). Also corresponding eigenvectors for A were found which form the matrix

$$S = \begin{bmatrix} \frac{-1+i}{2} & -\frac{1+i}{2} \\ 1 & 1 \end{bmatrix}.$$

From the preceding theory we may be sure that

$$S^{-1}AS = \begin{bmatrix} 3+i & 0 \\ 0 & 3-i \end{bmatrix}.$$

Triangularizable matrices

It has been seen that not every complex square matrix is diagonalizable. Compensating for this failure is the fact such a matrix is always similar to an upper triangular matrix, i. e., a matrix with 0's below the principal diagonal.

Let A be a square matrix over a field F. Then A is said to be *triangularizable over F* if there is an invertible matrix S over F such that $A = STS^{-1}$ or equivalently $S^{-1}AS = T$, where T is upper triangular. It will also be convenient to say that S *triangularizes A*. Note that the diagonal entries of the triangular matrix T will necessarily be the eigenvalues of A. This is because of Example (8.4.3) and the fact that similar matrices have the same eigenvalues. Thus a necessary condition for A to be triangularizable over F is that all its eigenvalues belong to F. In fact the converse is also true.

(8.4.8). *A square matrix A over a field F all of whose eigenvalues lie in F is triangularizable over F.*

Proof. We show by induction on n that A is triangularizable. If $n = 1$, there is nothing to prove, so let $n > 1$. Assume the result is true for $(n-1) \times (n-1)$ matrices.

By hypothesis A has at least one eigenvalue c in F, with associated eigenvector X say. Since $X \neq 0$, it is possible to adjoin vectors to X to produce a basis of F^n, say $\{X = X_1, X_2, \ldots, X_n\}$; here we have used (8.2.6). Left multiplication of the vectors of F^n by A gives rise to linear operator α on F^n. With respect to the basis $\{X_1, \ldots, X_n\}$, the linear operator α is represented by a matrix with the special form

$$B_1 = \begin{bmatrix} c & A_2 \\ 0 & A_1 \end{bmatrix}$$

where A_1 and A_2 are matrices over F and A_1 has $n-1$ rows and columns. The reason for the special form is that $\alpha(X_1) = AX_1 = cX_1$ since X_1 is an eigenvector of A with associated eigenvalue c. The matrices A and B_1 are similar since they represent the same linear operator α. Suppose that in fact $B_1 = S_1^{-1}AS_1$ where S_1 is an invertible $n \times n$ matrix.

Observe that the eigenvalues of A_1 are among those of B_1 and hence A. By induction on n there is an invertible matrix S_2 with $n-1$ rows and columns such that $B_2 = S_2^{-1}A_1S_2$ is upper triangular. Now write

$$S = S_1 \begin{bmatrix} 1 & 0 \\ 0 & S_2 \end{bmatrix}.$$

This is a product of invertible matrices, so it is invertible. An easy matrix computation shows that

$$S^{-1}AS = \begin{bmatrix} 1 & 0 \\ 0 & S_2^{-1} \end{bmatrix} (S_1^{-1}AS_1) \begin{bmatrix} 1 & 0 \\ 0 & S_2 \end{bmatrix} = \begin{bmatrix} 1 & 0 \\ 0 & S_2^{-1} \end{bmatrix} B_1 \begin{bmatrix} 1 & 0 \\ 0 & S_2 \end{bmatrix}.$$

From this we obtain

$$S^{-1}AS = \begin{bmatrix} 1 & 0 \\ 0 & S_2^{-1} \end{bmatrix} \begin{bmatrix} c & A_2 \\ 0 & A_1 \end{bmatrix} \begin{bmatrix} 1 & 0 \\ 0 & S_2 \end{bmatrix} = \begin{bmatrix} c & A_2S_2 \\ 0 & S_2^{-1}A_1S_2 \end{bmatrix} = \begin{bmatrix} c & A_2S_2 \\ 0 & B_2 \end{bmatrix} = T.$$

The matrix T is upper triangular, so the theorem is proved. $\qquad\qquad\square$

The preceding proof provides a method for triangularizing a matrix.

Example (8.4.6). Triangularize the matrix $A = \begin{bmatrix} 1 & 1 \\ -1 & 3 \end{bmatrix}$ over $\mathbb{C}$.

The characteristic polynomial of A is $t^2 - 4t + 4$, so both eigenvalues equal 2. Solving $(2I_2 - A)X = 0$, we find that all the eigenvectors of A are scalar multiples of $X_1 = \begin{bmatrix} 1 \\ 1 \end{bmatrix}$. Therefore by (8.4.6) the matrix A is *not* diagonalizable.

Let α be the linear operator on $\mathbb{C}^2$ arising from left multiplication by A. Adjoin a vector to X_2 to X_1 to get a basis $\mathcal{B}_2 = \{X_1, X_2\}$ of $\mathbb{C}^2$: for example let $X_2 = \begin{bmatrix} 0 \\ 1 \end{bmatrix}$. Denote by $\mathcal{B}_1$ the standard basis of $\mathbb{C}^2$. The change of basis $\mathcal{B}_2 \rightarrow \mathcal{B}_1$ has transition matrix $S = \begin{bmatrix} 1 & 0 \\ 1 & 1 \end{bmatrix}$, so $S_1 = S^{-1} = \begin{bmatrix} 1 & 0 \\ -1 & 1 \end{bmatrix}$ is the transition matrix of the change of basis $\mathcal{B}_1 \rightarrow \mathcal{B}_2$. Therefore by (8.3.13) the matrix that represents α with respect to the basis $\mathcal{B}_2$ is $S_1AS_1^{-1} = \begin{bmatrix} 2 & 1 \\ 0 & 2 \end{bmatrix} = T$. Hence $A = S_1^{-1}TS_1 = STS^{-1}$ and S triangularizes A.

To conclude the chapter we show how to solve a system of linear recurrences by using matrix diagonalization.

Example (8.4.7). In a population of rabbits and weasels it is observed that each year the number of rabbits is equal to four times the number of rabbits less twice the number of weasels in the previous year. The number of weasels in any year equals the sum of the numbers of rabbits and weasels in the previous year. If the initial numbers of rabbits and weasels were 100 and 10 respectively, find the numbers of each species after n years.

Let r_n and w_n denote the respective numbers of rabbits and weasels after n years. The information given translates into the two *linear recurrence relations*

$$\begin{cases} r_{n+1} = 4r_n - 2w_n \\ w_{n+1} = r_n + w_n \end{cases}$$

together with the initial conditions $r_0 = 100$, $w_0 = 10$. We have to solve this system of linear recurrence relations for r_n and w_n.

To see how eigenvalues enter into the problem, write the system of recurrences in matrix form. Put $X_n = \begin{bmatrix} r_n \\ w_n \end{bmatrix}$ and $A = \begin{bmatrix} 4 & -2 \\ 1 & 1 \end{bmatrix}$. Then the two recurrences are equivalent to the single matrix equation

$$X_{n+1} = AX_n,$$

while the initial conditions assert that $X_0 = \begin{bmatrix} 100 \\ 10 \end{bmatrix}$. These equations enable us to calculate successive vectors X_n; for $X_1 = AX_0$, $X_2 = A^2 X_0$ and in general $X_n = A^n X_0$.

In principle this provides a solution to the problem. However, it involves calculating powers of the matrix A. Fortunately A is diagonalizable since it has distinct eigenvalues 2 and 3. Corresponding eigenvectors are found to be $\begin{bmatrix} 1 \\ 1 \end{bmatrix}$ and $\begin{bmatrix} 2 \\ 1 \end{bmatrix}$; therefore the matrix $S = \begin{bmatrix} 1 & 2 \\ 1 & 1 \end{bmatrix}$ diagonalizes A, and

$$S^{-1}AS = \begin{bmatrix} 2 & 0 \\ 0 & 3 \end{bmatrix} = D.$$

It is now easy to compute powers since $A^n = (SDS^{-1})^n = SD^n S^{-1}$. Therefore $X_n = A^n X_0 = SD^n S^{-1} X_0$ and thus

$$X_n = \begin{bmatrix} 1 & 2 \\ 1 & 1 \end{bmatrix} \begin{bmatrix} 2^n & 0 \\ 0 & 3^n \end{bmatrix} \begin{bmatrix} -1 & 2 \\ 1 & -1 \end{bmatrix} \begin{bmatrix} 100 \\ 10 \end{bmatrix},$$

which leads to

$$X_n = \begin{bmatrix} 180 \cdot 3^n - 80 \cdot 2^n \\ 90 \cdot 3^n - 80 \cdot 2^n \end{bmatrix}.$$

The solution to the problem can now be read off:

$$r_n = 180 \cdot 3^n - 80 \cdot 2^n \quad \text{and} \quad w_n = 90 \cdot 3^n - 80 \cdot 2^n.$$

Notice that r_n and w_n both increase without limit as $n \to \infty$ since 3^n is the dominant term; however, $\lim_{n \to \infty} (\frac{r_n}{w_n}) = 2$. The conclusion is that, while both populations explode, in the long run there will be twice as many rabbits as weasels.

Exercises (8.4).

(1) Find the eigenvectors and eigenvalues of the following matrices:

$$\begin{bmatrix} 1 & 5 \\ 3 & 3 \end{bmatrix}; \quad \begin{bmatrix} 1 & 2 & -1 \\ 1 & 0 & 1 \\ 4 & -4 & 5 \end{bmatrix}; \quad \begin{bmatrix} 1 & 0 & 0 & 0 \\ 2 & 2 & 0 & 0 \\ 1 & 0 & 3 & 0 \\ 0 & 1 & -1 & 4 \end{bmatrix}.$$

(2) Prove that $\text{tr}(A + B) = \text{tr}(A) + \text{tr}(B)$ and $\text{tr}(cA) = c\,\text{tr}(A)$ where A and B are $n \times n$ matrices and c is a scalar.

(3) If A and B are non-singular $n \times n$ matrices, show that AB and BA have the same eigenvalues.

(4) Suppose that A is a square matrix with real entries and real eigenvalues. Prove that each eigenvalue of A has an associated *real* eigenvector.

(5) A real square matrix with distinct eigenvalues is diagonalizable over $\mathbb{R}$: true or false?

(6) Let $p(t)$ be the polynomial $t^n + a_{n-1}t^{n-1} + a_{n-2}t^{n-2} + \cdots + a_0$ over a field F. Show that $p(t)$ is the characteristic polynomial of the matrix

$$\begin{bmatrix} 0 & 0 & \cdots & 0 & -a_0 \\ 1 & 0 & \cdots & 0 & -a_1 \\ 0 & 1 & \cdots & 0 & -a_2 \\ . & . & \cdots & . & . \\ 0 & 0 & \cdots & 1 & -a_{n-1} \end{bmatrix}.$$

(This is called the *companion matrix* of $p(t)$):

(7) Find matrices which diagonalize the following matrices:

(i) $\begin{bmatrix} 1 & 5 \\ 3 & 3 \end{bmatrix}$; (ii) $\begin{bmatrix} 1 & 2 & -1 \\ 1 & 0 & 1 \\ 4 & -4 & 5 \end{bmatrix}$.

(8) For which values of a and b is the matrix $\begin{bmatrix} 0 & a \\ b & 0 \end{bmatrix}$ diagonalizable over $\mathbb{C}$?

(9) Prove that a complex 2×2 matrix is *not* diagonalizable if and only if it is similar to a matrix of the form $\begin{bmatrix} a & b \\ 0 & a \end{bmatrix}$ where $b \neq 0$.

(10) Let A be a diagonalizable matrix and assume that S is a matrix which diagonalizes A. Prove that a matrix T diagonalizes A if and only if it is of the form $T = CS$ where C is a matrix such that $AC = CA$.

(11) If A is a non-singular matrix with eigenvalues $c_1, \ldots, c_n$, show that the eigenvalues of A^{-1} are $c_1^{-1}, \ldots, c_n^{-1}$.

(12) Let α be a linear operator on a complex n-dimensional vector space V. Prove that there is a basis $\{v_1, \ldots, v_n\}$ of V such that $\alpha(v_i)$ is a linear combination of $v_1, v_2, \ldots, v_i$ for $i = 1, \ldots, n$.

(13) Let $\delta : P_n(\mathbb{R}) \to P_n(\mathbb{R})$ be the linear operator corresponding to differentiation. Show that all the eigenvalues of δ are zero. What are the eigenvectors?

(14) Let $c_1, \ldots, c_n$ be the distinct eigenvalues of a complex $n \times n$ matrix A. Prove that the eigenvalues of A^m are $c_1^m, \ldots, c_n^m$ where m is any positive integer. [Hint: the matrix is triangularizable by (8.4.8).]

(15) Prove that a square matrix and its transpose have the same eigenvalues.

(16) Use matrix diagonalization to solve the following system of linear recurrences:

$$\begin{cases} x_{n+1} = 2x_n + 10y_n \\ y_{n+1} = 2x_n + 3y_n \end{cases}$$

with the initial conditions $x_0 = 0$, $y_0 = 1$.

9 Introduction to Modules

After groups, rings and vector spaces, the most useful algebraic structures are probably modules. A module is an abelian group on which a ring acts subject to certain natural rules. Aside from their intrinsic interest as algebraic objects, modules have important applications to linear operators, canonical forms of matrices and representations of groups.

9.1 Elements of module theory

Let R be a ring and let M be an abelian group which is written additively. Then M is said to be a *left R-module* if there is a *left action* of R on M, i. e., a map from $R \times M$ to M, written $(r, a) \mapsto r \cdot a$, $(r \in R, a \in M)$, such that the following axioms are valid for all $r, s \in R$ and $a, b \in M$:

(i) $r \cdot (a + b) = r \cdot a + r \cdot b$;
(ii) $(r + s) \cdot a = r \cdot a + s \cdot a$;
(iii) $(rs) \cdot a = r \cdot (s \cdot a)$.

If the ring R has an identity element and if in addition
(iv) $1_R \cdot a = a$,

for all $a \in M$, the module M is called *unitary*. It will be a tacit assumption here that whenever a ring R has identity, an R-module is unitary.

A *right R-module* is defined in the analogous fashion via a *right action* of R on M. Sometimes it is convenient to indicate whether an R-module M is left or right by writing

$$_R M \quad \text{or} \quad M_R,$$

respectively.

It is usually not necessary to study left and right R-modules separately since one can pass to a module over the *opposite ring*

$$R^{op}$$

of R. This is the ring with the same underlying set and operation of addition as R, but with the *opposite multiplication*, denoted by $*$ where

$$r * s = sr, \quad (r, s \in R).$$

It is easy to check that R^{op} is a ring. Of course $R = R^{op}$ if R is a commutative ring. The relation between left and right modules is made clear by the next result.

https://doi.org/10.1515/9783110691160-009

(9.1.1). *Let R be a ring and M an R-module.*
(i) *If M is a left R-module, it is also right R^{op}-module with the right action $a \cdot r = r \cdot a$.*
(ii) *If M is a right R-module, it is also left R^{op}-module with the left action $r \cdot a = a \cdot r$.*

Proof. (i) The axioms for a right action have to be verified, the crucial one being

$$(a \cdot r) \cdot s = (r \cdot a) \cdot s = s \cdot (r \cdot a) = (sr) \cdot a = (r * s) \cdot a = a \cdot (r * s):$$

here $*$ denotes the ring operation in R^{op}. The other module axioms are easily checked.
□

This result allows us to concentrate on left modules.

Elementary properties

The simplest consequences of the module axioms are collected in the next result, which, as will usually be the case, is stated for left modules.

(9.1.2). *Let M be a left R-module and let $a \in M$, $r \in R$ and $n \in \mathbb{Z}$. Then:*
(i) $r \cdot 0_M = 0_M$;
(ii) $0_R \cdot a = 0_M$;
(iii) $n(r \cdot a) = (nr) \cdot a = r \cdot (na)$.

Proof. For (i) put $a = 0_M = b$ in module axiom (i): for (ii) put $r = 0_R = s$ in axiom (ii). The proof of (iii) requires a little more effort. If $n > 0$, the statements are quickly proved by induction on n. For $n = 0$ they follow at once from (i) and (ii).

Next consider the case $n = -1$. The elements $(-r) \cdot a$ and $r \cdot (-a)$ both equal $-(r \cdot a)$ since $(-r) \cdot a + r \cdot a = (-r + r) \cdot a = 0_R \cdot a = 0_M$ and $r \cdot (-a) + r \cdot a = r \cdot (-a + a) = r \cdot 0_M = 0_M$ by (i) and (ii).

Finally, let $n < 0$. Then $-n(r \cdot a) = (-n)r \cdot a = r \cdot (-na)$. Take the negative of each side and use the case $n = -1$ to get $n(r \cdot a) = (nr) \cdot a = r \cdot (na)$, as required. □

In future we will write 0 for both 0_R and 0_M.

Examples of modules

Next we list some standard sources of modules.
(i) Let R be an arbitrary ring. Define a left action of R on itself by using the ring product: thus $r \cdot s = rs$, $(r, s \in R)$. The ring axioms guarantee the validity of the module axioms. In a similar way R can be made into a right R-module by using the ring product. To distinguish when the ring is being regarded as a left or a right module,

we will often write

$$_R R \quad \text{and} \quad R_R$$

respectively.

(ii) Let F be a field. Then a left F-module is simply a vector space over F since the vector space axioms are just those for an F-module.

(iii) An abelian group A is a left $\mathbb{Z}$-module in which the action is $n \cdot a = na$, $n \in \mathbb{Z}$, $a \in A$.

Conversely, *if A is a $\mathbb{Z}$-module, the module action is $n \cdot a = na$.* To see this set $r = 1$ in (9.1.2)(iii), keeping in mind that A is a unitary module. Consequently, there is only one way to make an abelian group into a $\mathbb{Z}$-module.

These examples show that the module concept is a broad one, encompassing rings, abelian groups and vector spaces.

Bimodules

Let R and S be a pair of rings. An (R, S)-*bimodule* is an abelian group M which is simultaneously a left R-module and a right S-module, and in which the left and right actions are linked by the law

$$(r \cdot a) \cdot s = r \cdot (a \cdot s),$$

where $r \in R$, $s \in S$, $a \in M$. The notation

$$_R M_S$$

will be used to indicate an (R, S)-bimodule. For example, a ring R is an (R, R)-bimodule via the ring operations. Of course, if R is a commutative ring, $R = R^{op}$ and there is no difference between a left R-module, a right R-module and an (R, R)-bimodule.

Submodules

Groups have subgroups, rings have subrings and vector spaces have subspaces, so it is to be expected that submodules will play a role in module theory.

Let M be a left R-module. An R-*submodule* of M is a subgroup N of M which has the additional property

$$a \in N, \ r \in R \ \Rightarrow \ r \cdot a \in N.$$

Notice that N itself is an R-module. There is a corresponding definition for right modules. Here are some standard examples of submodules.

(i)   If R is a ring, the submodules of $_R R$ are the left ideals of R, while those of R_R are the right ideals.

(ii) Every module has the *zero submodule*, containing only the zero element, and the *improper submodule*, namely the module itself.

Submodules generated by subsets

Let R be a ring and M a left R-module. It follows from the definition of a submodule that *the intersection of a non-empty set of submodules of M is itself a submodule*. Now let X be a non-empty subset of M. There is at least one submodule of M containing X, namely M itself. Thus we can form the intersection of all the submodules that contain X, which is a submodule called the *submodule generated* by X. It is evidently the smallest submodule of M containing X.

It is natural to ask what the elements of this submodule look like; recall that similar questions arose for subgroups, subrings, ideals and subspaces. The answer in the case of a ring with identity is given next.

(9.1.3). *Let R be a ring with identity and M a left R-module. If X is a non-empty subset of M, the submodule of M generated by X consists of all elements of the form*

$$r_1 \cdot x_1 + r_2 \cdot x_2 + \cdots + r_n \cdot x_n$$

where $r_i \in R$, $x_i \in X$, $n \geq 0$.

Proof. Let N denote the set of all elements of the form $r_1 \cdot x_1 + r_2 \cdot x_2 + \cdots + r_n \cdot x_n$ with $r_i \in R$, $x_i \in X$, $n \geq 0$. (Note that when $n = 0$, the sum is to be interpreted as 0.) It is an easy verification that N is a submodule. Now $X \subseteq N$ since $x = 1 \cdot x \in N$ for all $x \in X$. Hence the submodule L generated by X is contained in N. On the other hand, $N \subseteq L$, since it is clear from their form that every element of N belongs to L. Therefore $L = N$. $\qquad\square$

If R is a ring with identity and X is a subset of a left R-module, the notation

$$R \cdot X$$

will be used to denote the submodule generated by X. (If R is a field, so that R-modules are vector spaces, the notation used in Section 8.1 for $R \cdot X$ was $\langle X \rangle$.)

An R-module M is said to be *finitely generated* if it can be generated by a finite subset X. An important special case is when $X = \{x\}$. In this situation, if R has an identity, we write $R \cdot x$ for $R \cdot X$; then M is called a *cyclic R-module*. For example, the cyclic submodules of $_R R$ are the principal left ideals of R, i. e., those of the form $Rx = \{r \cdot x \mid r \in R\}$.

Quotient modules and homomorphisms

Just as for groups, rings and vector spaces, we can form quotients of modules. Let N be a submodule of a left R-module M. Since N is a subgroup of the abelian group M, the quotient $M/N = \{a + N \mid a \in M\}$, consisting of all cosets of N in M, already has the structure of an abelian group. To make M/N into a left R-module a left action must be specified. The natural candidate is the rule

$$r \cdot (a + N) = r \cdot a + N, \quad (a \in M, r \in R).$$

As usual when an operation is to be defined on a quotient structure, the question arises as to whether it is well defined. Let $b \in a + N$, so that $b = a + c$ with $c \in N$. Then $r \cdot b = r \cdot a + r \cdot c \in r \cdot a + N$ since $r \cdot c \in N$. Hence $r \cdot a + N = r \cdot b + N$ and the left action has been well defined. The task of checking the validity of the module axioms is left to the reader. The module M/N is the *quotient module* of M by N.

It is to be expected that there are mappings between modules called module homomorphisms. Let M, N be two left modules over a ring R. An *R-module homomorphism* from M to N is a homomorphism of abelian groups

$$\alpha : M \to N$$

which has the additional property that

$$\alpha(r \cdot a) = r \cdot \alpha(a) \quad \text{for } r \in R, a \in M.$$

Thus the mapping α connects the module structures of M and N. A module homomorphism from M to itself is sometimes called an *endomorphism* of M.

A standard example is the *canonical homomorphism* ν from an R-module M to the quotient module M/N where N is a submodule of M. This is defined by $\nu(a) = a + N$. We already know from group theory that ν is a group homomorphism. To show that it is a module homomorphism simply observe that $\nu(r \cdot a) = r \cdot a + N = r \cdot (a + N) = r \cdot \nu(a)$.

(9.1.4). *Let M and N be left modules over a ring R and let $\alpha : M \to N$ be a module homomorphism. Then $\mathrm{Im}(\alpha)$ and $\mathrm{Ker}(\alpha)$ are submodules of N and M respectively.*

Of course group theory tells us that $\mathrm{Im}(\alpha)$ and $\mathrm{Ker}(\alpha)$ are subgroups of N and M. It is just a matter of verifying that they are submodules, another simple task that is left to the reader.

A module homomorphism which is bijective is called a *module isomorphism*. If there is a module isomorphism between two R-modules M and N, they are said to be *R-isomorphic*, in symbols

$$M \overset{R}{\cong} N.$$

It is an important observation that *the inverse of a module isomorphism is also a module isomorphism* – see Exercise (9.1.4).

The isomorphism theorems for modules

Just as in group theory there are theorems connecting module homomorphisms and quotient modules.

(9.1.5) (First Isomorphism Theorem). *Let $\alpha : M \to N$ be an R-module homomorphism. Then $M/\mathrm{Ker}(\alpha) \overset{R}{\simeq} \mathrm{Im}(\alpha)$*

(9.1.6) (Second Isomorphism Theorem). *Let M and N be submodules of an R-module. Then $M + N$ and $M \cap N$ are submodules and $(M + N)/N \overset{R}{\simeq} M/(M \cap N)$.*

(9.1.7) (Third Isomorphism Theorem). *Let L, M, N be submodules of an R-module such that $L \subseteq M \subseteq N$. Then M/L is a submodule of N/L and $(N/L)/(M/L) \overset{R}{\simeq} N/M$.*

Proof. We know from (4.3.4), (4.3.5) and (4.3.6) that each of the specified maps is an isomorphism of groups. To complete the proofs it is a question of verifying that the relevant maps are module homomorphisms. For example, take the case of (9.1.5). Let $\theta : M/\mathrm{Ker}(\alpha) \to \mathrm{Im}(\alpha)$ be defined by $\theta(a + \mathrm{Ker}(\alpha)) = \alpha(a)$, with $a \in M$. By (4.3.4) θ is a homomorphism of groups. Check that it is a homomorphism of R-modules. By definition $\theta(r \cdot (a + \mathrm{Ker}(\alpha))) = \theta(r \cdot a + \mathrm{Ker}(\alpha)) = \alpha(r \cdot a) = r \cdot \alpha(a) = r \cdot \theta(a + \mathrm{Ker}(\alpha))$.

In a similar way (9.1.6) and (9.1.7) can be established. $\qquad\square$

We mention, without writing down the details, that there is a module version of the Correspondence Theorem – cf. (4.2.2). This theorem describes the submodules of a quotient module M/N as having the form L/N where L is a submodule of M containing N.

The structure of cyclic modules

Sufficient machinery has been developed to permit a description of cyclic R-modules when R is a ring with identity.

(9.1.8). *Let R be a ring with identity.*
(i) *If M is a cyclic left R-module, then $M \overset{R}{\simeq} {}_R R/L$ where L is a left ideal of R.*
(ii) *Conversely, if L is a left ideal of R, then ${}_R R/L$ is the cyclic left R-module generated by $1_R + L$.*

Proof. Assume that M is cyclic and $M = R \cdot a$ where $a \in M$. Define a function $\alpha : {}_R R \to M$ by $\alpha(r) = r \cdot a$. Check that α is an R-module homomorphism. For example, let $r_1, r \in R$; then $\alpha(r_1 \cdot r) = \alpha(r_1 r) = (r_1 r) \cdot a = r_1 \cdot (r \cdot a) = r_1 \cdot \alpha(r)$. Also α is surjective since each element of M has the form $r \cdot a = \alpha(r)$ for some $r \in R$. Set $L = \mathrm{Ker}(\alpha)$ and note that

L is a left ideal by (9.1.4). Hence $_RR/L \overset{R}{\cong} \text{Im}(\alpha) = M$ by (9.1.5). The converse statement is obvious. □

The kernel of the function α in the proof of (9.1.8)(i) is the set $\{r \in R \mid r \cdot a = 0\}$: this left ideal of R is called the *annihilator of a in R* and is denoted by

$$\text{Ann}_R(a).$$

Since cyclic left R-modules have been seen to correspond to left ideals of the ring R, it is to be expected that module theory will be more complicated for rings with many ideals. The simplest situation is, of course, for fields, which have no proper non-zero ideals: in this case we are dealing with vector spaces over a field and every cyclic module is a 1-dimensional space isomorphic with the field itself.

Direct sums of submodules

Just as for vector spaces, there is the notion of a direct sum of submodules. Let M be a module with a family of submodules $\{M_\lambda \mid \lambda \in \Lambda\}$. Suppose that

$$M_\lambda \cap \sum_{\mu \neq \lambda} M_\mu = 0$$

for all $\lambda \in \Lambda$. Then the M_λ generate their (internal) *direct sum*, which is written

$$\bigoplus_{\lambda \in \Lambda} M_\lambda.$$

This is a subgroup of M, as we know from Section 4.2, (where the multiplicative notation was used). It is evidently also a submodule. In the case where Λ is finite and $\Lambda = \{1, 2, \ldots, n\}$, we write the direct sum as

$$M_1 \oplus M_2 \oplus \cdots \oplus M_n.$$

It is also possible to form the *external direct sum* of a set of modules – see Section 4.2 where external direct products of groups were defined. We will encounter mainly the situation is where there are finitely many modules $\{M_1, M_2, \ldots, M_n\}$. The external direct sum of these modules is the set product $M_1 \times M_2 \times \cdots \times M_n$ where elements are added componentwise and the action of the ring is on components. The external direct sum is also denoted by $M_1 \oplus M_2 \oplus \cdots \oplus M_n$: we will sometimes write

$$a_1 \oplus a_2 \oplus \cdots \oplus a_n$$

for $(a_1, a_2, \ldots, a_n)$ to distinguish the direct sum from the set product. An external direct sum is isomorphic with an internal direct sum, as was seen for groups in Section 4.2.

External direct sums can be extended to the case where there are infinitely many modules by using choice functions, just as for groups. Thus we obtain restricted direct sums and unrestricted direct sums of modules. One can also form the associated canonical injections and canonical projections. The definitions and results given in Section 4.2 in the case of groups apply equally to modules, allowing for the additive notation used for modules. It is only necessary to verify that any homomorphisms involved are module homomorphisms, usually a routine task.

Simple modules

Let M be a module over a ring R. Then M is said to be a *simple module* if M is non-zero and it has no proper non-zero submodules. The next result characterizes simple modules in terms of left ideals which are maximal, i. e., maximal proper left ideals.

(9.1.9). *Let R be a ring with identity. If M is a simple R-module, then $M \overset{R}{\simeq} R/L$ where L is a maximal left ideal of R. Conversely, such an R/L is a simple R-module.*

Proof. Let a be a non-zero element of M. Then $Ra \neq 0$ as $a = 1_R \cdot a$. Thus $Ra = M$ since M is a simple R-module. Hence M is a cyclic module and by (9.1.8) we have $M \overset{R}{\simeq} R/L$ where $L = \mathrm{Ann}_R(a)$ is the annihilator of a. That L must be a maximal left ideal of R follows from the Correspondence Theorem for modules mentioned after (9.1.7). $\qquad \square$

If R is a ring with identity, then by (6.2.5) it has maximal left ideals and therefore by (9.1.9) simple R-modules exist. On the other hand, it was observed in Section 6.2 that a ring without an identity element may not have any maximal left ideals and thus may not possess simple modules. Thus we will always assume the ring has identity in discussions of simple modules.

Semisimple modules

A module is called *semisimple* if it is the direct sum of simple submodules. For example, if F is a field, any F-module, i. e., F-vector space, is semisimple since it has a basis by (8.2.7) and simple F-modules are just 1-dimensional vector spaces. Semisimple modules have properties which make them interesting *per se*; in addition they arise naturally in ring theory and representation theory of groups.

(9.1.10). *Let M be a module over a ring R with identity and assume that M is generated by simple submodules. Then the following statements are valid.*
(i) *the module M is semisimple, say $M = \bigoplus_{i \in I} S_i$ where the S_i are simple modules;*
(ii) *every quotient of M is semisimple;*
(iii) *a simple quotient of M is isomorphic with one of the simple submodules S_i in (i).*

Proof. (i) Let M be generated by a set of simple submodules $\{S_j \mid j \in J\}$, where J is non-empty. Let $\emptyset \neq K \subseteq J$ and observe that the subgroup $\langle S_k \mid k \in K \rangle$ is actually an R-submodule. We will call K independent if $\langle S_k \mid k \in K \rangle$ is the direct sum of the S_k. Obviously any 1-element subset of J is independent. Now the set of all independent subsets of J is partially ordered by set inclusion. Furthermore, the union of a chain of independent subsets is independent. Therefore by Zorn's Lemma there is a maximal independent subset I of J. Put $N = \langle S_i \mid i \in I \rangle$, which is the direct sum of these S_i. If $N = M$, then M is semisimple, so assume that $N \neq M$. Hence there must exist $i \in I$ such that $S_i \nsubseteq N$. Since S_i is simple and $N \cap S_i$ is a submodule, it follows that $N \cap S_i = 0$ and $N + S_i = N \oplus S_i$. But this implies that the set $I \cup \{i\}$ is independent, contradicting the maximality of I.

(ii) If N is a submodule of M, then clearly M/N is a sum – although perhaps not a direct sum – of simple submodules. However, we can apply (i) to show that M/N is semisimple.

(iii) Let M/N be a simple quotient of M. Then $S_i \nsubseteq N$ for some i since $N \neq M$. Thus $N \cap S_i = 0$ and $M = N + S_i$ since N is a maximal submodule of M. Hence $M = N \oplus S_i$ and $M/N \simeq S_i$. $\qquad\square$

A submodule N of a module M is said to be a *direct summand* of M if there is a submodule L such that $M = N \oplus L$. It is a notable property of semisimple modules that every submodule is a direct summand.

(9.1.11). *Let N be a submodule of a semisimple module M over a ring with identity. Then N is a direct summand of M and hence N is semisimple.*

Proof. Clearly we may assume that $N \neq M$. Let $M = \bigoplus_{i \in I} M_i$ where each M_i is a simple submodule. Define S to be the set of all subsets J of I such that $N \cap \langle M_j \mid j \in J \rangle = 0$. Notice that S is not empty: for there exists $i_0 \in I$ such that $M_{i_0} \nsubseteq N$ and hence $N \cap M_{i_0} = 0$. Also S is partially ordered by set inclusion and the union of a chain in S belongs to S. Apply Zorn's Lemma and deduce the existence of a maximal element of S, say J. Put $L = \langle M_j \mid j \in J \rangle$. Then $N + L = N \oplus L$. Assume that $N + L \neq M$; thus there is an $i \in I$ such that $M_i \nsubseteq N + L$. Therefore $M_i \cap (N + L) = 0$, since M_i is simple. Next $N \cap (M_i \oplus L) \neq 0$ by maximality of L. Thus there exists $n \neq 0$ in N such that $n = m + \ell$ where $m \in M_i$, $\ell \in L$; it follows that $m = n - \ell \in M_i \cap (N + L) = 0$, a contradiction. Therefore $M = N \oplus L$. Finally, $N \overset{R}{\simeq} M/L$, so N is semisimple by (9.1.10). $\qquad\square$

The next result is the converse of (9.1.11): it identifies semisimple modules as precisely those modules in which every submodule is a direct summand.

(9.1.12). *Let M be a module over a ring with identity. Assume that every submodule of M is a direct summand. Then M is semisimple.*

Proof. The main step in the proof is to show that if N is a non-zero submodule of M, then N contains a simple submodule. Let $0 \neq a \in N$. Apply Zorn's Lemma to show that there is a submodule L of N which is maximal subject to $a \notin L$. By hypothesis $M = L \oplus P$ for some submodule P. Intersecting with N, we obtain $N = N \cap (L \oplus P) = L \oplus P_1$ where $P_1 = N \cap P$. If P_1 is simple, we are done, so assume this is not the case and let Q_1 be a proper non-zero submodule of P_1. Then Q_1 is a direct summand of M and hence of P_1 by intersecting with P_1. Hence there is an expression $P_1 = Q_1 \oplus Q_2$ for some submodule $Q_2 \neq 0$. Thus L is properly contained in both $L \oplus Q_1$ and $L \oplus Q_2$. Therefore by maximality of L we conclude that $a \in L \oplus Q_i$ for $i = 1, 2$. However, $(L \oplus Q_1) \cap (L \oplus Q_2) = L$, since $L \cap P_1 = 0$. This yields the contradiction $a \in L$. It follows that N has a simple submodule.

Next let S be the submodule generated by all the simple submodules of M; then S is semisimple by (9.1.10). By hypothesis $M = S \oplus T$ for some submodule T. If $T \neq 0$, it has a simple submodule T_0 by the first part of the proof. But then $T_0 \leq S \cap T = 0$. By this contradiction $T = 0$ and $M = S$, so M is semisimple. $\square$

Finiteness conditions on modules

Modules are frequently studied in conjunction with finiteness restrictions on their submodules.

(9.1.13). *Let S be a non-empty set of submodules of a module. Then the following statements about S are equivalent.*

(i) *The set S satisfies the* ascending chain condition, *i. e., there does not exist an infinite ascending chain of submodules $M_1 \subset M_2 \subset \cdots$ with $M_i \in S$.*

(ii) *The set S satisfies the* maximal condition, *which asserts that every non-empty subset of S has a maximal element, i. e., an element which is not properly contained in any other element of S.*

The corresponding result for finiteness conditions on ideals in a ring was proved in (6.4.1). The proof of (9.1.13) is very similar. A module for which the set of all submodules satisfies the equivalent conditions in (9.1.13) is said to be *noetherian*. Notice that if R is a ring, then $_RR$ is a noetherian R-module if and only if R is a left noetherian ring – see Section 6.4.

The next result provides some insight into the nature of the noetherian condition for modules.

(9.1.14). *Let M be a module. Then M is noetherian if and only if every submodule of M is finitely generated.*

Again there was a similar result (6.4.2) for rings and ideals: the proof of (9.1.14) is nearly identical.

A noetherian module is always finitely generated, as (9.1.14) shows, but the converse is false: finitely generated modules need not be noetherian – see Exercise (9.1.9). Therefore the next result is of interest.

(9.1.15). *Let R is a left noetherian ring with identity and let M be a finitely generated R-module. Then M is noetherian.*

Proof. By hypothesis there exist elements $a_1, a_2, \ldots, a_k$ such that $M = R \cdot a_1 + R \cdot a_2 + \cdots + R \cdot a_k$. Since $R \cdot a \overset{R}{\simeq} {}_R R/\text{Ann}_R(a)$ by (9.1.8) and ${}_R R$ is noetherian, we see that $R \cdot a$ is a noetherian R-module. Thus the result is true when $k = 1$. Let $k > 1$ and argue by induction on k; then $N = R \cdot a_2 + \cdots + R \cdot a_k$ is noetherian. Next $M = R \cdot a_1 + N$ and $M/N \overset{R}{\simeq} R \cdot a_1/(R \cdot a_1) \cap N$ by (9.1.6); this quotient is noetherian since $R \cdot a_1$ is noetherian. Finally, since M/N and N are both noetherian, it follows that M is noetherian by Exercise (9.1.10). □

This result provides many examples of noetherian modules. Recall from (6.4.5) that a finitely generated commutative ring with identity is noetherian. Therefore by (9.1.15) *a finitely generated module over a finitely generated commutative ring with identity is noetherian.*

Bases and free modules

The concept of a basis of a vector space extends in a natural way to modules. Let M be a left module over a ring R with identity. A non-empty subset X of M is called an *R-basis of M* if the following hold:

(i) $M = R \cdot X$, i. e., X generates M as an R-module.
(ii) X is *R-linearly independent*, i. e., if $r_1 \cdot x_1 + r_2 \cdot x_2 + \cdots + r_k \cdot x_k = 0$ with $r_i \in R$ and distinct $x_i \in X$, then $r_1 = r_2 = \cdots = r_k = 0$.

It is easy to see that these properties taken together are equivalent to every element of the module having a unique expression as an R-linear combination of elements of X: cf. (8.2.5).

Unlike vector spaces, a module need not have a basis. Indeed there are abelian groups without non-trivial elements of finite order that have no bases.

Example (9.1.1). The additive group $\mathbb{Q}$ of rational numbers does not have a basis.

For suppose that $\mathbb{Q}$ has a basis X. If X contains two different elements $\frac{m_1}{n_1}, \frac{m_2}{n_2}$, then

$$m_2 n_1 \frac{m_1}{n_1} - m_1 n_2 \frac{m_2}{n_2} = 0,$$

contradicting linear independence. Therefore X has a single element, which is certainly false since Q is not cyclic.

Let R be a ring with identity and M a left R-module. If M has a basis X, then it is called a *free module on X*. If R is a field, all non-zero modules are free since every non-zero vector space has a basis, but, as has been seen, not every $\mathbb{Z}$-module has a basis. Free $\mathbb{Z}$-modules are called *free abelian groups*.

We will investigate the properties of free modules next. Let M be a free R-module with a basis X. Then $M = \sum_{x \in X} R \cdot x$ and also $(R \cdot x) \cap \sum_{y \in X - \{x\}} R \cdot y = 0$ by uniqueness of expression as a linear combination of basis elements. Hence $M = \bigoplus_{x \in X} R \cdot x$. Next $R \cdot x$ is clearly a cyclic module, so $R \cdot x \simeq {}_R R/L$ where $L = \operatorname{Ann}_R(x)$ by (9.1.8). If $r \in L$, then $0 = r \cdot x = 0 \cdot x$, from which it follows by uniqueness of expression that $r = 0$ and $L = 0$. Thus $R \cdot x \overset{R}{\simeq} {}_R R$. These conclusions are summed up in:

(9.1.16). *Let R be a ring with identity and let M be a free R-module with a basis X. Then $M = \bigoplus_{x \in X} M_x$ where $M_x \overset{R}{\simeq} {}_R R$.*

The significance of free modules in module theory becomes clear from the next result, which states that every module is isomorphic with a quotient of a free module.

(9.1.17). *Let R be a ring with identity and let M be a left R-module which is generated by a subset $X = \{x_\lambda \mid \lambda \in \Lambda\}$. If F is a free left R-module with basis $\bar{X} = \{\bar{x}_\lambda \mid \lambda \in \Lambda\}$, there is a surjective R-module homomorphism $\theta : F \to M$ such that $\theta(\bar{x}_\lambda) = x_\lambda$ for all $\lambda \in \Lambda$. Thus $M \overset{R}{\simeq} F/\operatorname{Ker}(\theta)$.*

Proof. If $f \in F$, there is a *unique* expression $f = r_1 \cdot \bar{x}_{\lambda_1} + r_2 \cdot \bar{x}_{\lambda_2} + \cdots + r_n \cdot \bar{x}_{\lambda_n}$ with $r_i \in R$ and distinct $x_{\lambda_i} \in \bar{X}$. Define $\theta(f) = r_1 \cdot x_{\lambda_1} + r_2 \cdot x_{\lambda_2} + \cdots + r_n \cdot x_{\lambda_n}$. Then θ is a surjective module homomorphism from F to M. $\square$

Next comes a useful property of free modules.

(9.1.18). *Let R be a ring with identity and M a left R-module with a submodule N such that M/N is free. Then there is a submodule F such that $M = N \oplus F$ and $F \overset{R}{\simeq} M/N$.*

Proof. Let $X = \{x_\lambda + N \mid \lambda \in \Lambda\}$ be an R-basis of M/N and let F be the submodule of M generated by all the elements x_λ. Certainly $M = N + F$. Suppose that $f \in N \cap F$. Then $f = r_1 \cdot x_{\lambda_1} + \cdots + r_n \cdot x_{\lambda_n}$ where $r_i \in R$ and the x_{λ_i} are distinct. Hence

$$r_1 \cdot (x_{\lambda_1} + N) + \cdots + r_n \cdot (x_{\lambda_n} + N) = f + N = 0_{M/N}.$$

Since X is a basis of M/N, it follows that $r_i = 0$ for all i and $f = 0$. Therefore $N \cap F = 0$ and $M = N \oplus F$. $\square$

Finally, we address the question of the cardinality of a basis in a free module. Recall from (8.2.8) that any two bases of a vector space have the same cardinal, which is termed the dimension of the space. In general it is possible for a free module to have bases of different cardinalities. However, the situation is completely different in the case of modules over commutative rings.

(9.1.19). *Let M be a free module over a commutative ring R with identity. Then any two bases of M have the same cardinality.*

Proof. Since R has identity, it has a maximal proper ideal S. Then $K = R/S$ is a field by (6.3.7) since R is commutative. Let N be the subgroup of M generated by all elements of the form $s \cdot a$ where $s \in S$, $a \in M$. Then N is a submodule and $\bar{M} = M/N$ is a K-vector space via the action $(r + S) \cdot (a + N) = r \cdot a + N$, where $r \in R$, $a \in M$: here it is necessary to check that this action is well defined.

Next let $X = \{x_\lambda \mid \lambda \in \Lambda\}$ be a basis of M: we will show that $\bar{X} = \{x_\lambda + N \mid \lambda \in \Lambda\}$ is a K-basis of $\bar{M}$. Clearly $\bar{X}$ generates $\bar{M}$, so it remains to establish K-linear independence. Suppose that $\sum_{i=1}^{n}(r_{\lambda_i} + S) \cdot (x_{\lambda_i} + N) = 0_{\bar{M}}$ where the λ_i are distinct and $r_{\lambda_i} \in R$. Then $\sum_{i=1}^{n} r_{\lambda_i} \cdot x_{\lambda_i} \in N$. After adjusting the notation, we arrive at a relation of the form

$$\sum_{i=1}^{n} r_{\lambda_i} \cdot x_{\lambda_i} = \sum_{i=1}^{n} s_{\lambda_i} \cdot x_{\lambda_i}$$

for certain $s_{\lambda_i} \in S$. Since the x_{λ_i} are linearly independent, it follows that $r_{\lambda_i} = s_{\lambda_i} \in S$ and $r_{\lambda_i} + S = 0_K$. Therefore the $x_{\lambda_i} + N$ are linearly independent, so they form a basis of the K-space $\bar{M}$. Hence $|X| = |\bar{X}| = \dim_K(\bar{M})$, which shows that all R-bases of M have this cardinality. $\qquad\square$

The cardinality of a basis of a free module F, when this is unique, is called the *rank* of F, in symbols rank(F). A zero module is regarded as a free module of rank 0.

Homomorphism groups

Let M and N be left modules over a ring R. The set of all R-module homomorphisms from M to N is written

$$\mathrm{Hom}_R(M, N).$$

This set can be endowed with the structure of an abelian group in which the group operation is defined as follows. If $\alpha, \beta \in \mathrm{Hom}_R(M, N)$, then $\alpha + \beta : M \to N$ is given by the rule

$$(\alpha + \beta)(a) = \alpha(a) + \beta(a)$$

where $a \in M$. It is a simple verification that $\alpha + \beta = \beta + \alpha \in \mathrm{Hom}_R(M, N)$. The identity element is the zero mapping $a \mapsto 0_N$ and the negative of α is $-\alpha$ where $(-\alpha)(a) = -(\alpha(a))$. The group axioms are quickly verified.

A homomorphism from an R-module M to itself is called an *endomorphism* of M. The endomorphisms of an R-module M form a ring in which the product operation is functional composition, as a simple check shows. The ring of endomorphisms of an R-module M is denoted by

$$\mathrm{End}_R(M).$$

When $R = F$ is a field, $\text{Hom}_F(V, W) = L(V, W)$, the set of F-linear mappings from V to W, and $\text{End}_F(V) = L(V)$, is the set of linear operators on the F-vector space V. In particular $\text{Hom}_F(V, W)$ is an F-vector space. In general $\text{Hom}_R(M, N)$ is not an R-module, but it can inherit a module structure from M or N. This shown by the next result.

(9.1.20). *Let $_R M_S$ and $_R N_T$ be bimodules with respect to rings R, S, T as indicated. Then $\text{Hom}_R(M, N)$ is an (S, T)-bimodule in which the module actions of S and T are given by*

$$(s \cdot \alpha)(a) = \alpha(a \cdot s) \quad \text{and} \quad (\alpha \cdot t)(a) = \alpha(a) \cdot t$$

where $a \in M, s \in S, t \in T$.

Proof. We will check the module axioms for the first action, leaving the second action to the reader. Let $\alpha \in \text{Hom}_R(M, N)$, $a_i, a \in M$, $r \in R$, $s \in S$; then $(s \cdot \alpha)(a_1 + a_2) = \alpha((a_1 + a_2) \cdot s) = \alpha(a_1 \cdot s + a_2 \cdot s) = \alpha(a_1 \cdot s) + \alpha(a_2 \cdot s) = (s \cdot \alpha)(a_1) + (s \cdot \alpha)(a_2)$. Also $(s \cdot \alpha)(r \cdot a) = \alpha((r \cdot a) \cdot s) = \alpha(r \cdot (a \cdot s)) = r \cdot (\alpha(a \cdot s)) = r \cdot ((s \cdot \alpha)(a))$. Hence $s \cdot \alpha \in \text{Hom}_R(M, N)$.

Next it must be proved that $s \cdot (\alpha_1 + \alpha_2) = s \cdot \alpha_1 + s \cdot \alpha_2$, $(s_1 + s_2) \cdot \alpha = s_1 \cdot \alpha + s_2 \cdot \alpha$ and $s_1 \cdot (s_2 \cdot \alpha) = (s_1 s_2) \cdot \alpha$, where $s, s_i \in S, \alpha, \alpha_i \in \text{Hom}_R(M, N)$. Let us take the third statement, leaving the others to the reader. If $a \in M$, we have $(s_1 \cdot (s_2 \cdot \alpha))(a) = (s_2 \cdot \alpha)(a \cdot s_1) = \alpha((a \cdot s_1) \cdot s_2) = \alpha(a \cdot (s_1 s_2)) = ((s_1 s_2) \cdot \alpha)(a)$, as required.

Finally, check the bimodule property. Let $\alpha \in \text{Hom}_R(M, N)$, $s \in S$, $t \in T$; then $((s \cdot \alpha) \cdot t)(a) = ((s \cdot \alpha)(a)) \cdot t = (\alpha(a \cdot s)) \cdot t = (\alpha \cdot t)(a \cdot s) = (s \cdot (\alpha \cdot t))(a)$ for all $a \in M$. Therefore $(s \cdot \alpha) \cdot t = s \cdot (\alpha \cdot t)$. $\square$

Of course, if we just have $_R M_S$ or $_R N_T$, then $\text{Hom}_R(M, N)$ is merely a left S-module or a right T-module respectively. Next comes a useful result applicable to the computation of homomorphism groups.

(9.1.21). *Let R, S be rings with identity elements and let $_R M_S$ be a bimodule as indicated. Then*

$$\text{Hom}_R(R, M) \overset{R,S}{\cong} M.$$

Proof. First of all notice that $\text{Hom}_R(R, M)$ is an (R, S)-bimodule. A function $\Phi : \text{Hom}_R(R, M) \rightarrow M$ is defined by $\Phi(\theta) = \theta(1_R)$. Verify that Φ is a homomorphism of (R, S)-bimodules. Next we show that Φ is injective. Suppose that $\Phi(\theta) = 0$, so $\theta(1_R) = 0$; then $\theta(r) = \theta(r \cdot 1_R) = r \cdot \theta(1_R) = 0$ for all $r \in R$ and $\theta = 0$. Finally, Φ is surjective. For, let $a \in M$ and define $\theta_a : \text{Hom}_R(R, M) \rightarrow M$ by $\theta_a(r) = r \cdot a$. Then $\Phi(\theta_a) = \theta_a(1_R) = 1_R \cdot a = a$, showing that Φ is surjective and hence is an isomorphism. $\square$

Schur's Lemma

The following result is of importance in calculating $\mathrm{Hom}_R(M, N)$ when both modules are simple; it is traditionally known as Schur's Lemma.[1]

(9.1.22). *Let $_RM$ and $_RN$ be simple modules over a ring R.*
(i) *If M and N are not isomorphic, then $\mathrm{Hom}_R(M, N) = 0$.*
(ii) *If M and N are isomorphic, then $\mathrm{Hom}_R(M, N) \simeq \mathrm{End}_R(M)$ is a division ring.*

Proof. Let $\alpha : M \to N$ be an R-homomorphism. Since $\mathrm{Ker}(\alpha)$ and $\mathrm{Im}(\alpha)$ are submodules of M and N respectively, either $\alpha = 0$ or else $\mathrm{Ker}(\alpha) = 0$ and $\mathrm{Im}(\alpha) = N$, that is to say, α is an isomorphism. Therefore $\mathrm{Hom}_R(M, N) = 0$ if M and N are not isomorphic. If M and N are isomorphic, then $\mathrm{Hom}_R(M, N) \simeq \mathrm{Hom}_R(M, M) = \mathrm{End}_R(M)$ and every non-zero element of the latter has an inverse: thus $\mathrm{End}_R(M)$ is a division ring. □

Induced and coinduced mappings

When a homomorphism α between modules is given, there is an induced homomorphism α_* and a coinduced homomorphism α^* between homomorphism groups, as explained below.

(9.1.23). *Let A, B, M be left modules over a ring R and let $\alpha : A \to B$ be a module homomorphism. Then the following are true.*
(i) *There is a group homomorphism $\alpha_* : \mathrm{Hom}_R(M, A) \to \mathrm{Hom}_R(M, B)$ such that $\alpha_*(\theta) = \alpha\theta$ for $\theta \in \mathrm{Hom}_R(M, A)$.*
(ii) *There is a group homomorphism $\alpha^* : \mathrm{Hom}_R(B, M) \to \mathrm{Hom}_R(A, M)$ such that $\alpha^*(\phi) = \phi\alpha$ for $\phi \in \mathrm{Hom}_R(M, B)$.*

Proof. Only (ii) will be proved. Let $\phi \in \mathrm{Hom}_R(B, M)$. Certainly $\alpha^*(\phi) = \phi\alpha$ is a function from A to M. We check that it is an R-module homomorphism. Let $a, a_i \in A$ and $r \in R$. Firstly $\alpha^*(\phi)(a_1 + a_2) = \phi\alpha(a_1 + a_2) = \phi(\alpha(a_1) + \alpha(a_2)) = \phi\alpha(a_1) + \phi\alpha(a_2) = \alpha^*(\phi)(a_1) + \alpha^*(\phi)(a_2)$. Then $(\alpha^*(\phi))(r \cdot a) = \phi\alpha(r \cdot a) = \phi(r \cdot \alpha(a)) = r \cdot (\phi\alpha(a)) = r \cdot (\alpha^*(\phi)(a))$. Hence $\alpha^*(\phi) \in \mathrm{Hom}_R(B, M)$.

Finally, we prove that α^* is a group homomorphism. Let $\phi_i \in \mathrm{Hom}_R(B, M)$. Then $\alpha^*(\phi_1 + \phi_2) = (\phi_1 + \phi_2)\alpha = \phi_1\alpha + \phi_2\alpha = \alpha^*(\phi_1) + \alpha^*(\phi_2)$, as required. □

The induced and coinduced homomorphisms just introduced have notable properties when applied to composites. In the next result the $*$ notation for these homomorphisms is used.

(9.1.24). *Let A, B, C, M be left modules over a ring R and let $\alpha : A \to B$ and $\beta : B \to C$ be R-module homomorphisms. Then (i) $(\beta\alpha)_* = \beta_*\alpha_*$ and (ii) $(\beta\alpha)^* = \alpha^*\beta^*$.*

---

[1] Issai Schur (1875–1941).

Proof. For example, to prove (ii) let $\phi \in \mathrm{Hom}_R(C, M)$. Then $(\beta\alpha)^*(\phi) = \phi(\beta\alpha) = (\phi\beta)\alpha = \beta^*(\phi)\alpha = \alpha^*(\beta^*(\phi)) = \alpha^*\beta^*(\phi)$ and hence $(\beta\alpha)^* = \alpha^*\beta^*$. $\qquad\square$

Exact sequences

An *exact sequence* of modules over a ring R is a chain of R-modules and R-module homomorphisms

$$\cdots \longrightarrow A_{i-1} \xrightarrow{\alpha_{i-1}} A_i \xrightarrow{\alpha_i} A_{i+1} \longrightarrow \cdots, \quad (i \in I),$$

where I is a linearly ordered set, such that $\mathrm{Im}(\alpha_{i-1}) = \mathrm{Ker}(\alpha_i)$ for all i. Here the chain can be finite or infinite in either direction. We note some important special types of exact sequences:

$$0 \to A \xrightarrow{\alpha} B \xrightarrow{\beta} C \quad \text{and} \quad A \xrightarrow{\alpha} B \xrightarrow{\beta} C \to 0.$$

In the first sequence exactness at A means that $\mathrm{Ker}(\alpha) = 0$, i. e., α is injective: in the second sequence exactness at C shows that $\mathrm{Im}(\beta) = C$, i. e., β is surjective. The combination of the two types

$$0 \to A \xrightarrow{\alpha} B \xrightarrow{\beta} C \to 0$$

is called a *short exact sequence*: in this case $A \overset{R}{\simeq} \mathrm{Im}(\alpha) = \mathrm{Ker}(\beta)$ and $B/\mathrm{Ker}(\beta) \overset{R}{\simeq} C$. Note that the maps $0 \to A$ and $C \to 0$ are necessarily zero maps.

The Hom construction has the critical property of preserving exactness of sequences on the left.

(9.1.25) (Left exactness of Hom). *Let M be a left R-module where R is an arbitrary ring. Then the following hold.*

(i) *If $0 \to A \xrightarrow{\alpha} B \xrightarrow{\beta} C$ is an exact sequence of left R-modules, the induced sequence of abelian groups and homomorphisms*

$$0 \to \mathrm{Hom}_R(M, A) \xrightarrow{\alpha_*} \mathrm{Hom}_R(M, B) \xrightarrow{\beta_*} \mathrm{Hom}_R(M, C)$$

is exact.

(ii) *If $A \xrightarrow{\alpha} B \xrightarrow{\beta} C \to 0$ is an exact sequence of left R-modules, the coinduced sequence of abelian groups and homomorphisms*

$$0 \to \mathrm{Hom}_R(C, M) \xrightarrow{\beta^*} \mathrm{Hom}_R(B, M) \xrightarrow{\alpha^*} \mathrm{Hom}_R(A, M)$$

is exact.

Proof. Only (i) will be proved, the proof of (ii) being similar. Firstly, α_* is injective. For suppose that $\alpha_*(\theta) = 0$, i. e., $\alpha\theta = 0$. Since α is injective, it follows that $\theta = 0$ and hence the sequence is exact at $\mathrm{Hom}_R(M, A)$.

Now for exactness at $\mathrm{Hom}_R(M, B)$, i. e., $\mathrm{Ker}(\beta_*) = \mathrm{Im}(\alpha_*)$. Since $\mathrm{Im}(\alpha) = \mathrm{Ker}(\beta)$, we have $\beta_*\alpha_* = (\beta\alpha)_* = 0_* = 0$ by (9.1.24). Hence $\mathrm{Im}(\alpha_*) \subseteq \mathrm{Ker}(\beta_*)$. Next let $\phi \in \mathrm{Ker}(\beta_*)$, so that we have $0 = \beta_*(\phi) = \beta\phi$. If $m \in M$, then $\beta\phi(m) = 0$, so $\phi(m) \in \mathrm{Ker}(\beta) = \mathrm{Im}(\alpha)$. Hence $\phi(m) = \alpha(a)$ for some $a \in A$. In fact the element a is unique: for, if also $\phi(m) = \alpha(a')$, then $a = a'$ by injectivity of α. This allows us to define unambiguously a function $\theta : M \to A$ by $\theta(m) = a$ where $\phi(m) = \alpha(a)$. It is easy to see that θ is an R-module homomorphism. Next $(\alpha_*(\theta))(m) = \alpha\theta(m) = \alpha(a) = \phi(m)$ for all $m \in M$. Therefore $\alpha_*(\theta) = \phi$ and $\phi \in \mathrm{Im}(\alpha_*)$, so that $\mathrm{Ker}(\beta_*) = \mathrm{Im}(\alpha_*)$, as was to be proved. $\square$

Exercises (9.1).

(1) Let L, M, N be submodules of an R-module such that $N \subseteq M$. Prove the following statements.

   (i) $(L \cap M)/(L \cap N)$ is R-isomorphic with a submodule of M/N.

   (ii) $(L + M)/(L + N)$ is R-isomorphic with a quotient of M/N.

(2) Let L, M, N be submodules of an R-module such that $N \subseteq M$. If $L + M = L + N$ and $L \cap M = L \cap N$, prove that $M = N$.

(3) Let X be a non-empty subset of an R-module M. If the ring R does *not* have an identity element, what is the general form of an element of the submodule of M generated by X?

(4) If $\alpha : M \to N$ is a module isomorphism, show that $\alpha^{-1} : N \to M$ is also a module isomorphism.

(5) State and prove the Correspondence Theorem for modules.

(6) Let R be a commutative ring with identity. Prove that R is a field if and only if every non-zero cyclic R-module is isomorphic with R.

(7) Let R, S be rings and let $_RM_S$ be a bimodule as indicated. If R has an identity element, prove that $\mathrm{Hom}_R(_RR, \, _RM_S) \overset{S}{\cong} M$.

(8) Establish (9.1.25)(ii).

(9) Give an example of a finitely generated module that is not noetherian. [Hint: if R is a ring with identity, then $_RR$ is a finitely generated R-module.]

(10) Let M be a module with a submodule N. If N and M/N are noetherian, prove that M is noetherian.

(11) Let M be an R-module with a submodule N such that $M \overset{R}{\cong} M/N$. If M is noetherian, prove that $N = 0$. [Hint: use the ascending chain form of the noetherian property.]

(12) Let u, v be elements of a principal ideal domain R such that $\gcd\{u, v\} = 1$. Prove that $R/Ru \oplus R/Rv \overset{R}{\cong} R/Ruv$. Then extend the result to n relatively prime elements $u_1, u_2, \ldots, u_n$. [Hint: show that the assignment $r + (uv) \mapsto (r + (u), r + (v))$, $r \in R$ determines an isomorphism by using the Chinese Remainder Theorem for R.]

(13) Explain how to define the unrestricted and restricted direct sums of an infinite set of modules by using choice functions as in Section 4.2.

(14) An exact sequence of R-modules and homomorphisms $0 \to A \xrightarrow{\alpha} B \xrightarrow{\beta} C \to 0$ is said to *split* if there is a module homomorphism $\gamma : C \to B$ such that $\beta\gamma$ is the identity map on C. Prove that in this event $B = \mathrm{Im}(\alpha) \oplus \mathrm{Im}(\gamma) \overset{R}{\simeq} A \oplus C$.

(15) Show that an exact sequence $0 \to A \to B \to F \to 0$ always splits if F is a free module.

(16) Prove that the exact sequence $0 \to \mathbb{Z} \to \mathbb{Q} \to \mathbb{Q}/\mathbb{Z} \to 0$ in which all the maps are the natural ones does not split.

(17) Establish the following analogue of (9.1.13). Let $\mathcal{S}$ be a non-empty set of submodules of a module. Then the following statements about $\mathcal{S}$ are equivalent.

   (i)   The set $\mathcal{S}$ satisfies the descending chain condition, i. e., there does not exist an infinite descending chain of submodules $M_1 \supset M_2 \supset \cdots$ with $M_i \in \mathcal{S}$.

   (ii) The set $\mathcal{S}$ satisfies the minimal condition, which asserts that every non-empty subset of $\mathcal{S}$ has a minimal element, i. e., an element which does not properly contain any other element of $\mathcal{S}$.

   (If the set of all submodules satisfies these finiteness conditions, the module is said to be *artinian*.)

(18) Let A, B, C be modules over a ring R. Prove that $\mathrm{Hom}_R(A \oplus B, C) \simeq \mathrm{Hom}_R(A, C) \oplus \mathrm{Hom}_R(B, C)$ and $\mathrm{Hom}_R(A, B \oplus C) \simeq \mathrm{Hom}_R(A, B) \oplus \mathrm{Hom}_R(A, C)$.

(19) Let R be a ring with identity and let F be a free R-module with a basis X. If $\alpha : X \to M$ is a map from X into some R-module M, prove that there is a unique R-module homomorphism $\beta : F \to M$ such that $\beta\mu = \alpha$, where $\mu : X \to F$ is the inclusion map.

(20) Let R be a ring with identity and let F be an R-module with a set of generators X. Assume that the property described in Exercise (9.1.19) holds for F and X. Prove that X is a basis of F, so that F is a free module. [Hint: form a free module M on X and then a module homomorphism from F to M; from this deduce that X is R-linearly independent in F.]

9.2 Modules over principal ideal domains

In this section we restrict attention to modules over commutative rings. The main objective is to determine the structure of finitely generated modules over PID's. This is one of the central results of abstract algebra and it has applications to finitely generated abelian groups, linear operators on finite dimensional vector spaces and canonical forms of matrices.

Torsion elements

Let R be a commutative ring with identity and let M be an R-module. Recall that we need not distinguish between left and right modules. An element a of M is called an

R-torsion element if there exists $r \neq 0$ in R such that $r \cdot a = 0$. Equivalently, the annihilator $\mathrm{Ann}_R(a)$ is a non-zero ideal of R. If every element of M is a torsion element, M is called an *R-torsion module*. On the other hand, if 0 is the only torsion element of M, the module is said to be *R-torsion-free*. (The terminology comes from topology.)

For example, a torsion element of a $\mathbb{Z}$-module, i.e., an abelian group, is an element of finite order and a torsion-free $\mathbb{Z}$-module is an abelian group in which every non-trivial element has infinite order.

(9.2.1). *Let R be an integral domain and M an R-module. Then the torsion elements of M form a submodule T, the torsion submodule, such that M/T is torsion-free.*

Proof. Let a and b be torsion elements of M; thus there exist $r, s \neq 0$ in R such that $r \cdot a = 0 = s \cdot b$. Since R is an integral domain, $rs \neq 0$. Now $rs \cdot (a \pm b) = s \cdot (r \cdot a) \pm r \cdot (s \cdot b) = 0$, which shows that $a \pm b \in T$. Next let $u \in R$; then $r \cdot (u \cdot a) = u \cdot (r \cdot a) = u \cdot 0 = 0$, so $u \cdot a \in T$. Hence T is a submodule.

Now suppose that $a + T$ is a torsion element of M/T. Then $r \cdot (a + T) = 0_{M/T} = T$ for some $r \neq 0$ in R, that is, $r \cdot a \in T$. Therefore $s \cdot (r \cdot a) = 0$ for some $s \neq 0$ in R. Hence $(sr) \cdot a = 0$ and $sr \neq 0$, from which it follows that $a \in T$ and $a + T = T = 0_{M/T}$. $\square$

p-Torsion modules

Next the concept of a torsion module will be refined. Let p denote an irreducible element of an integral domain R: thus the only divisors of p are associates and units. An element a of an R-module M is termed a *p-torsion element* if $p^i \cdot a = 0$ for some $i > 0$. If every element of M is p-torsion, then M is called a *p-torsion* module.

(9.2.2). *Let M be a module over an integral domain R and let p be an irreducible element of R. Then the following statements hold.*
(i) *The p-torsion elements form a submodule M_p of M, (called the p-torsion submodule).*
(ii) *If R is a principal ideal domain, a non-zero element a in M is a p-torsion element if and only if $\mathrm{Ann}_R(a) = (p^i)$ for some $i > 0$.*

Proof. The proof of (i) is a simple exercise. As for (ii), let $I = \mathrm{Ann}_R(a)$; then $I = (s)$ where $s \in R$ is a non-zero, non-unit, since R is a PID. If a is a p-torsion element, $p^j \in I = (s)$ for some $j > 0$ and hence s divides p^j. Since R is a UFD by (7.3.2), it follows that $s = p^i u$ where $0 < i \leq j$ and u a unit of R. Therefore $I = (s) = (p^i)$. The converse is clear. $\square$

The first really significant result about torsion modules over a PID is:

(9.2.3) (The Primary Decomposition Theorem). *Let M be a torsion module over a principal ideal domain R and let P be a complete set of irreducible elements for R. Then M is the direct sum of the p-torsion components M_p for $p \in P$.*

Proof. Let $0 \neq a \in M$. Since M is a torsion module, there exists $r \neq 0$ in R such that $r \cdot a = 0$. Note that r cannot be a unit of R since otherwise $a = 0$. Write $r = up_1^{e_1}p_2^{e_2}\cdots p_k^{e_k}$ where the p_i are distinct elements of P, $e_i > 0$ and u is a unit of R. Let r_i denote the product that remains when the factor $p_i^{e_i}$ is deleted from r. Then $r_1, r_2, \ldots, r_k$ are relatively prime since they have no common irreducible factors. By (7.2.3) applied repeatedly, there exist $s_i \in R$ such that $r_1s_1 + r_2s_2 + \cdots + r_ks_k = 1$. Consequently $a = 1 \cdot a = (r_1s_1) \cdot a + (r_2s_2) \cdot a + \cdots + (r_ks_k) \cdot a$. Now $p_i^{e_i} \cdot ((r_is_i) \cdot a) = (s_ip_i^{e_i}r_i) \cdot a = s_i \cdot (r \cdot a) = 0$. Hence $(r_is_i) \cdot a \in M_{p_i}$, from which it follows that M is the sum of the submodules M_p with $p \in P$.

To complete the proof it must be shown that the sum is direct. Suppose that $b \in M_p \cap \sum_{q \in P - \{p\}} M_q$. Then $p^m \cdot b = 0$ for some $m > 0$. Also there is an expression $b = b_1 + b_2 + \cdots + b_\ell$ with $b_i \in M_{q_i}$, $q_i \in P - \{p\}$ where the q_i are distinct. Thus $q_i^{m_i} \cdot b_i = 0$ for some $m_i > 0$, and hence $q \cdot b = 0$ where $q = q_1^{m_1}q_2^{m_2}\cdots q_\ell^{m_\ell}$. Since none of the q_i can equal p, the elements q and p^m are relatively prime and hence by (7.2.3) there exist $u, v \in R$ such that $p^mu + qv = 1$. Therefore

$$b = 1 \cdot b = (p^mu + qv) \cdot b = u \cdot (p^m \cdot b) + v \cdot (q \cdot b) = 0,$$

and it follows that $M_p \cap \sum_{q \in \pi - \{p\}} M_q = 0$, so the sum is direct. $\qquad\square$

What the Primary Decomposition Theorem does is reduce the study of torsion modules over a PID to the case of p-torsion modules.

Submodules of free modules

Before we can proceed further with the study of finitely generated modules over PID's, we need to gain a better understanding of free modules. As a first step let us consider submodules of free modules and show these are also free. For simplicity we will discuss only free modules of finite rank, although the results are true in the infinite case as well.

(9.2.4). *Let S be a submodule of a finitely generated free module F over a principal ideal domain R. Then S is a free module with rank less than or equal to the rank of F.*

Proof. By hypothesis F has finite rank, say r. If $S = 0$, it is free with rank 0, so we can assume that $S \neq 0$ and thus $r > 0$. Suppose first that $r = 1$, so that $F \stackrel{R}{\simeq} R$. Identifying F with R, we see that S is an ideal of R and thus $S = (s)$ for some s, since R is a PID. The assignment $x \mapsto xs$, $(x \in R)$, determines a surjective R-module homomorphism from R to S. It is also injective because R is a domain, so it is a module isomorphism and $S \stackrel{R}{\simeq} R$. Thus S is a free module of rank 1.

Next assume that $r > 1$ and let $\{x_1, x_2, \ldots, x_r\}$ be a basis of F. Define F_i to be the submodule of F generated by $x_1, x_2, \ldots, x_i$, so we have the chain of submodules of F

$$0 = F_0 \subset F_1 \subset F_2 \subset \cdots \subset F_r = F.$$

Clearly F_i is free with basis $\{x_1, x_2, \ldots, x_i\}$ and rank i. Define $S_i = S \cap F_i$, a submodule of F; then there is a chain of submodules $0 = S_0 \subseteq S_1 \subseteq S_2 \subseteq \cdots \subseteq S_r = S$. By (9.1.6)

$$S_{i+1}/S_i = S \cap F_{i+1}/S \cap F_i \overset{R}{\simeq} ((S \cap F_{i+1}) + F_i)/F_i \subseteq F_{i+1}/F_i.$$

Since $F_{i+1}/F_i \overset{R}{\simeq} R$, either $S_i = S_{i+1}$ or $S_{i+1}/S_i \overset{R}{\simeq} R$ by the rank 1 case. By (9.1.18) there is a submodule T_{i+1} such that $S_{i+1} = S_i \oplus T_{i+1}$. From this it follows that $S = T_1 \oplus T_2 \oplus \cdots \oplus T_r$. In addition $T_{i+1} \overset{R}{\simeq} S_{i+1}/S_i$ and hence either $T_{i+1} = 0$ or $T_{i+1} \overset{R}{\simeq} R$. Therefore S is a free module with rank at most r. $\qquad\square$

An important consequence of the last result is:

Corollary (9.2.5). *Let R be a principal ideal domain and let M be an R-module which can be generated by n elements. If N is a submodule of M, then N can be generated by n or fewer elements.*

Proof. From (9.1.17) it follows that $M \overset{R}{\simeq} F/L$ where F is a free module of rank n and L is a submodule. By the Correspondence Theorem for modules, $N \overset{R}{\simeq} S/L$ for some submodule S of F containing L. By (9.2.4) S can be generated by n or fewer elements, from which it follows that N also has this property. $\qquad\square$

We are now equipped with sufficient knowledge of free modules over PID's to determine the structure of finitely generated, torsion-free modules.

(9.2.6). *Let M be a finitely generated torsion-free module over a principal ideal domain R. Then M is a free module.*

Proof. It may be assumed that $M \neq 0$. Suppose that M is generated by non-zero elements $a_1, a_2, \ldots, a_n$. If $n = 1$, then $M = R \cdot a_1$, so that $M \overset{R}{\simeq} R/\mathrm{Ann}_R(a_1)$ by (9.1.8). However, $\mathrm{Ann}_R(a_1) = 0$ since $a_1 \neq 0$ and M is torsion-free. Hence $M \overset{R}{\simeq} R$ and M is a free module of rank 1.

Let $n > 1$ and use induction on n. For convenience let us write $a = a_1$. Denote by $N/(R \cdot a)$ the torsion-submodule of $M/(R \cdot a)$. By (9.2.1) the module M/N is torsion-free, and clearly it can be generated by $n - 1$ elements. Therefore by induction hypothesis M/N is free and (9.1.18) shows that there is a submodule L such that $M = N \oplus L$; moreover, $L \overset{R}{\simeq} M/N$, so L is free. Thus it is enough to prove that N is a free module.

By (9.2.5) the submodule N can be finitely generated, say by $b_1, b_2, \ldots, b_k$. Since $b_i \in N$, there exists $r_i \neq 0$ in R such that $r_i \cdot b_i \in R \cdot a$. Writing $r = r_1 r_2 \cdots r_k \neq 0$, we have $r \cdot b_i \in R \cdot a$ for $i = 1, 2, \ldots, k$, which implies that $r \cdot N \subseteq R \cdot a$. But $R \cdot a \overset{R}{\simeq} R$ since $\mathrm{Ann}_R(a) = 0$, so $r \cdot N$ is free by the case $n = 1$. Finally, $N \overset{R}{\simeq} r \cdot N$ via the map $b \mapsto r \cdot b$ since M is torsion-free. Consequently N is a free module. $\qquad\square$

Corollary (9.2.7). *Let M be a finitely generated module over a principal ideal domain R and let T denote the torsion submodule of M. Then $M = T \oplus F$ where F is a free module of finite rank.*

Proof. By (9.2.1) the quotient M/T is torsion-free and it is evidently finitely generated. Hence M/T is free by (9.2.6). From (9.1.18) we deduce that $M = T \oplus F$ where $F \overset{R}{\simeq} M/T$, so F is free. □

Combining (9.2.7) with the Primary Decomposition Theorem (9.2.3), we see that the remaining obstacle to determining the structure of finitely generated modules over a PID is the case of a finitely generated p-torsion module. This is overcome in the next major result.

(9.2.8). *Let M be a finitely generated module over a principal ideal domain R. Assume that M is a p-torsion module for some irreducible element p of R. Then M is a direct sum of finitely many cyclic p-torsion modules.*

Notice that by (9.1.8) and (9.2.2) a cyclic p-torsion R-module is isomorphic with $R/(p^i)$ for some $i > 0$. Thus (9.2.8) shows that the module M is completely determined by certain powers of irreducible elements of R.

The proof of (9.2.8) is one of the harder ones in this book. The reader is advised to look out for the main ideas in the proof and try not to get bogged down in the details.

Proof of (9.2.8). We can suppose that $M \neq 0$; let it be generated by non-zero elements $b_1, b_2, \ldots, b_k$. Then $p^{e_i} \cdot b_i = 0$ where $e_i > 0$. Let e be the largest of the e_i, so that $p^e \cdot b_i = 0$ for all i and thus $p^e \cdot M = 0$. Choose e to be the smallest positive integer with this property. Hence there exists $a \in M$ such that $p^{e-1} \cdot a \neq 0$, and thus $\text{Ann}_R(a) = (p^e) = \text{Ann}_R(M)$.

The main step in the proof is to establish the following statement.

For any $a \in M$ such that $\text{Ann}_R(a) = (p^e) = \text{Ann}_R(M)$, the cyclic submodule $R \cdot a$ is a direct summand of M. $\qquad (*)$

Let us assume the statement $(*)$ is false: a series of contradictions will then ensue. By (9.2.5) every submodule of M is finitely generated and hence M is noetherian by (9.1.14). We claim that M contains a submodule M_0 which is maximal subject to having the following properties:

(i) $\bar{M} = M/M_0$ has an element $\bar{a}$ such that $\text{Ann}_R(\bar{a}) = (p^e) = \text{Ann}(\bar{M})$;
(ii) $R \cdot \bar{a}$ is *not* a direct summand of $\bar{M}$.

Certainly there are submodules with these properties, for example the zero submodule qualifies with a in place of $\bar{a}$. The maximal condition on submodules guarantees that there is a maximal one. Since we are only looking for a contradiction, we can just well work with the module $\bar{M}$: thus we will assume that $M_0 = 0$ and $M = \bar{M}$. Consequently, the statement $(*)$ is true for every proper quotient of M, but false for M itself.

Suppose first that there exists $b \in M - R \cdot a$ such that $p \cdot b = 0$. Notice that $R \cdot b$ is a module over the field $R/(p)$ since $p \cdot b = 0$; thus it is a 1-dimensional vector space over $R/(p)$. Therefore $(R \cdot a) \cap (R \cdot b)$, being a subspace of $R \cdot b$, is either 0 or $R \cdot b$. In the second case $R \cdot b \subseteq R \cdot a$ and $b \in R \cdot a$, contrary to the choice of b. Thus $(R \cdot a) \cap (R \cdot b) = 0$. Next $p^{e-1} \cdot (a + R \cdot b) = p^{e-1} \cdot a + R \cdot b$, which cannot equal $0_{M/R \cdot b}$, since otherwise $p^{e-1} \cdot a \in (R \cdot a) \cap (R \cdot b) = 0$, another contradiction. Therefore $p^{e-1} \cdot (a + R \cdot b) \neq 0_{M/R \cdot b}$ and $\mathrm{Ann}_R(a + R \cdot b) = (p^e) = \mathrm{Ann}_R(M/R \cdot b)$. This means that the module $M/R \cdot a$ and the element $a + R \cdot a$ satisfy the hypotheses of $(*)$ above. Since $M/(R \cdot b)$ is a proper quotient of M, there is a direct decomposition $M/(R \cdot b) = R \cdot (a + R \cdot b) \oplus N/(R \cdot b)$. Consequently $M = (R \cdot a) + N$, while $(R \cdot a) \cap N \subseteq (R \cdot a) \cap (R \cdot b) = 0$ and $M = R \cdot a \oplus N$, contradicting the fact that $(*)$ is false for M.

From the discussion of the previous paragraph, we conclude that $R \cdot a$ contains all elements b of M such that $p \cdot b = 0$. Let $c \in M - (R \cdot a)$ be chosen such that $\mathrm{Ann}_R(c) = (p^k)$ with k minimal. Then $1 < k \leq e$ since $p \cdot c$ cannot equal 0. Next $0 = p^k \cdot c = p^{k-1} \cdot (p \cdot c)$, and by minimality of k we have $p \cdot c \in R \cdot a$: now write $p \cdot c = r \cdot a$ with $r \in R$. Thus $0 = p^k \cdot c = p^{k-1} \cdot (p \cdot c) = p^{k-1} \cdot (r \cdot a) = (p^{k-1}r) \cdot a$, from which it follows that p^e divides $p^{k-1}r$. Since $k - 1 < e$, we deduce that p divides r. Write $r = pr'$ with $r' \in R$. Then $p \cdot c = r \cdot a = (pr') \cdot a$ and hence $p \cdot (c - r' \cdot a) = 0$. Consequently $c - r' \cdot a \in R \cdot a$ and hence $c \in R \cdot a$. This contradiction finally establishes the truth of the statement $(*)$ above.

From this point it is but a short step to finish the proof. Writing a_1 for a, we have shown that $M = R \cdot a_1 \oplus M_1$ for some finitely generated submodule M_1. Either $M_1 = 0$, in which event $M = R \cdot a_1$ and we are done, or else $M_1 \neq 0$ and the same argument may be applied to M_1, yielding $M_1 = R \cdot a_2 \oplus M_2$ and $M = R \cdot a_1 \oplus R \cdot a_2 \oplus M_2$ for a suitable element a_2 and finitely generated submodule M_2. The argument may be repeated if M_2 is non-zero, and so on. Because the ascending chain condition is valid in the noetherian module M, we will eventually reach a direct decomposition $M = R \cdot a_1 \oplus R \cdot a_2 \oplus \cdots \oplus R \cdot a_n$, and the theorem is proved. $\square$

The structure theorem for finitely generated modules over a PID can now be stated.

(9.2.9). *Let M be a finitely generated module over a principal ideal domain R. Then M is the direct sum of finitely many cyclic R-modules. More precisely*

$$M = F \oplus M_1 \oplus M_2 \oplus \cdots \oplus M_k$$

where F is a free module of finite rank $r \geq 0$ and

$$M_i = M_i(1) \oplus M_i(2) \oplus \cdots \oplus M_i(\ell_i), \quad i = 1, 2, \ldots, k,$$

where $M_i(j)$ is the direct sum of n_{ij} isomorphic copies of $R/(p_i^j)$, $(j = 1, 2, \ldots, \ell_i)$, $n_{ij} \geq 0$, $n_{i\ell_i} > 0$ and the p_i are distinct elements in a complete set of irreducible elements for R.

Proof. From (9.2.7) we have $M = F \oplus T$ where T is the torsion submodule of M and F is a finitely generated torsion-free module. By (9.2.6) the submodule F is free. Next T is finitely generated since M is noetherian, so by (9.2.3) $T = M_1 \oplus M_2 \oplus \cdots \oplus M_k$ where $M_i \neq 0$ is the p_i-torsion submodule of M and the p_i are distinct elements in a complete set of irreducibles. Finally, by (9.2.8) M_i is a direct sum of cyclic p_i-torsion modules each of which is isomorphic with some $R/(p_i^j)$. By grouping together isomorphic cyclic modules in the direct sum, we obtain the desired result. □

While the last theorem gives a clear picture of the structure of the module M, it leaves a natural question open, namely, what is the significance of the data r, k, p_i, ℓ_i, n_{ij}? The module M will usually have many direct decompositions of the type in (9.2.9), so the question arises as to whether different sets of data could arise from different decompositions. In other words we are asking if r, k, p_i, ℓ_i, n_{ij} are true invariants of the module M. The answer is supplied by the result that follows.

(9.2.10). *Let M be a finitely generated module over a principal ideal domain R and suppose that M has two direct decompositions into cyclic submodules of the type in (9.2.9),*

$$M = F \oplus M_1 \oplus M_2 \oplus \cdots \oplus M_k = \bar{F} \oplus \bar{M}_1 \oplus \bar{M}_2 \oplus \cdots \oplus \bar{M}_{\bar{k}},$$

with corresponding data r, k, p_i, ℓ_i, n_{ij} and $\bar{r}$, $\bar{k}$, $\bar{p}_i$, $\bar{\ell}_i$, $\bar{n}_{ij}$. Then $r = \bar{r}$, $k = \bar{k}$, $p_i = \bar{p}_i$, $\ell_i = \bar{\ell}_i$, $n_{ij} = \bar{n}_{ij}$, after possible reordering of the $\bar{M}_i$.

Proof. In the first place the torsion submodule of M is evidently

$$T = M_1 \oplus M_2 \oplus \cdots \oplus M_k = \bar{M}_1 \oplus \bar{M}_2 \oplus \cdots \oplus \bar{M}_{\bar{k}}.$$

Hence $F \overset{R}{\simeq} M/T \overset{R}{\simeq} \bar{F}$ and by (9.1.19) we deduce that $r = \bar{r}$. Also the p_i and $\bar{p}_i$ are the irreducible elements with non-trivial torsion components in M. Thus $k = \bar{k}$ and the $\bar{p}_i$ can be relabelled so that $p_i = \bar{p}_i$. Consequently we can assume that M itself is a p-torsion module for some irreducible element p, and that

$$M = M(1) \oplus M(2) \oplus \cdots \oplus M(\ell) = \bar{M}(1) \oplus \bar{M}(2) \oplus \cdots \oplus \bar{M}(\bar{\ell}),$$

where $M(j)$ and $\bar{M}(j)$ are direct sums of n_j and $\bar{n}_j$ copies of $R/(p^j)$ respectively. Note that n_ℓ, $\bar{n}_{\bar{\ell}} > 0$. Our task is to prove that $n_j = \bar{n}_j$ and $\ell = \bar{\ell}$.

We introduce the useful notation $M[p] = \{a \in M \mid p \cdot a = 0\}$: notice that $M[p]$ an R-submodule of M, indeed it is a vector space over the field $R/(p)$. Observe also that

$$p^m \cdot (R/(p^j) \overset{R}{\simeq} R/(p^{j-m}) \quad \text{if } m < j,$$

while $p^m \cdot (R/(p^j)) = 0$ if $m \geq j$.

A consequence of these observations is that $p^m \cdot M(j) = 0$ if $m \geq j$ and $p^m \cdot M(j)$ is the direct sum of n_j copies of $R/(p^{j-m})$ if $m < j$. Therefore $(p^m \cdot M)[p]$ is an $R/(p)$-vector space with dimension $n_{m+1} + n_{m+2} + \cdots + n_\ell$. Of course, the same argument may be applied to the second direct decomposition. Now clearly $(p^m \cdot M)[p]$ depends only on the module M, not on any particular direct decomposition of it. Therefore, on equating dimensions, we obtain the system of linear equations

$$n_{m+1} + n_{m+2} + \cdots + n_\ell = \bar{n}_{m+1} + \bar{n}_{m+2} + \cdots + \bar{n}_{\bar{\ell}}$$

for $m = 1, 2, \dots$. Since $n_\ell, \bar{n}_{\bar{\ell}} > 0$, it follows that $\ell = \bar{\ell}$. By solving back this linear system from the final equation, we find that $n_j = \bar{n}_j$, for $j = 1, 2, \dots, \ell$. $\qquad\square$

Elementary divisors and invariant factors

If M is a finitely generated module over a PID R, the invariants p_i^j for which $R/(p_i^j)$ is isomorphic with one of the direct summands of M in (9.2.9) are called the *elementary divisors* of M. The torsion submodule is determined by the elementary divisors together with their multiplicities. The elementary divisors are *invariants* of the module and do not depend on a particular direct decomposition.

Let us suppose that the elementary divisors are arranged to form a rectangular array as shown below,

$$
\begin{array}{cccc}
p_1^{r_{11}} & p_1^{r_{12}} & \cdots & p_1^{r_{1\ell}} \\
p_2^{r_{21}} & p_2^{r_{22}} & \cdots & p_2^{r_{2\ell}} \\
\cdot & \cdot & \cdots & \cdot \\
p_k^{r_{k1}} & p_k^{r_{k2}} & \cdots & p_k^{r_{k\ell}}
\end{array}
$$

where $0 \leq r_{i1} \leq r_{i2} \leq \cdots \leq r_{i\ell}$, at least one element in each row and column is different from 1, and ℓ is the maximum of $\ell_1, \ell_2, \dots, \ell_k$. Here in order to ensure that all the rows of the array have the same length, it may be necessary to introduce several 1's at the beginning of a row.

Now define

$$s_j = p_1^{r_{1j}} p_2^{r_{2j}} \cdots p_k^{r_{kj}}, \quad j = 1, 2, \dots, \ell,$$

the product of the elements in column j. The ring elements $s_1, s_2, \dots, s_\ell$, which cannot be units, are called the *invariant factors* of M. These are also invariants of the module since they are expressed in terms of the elementary divisors. The invariant factors have the noteworthy divisibility properties

$$s_1 \mid s_2 \mid \cdots \mid s_\ell$$

since $r_{ij} \leq r_{ij+1}$.

We remark that if $u, v \in R$ are relatively prime, then $R/(u) \oplus R/(v) \overset{R}{\simeq} R/(uv)$, which is Exercise (9.1.12). This observation allows us to combine all the cyclic modules associated with entries in the jth column of the array of elementary divisors into a single cyclic submodule $R/(s_j)$. In this way we obtain an alternative form of (9.2.9).

(9.2.11). *Let M be a finitely generated module over a principal ideal domain R. Then*

$$M \overset{R}{\simeq} F \oplus R/(s_1) \oplus R/(s_2) \oplus \cdots \oplus R/(s_\ell)$$

where F is a free module of finite rank and the s_j are the invariant factors of M.

Here is an example with $R = \mathbb{Z}$ to illustrate the procedure for finding the invariant factors when the elementary divisors are known.

Example (9.2.1). Consider the finite abelian group

$$A = \mathbb{Z}_2 \oplus \mathbb{Z}_2 \oplus \mathbb{Z}_2 \oplus \mathbb{Z}_3 \oplus \mathbb{Z}_5 \oplus \mathbb{Z}_{5^2}.$$

The elementary divisors of A are quickly identified from the direct decomposition as $2, 2, 2, 3, 5, 5^2$. Arrange these to form an array with 1's inserted appropriately,

$$
\begin{array}{ccc}
2 & 2 & 2 \\
1 & 1 & 3 \\
1 & 5 & 5^2
\end{array}
$$

Forming the products of the columns, we find the invariant factors to be 2, 10, 150. Therefore $A \simeq \mathbb{Z}_2 \oplus \mathbb{Z}_{10} \oplus \mathbb{Z}_{150}$.

Presentations of modules

Let R be a PID and M a finitely generated R-module generated by elements $a_1, a_2, \ldots, a_n$. Suppose that F is a free R-module with basis $\{x_1, x_2, \ldots, x_n\}$. Then by (9.1.17) there is a surjective R-module homomorphism $\theta : F \to M$ such that $\theta(x_i) = a_i$ for $i = 1, \ldots, n$. Thus $M \overset{R}{\simeq} F/N$ where $N = \mathrm{Ker}(\theta)$. By (9.2.5) N is a finitely generated R-module, say with generators $y_1, y_2, \ldots, y_m$, where $m \leq n$, and there are expressions $y_j = \sum_{k=1}^{n} u_{jk} \cdot x_k$ with $u_{jk} \in R$.

Conversely, suppose we start with a free R module F with basis $\{x_1, x_2, \ldots, x_n\}$ and elements $y_1, y_2, \ldots, y_m$ of F where $y_j = \sum_{k=1}^{n} u_{jk} \cdot x_k, u_{jk} \in R$. Let $N = R \cdot \{y_1, y_2, \ldots, y_m\}$ and put $M = F/N$. Then M is a finitely generated R-module which may be written in the form

$$M = \langle x_1, x_2, \ldots, x_n \mid y_1, y_2, \ldots, y_m \rangle.$$

This called a *presentation* of the R-module M: the x_i are the *generators* and the y_j are the *relators* of the presentation. We should think of the generators $x_1, x_2, \ldots, x_n$ as being subject to the *relations* $y_1 = 0, y_2 = 0, \ldots, y_m = 0$. The presentation is determined by the matrix

$$U = [u_{ij}]_{m,n} \in M_{m,n}(R),$$

which is called the *presentation matrix*.

Since every finitely generated R-module has a presentation which determines it up to isomorphism, a natural question arises: given a presentation, how can one discover the structure of the module? We will answer the question in the case of modules over a Euclidean domain by describing a procedure which, when applied to a presentation matrix, gives the invariant factors and hence the structure of the module determined by the presentation.

The key observation is that there three types of matrix operation that can be applied to a presentation matrix U without changing the isomorphism type of the associated module M. These are:

(I)   Interchange of two rows or columns.

(II) Addition of an R-multiple of one row to another.

(III) Addition of an R-multiple of one column to another.

Clearly interchange of two rows merely changes the order of the relators and of two columns the order of generators. Adding a multiple of row i to row j produces a new relator which is a consequence of the relator associated with row j and which also implies it.

Justification of the column operation (III) requires a little more thought. Suppose we add c times column i to column j where $c \in R$. This amounts to replacing the generator x_i by a new generator $x_i' = x_i - c \cdot x_j$ and making the substitution in the relators, as can be seen from the equation

$$u_{ri} \cdot x_i' + (u_{rj} + cu_{ri}) \cdot x_j = u_{ri} \cdot x_i + u_{rj} \cdot x_j.$$

For it shows that the new matrix represents a presentation in generators $x_1, x_2, \ldots, x_i', \ldots, x_n$ with relations equivalent to the original ones. The important point to keep in mind is that while these operations change the presentation, they do not change the isomorphism type of the corresponding module.

If a matrix V is obtained from a matrix $U \in M_{m,n}(R)$ by means of a finite sequence of operations of types (I), (II), (III) above, then V is said to be *R-equivalent* to U, in symbols

$$U \stackrel{R}{\equiv} V.$$

This is obviously an equivalence relation on matrices. The critical result needed is the following.

(9.2.12). *Let R be a Euclidean domain and U an $m \times n$ matrix with entries in R. Then U is R-equivalent to an $m \times n$ diagonal matrix*

$$V = \text{diag}(d_1, d_2, \ldots, d_k, 0, \ldots, 0)$$

where $0 \neq d_i \in R, k \geq 0$ and $d_1 | d_2 | \cdots | d_k$.

The point here is that the matrix V in (9.2.12) has $d_1, d_2, \ldots, d_k$ on the principal diagonal and zeroes elsewhere.

Proof of (9.2.12). Let $\delta : R - \{0\} \to \mathbb{N}$ be the associated function for the Euclidean domain R and recall that R is a PID by (7.2.1). We can assume that $U \neq 0$. To initiate the procedure move a non-zero entry b_1 to the $(1, 1)$ position by using row and column interchanges. Suppose that b_1 does not divide some entry c in row 1 or column 1: let us say the latter. Using the division algorithm for R, write $c = b_1 q + b_2$ where $q, b_2 \in R$ and $\delta(b_2) < \delta(b_1)$. Subtract q times row 1 from the row containing c, the effect of which is to replace c by b_2. Then move b_2 up to the $(1, 1)$ position.

If b_2 does not divide some entry in row 1 or column 1, repeat the procedure. Continuation of this process yields a sequence of elements $b_1, b_2, \ldots$, in R such that $\delta(b_1) > \delta(b_2) > \cdots$. Since the δ_i are non-negative integers, the process must terminate and when this happens, we will have a matrix R-equivalent to U with an element a_1 in the $(1, 1)$ position which divides every entry in row 1 and column 1. By further row and column subtractions we can clear out all the entries in row 1 and column 1 except the $(1, 1)$ entry to obtain a matrix of the form

$$\begin{bmatrix} a_1 & 0 \\ 0 & U_1 \end{bmatrix}$$

which is R-equivalent to U; here of course U_1 is an $(m-1) \times (n-1)$ matrix. By induction on m the matrix U_1 is R-equivalent to a matrix $\text{diag}(a_2, a_3, \ldots, a_k, 0 \ldots, 0)$ and therefore

$$U \overset{R}{\equiv} D = \text{diag}(a_1, a_2, a_3, \ldots, a_k, 0 \ldots, 0).$$

Suppose that a_1 does not divide a_2. Let $d = va_1 + wa_2$ be a gcd of a_1 and a_2 with $v, w \in R$. Then, using the operations of types (I), (II), (III), we obtain

$$\begin{bmatrix} a_1 & 0 \\ 0 & a_2 \end{bmatrix} \overset{R}{\equiv} \begin{bmatrix} a & va_1 + wa_2 \\ 0 & a_2 \end{bmatrix} = \begin{bmatrix} a_1 & d \\ 0 & a_2 \end{bmatrix} \overset{R}{\equiv} \begin{bmatrix} d & a_1 \\ a_2 & 0 \end{bmatrix} \overset{R}{\equiv} \begin{bmatrix} d & 0 \\ 0 & \frac{a_1 a_2}{d} \end{bmatrix}.$$

Note that d divides $\frac{a_1 a_2}{d}$. Use this routine to replace a_1 by d in the diagonal matrix D. Repeating the procedure for $a_3, \ldots, a_k$, we get $U \overset{R}{\equiv} \text{diag}(d_1, \bar{a}_2, \ldots, \bar{a}_k, 0, \ldots, 0)$ where d_1 is a gcd, and hence a linear combination, of $a_1, a_2, a_3, \ldots, a_k$, and d_1 divides each of

$\bar{a}_2, \ldots, \bar{a}_k$. By induction we conclude that

$$U \overset{R}{\equiv} \text{diag}(d_1, d_2, \ldots, d_k, 0 \ldots, 0)$$

where $d_2 | d_3 | \cdots | d_k$ and d_2 is an R-linear combination of $\bar{a}_2, \bar{a}_3, \ldots, \bar{a}_k$. Hence $d_1 | d_2$ since $\bar{a}_2, \bar{a}_3, \ldots, \bar{a}_k$ are divisible by d_1. This completes the proof. $\qquad\square$

The diagonal matrix V in (9.2.12) is called the *Smith canonical form*[2] of U. Let us apply this method to the presentation matrix U for a finitely generated module $M = F/N$ over a Euclidean domain R where F is a free R-module. Then $U \overset{R}{\equiv} V$ where $V = \text{diag}(d_1, d_2, \ldots, d_k, 0, \ldots, 0)$, $0 \neq d_i \in R$ and $d_1 | d_2 | \cdots | d_k$. The matrix V is the Smith canonical form of U; it gives a new presentation of M which is much simpler in form, having generators $x'_1, x'_2, \ldots, x'_n$ and relators $d_1 x'_1, d_2 x'_2, \ldots, d_k x'_k$. From this presentation we read off that

$$M \overset{R}{\simeq} R/(d_1) \oplus R/(d_2) \oplus \cdots \oplus R/(d_k) \oplus \underbrace{R \oplus \cdots \oplus R}_{n-k}.$$

Thus $n - k$ is the number of cyclic summands isomorphic with R, while the non-unit d_i's are the invariant factors.

Example (9.2.2). Let A be the abelian group with generators x, y, z and relations

$$3x + 4y + 3z = 0, \quad 6x + 4y + 6z = 0, \quad 3x + 8y + 3z = 0.$$

In this example $R = \mathbb{Z}$ and the presentation matrix is

$$U = \begin{bmatrix} 3 & 4 & 3 \\ 6 & 4 & 6 \\ 3 & 8 & 3 \end{bmatrix}.$$

Following the steps in the algorithm in (9.2.12), we find that

$$U \overset{\mathbb{Z}}{\equiv} \begin{bmatrix} 1 & 0 & 0 \\ 0 & 12 & 0 \\ 0 & 0 & 0 \end{bmatrix} = V,$$

which is the Smith canonical form of U. Hence $A \simeq \mathbb{Z}_1 \oplus \mathbb{Z}_{12} \oplus \mathbb{Z}$, i. e.,

$$A \simeq \mathbb{Z}_{12} \oplus \mathbb{Z} \simeq \mathbb{Z}_3 \oplus \mathbb{Z}_4 \oplus \mathbb{Z}.$$

The single invariant factor is 12 and the elementary divisors are 3, 4.

---

2 Henry John Stephen Smith (1826–1883).

The description of finite abelian groups afforded by the preceding theory is precise enough for us to make an exact count of the groups of given order.

(9.2.13). *Let $n > 1$ be an integer and write $n = p_1^{e_1} p_2^{e_2} \cdots p_k^{e_k}$ where $e_i > 0$ and the p_i are distinct primes. Then the number of isomorphism types of abelian groups of order n is*

$$\lambda(e_1)\lambda(e_2) \cdots \lambda(e_k)$$

where $\lambda(i)$ is the number of partitions of i.

Proof. First let A be an abelian group of order $p^e > 1$ where p is a prime. By (9.2.8) the group A is the direct sum of ℓ_1 copies of $\mathbb{Z}_p$, ℓ_2 copies of $\mathbb{Z}_{p^2}$, etc, where $\ell_i \geq 0$ and $e = \ell_1 + 2\ell_2 + 3\ell_3 + \cdots$. Thus we have a partition of e into ℓ_1 1-subsets, ℓ_2 2-subsets, etc. Conversely, every partition of e leads to an abelian group of order p^e and different partitions yield non-isomorphic groups since the invariant factors are different. Therefore the number of possible isomorphism types for A is $\lambda(e)$.

Now let A be an abelian group of order $n = p_1^{e_1} p_2^{e_2} \cdots p_k^{e_k}$; then $A = A_{p_1} \oplus A_{p_2} \oplus \cdots \oplus A_{p_k}$ where A_i is the p_i-torsion component and $|A_i| = p_i^{e_i}$. There are $\lambda(e_i)$ possible isomorphism types for A_{p_i}, so the number of isomorphism types for A is $\lambda(e_1)\lambda(e_2) \cdots \lambda(e_k)$. $\square$

Example (9.2.3). Find all abelian groups of order 600.

Since $600 = 2^3 \cdot 3 \cdot 5^2$, the number of abelian groups of order 600 is $\lambda(3)\lambda(1)\lambda(2) = 3 \times 1 \times 2 = 6$. The isomorphism types are determined by the partitions of 3 and 2, namely $3 = 1 + 2 = 1 + 1 + 1$ and $2 = 1 + 1$. Hence the six isomorphism types are:

$$\mathbb{Z}_8 \oplus \mathbb{Z}_3 \oplus \mathbb{Z}_{5^2}, \quad \mathbb{Z}_8 \oplus \mathbb{Z}_3 \oplus \mathbb{Z}_5 \oplus \mathbb{Z}_5, \quad \mathbb{Z}_2 \oplus \mathbb{Z}_4 \oplus \mathbb{Z}_3 \oplus \mathbb{Z}_{5^2},$$

$$\mathbb{Z}_2 \oplus \mathbb{Z}_4 \oplus \mathbb{Z}_3 \oplus \mathbb{Z}_5 \oplus \mathbb{Z}_5, \quad \mathbb{Z}_2 \oplus \mathbb{Z}_2 \oplus \mathbb{Z}_2 \oplus \mathbb{Z}_3 \oplus \mathbb{Z}_{5^2}, \quad \mathbb{Z}_2 \oplus \mathbb{Z}_2 \oplus \mathbb{Z}_2 \oplus \mathbb{Z}_3 \oplus \mathbb{Z}_5 \oplus \mathbb{Z}_5.$$

Notice that $\mathbb{Z}_8 \oplus \mathbb{Z}_3 \oplus \mathbb{Z}_{5^2}$ is the cyclic group of order 600.

Of course the task of counting the non-abelian groups of given finite order is much more formidable.

Exercises (9.2).

(1) Let R be a domain with field of fractions F and $R \subseteq F$. Regard F as an R-module via the field operations. Prove that F is torsion-free and F/R is a torsion module.

(2) Let $R = \mathbb{Z}_6$, the ring of congruence classes modulo 6. Find the torsion elements in the module $_R R$ and conclude that the torsion elements in a module do not always form a submodule.

(3) Let $p_1, p_2, \ldots$ be the sequence of primes and let $\langle a_i \rangle$ be an additively written group of order p_i. Define A to be the set of all sequences $(x_1, x_2, \ldots)$ where $x_i \in \langle a_i \rangle$. Make A into an abelian group by adding components. Thus A is the unrestricted direct sum of the $\mathbb{Z}_{p_i}$.

(i) Show that the torsion subgroup T consists of the sequences in which all but a finite number of components are 0.

(ii) Prove that $\bar{A} = A/T$ has the property $\bar{A} = p\bar{A}$ for all primes p.

(iii) Prove that $\bigcap_p pA = 0$.

(iv) Deduce from (ii) and (iii) that T is *not* a direct summand of A.

(4) Find the elementary divisors and invariant factors of the group $\mathbb{Z}_4 \oplus \mathbb{Z}_{30} \oplus \mathbb{Z}_{35}$.

(5) Show that there are six isomorphism types of abelian groups of order 1350.

(6) Let A be a torsion-free abelian group and define $D = \bigcap_{n=1,2,\ldots} nA$. Prove that (i) A/D is torsion-free and (ii) $D = nD$ for all $n > 0$.

(7) A finitely generated abelian group A is given by a presentation with generators x, y, z, u and relators $x - y - z - u$, $3x + y - z + u$, $2x + 3y - 2z + t$. Find the invariant factors and hence the structure of A.

(8) Let A be a finite abelian group and denote by $v_n(A)$ the number of elements of A which have order exactly n.

(i) If $n = p_1^{e_1} p_2^{e_2} \cdots p_k^{e_k}$ with distinct primes p_i, show that $v_n(A) = v_{p_1^{e_1}}(A) v_{p_2^{e_2}}(A) \cdots v_{p_k^{e_k}}(A)$.

(ii) Let A be a finite abelian p-group. Define $A[p^i] = \{a \in A \mid p^i a = 0\}$. Prove that $v_{p^e}(A) = |A[p^e]| - |A[p^{e-1}]|$ for $e \geq 1$.

(9) Let A be a finite abelian p-group. Assume that A is the direct sum of r_i cyclic groups of order p^i where $i = 1, 2 \ldots, \ell$. Prove that $|A[p^i]| = p^{s_i}$ where $s_i = r_1 + 2r_2 + \cdots + (i - 1)r_{i-1} + i(r_i + r_{i+1} + \cdots + r_\ell)$ for $1 \leq i \leq \ell$.

(10) Let A and B be finite abelian groups. If $v_n(A) = v_n(B)$ for all positive integers n, prove that $A \simeq B$. [Hint: use Exercises (9.2.8) and (9.2.9).]

9.3 Applications to linear operators

One of the most convincing applications of the theory of modules over a PID is to the study of linear operators on a finite dimensional vector space. The relationship between modules and linear operators is not obvious, so some explanation is called for.

Let V be a finite dimensional vector space over a field F with $n = \dim(V) > 0$ and let α be a fixed linear operator on V. Set $R = F[t]$, the ring of polynomials in t over F, and recall that R is a PID by (7.2.2). The basic idea is to make V into an R-module by defining

$$f \cdot v = f(\alpha)(v), \quad (f \in R, v \in V).$$

The notation here is as follows: if $f = a_0 + a_1 t + \cdots + a_m t^m \in R$, then $f(\alpha)$ is the linear operator $a_0 1 + a_1 \alpha + \cdots + a_m \alpha^m$. (Here 1 is the identity linear operator on V.) It is straightforward to check the validity the module axioms for the action specified.

Next the properties of the R-module V will be investigated. Let $v \in V$; since $\dim(V) = n$, the subset

$$\{v, \alpha(v), \alpha^2(v), \ldots, \alpha^n(v)\}$$

must be linearly dependent by (8.2.3). Hence there exist elements $a_0, a_1, \ldots, a_m$ of F, not all equal to zero, such that $a_0 v + a_1 \alpha(v) + \cdots + a_n \alpha^n(v) = 0$. Put $g = a_0 + a_1 t + \cdots + a_n t^n \in R$, noting that $g \neq 0$. Then $g \cdot v = g(\alpha)(v) = 0$, so V is a torsion R-module. In fact more is true. Let $\{v_1, v_2, \ldots, v_n\}$ be a basis of V. Then there exist $g_i \neq 0$ in R such that $g_i \cdot v_i = 0$ for $i = 1, 2, \ldots, n$. Put $h = g_1 g_2 \cdots g_n \neq 0$; then $h \cdot v_i = 0$ for all i and thus $h \cdot v = 0$ for all $v \in V$, i.e., $h(\alpha) = 0$. It follows that $\mathrm{Ann}_R(V) \neq 0$.

Since R is a PID, $\mathrm{Ann}_R(V) = (f)$ for some $f \in R$ and clearly we may choose the polynomial f to be monic. Thus a polynomial g belongs to $\mathrm{Ann}_R(V)$ if and only if f divides g, and consequently f is the unique monic polynomial of smallest degree such that $f(\alpha) = 0$. These conclusions are summed up in:

(9.3.1). *Let α be a linear operator on a finite dimensional vector space V over a field F. Then there is a unique monic polynomial f in $F[t]$ of smallest degree such that $f(\alpha) = 0$. Moreover, $g(\alpha) = 0$ if and only if f divides g in $F[t]$.*

The polynomial f is called the *minimum polynomial* of α. The next step forward is to apply the Primary Decomposition Theorem (9.2.3) to the torsion module V. According to this result there is a direct decomposition

$$V = V_1 \oplus V_2 \oplus \cdots \oplus V_k$$

where $V_i \neq 0$ is the p_i-torsion submodule of V and $p_1, p_2, \ldots, p_k$ are distinct monic irreducible elements of $R = F[t]$. There are only finitely many such V_i since V is finite dimensional. The restriction of α to V_i is a linear operator α_i, which has minimum polynomial of the form $p_i^{e_i}$. If $g \in R$, then $g(\alpha) = 0$ if and only if $g(\alpha_i) = 0$, i.e., $p_i^{e_i} | g$, for all i. It follows that the minimum polynomial of α is $f = p_1^{e_1} p_2^{e_2} \cdots p_k^{e_k}$. Thus we have proved the following theorem.

(9.3.2). *Let α be a linear operator on a finite dimensional vector space V over a field F, and suppose that the minimum polynomial of α is $f = p_1^{e_1} p_2^{e_2} \cdots p_k^{e_k}$ where the p_i are distinct monic irreducibles in $F[t]$ and $e_i > 0$. Then $V = V_1 \oplus V_2 \oplus \cdots \oplus V_k$ where V_i is the p_i-torsion submodule of V. Moreover, $p_i^{e_i}$ is the minimum polynomial of $\alpha_i = \alpha|_{V_i}$.*

The case of an algebraically closed field

Up to this point the field has been arbitrary. However, important simplifications occur for an algebraically closed field F: for then an irreducible polynomial over F has degree 1. In particular these simplifications apply to the complex field $\mathbb{C}$ by the Fundamental Theorem of Algebra – for this see (12.3.6).

Consider the situation of (9.3.2) when F is algebraically closed and $p_i = t - a_i$ with $a_i \in F$. The minimum polynomial of α is then

$$f = (t - a_1)^{e_1}(t - a_2)^{e_2} \cdots (t - a_k)^{e_k}.$$

Let $V = V_1 \oplus V_2 \oplus \cdots \oplus V_k$ be the primary decomposition of the $F[t]$-module V, with V_i the p_i-torsion component. Thus $\alpha_i = \alpha|_{V_i}$ has minimum polynomial $(t - a_i)^{e_i}$ and $(\alpha_i - a_i 1)^{e_i} = 0$. This means that $\alpha_i - a_i 1$ is a *nilpotent* linear operator on V_i, i. e., some positive power of it equals to zero.

Define two new linear operators δ, ν on V by $\delta|_{V_i} = a_i 1$, for $i = 1, 2, \ldots, k$, and $\nu = \alpha - \delta$. Then $\nu_i = \nu|_{V_i} = \alpha_i - a_i 1$ and hence $\nu_i^{e_i} = 0$, which implies that $\nu^e = 0$ where e is the largest of $e_1, e_2, \ldots, e_k$. Thus ν is a nilpotent linear operator on V. Notice that δ_i, being multiplication by a_i, commutes with ν_i, from which it follows that $\delta\nu = \nu\delta$.

The important feature of the linear operator δ is that it is diagonalizable, since δ acts on V_i by multiplication by a_i. This leads to the following result.

(9.3.3). *Let V be a finite dimensional vector space over an algebraically closed field F and let α be a linear operator on V. Then there are linear operators δ, ν on V such that $\alpha = \delta + \nu$ and $\delta\nu = \nu\delta$, where δ is diagonalizable and ν is nilpotent.*

Keep in mind that (9.3.3) can be applied to an $n \times n$ matrix A over F if we take α to be the linear operator on F^n in which $X \mapsto AX$. The statement then takes the form that $A = D + N$ and $DN = ND$ where D is a diagonalizable and N a nilpotent matrix.

Example (9.3.1). Let $A = \left[\begin{smallmatrix} -7 & 27 \\ -3 & 11 \end{smallmatrix} \right]$. The characteristic polynomial of A is $(t - 2)^2$. The minimum polynomial is also $(t - 2)^2$, either by direct matrix multiplication or by (9.3.5) below. Thus $k = 1$ and $V = V_1$ in the previous notation; hence $D = 2I_2$. Put $N = A - D = \left[\begin{smallmatrix} -9 & 27 \\ -3 & 9 \end{smallmatrix} \right]$, so that $A = D + N$ and $N^2 = 0$; also note that $DN = ND$.

Rational canonical form

It is time to apply the full force of the structure theorem for modules over a PID to a linear operator α on an n-dimensional vector space V over an arbitrary field F. Bear in mind that V is a torsion module over $R = F[t]$ via the ring action $f \cdot v = f(\alpha)(v)$. Then by (9.2.11)

$$V = V_1 \oplus V_2 \oplus \cdots \oplus V_\ell$$

where $V_i = R \cdot v_i \stackrel{R}{\simeq} R/(s_i)$ and $\mathrm{Ann}_R(v_i) = (s_i)$. Here $s_1, s_2, \ldots, s_\ell$ are the invariant factors, which satisfy $s_1 | s_2 | \ldots | s_\ell$. Recall that these s_i can be chosen to be monic. Let $\alpha_i = \alpha|_{V_i}$. If $g \in R$, then $g(\alpha) = 0$ if and only if $g(\alpha_i) = 0$, that is, $g \in (s_i)$ for $i = 1, 2 \ldots, \ell$. The divisibility property of the s_i implies that this happens precisely when s_ℓ divides g. Consequently, *the final invariant factor s_ℓ is the minimum polynomial of α.*

Next we will show that $\dim_F(R/(s_i)) = \deg(s_i)$. Write

$$s_i = t^{n_i} + a_{in_i-1}t^{n_i-1} + \cdots + a_{i1}t + a_{i0}, \quad (a_{ij} \in R).$$

If $g \in R$, then $g = qs_i + r$ where $q, r \in R$ and $r_i = 0$ or $\deg(r_i) < \deg(s_i) = n_i$. Then $g + (s_i) = r_i + (s_i)$, so that $\dim_F(R/(s_i)) \leq n_i$. Suppose that $1 + (s_i), t + (s_i), \ldots, t^{n_i-1} + (s_i)$ are linearly dependent and $b_0 1 + b_1 t + \cdots + b_{n_i-1}t^{n_i-1} + (s_i) = 0_{R/(s_i)}$ where not all the $b_i \in F$ are zero. Let $g = b_0 + b_1 t + \cdots + b_{n_i-1}t^{n_i-1}$; thus $g \neq 0$. Since $g + (s_i) = 0_{R/(s_i)}$, we have $g \in (s_i)$ and s_i divides g. But $\deg(g) < \deg(s_i)$, which can only mean that $g = 0$. By this contradiction $1 + (s_i), t + (s_i), \ldots, t^{n_i-1} + (s_i)$ are linearly independent and these elements form an F-basis of $R/(s_i)$. Hence $\dim(V_i) = \dim(R/(s_i)) = n_i$ and $\dim(V) = \sum_{i=1}^{\ell} n_i$.

Since $V_i \overset{R}{\simeq} R/(s_i)$ via the assignment $r \cdot v_i \mapsto r + (s_i)$, the subspace V_i has the basis $\{v_i, \alpha(v_i), \alpha^2(v_i), \ldots, \alpha^{n_i-1}(v_i)\}$. Let us identify the matrix which represents α_i with respect to this ordered basis. Now $\alpha(\alpha^j(v_i)) = \alpha^{j+1}(v_i)$ if $0 \leq j < n_i - 1$ and

$$\alpha(\alpha^{n_i-1}(v_i)) = \alpha^{n_i}(v_i) = -a_{i0}v_i - a_{i1}\alpha(v_i) - \cdots - a_{in_i-1}\alpha^{n_i-1}(v_i)$$

since $s_i(\alpha_i) = 0$. Therefore α_i is represented by the $n_i \times n_i$ matrix

$$R_i = \begin{bmatrix} 0 & 0 & \ldots & 0 & -a_{i0} \\ 1 & 0 & \ldots & 0 & -a_{i1} \\ 0 & 1 & \ldots & 0 & -a_{i2} \\ \cdot & \cdot & \ldots & \cdot & \cdot \\ 0 & 0 & \ldots & 1 & -a_{in_i-1} \end{bmatrix}.$$

This is the *companion matrix* of the polynomial s_i – see Exercise (8.4.6). Note that s_i is the minimum polynomial of α_i and hence of R_i.

Now form the union of the chosen bases of the V_i to obtain a basis of V with respect to which α is represented by the block matrix

$$C = \begin{bmatrix} R_1 & 0 & \ldots & 0 \\ 0 & R_2 & \ldots & 0 \\ \cdot & \cdot & \ldots & \cdot \\ 0 & 0 & \ldots & R_\ell \end{bmatrix}.$$

This is called the *rational canonical form* of α.

Recall that the characteristic polynomial of α is $\det(tI_n - C)$. Now

$$\det(tI_n - C) = \begin{vmatrix} tI_{n_1} - R_1 & 0 & \ldots & 0 \\ 0 & tI_{n_2} - R_2 & \ldots & 0 \\ \cdot & \cdot & \ldots & \cdot \\ 0 & 0 & \ldots & tI_{n_\ell} - R_\ell \end{vmatrix},$$

which is equal to the product $\det(tI_{n_1} - R_1)\det(tI_{n_2} - R_2) \cdots \det(I_{n_\ell} - R_\ell)$. Also

$$\det(tI_{n_i} - R_i) = \begin{vmatrix} t & 0 & 0 & \cdots & 0 & a_{i0} \\ -1 & t & 0 & \cdots & 0 & a_{i1} \\ 0 & -1 & t & \cdots & 0 & a_{i2} \\ . & . & . & \cdots & . & . \\ 0 & 0 & 0 & \cdots & -1 & t + a_{in_i} \end{vmatrix},$$

which by direct determinantal expansion equals $a_{i0} + a_{i1}t + \cdots + a_{in_i-1}t^{n_i-1} + t^{n_i} = s_i$. Therefore $\det(tI_n - C) = s_1 s_2 \cdots s_\ell$.

These conclusions are summed up in the following fundamental result.

(9.3.4) (Rational canonical form). *Let α be a linear operator on a finite dimensional vector space V over an arbitrary field. Then the following statements hold.*
(i) *α can be represented with respect to a suitable basis of V by a matrix in rational canonical form.*
(ii) *The final invariant factor of α is the minimum polynomial.*
(iii) *The product of the invariant factors of α equals the characteristic polynomial.*

Corollary (9.3.5) (The Cayley-Hamilton Theorem). *The minimum polynomial of a linear operator divides the characteristic polynomial and these polynomials have the same irreducible factors.*

This follows directly from (9.3.4). The preceding very powerful results have been stated for a linear operator. Of course, they apply equally to an $n \times n$ matrix A over a field F. Thus by (9.3.4) every square matrix is similar to a matrix in rational canonical form and also the Cayley-Hamilton Theorem is valid.

Nilpotent linear operators

Rational canonical form is particularly effective when applied to a nilpotent linear operator α on an n-dimensional vector space V over an arbitrary field F. Since $\alpha^k = 0$ for some $k > 0$, the minimum polynomial must divide t^k and thus has the form t^m where $m \le k$. The invariant factors satisfy $s_1 | s_2 | \cdots | s_\ell = t^m$ by (9.3.4). Hence $s_i = t^{n_i}$ where $n_1 \le n_2 \le \cdots \le n_\ell = m$. The characteristic polynomial of α equals $s_1 s_2 \cdots s_\ell = t^n$ and thus $n = \sum_{i=1}^{\ell} n_i$.

The companion matrix of s_i is the $n_i \times n_i$ matrix

$$R_i = \begin{bmatrix} 0 & 0 & \cdots & 0 & 0 \\ 1 & 0 & \cdots & 0 & 0 \\ 0 & 1 & \cdots & 0 & 0 \\ . & . & \cdots & . & . \\ 0 & 0 & \cdots & 1 & 0 \end{bmatrix}$$

and the rational canonical form of α is the block matrix formed by $R_1, R_2, \ldots, R_\ell$. This is a triangular matrix with zeros on and above the principal diagonal, a type of matrix called *lower zero triangular*. Applying this in matrix form, we deduce:

(9.3.6). *A nilpotent matrix is similar to a lower zero triangular matrix.*

Rational canonical form allows us to make an exact count of the similarity types of nilpotent $n \times n$ matrix.

(9.3.7). *The number of similarity types of nilpotent $n \times n$ matrices over any field equals $\lambda(n)$ where λ is the partition function.*

Proof. Let A be an $n \times n$ nilpotent matrix. Let m_i denote the number of rational blocks with exactly i 1's on the subdiagonal. Thus $m_i \geq 0$ and $0 \leq i \leq n - 1$. Then $n = \sum_{i=0}^{n-1}(i+1)m_i$, so that we have a partition of n. Conversely, each partition of n allows us to assemble a nilpotent matrix, the rational blocks coming from the subsets in the partition. Moreover, different partitions give rise to non-similar matrices by uniqueness of the invariant factors. □

Example (9.3.2). Since $\lambda(3) = 3$, there are three similarity types of nilpotent 3×3 matrices, corresponding to the partitions of 3, which are $1 + 1 + 1$, $1 + 2$, 3. The respective types of matrix are

$$
\begin{bmatrix} 0 & 0 & 0 \\ 0 & 0 & 0 \\ 0 & 0 & 0 \end{bmatrix}, \quad
\begin{bmatrix} 0 & 0 & 0 \\ 0 & 0 & 0 \\ 0 & 1 & 0 \end{bmatrix}, \quad
\begin{bmatrix} 0 & 0 & 0 \\ 1 & 0 & 0 \\ 0 & 1 & 0 \end{bmatrix}.
$$

Jordan canonical form

Let α be a linear operator on an n-dimensional vector space V over a field F and let f denote the minimum polynomial of α. Assume that f splits into linear factors over F, which by the Cayley-Hamilton Theorem amounts to requiring all eigenvalues of α to be in F, which will certainly be true if F is algebraically closed.

In this case there is a simpler canonical form for α called *Jordan normal form*. Write

$$
f = (t - a_1)^{e_1}(t - a_2)^{e_2} \cdots (t - a_k)^{e_k}
$$

where $e_i > 0$ and the a_i are distinct elements of the field F. By (9.3.5) the roots of f are the roots of the characteristic polynomial, so $a_1, a_2, \ldots, a_k$ are the distinct eigenvalues of α. By the Primary Decomposition Theorem $V = V_1 \oplus V_2 \oplus \cdots \oplus V_k$ where V_i is the $p_i = (t - a_i)$-torsion submodule of V. Write $n_i = \dim(V_i)$, so that $n = \sum_{i=1}^{k} n_i$. Then $\alpha_i = \alpha|_{V_i}$ has $(t - a_i)^{e_i}$ as its minimum polynomial by (9.3.2); thus $(\alpha_i - a_i 1_{n_i})^{e_i} = 0$ and $\alpha_i - a_i 1_{n_i}$ is a nilpotent linear operator on V_i. By the discussion of nilpotent linear operators above, $\alpha_i - a_i 1_{n_i}$ is represented with respect to a suitable basis of V_i by a matrix consisting of ℓ_{ij} blocks of size n_{ij} with the form

$$\begin{bmatrix} 0 & 0 & \cdots & 0 & 0 \\ 1 & 0 & \cdots & 0 & 0 \\ 0 & 1 & \cdots & 0 & 0 \\ \cdot & \cdot & \cdots & \cdot & \cdot \\ 0 & 0 & \cdots & 1 & 0 \end{bmatrix}$$

for $j = 1, 2, \ldots, e_i$. Here $n_{i1} \le n_{i2} \le \cdots \le n_{ie_i}$ and $\sum_{j=1}^{e_i} \ell_{ij} n_{ij} = \dim(V_i) = n_i$. Consequently, α_i is represented by a matrix consisting of ℓ_{ij} blocks J_{ij} of size n_{ij} with the form

$$J_{ij} = \begin{bmatrix} a_i & 0 & \cdots & 0 & 0 & 0 \\ 1 & a_i & 0 & 0 & \cdots & 0 \\ 0 & 1 & a_i & 0 & \cdots & 0 \\ \cdot & \cdot & \cdots & \cdot & \cdot & \cdot \\ 0 & 0 & \cdots & 0 & 1 & a_i \end{bmatrix}.$$

Such matrices are called *Jordan blocks* and they are unique up to order since they are determined by the elementary divisors of α. Therefore we can state:

(9.3.8) (Jordan canonical form). *Let α be a linear operator on a finite dimensional vector space over a field F. Assume that the minimum polynomial of α splits into linear factors over F. Then α can be represented with respect to a suitable basis by a matrix with Jordan blocks on the diagonal which are unique up to order.*

The matrix form of (9.3.8) asserts that an $n \times n$ matrix A whose minimum polynomial is a product of linear factors over F is similar to a matrix with Jordan blocks on the diagonal. Therefore, in particular, A *is similar to a lower triangular matrix over F*, i. e., with zeros above the diagonal – cf. (8.4.8).

Example (9.3.3). Find all similarity types of complex 3×3 matrices A which satisfy the equation $A(A - 2I)^2 = 0$.

From the information furnished the minimum polynomial f of A divides $t(t-2)^2$. Hence there are five possibilities for f, which are listed below with the corresponding Jordan canonical form J of A:

(i) $f = t$: in this case $A = J = 0$.

(ii) $f = t - 2$: $J = 2I_3$.

(iii) $f = (t-2)^2$: $J = \begin{bmatrix} 2 & 0 & 0 \\ 1 & 2 & 0 \\ 0 & 0 & 2 \end{bmatrix}$.

(iv) $f = t(t-2)$: $J = \begin{bmatrix} 0 & 0 & 0 \\ 0 & 2 & 0 \\ 0 & 0 & 2 \end{bmatrix}$ or $\begin{bmatrix} 0 & 0 & 0 \\ 0 & 0 & 0 \\ 0 & 0 & 2 \end{bmatrix}$.

(v) $f = t(t-2)^2$: $J = \begin{bmatrix} 0 & 0 & 0 \\ 0 & 2 & 0 \\ 0 & 1 & 2 \end{bmatrix}$.

Hence there are six types of matrix up to similarity.

Computing the invariant factors

We end the chapter by describing an algorithm for calculating the invariant factors of a linear operator or matrix: it will be stated for matrices. The basis is the method for calculating Smith canonical form developed in (9.2.12).

(9.3.9). *Let A be an $n \times n$ matrix over a field F. Then the Smith canonical form of the matrix $tI - A$ is $\mathrm{diag}(1, 1, \ldots, 1, s_1, s_2, \ldots, s_\ell)$ where $s_1, s_2, \ldots, s_\ell$ are the invariant factors of A.*

Proof. Let S denote the rational canonical form of A. Then $S = XAX^{-1}$ for some non-singular matrix X over F. It follows that S and A have the same invariant factors since they represent the same linear operator on F^n, but with respect to different bases. Also $tI - S = X(tI - S)X^{-1}$, so by the same reasoning $tI - S$ and $tI - A$ have the same Smith canonical form. Therefore we may assume that $A = S$, i.e., A is in rational canonical form.

Let $R_1, R_2, \ldots, R_\ell$ be the blocks in the rational canonical form of A corresponding to the invariant factors $s_1 | s_2 | \ldots | s_\ell$. It is enough to prove that the Smith canonical form of $tI - R_i$ is $\mathrm{diag}(1, 1, \ldots, 1, s_i)$; for then $tI - A$ will have $\mathrm{diag}(1, 1, \ldots, 1, s_1, s_2, \ldots, s_\ell)$ as its Smith canonical form. Let $s_i = a_{i0} + a_{i1}t + \cdots + a_{in_i-1}t^{n_i-1} + t^{n_i}$; thus

$$R_i = \begin{bmatrix} 0 & 0 & \ldots & 0 & -a_{i0} \\ 1 & 0 & \ldots & 0 & -a_{i1} \\ 0 & 1 & \ldots & 0 & -a_{i2} \\ . & . & \ldots & . & . \\ 0 & 0 & \ldots & 1 & -a_{in_i-1} \end{bmatrix}.$$

Since $F[t]$ is a Euclidean domain, we can transform the matrix

$$tI - R_i = \begin{bmatrix} t & 0 & 0 & \ldots & 0 & a_{i0} \\ -1 & t & 0 & \ldots & 0 & a_{i1} \\ 0 & -1 & t & \ldots & 0 & a_{i2} \\ . & . & . & \ldots & . & . \\ 0 & 0 & . & \ldots & -1 & t + a_{in_i-1} \end{bmatrix}$$

into Smith canonical form using the method of (9.2.12). This is readily seen to be $\mathrm{diag}(1, 1, \ldots, 1, s_i)$, as the reader should verify, at least for $n_i \leq 3$. (Note the absence of zeros since V is a torsion module.) The required result now follows. $\square$

Example (9.3.4). Consider the rational matrix

$$A = \begin{bmatrix} 0 & 4 & 1 \\ -1 & -4 & 2 \\ 0 & 0 & -2 \end{bmatrix}.$$

Apply suitable row and column operations to put the matrix

$$tI - A = \begin{bmatrix} t & -4 & -1 \\ 1 & t+4 & -2 \\ 0 & 0 & t+2 \end{bmatrix}$$

into Smith canonical form

$$\begin{bmatrix} 1 & 0 & 0 \\ 0 & 1 & 0 \\ 0 & 0 & (t+2)^3 \end{bmatrix}.$$

Hence there is just one invariant factor $s_1 = (t+2)^3$. The rational canonical form of A can now be written down immediately as

$$\begin{bmatrix} 0 & 0 & -8 \\ 1 & 0 & -12 \\ 0 & 1 & -6 \end{bmatrix}.$$

The minimum polynomial is $(t+2)^3$, so the Jordan canonical form is

$$\begin{bmatrix} -2 & 0 & 0 \\ 1 & -2 & 0 \\ 0 & 1 & -2 \end{bmatrix}.$$

Exercises (9.3).

(1) Find all similarity types of 3×3 rational matrices A which satisfy the equation $A^4 = A^5$.

(2) Find the invariant factors and rational canonical form of the rational matrix

$$\begin{bmatrix} 2 & 3 & 1 \\ 1 & 2 & 1 \\ 0 & 0 & -4 \end{bmatrix}.$$

(3) Find the Jordan canonical form and minimum polynomial of the rational matrix

$$\begin{bmatrix} 3 & 1 & 0 \\ -1 & 1 & 0 \\ 0 & 0 & 2 \end{bmatrix}.$$

(4) Let A be an $n \times n$ matrix over $\mathbb{Q}$ and let p be a prime. Assume that $A^p = I$. Prove that the number of similarity types of A is $[\frac{n}{p-1}] + 1$. [Hint: recall from Example (7.4.6) that the polynomial $1 + t + t^2 + \cdots + t^{p-1}$ is irreducible over $\mathbb{Q}$.]

(5) Prove that a square matrix over a field is similar to its transpose. (You may assume the field contains all roots of the minimum polynomial of A).

(6) Prove that every square matrix is similar to an upper triangular matrix.

(7) Let A be a non-singular $n \times n$ matrix over an algebraically closed field F and let $J_1, J_2, \ldots, J_k$ be the blocks in the Jordan canonical form of A. Prove that A has finite order if and only if each J_i has finite order and in that case the order of A is $\mathrm{lcm}\{|J_1|, |J_2|, \ldots, |J_k|\}$.

(8) Let J be an $n \times n$ Jordan block over a field F with diagonal elements equal to $a \neq 0$. If $n > 1$, prove that J has finite order if and only if a has finite order in F^* and $p = \mathrm{char}(F) \neq 0$, and that in this event $|J|$ divides $|a| \cdot p^{n-1}$.

(9) Let A be a non-singular $n \times n$ matrix over an algebraically closed field of characteristic 0. Let $a_1, a_2, \ldots, a_n$ be the eigenvalues of A. Prove that A has finite order if and only if each a_i has finite order and then $|A| = \mathrm{lcm}\{|a_1|, |a_2|, \ldots, |a_\ell|\}$.

(10) Find the Jordan canonical form of the matrix

$$A = \begin{bmatrix} 0 & 0 & 1 \\ 1 & 0 & 4 \\ 0 & 1 & 3 \end{bmatrix}$$

over $GF(7)$, the field of seven elements. Then use it to prove that $|A| = 7$.

10 The Structure of Groups

In this chapter we will pursue the study of groups at a deeper level. A common method of investigation in algebra is to break up a complex structure into simpler substructures. The hope is that by repeated application of this procedure one will eventually arrive at substructures that are easy to understand. It may then be possible in some sense to synthesize these substructures to reconstruct the original structure. While it is rare for the procedure just described to be brought to such a perfect state of completion, the analytic-synthetic method can yield valuable information and suggest new directions. We will consider instances where this procedure can be employed in group theory.

10.1 The Jordan–Hölder Theorem

A basic concept in group theory is that of a *finite series in a group G*. By this is meant a finite chain of subgroups $S = \{G_i \mid i = 0, 1, \ldots, n\}$ leading from the identity subgroup to G, with each term normal in its successor, that is, a chain of the form

$$1 = G_0 \lhd G_1 \lhd \cdots \lhd G_n = G.$$

The G_i are the *terms* of the series and the quotient groups G_{i+1}/G_i are the *factors*. The *length* of the series is defined to be the number of non-trivial factors. Keep in mind that G_i may not be normal in G since normality is not a transitive relation – see Exercise (4.2.6).

A subgroup H which appears in a series in a group G is called a *subnormal subgroup*; clearly this is equivalent to there being a chain of normality relations leading from H to G,

$$H = H_0 \lhd H_1 \lhd \cdots \lhd H_m = G.$$

A partial order on the set of series in a group G is defined as follows. A series S is called a *refinement* of a series T if every term of T is also a term of S. If S has at least one term that is not a term of T, then S is a *proper refinement* of T. It is easy to see that the relation of being a refinement is a partial order on the set of all series in G.

Example (10.1.1). The symmetric group S_4 has the series $1 \lhd V \lhd A_4 \lhd S_4$ where V is the Klein 4-group. This is a refinement of the series $1 \lhd A_4 \lhd S_4$.

Isomorphic series

Two series S and T in a group G are called *isomorphic* if there is a bijection from the set of non-trivial factors of S to the set of non-trivial factors of T such that corresponding

https://doi.org/10.1515/9783110691160-010

factors are isomorphic groups. Isomorphic series must have the same length, but the isomorphic factors may occur at different points in the series.

Example (10.1.2). In $\mathbb{Z}_6$ the series $0 \vartriangleleft \langle [2] \rangle \vartriangleleft \mathbb{Z}_6$ and $0 \vartriangleleft \langle [3] \rangle \vartriangleleft \mathbb{Z}_6$ are isomorphic since $\langle [2] \rangle \simeq \mathbb{Z}_6 / \langle [3] \rangle$ and $\langle [3] \rangle \simeq \mathbb{Z}_6 / \langle [2] \rangle$.

The foundation for the theory of series in groups is the following technical result. It can be viewed as a generalization of the Second Isomorphism Theorem.

(10.1.1) (Zassenhaus's[1] Lemma). *Let A_1, A_2, B_1, B_2 be subgroups of a group such that $A_1 \vartriangleleft A_2$ and $B_1 \vartriangleleft B_2$. Define $D_{ij} = A_i \cap B_j$, $(i, j = 1, 2)$. Then $A_1 D_{21} \vartriangleleft A_1 D_{22}$ and $B_1 D_{12} \vartriangleleft B_1 D_{22}$. Furthermore*

$$A_1 D_{22} / A_1 D_{21} \simeq B_1 D_{22} / B_1 D_{12}.$$

Proof. The Hasse diagram below displays all the relevant subgroups.

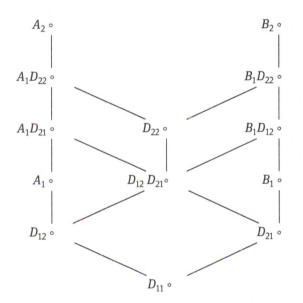

From $B_1 \vartriangleleft B_2$ we obtain $D_{21} \vartriangleleft D_{22}$ by intersecting with A_2. Since $A_1 \vartriangleleft A_2$, it follows that $A_1 D_{21} \vartriangleleft A_1 D_{22}$ on applying the canonical homomorphism $A_2 \rightarrow A_2 / A_1$. Similarly $B_1 D_{12} \vartriangleleft B_1 D_{22}$. Now we invoke (4.3.5) with $H = D_{22}$ and $N = A_1 D_{21}$ to give $HN/N \simeq H/H \cap N$. But $HN = A_1 D_{22}$ and $H \cap N = D_{22} \cap (A_1 D_{21}) = D_{12} D_{21}$ by (4.1.11). The conclusion is that $A_1 D_{22} / A_1 D_{21} \simeq D_{22} / D_{12} D_{21}$. By the same argument $B_1 D_{22} / B_1 D_{12} \simeq D_{22} / D_{12} D_{21}$, from which the result follows. □

---

1 Hans Zassenhaus (1912–1991).

The main use of Zassenhaus's Lemma is to prove a theorem about refinements: its statement is remarkably simple.

(10.1.2) (The Schreier[2] Refinement Theorem). *Any two series in a group have isomorphic refinements.*

Proof. Let $1 = H_0 \lhd H_1 \lhd \cdots \lhd H_l = G$ and $1 = K_0 \lhd K_1 \lhd \cdots \lhd K_m = G$ be two series in a group G. Define subgroups $H_{ij} = H_i(H_{i+1} \cap K_j)$ for $0 \le i \le l - 1$, $0 \le j \le m$ and $K_{ij} = K_j(H_i \cap K_{j+1})$ for $0 \le i \le l$, $0 \le j \le m - 1$. Apply (10.1.1) with $A_1 = H_i$, $A_2 = H_{i+1}$, $B_1 = K_j$ and $B_2 = K_{j+1}$; the conclusion is that $H_{ij} \lhd H_{ij+1}$ and $K_{ij} \lhd K_{i+1j}$, and also that $H_{ij+1}/H_{ij} \simeq K_{i+1j}/K_{ij}$. Therefore the series $\{H_{ij} \mid i = 0, 1, \ldots, l - 1, j = 0, 1, \ldots m\}$ and $\{K_{ij} \mid i = 0, 1, \ldots, l, j = 0, 1, \ldots, m - 1\}$ are isomorphic refinements of $\{H_i \mid i = 0, 1, \ldots, l\}$ and $\{K_j \mid j = 0, 1, \ldots, m\}$ respectively. $\square$

Composition series

A series which has no proper refinements is called a *composition series* and its factors are *composition factors*. If G is a finite group, we can start with any series, for example $1 \lhd G$, and keep refining it until a composition series is reached. Thus *every finite group has a composition series*. However, not every infinite group has a composition series, as is shown by (10.1.6) below.

A composition series can be recognized from the nature of its factors.

(10.1.3). *A series is a composition series if and only if all its factors are simple groups.*

Proof. Let X/Y be a factor of a series in a group G. If X/Y is not simple, there is a subgroup W such that $Y < W < X$ and $W \lhd X$; here the Correspondence Theorem (4.2.2) has been invoked. Adjoining W to the given series, we obtain a new series which is a proper refinement, with the terms $Y \lhd W \lhd X$ replacing $Y \lhd X$.

Conversely, if a series in G has a proper refinement, there must be two consecutive terms $Y \lhd X$ of the original series with additional terms of the refined series between them. Hence there is a subgroup W in the refined series such that $Y < W < X$ and $W \lhd X$. But then W/Y is a proper non-trivial subgroup of X/Y, so the latter cannot be simple. Hence the result is proved. $\square$

The main result about composition series is a celebrated theorem associated with the names of two prominent 19th Century algebraists, Camille Jordan (1838–1922) and Otto Hölder (1859–1937).

(10.1.4) (The Jordan–Hölder Theorem). *Let $\mathcal{S}$ be a composition series in a group G and suppose that $\mathcal{T}$ is any series in G. Then $\mathcal{T}$ has a refinement which is isomorphic with $\mathcal{S}$.*

---

2 Otto Schreier (1901–1929).

The most important case is when $\mathcal{T}$ itself is a composition series and the conclusion is that $\mathcal{T}$ is isomorphic with $\mathcal{S}$. Thus we obtain:

Corollary (10.1.5). *Any two composition series in a group are isomorphic.*

Proof of (10.1.4). By the Refinement Theorem (10.1.2) the series $\mathcal{S}$ and $\mathcal{T}$ have isomorphic refinements. But $\mathcal{S}$ is a composition series, so it is isomorphic with a refinement of $\mathcal{T}$. $\qquad\square$

Example (10.1.3). Consider the symmetric group S_4. It has a series

$$1 \triangleleft C \triangleleft V \triangleleft A_4 \triangleleft S_4$$

where $|C| = 2$ and V is the Klein 4-group. Now C, V/C and S_4/A_4 all have order 2, while A_4/V has order 3, so all factors of the series are simple. By (10.1.3) the series is a composition series with composition factors $\mathbb{Z}_2$, $\mathbb{Z}_2$, $\mathbb{Z}_3$, $\mathbb{Z}_2$.

The next result demonstrates that not every group has a composition series.

(10.1.6). *An abelian group A has a composition series if and only if it is finite.*

Proof. Only necessity is in doubt, so assume that A has a composition series. Each factor of the series is simple and abelian, and thus has no proper non-trivial subgroups. By (4.1.9) the factors have prime orders and therefore A is finite. $\qquad\square$

Example (10.1.4) (Composition series in $\mathbb{Z}_n$). Let n be an integer greater than 1. The group $\mathbb{Z}_n$ has a composition series with each factor of prime order. Since the product of the orders of the composition factors is equal to n, the group order, it follows that n is a product of primes, which is the first part of the Fundamental Theorem of Arithmetic. In fact we can also obtain the uniqueness part.

Suppose that $n = p_1 p_2 \cdots p_k$ is an expression for n as a product of primes. Define H_i to be the subgroup of $\mathbb{Z}_n$ generated by the congruence class $[p_{i+1} p_{i+2} \cdots p_k]$ where $0 \le i < k$ and let $H_k = \mathbb{Z}_n$. Then

$$0 = H_0 \triangleleft H_1 \triangleleft \cdots \triangleleft H_{k-1} \triangleleft H_k = \mathbb{Z}_n$$

is a series in $\mathbb{Z}_n$. Now clearly $|H_i| = p_1 p_2 \cdots p_i$ and hence $|H_{i+1}/H_i| = p_{i+1}$. Thus we have constructed a composition series in $\mathbb{Z}_n$ with factors of orders $p_1, p_2, \ldots, p_k$.

If $n = q_1 q_2 \cdots q_l$ is another expression for n as product of primes, there is a corresponding composition series with factors of orders $q_1, q_2, \ldots, q_l$. By the Jordan–Hölder Theorem these composition series are isomorphic. Consequently, $k = \ell$ and the q_j's must be the p_i's in some order. Thus we have recovered the Fundamental Theorem of Arithmetic from the Jordan–Hölder Theorem.

Some simple groups

The investigation so far shows that in a sense a finite group decomposes into a number of simple groups, namely its composition factors. The only simple groups we currently know are the groups of prime order and the alternating group A_5 – see (5.3.10). It is definitely time to expand this list, which we do by proving:

(10.1.7). *The alternating group A_n is simple if and only if $n \neq 1$, 2 or 4.*

The proof uses the following property of 3-cycles.

(10.1.8). *If $n \geq 3$, the alternating group A_n is generated by 3-cycles.*

Proof. First of all note that 3-cycles are even and hence belong to A_n. Next each element of A_n is the product of an even number of transpositions by (3.1.4). Finally, note the equations $(ac)(ab) = (abc)$ and $(ab)(cd) = (adb)(adc)$, where a, b, c, d are all different; these demonstrate that every element of A_n is a product of 3-cycles. □

Proof of (10.1.7). In the first place A_4 has a normal subgroup of order 4, so it cannot be simple. Also A_1 and A_2 have order 1, so these are also excluded. However, A_3 is simple because its order is 3. Thus we can assume that $n \geq 5$ and aim to show that A_n is simple. If this is false, there is a proper, non-trivial normal subgroup N. The proof analyzes the possible forms of elements of N.

Assume first that N contains a 3-cycle (abc). If $(a'b'c')$ is another 3-cycle and π in S_n sends a, b, c to a', b', c' respectively, then $\pi(abc)\pi^{-1} = (a'b'c')$. If π is even, it follows that $(a'b'c') \in N$. If, on the other hand, π is odd, it can be replaced by the even permutation $\pi \circ (ef)$ where e, f are different from a, b, c – notice that this uses $n \geq 5$. We will still have $\pi(abc)\pi^{-1} = (a'b'c')$ since $(ef)(abc)(ef) = (abc)$. Consequently N contains all 3-cycles and by (10.1.8) $N = A_n$, a contradiction. Hence N cannot contain a 3-cycle.

Assume next that N contains a permutation π whose disjoint cycle decomposition involves a cycle of length at least 4, say

$$\pi = (a_1 a_2 a_3 a_4 \cdots) \cdots$$

where the final dots indicate the possible presence of further disjoint cycles. Now N also contains the conjugate of π

$$\pi' = (a_1 a_2 a_3)\pi(a_1 a_2 a_3)^{-1} = (a_2 a_3 a_1 a_4 \cdots) \cdots .$$

Therefore N contains $\pi'\pi^{-1} = (a_1 a_2 a_4)$: here the point to note is that the other cycles cancel. Since this conclusion is untenable, elements in N must have disjoint cycle decompositions involving cycles of length at most 3. Furthermore, such elements cannot involve just one 3-cycle, otherwise by squaring we would obtain a 3-cycle in N.

Assume next that N contains a permutation with at least two disjoint 3-cycles, say $\pi = (abc)(a'b'c')\cdots$. Then N contains the conjugate

$$\pi' = (a'b'c)\pi(a'b'c)^{-1} = (aba')(cc'b')\cdots,$$

and hence it contains $\pi\pi' = (aca'bb')\cdots$, which has been seen to be impossible. Therefore each non-trivial element of N must be the product of an even number of disjoint transpositions.

If $\pi = (ab)(a'b') \in N$, then N contains $\pi' = (acb)\pi(acb)^{-1} = (ac)(a'b')$ for any c unaffected by π. But then N will contain $\pi\pi' = (acb)$, which is false. Consequently, if $1 \neq \pi \in N$, then $\pi = (a_1b_1)(a_2b_2)(a_3b_3)(a_4b_4)\cdots$, with at least four transpositions. It follows that N also contains

$$\pi' = (a_3b_2)(a_2b_1)\pi(a_2b_1)(a_3b_2) = (a_1a_2)(a_3b_1)(b_2b_3)(a_4b_4)\cdots$$

and hence N contains $\pi\pi' = (a_1b_2a_3)(a_2b_1b_3)$, a final contradiction. □

As a consequence of (10.1.8) there are infinitely many simple alternating groups. The simplicity of A_n will now be used to determine the composition series of S_n.

(10.1.9). *If $n = 3$ or $n \geq 5$, then $1 \triangleleft A_n \triangleleft S_n$ is the unique composition series of S_n.*

Proof. In the first place this is a composition series since A_n and $S_n/A_n \simeq \mathbb{Z}_2$ are simple. Suppose that N is a non-trivial, proper normal subgroup of S_n. We will show that $N = A_n$, which will settle the matter. First note that $N \cap A_n \triangleleft A_n$, so that either $N \cap A_n = 1$ or $A_n \leq N$ since A_n is simple. Now $|S_n : A_n| = 2$, so if $A_n \leq N$, then $N = A_n$. Suppose that $N \cap A_n = 1$. Then $NA_n = S_n$ and $|N| = |NA_n/A_n| = |S_n/A_n| = 2$. Thus N contains a single non-identity element π, (necessarily an odd permutation). Since $N \triangleleft S_n$, the permutation π belongs to the center of S_n; however $Z(S_n) = 1$ by Exercise (4.2.10), so a final contradiction is reached. □

Projective linear groups

We mention in passing another infinite family of finite simple groups. Let F be any field. It is not difficult to prove by direct matrix calculations that the center of the general linear group $\mathrm{GL}_n(F)$ is just the subgroup of all scalar matrices fI_n where $f \in F$, (cf. Exercise (4.2.12)). The *projective general linear group of degree n over F* is defined to be

$$\mathrm{PGL}_n(F) = \mathrm{GL}_n(F)/Z(\mathrm{GL}_n(F)).$$

Recall that $\mathrm{SL}_n(F)$ is the special linear group consisting of all matrices in $\mathrm{GL}_n(F)$ with determinant equal to 1. The center of $\mathrm{SL}_n(F)$ can be shown to be $Z(\mathrm{GL}_n(F)) \cap \mathrm{SL}_n(F)$. Therefore by (4.3.5)

$$\mathrm{SL}_n(F)Z(\mathrm{GL}_n(F))/Z(\mathrm{GL}_n(F)) \simeq \mathrm{SL}_n(F)/Z(\mathrm{SL}_n(F)).$$

The latter is called the *projective special linear group*

$$\mathrm{PSL}_n(F).$$

The projective special linear groups are usually simple, as the following result shows.

(10.1.10). *Let F be a field and let $n > 1$. Then $\mathrm{PSL}_n(F)$ is simple if and only if $n \geq 3$ or $n = 2$ and F has more than three elements.*

This result can be proved by direct, if tedious, matrix calculations – see for example [15]. If F is a finite field, its order is a prime power q by (8.2.17). Moreover, by (11.3.5) below, there is up to isomorphism just one field of order q. If F is a field of order q, it is better notation to write

$$\mathrm{GL}_n(q),\ \mathrm{PGL}_n(q),\ \mathrm{PSL}_n(q)$$

instead of $\mathrm{GL}_n(F)$, $\mathrm{PGL}_n(F)$, $\mathrm{PSL}_n(F)$.

It is not hard to compute the orders of these groups. In the first place, $|Z(\mathrm{GL}_n(F))| = |F^*| = q - 1$, where $F^* = U(F) = F - 0$, and also $|Z(\mathrm{SL}_n(F))| = \gcd\{n, q - 1\}$. For the last statement we need to know that F^* is cyclic: for a proof see (11.3.6) below. The orders of the projective groups can now be read off. A simple count of the non-singular $n \times n$ matrices over F reveals that

$$|\mathrm{GL}_n(q)| = (q^n - 1)(q^n - q) \cdots (q^n - q^{n-1}),$$

while $|\mathrm{SL}_n(q)| = |\mathrm{GL}_n(q)|/(q-1)$. Thus we have formulas for the orders of the projective groups.

(10.1.11).
(i) $|\mathrm{PGL}_n(q)| = |\mathrm{GL}_n(q)|/(q - 1)$;
(ii) $|\mathrm{PSL}_n(q)| = |\mathrm{SL}_n(q)|/\gcd\{n, q - 1\}$.

For example, $\mathrm{PSL}_2(5)$ is a simple group of order 60. In fact there is only one simple group of this order – see Exercise (10.2.18) – so $\mathrm{PSL}_2(5)$ must be isomorphic with A_5. But $\mathrm{PSL}_2(7)$ of order 168 and $\mathrm{PSL}_2(8)$ of order 504 are simple groups that are not of alternating type.

Projective groups and projective space

We indicate briefly how the projective groups arise in geometry. Let V be an $(n + 1)$-dimensional vector space over a field F and let V^* denote the set of all non-zero vectors in V. An equivalence relation $\sim$ on V^* is introduced by the following rule: $u \sim v$ if and only if $u = fv$ for some $f \neq 0$ in F. Let $[v]$ be the equivalence class of the vector v, so this is just the set of non-zero multiples of v. The set

$$\tilde{V} = \{[v] \mid v \in V^*\}$$

is called *n-dimensional projective space* over F.

Next let α be a bijective linear operator on V. Then there is an induced mapping $\tilde{\alpha} : \tilde{V} \to \tilde{V}$ defined by the rule

$$\tilde{\alpha}([v]) = [\alpha(v)].$$

Here $\tilde{\alpha}$ is called a *collineation* on $\tilde{V}$. It is not hard to see that the collineations on $\tilde{V}$ form a group $\mathrm{PGL}(\tilde{V})$ with respect to functional composition.

It is also straightforward to verify that the assignment $\alpha \mapsto \tilde{\alpha}$ gives rise to a surjective group homomorphism from $\mathrm{GL}(V)$, the group of invertible linear operators on V, to $\mathrm{PGL}(\tilde{V})$, with kernel equal to the subgroup of all scalar linear operators. Therefore $\mathrm{PGL}(\tilde{V}) \simeq \mathrm{PGL}_n(F)$, while $\mathrm{PSL}_n(F)$ corresponds to the subgroup of collineations arising from matrices with determinant equal to 1.

The classification of finite simple groups

The projective special linear groups form one of a number of infinite families of finite simple groups known collectively as *the simple groups of Lie type*. They arise as groups of automorphisms of simple Lie algebras. In addition to the alternating groups and the groups of Lie type, there are 26 isolated simple groups, the so-called *sporadic simple groups*. The smallest of these, the *Mathieu[3] group M_{11}*, has order 7920, while the largest one, the so-called *Monster*, has order

$$2^{46} \cdot 3^{20} \cdot 5^9 \cdot 7^6 \cdot 11^2 \cdot 13^3 \cdot 17 \cdot 19 \cdot 23 \cdot 29 \cdot 31 \cdot 41 \cdot 47 \cdot 59 \cdot 71,$$

or approximately 8.08×10^{53}.

It is now widely accepted that the alternating groups, the simple groups of Lie type and the sporadic simple groups account for all the finite non-abelian simple groups. While a complete proof of this result has yet to appear, it is the subject of a multi-volume work currently in preparation. The classification of finite simple groups is a synthesis of the work of many mathematicians and is by any standard one of the greatest scientific achievements of all time.

To conclude this section, consider how far we have come in trying to understand the structure of finite groups. If the aim is to construct all finite groups, the Jordan–Hölder Theorem shows that two steps are necessary:
(i) find all finite simple groups;

---

3 Émile Léonard Mathieu (1835–1890).

(ii) construct all possible group extensions of a given finite group N by a finite simple group S.

In step (ii) we have to construct all groups G with a normal subgroup M such that $M \simeq N$ and $G/M \simeq S$.

Let us accept that step (i) has been accomplished. A formal description of the extensions arising in (ii) is possible, but the general problem of deciding when two of the constructed groups are isomorphic is intractable. Thus the practicality of the scheme is questionable. However, this does not mean that the enterprize was not worthwhile since a vast amount of knowledge about finite groups has been accumulated during the course of the program.

Exercises (10.1).

(1) Show that isomorphic groups have the same composition factors.
(2) Find two non-isomorphic groups with the same composition factors.
(3) Show that S_3 has a unique composition series, but S_4 has exactly three composition series.
(4) Let G be a finite group and let $N \triangleleft G$. How are the composition factors of G related to those of N and G/N?
(5) Suppose that G is a group generated by normal subgroups $N_1, N_2, \ldots, N_k$ each of which is simple. Prove that G is the direct product of certain of the N_i. [Hint: Choose r maximal subject to the existence of normal subgroups $N_{i_1}, \ldots, N_{i_r}$ which generate their direct product. Then show that the direct product equals G.]
(6) Let G be as in the previous exercise. If $N \triangleleft G$, prove that N is a direct factor of G. [Hint: write $G = N_1 \times N_2 \times \cdots \times N_s$. Choose r maximal subject to $N, N_{i_1}, \ldots, N_{i_r}$ generating their direct product; then prove that this direct product equals G.]
(7) Let G be a group with a series in which each factor is either infinite cyclic or finite. Prove that any other series in G of this type has the same number of infinite factors, but not necessarily the same number of finite ones. [Hint: use (10.1.2).]
(8) Suppose that G is a group with a composition series. Prove that G satisfies the *ascending and descending chain conditions for subnormal subgroups*, i. e., there cannot exist an infinite ascending chain $H_1 < H_2 < H_3 < \cdots$ or an infinite descending chain $H_1 > H_2 > H_3 > \cdots$ where the H_i are subnormal subgroups of G. (For more on chain conditions see Exercise (3.3.10).)
(9) Prove that a group G which satisfies both the ascending and descending chain conditions on subnormal subgroups has a composition series. [Hint: start by choosing a minimal non-trivial subnormal subgroup of G.]
(10) Let D_n denote the subgroup of S_n generated by all the derangements where $n > 1$. Prove that $D_n = S_n$ if $n \neq 3$, but $D_3 = A_3$. Conclude that if $n \neq 3$, every permutation is a product of derangements. [Hint: first prove that $D_n \triangleleft S_n$ and that if $n \neq 3$, odd derangements exist. Deal first with the case $n = 4$. Then note that if $n > 4$, then $D_n = S_n$ by (10.1.9).]

10.2 Solvable and nilpotent groups

In this section we will discuss certain types of group which are wide generalizations of abelian groups, but which retain vestiges of commutativity. The basic concept is that of a *solvable group*, which is defined to be a group with a series all of whose factors are abelian. The terminology derives from the classical problem of solving algebraic equations by radicals, which is discussed in detail in Section 12.4. The length of a shortest series with abelian factors is called the *derived length* of the solvable group. Thus abelian groups are the solvable groups with derived length at most 1. Solvable group with derived length 2 or less are called *metabelian*.

Finite solvable groups are easily characterized in terms of their composition factors.

(10.2.1). *A finite group is solvable if and only if its composition factors have prime orders. In particular a simple group is solvable if and only if it has prime order.*

Proof. Let G be a finite solvable group; thus G has a series S with abelian factors. Refine S to a composition series of G. The factors of this series are simple and they are also abelian since they are isomorphic with quotients of subgroups of abelian groups. By (4.1.9) a simple abelian group has prime order. Hence composition factors of G have prime orders. The converse is an immediate consequence of the definition of solvability. $\square$

Solvability is well-behaved with respect to the formation of subgroups, quotient groups and extensions.

(10.2.2).
(i) *If G is a solvable group, then every subgroup and every quotient group of G is solvable.*
(ii) *Let G be a group with a normal subgroup N such that N and G/N are solvable. Then G is solvable.*

Proof. (i) Let $1 = G_0 \lhd G_1 \lhd \cdots \lhd G_n = G$ be a series with abelian factors and let H be a subgroup of G. Then

$$1 = G_0 \cap H \lhd G_1 \cap H \lhd \cdots \lhd G_n \cap H = H$$

is a series in H. Let $x, y \in G_{i+1} \cap H$. Then the commutator $[x, y] = xyx^{-1}y^{-1}$ belongs to G_i, because G_{i+1}/G_i is abelian, and clearly $[x, y] \in H$. Therefore $[x, y] \in G_i \cap H$ and $G_{i+1} \cap H / G_i \cap H$ is abelian, which shows that H is a solvable group.

Next let $N \lhd G$. Then G/N has the series

$$1 = G_0 N/N \lhd G_1 N/N \lhd \cdots \lhd G_n N/N = G/N.$$

Also $(G_{i+1}N/N)/(G_iN/N) \simeq G_{i+1}N/G_iN$ by (4.3.6). The assignment $xG_i \mapsto xG_iN$ determines a well defined, surjective homomorphism from G_{i+1}/G_i to $G_{i+1}N/G_iN$. Since G_{i+1}/G_i is abelian, the group $G_{i+1}N/G_iN$ is abelian and hence G/N is solvable.

(ii) The easy proof is left to the reader as an exercise.   □

The derived chain

Recall from Section 4.2 that the derived subgroup G' of a group G is the subgroup generated by all the commutators in G,

$$G' = \langle [x,y] \mid x,y \in G \rangle.$$

The *derived chain* $G^{(i)}$, $i = 0, 1, 2, \ldots$, is defined to be the descending sequence of subgroups formed by repeatedly taking derived subgroups: thus

$$G^{(0)} = G, \quad G^{(i+1)} = (G^{(i)})'.$$

Note that $G^{(i)} \triangleleft G$ and $G^{(i)}/G^{(i+1)}$ is an abelian group.

The important properties of the derived chain are that in a solvable group it reaches the identity subgroup and of all series with abelian factors it has shortest length.

(10.2.3). *Let* $1 = G_0 \triangleleft G_1 \triangleleft \cdots \triangleleft G_k = G$ *be a series with abelian factors in a solvable group* G. *Then* $G^{(i)} \leq G_{k-i}$ *for* $0 \leq i \leq k$. *In particular* $G^{(k)} = 1$, *so that the length of the derived chain equals the derived length of* G.

Proof. The containment is certainly true when $i = 0$. Assume that it is true for i. Since G_{k-i}/G_{k-i-1} is abelian, $G^{(i+1)} = (G^{(i)})' \leq (G_{k-i})' \leq G_{k-i-1}$, as required. On setting $i = k$, we find that $G^{(k)} = 1$.   □

Notice the consequence: a solvable group has a *normal series*, i. e., one in which every term is normal, with abelian factors: indeed the derived series is of this type.

It is sometimes possible to deduce solvability of a finite group by inspecting its order. Some group orders for which this can be done are given in the next result.

(10.2.4). *Let* p, q, r *be primes. Then any group whose order has the form* p^m, p^2q^2, p^mq *or* pqr *is solvable.*

Proof. First observe that in each case it is enough to show that there are no non-abelian simple groups with the order. For once this fact has been established, by applying it to the composition factors the general case will follow. If G is a simple group of order $p^m \neq 1$, then $Z(G) \neq 1$ by (5.3.6) and $Z(G) \triangleleft G$, so $G = Z(G)$ and G is abelian. Before proceeding further, recall that n_p denotes the number of Sylow p-subgroups in a finite group.

Next consider the case of a simple group G with order $p^m q$. We can of course assume that $p \neq q$. Then by Sylow's theorem $n_p \equiv 1 \pmod{p}$ and $n_p \mid q$, so that $n_p = q$, since $n_p = 1$ would mean that there is a normal Sylow p-subgroup.

Choose two distinct Sylow p-subgroups P_1 and P_2 whose intersection $I = P_1 \cap P_2$ has largest order. First of all suppose that $I = 1$. Then each pair of distinct Sylow p-subgroups intersects in 1, which makes it easy to count the number of non-trivial elements with order a power of p; indeed this number is $q(p^m - 1)$ since there are q Sylow p-subgroups. This leaves $p^m q - q(p^m - 1) = q$ elements of order prime to p. These elements must form a single Sylow q-subgroup, which is therefore normal in G, contradicting the simplicity of the group G. It follows that $I \neq 1$.

By Exercise (5.3.14), or (10.2.7) below, $I < N_i = N_{P_i}(I)$ for $i = 1, 2$. Thus $I \lhd J = \langle N_1, N_2 \rangle$. Suppose for the moment that J is a p-group. By Sylow's Theorem J is contained in some Sylow subgroup P_3 of G. But $P_1 \cap P_3 \geq P_1 \cap J > I$ since $N_1 \leq P_1 \cap J$, which contradicts the maximality of the intersection I. Therefore J is not a p-group.

By Lagrange's Theorem $|J|$ divides $|G| = p^m q$ and it is not a power of p, from which it follows that q must divide $|J|$. Let Q be a Sylow q-subgroup of J. By (4.1.12)

$$|P_1 Q| = \frac{|P_1| \cdot |Q|}{|P_1 \cap Q|} = \frac{p^m q}{1} = |G|,$$

and thus $G = P_1 Q$. Now let $g \in G$ and write $g = ab$ where $a \in P_1$, $b \in Q$. Note that $bIb^{-1} = I$ since $I \lhd J$ and $Q \leq J$. Hence $gIg^{-1} = a(bIb^{-1})a^{-1} = aIa^{-1} \leq P_1 < G$. But this means that $\bar{I} = \langle gIg^{-1} \mid g \in G \rangle \leq P_1 < G$ and also $1 \neq \bar{I} \lhd G$, a final contradiction.

The remaining group orders are left as exercises with appropriate hints – see Exercises (10.2.5) and (10.2.6). □

We mention two much deeper arithmetic criteria for a finite group to be solvable. The first states that *a group of order $p^m q^n$ is solvable if p and q are primes*. This is the celebrated *Burnside p-q Theorem*. It is best proved by using group characters – see (14.4.3).

An even more difficult result is the *Odd Order Theorem*, which asserts that *every group of odd order is solvable*. This famous theorem is due W. Feit[4] and J. G. Thompson: the original proof, published in 1963, was over 250 pages long. These results indicate that finite solvable groups are relatively common.

Nilpotent groups

Nilpotent groups form an important subclass of the class of solvable groups. A group G is said to be *nilpotent* if it has a *central series*, by which is meant a series of *normal* subgroups $1 = G_0 \lhd G_1 \lhd G_2 \lhd \cdots \lhd G_n = G$ such that G_{i+1}/G_i is contained in the center of G/G_i for all i. The length of a shortest central series is called the *nilpotent class* of G.

---

4 Walter Feit (1930–2004).

Abelian groups are just the nilpotent groups with class ≤ 1. Clearly every nilpotent group is solvable, but S_3 is a solvable group that is not nilpotent since its center is trivial.

The great source of finite nilpotent groups is the class of groups of prime power order.

(10.2.5). *Let G be a group of order p^m where p is a prime. Then G is nilpotent, and if $m > 1$, the nilpotent class of G is at most $m - 1$.*

Proof. Define a sequence of subgroups $\{Z_i\}$ by repeatedly forming centers. Thus $Z_0 = 1$ and $Z_{i+1}/Z_i = Z(G/Z_i)$. By (5.3.6), if $Z_i \neq G$, then $Z(G/Z_i) \neq 1$ and $Z_i < Z_{i+1}$. Since G is finite, there is a smallest integer n such that $Z_n = G$, and clearly $n \leq m$. Suppose that $n = m$. Then $|Z_{m-2}| \geq p^{m-2}$ and thus $|G/Z_{m-2}| \leq p^m/p^{m-2} = p^2$, which means that G/Z_{m-2} is abelian by (5.3.7). This yields the contradiction $Z_{m-1} = G$; therefore $n \leq m - 1$. □

The foregoing proof suggests a general construction, the *upper central chain* of a group G. This is the ascending chain of subgroups defined by repeatedly forming centers,

$$Z_0(G) = 1, \quad Z_{i+1}(G)/Z_i(G) = Z(G/Z_i(G)).$$

Thus $1 = Z_0 \leq Z_1 \leq \cdots$ and $Z_i \triangleleft G$. If G is finite, this chain will certainly terminate, although it may not reach G. The significance of the upper central chain for nilpotency is shown by the next result.

(10.2.6). *Let $1 = G_0 \triangleleft G_1 \triangleleft \cdots \triangleleft G_k = G$ be a central series in a nilpotent group G. Then $G_i \leq Z_i(G)$ for $0 \leq i \leq k$. In particular, $Z_k(G) = G$ and the length of the upper central chain equals the nilpotent class of G.*

Proof. We argue that $G_i \leq Z_i(G)$ by induction on i, which is certainly the case for $i = 0$. If it is true for i, then, since $G_{i+1}/G_i \leq Z(G/G_i)$, we have

$$G_{i+1}Z_i(G)/Z_i(G) \leq Z(G/Z_i(G)) = Z_{i+1}(G)/Z_i(G).$$

Thus $G_{i+1} \leq Z_{i+1}(G)$, which completes the induction. Consequently $G = G_k \leq Z_k(G)$ and $G = Z_k(G)$. □

Example (10.2.1). Let p be a prime and let $n > 1$. Denote by $U_n(p)$ the group of all $n \times n$ upper unitriangular matrices over the field $\mathbb{Z}_p$, i.e., matrices which have 1's on the diagonal and 0's below it. Counting the matrices of this type by enumerating possible superdiagonals, we find that $|U_n(p)| = p^{n-1} \cdot p^{n-2} \cdots p \cdot 1 = p^{n(n-1)/2}$. Therefore $U_n(p)$ is a nilpotent group, and in fact its class is $n - 1$, (see Exercise (10.2.11)).

Characterizations of finite nilpotent groups

There are several different descriptions of finite nilpotent groups, which shed light on the nature of the property of nilpotency.

(10.2.7). *Let G be a finite group. Then the following statements are equivalent:*
(i) *G is nilpotent;*
(ii) *every subgroup of G is subnormal;*
(iii) *every proper subgroup of G is smaller than its normalizer;*
(iv) *G is the direct product of its Sylow subgroups.*

Proof. (i) implies (ii). Let $1 = G_0 \lhd G_1 \lhd \cdots \lhd G_n = G$ be a central series and let H be a subgroup of G. Then $G_{i+1}/G_i \leq Z(G/G_i)$, so $HG_i/G_i \lhd HG_{i+1}/G_i$. Hence there is a chain of subgroups $H = HG_0 \lhd HG_1 \lhd \cdots \lhd HG_n = G$ and H is subnormal in G.

(ii) implies (iii). Let $H < G$; then H is subnormal in G, so there is a chain $H = H_0 \lhd H_1 \lhd \cdots \lhd H_m = G$. There is a least $i > 0$ such that $H \neq H_i$, and then $H = H_{i-1} \lhd H_i$. Therefore $H_i \leq N_G(H)$ and $N_G(H) \neq H$.

(iii) implies (iv). Let P be a Sylow p-subgroup of G. If P is not normal in G, then $N_G(P) < G$, and hence $N_G(P)$ is smaller than its normalizer. But this contradicts Exercise (5.3.15). Therefore $P \lhd G$ and P must be the unique Sylow p-subgroup, which will be written G_p.

Evidently $G_p \lhd G$ and $G_p \cap \langle G_q \mid q \neq p \rangle = 1$ since orders of elements from the intersecting subgroups are relatively prime. Clearly G is generated by its Sylow subgroups, so G is the direct product of the G_p.

(iv) implies (i). This follows quickly from the fact that a finite p-group is nilpotent. □

The unique Sylow p-subgroup G_p is called the *p-component* of the nilpotent group G.

The Frattini[5] subgroup

A very intriguing subgroup that can be formed in any group G is the *Frattini subgroup*

$$\phi(G).$$

This is defined to be the intersection of all the maximal subgroups of G. Here a *maximal subgroup* is a proper subgroup which is not contained in any larger proper subgroup. If G has no maximal subgroups, as is the case if G is trivial and might happen if G is infinite, then $\phi(G)$ is defined to be G. Note that $\phi(G)$ is normal in G. For example, S_3 has one maximal subgroup of order 3 and three of order 2: these intersect in 1, so $\phi(S_3) = 1$.

---

[5] Giovanni Frattini (1852–1925).

There is another, very different way of describing the Frattini subgroup which involves the notion of a non-generator. An element g of a group G is called a *non-generator* if $G = \langle g, X \rangle$ always implies that $G = \langle X \rangle$ where X is a non-empty subset of G. Thus a non-generator can be omitted from any generating set for G.

(10.2.8). *If G is an arbitrary group, then $\phi(G)$ is the set of all non-generators of G.*

Proof. Let g be a non-generator of G and assume that g is not in $\phi(G)$. Then there is at least one maximal subgroup of G which does not contain g, say M. Thus M is definitely smaller than $\langle g, M \rangle$, which implies that $G = \langle g, M \rangle$ since M is maximal. Therefore $G = M$ by the non-generator property, which is impossible since maximal subgroups are proper.

Conversely, let $g \in \phi(G)$ and suppose that $G = \langle g, X \rangle$, but $G \neq \langle X \rangle$. Apply Zorn's Lemma to the set of subgroups of G that contain X, but not g; hence this set has a maximal element, say M. Any subgroup that properly contains M must contain g by maximality of M, and hence must equal G. This means that M is actually a maximal subgroup of G. But then $g \in \phi(G) \leq M$, a contradiction. It follows that g is a non-generator. $\qquad\square$

Next we establish an important result that connects the Frattini subgroup to nilpotency.

(10.2.9). *If G is a finite group, then $\phi(G)$ is nilpotent.*

Proof. The proof depends on a useful trick known as *the Frattini argument*. Write $F = \phi(G)$ and let P be a Sylow p-subgroup of F. If $g \in G$, then $gPg^{-1} \leq F$ since $F \lhd G$: also $|gPg^{-1}| = |P|$. Therefore gPg^{-1} is a Sylow p-subgroup of F, and as such it must be conjugate to P in F by Sylow's Theorem. Thus $gPg^{-1} = xPx^{-1}$ for some x in F. This implies that $x^{-1}gP(x^{-1}g)^{-1} = P$, i. e., $x^{-1}g \in N_G(P)$ and $g \in FN_G(P)$. Thus the conclusion of the Frattini argument is that $G = FN_G(P)$. Now the non-generator property comes into play, allowing us to omit the elements of F one at a time, until eventually we get $G = N_G(P)$, i. e., $P \lhd G$. In particular $P \lhd F$, so that all the Sylow subgroups of F are normal and F is nilpotent by (10.2.7). $\qquad\square$

The Frattini subgroup of a finite p-group

The Frattini subgroup plays an especially significant role in the theory of finite p-groups. Suppose that G is a finite p-group. If M is a maximal subgroup of G, then, since G is nilpotent, M is subnormal and hence is normal in G. Furthermore G/M cannot have proper non-trivial subgroups by maximality of M. Thus $|G/M| = p$. Define the *pth power* of the group G to be

$$G^p = \langle g^p \mid g \in G \rangle.$$

Then $G^p G' \leq M$ for all M and $G^p G' \leq \phi(G)$.

On the other hand, $G/G^p G'$ is a finite abelian group in which every pth power is the identity, i. e., it is an elementary abelian p-group. By (8.2.16) such a group is a direct product of groups of order p. This fact enables us to construct maximal subgroups of $G/G^p G'$ by omitting all but one factor from the direct product. The resulting maximal subgroups of $G/G^p G'$ clearly intersect in the identity subgroup, which shows that $\phi(G) \leq G^p G'$. We have therefore proved:

(10.2.10). *If G is a finite p-group, then $\phi(G) = G^p G'$.*

Next suppose that $V = G/G^p G'$ has order p^d; thus d is the dimension of V as a vector space over the field $\mathbb{Z}_p$. Consider an arbitrary set X of generators for G. Now the subset $\{xG^p G' \mid x \in X\}$ clearly generates V as a vector space. By Exercise (8.2.10) there is a subset Y of X such that $\{yG^p G' \mid y \in Y\}$ is a basis of V. Of course $|Y| = d$. We claim that Y generates G. Certainly we have that $G = \langle Y, G^p G' \rangle = \langle Y, \phi(G) \rangle$. The non-generator property of $\phi(G)$ shows that $G = \langle Y \rangle$.

Summing up these conclusions, we have the following basic result on finite p-groups.

(10.2.11). *Let G be a finite p-group and let $G/\phi(G)$ have order p^d. Then every set of generators of G has a d-element subset that generates G. In particular G can be generated by d and no fewer elements.*

Example (10.2.2). A group G is constructed as the semidirect product of a cyclic group $\langle a \rangle$ of order 2^n with a Klein 4-group $V = \langle x, y \rangle$ where $n \geq 3$, $xax^{-1} = a^{-1}$ and $yay^{-1} = a^{1+2^{n-1}}$. Thus $|G| = 2^{n+2}$. Observe that $G' = \langle a^2 \rangle$ and thus G/G' is elementary abelian of order 8. Hence $\phi(G) = G^2 G' = \langle a^2 \rangle$. By (10.2.11) the group G can be generated by 3 and no fewer elements, and in fact $G = \langle a, x, y \rangle$.

Exercises (10.2).

(1) Let $M \triangleleft G$ and $N \triangleleft G$ where G is any group. If M and N are solvable, prove that MN is solvable.

(2) Let $M \triangleleft G$ and $N \triangleleft G$ for any group G. If G/M and G/N are solvable, prove that $G/M \cap N$ is solvable.

(3) Explain why a solvable group with a composition series is necessarily finite.

(4) Let G be a finite group with two non-trivial elements a and b such that $|a|, |b|, |ab|$ are relatively prime in pairs. Prove that G cannot be solvable. [Hint: put $H = \langle a, b \rangle$ and show that H/H' has order 1.]

(5) Prove that if p, q, r are primes, then every group of order pqr is solvable. [Hint: assume that G is a simple group of order pqr where $p < q < r$ and show that $n_r = pq$, $n_q \geq r$ and $n_p \geq q$. Now count elements of elements of G to obtain a contradiction.]

(6) Prove that if p and q are primes, then every group of order $p^2 q^2$ is solvable. [Hint: follow the method of proof for groups of order $p^m q$ in (10.2.4). Assume G is simple. Deal first with the case where each pair of Sylow p-subgroups intersects in 1. Then

choose two Sylow subgroups P_1 and P_2 such that $I = P_1 \cap P_2$ has order p and note that $I \lhd J = \langle P_1, P_2 \rangle$.]

(7) Establish the commutator identities $[x, y^{-1}] = y^{-1}[x, y]^{-1}y$ and $[x, yz] = [x, y]y[x, z]y^{-1}$.

(8) Let G be a group and let $z \in Z_2(G)$. Prove that the assignment $x \mapsto [z, x]$ determines a homomorphism from G to $Z(G)$ whose kernel contains $G'Z(G)$.

(9) Let G be a group such that $Z_1(G) < Z_2(G)$. Use Exercise (10.2.8) to show that $G > G'$.

(10) Find the upper central series of the group $G = \mathrm{Dih}(2^m)$ where $m \geq 2$. Hence compute the nilpotent class of G.

(11) Let $n > 1$ and let $G = U_n(p)$, the group of $n \times n$ upper unitriangular matrices over $\mathbb{Z}_p$. Define G_i to be the subgroup of all elements of G in which the first i superdiagonals consist of 0's, where $0 \leq i < n$. Show that the G_i are terms of a central series of G. Hence find an upper bound for the nilpotent class of G. (For a greater challenge find the exact value of the nilpotent class).

(12) Let G be a nilpotent group with a non-trivial normal subgroup N. Prove that $N \cap Z(G) \neq 1$.

(13) Let A be a maximal abelian normal subgroup (i. e., a largest abelian normal subgroup) of a nilpotent group G. Prove that $C_G(A) = A$. [Hint: assume this is false and apply Exercise (10.2.12) to $C_G(A)/A \lhd G/A$.]

(14) If every abelian normal subgroup of a nilpotent group is finite, prove that the group is finite.

(15) The *lower central chain* $\{\gamma_i(G)\}$ of group G is defined by $\gamma_1(G) = G$ and $\gamma_{i+1}(G) = [\gamma_i(G), G]$. If G is a nilpotent group, prove that the lower central sequence reaches 1 and its length equals the nilpotent class of G. (If H, K are subgroups of a group, then $[H, K]$ denotes the subgroup generated by all commutators $[h, k]$, $h \in H$, $k \in K$.)

(16) Find the Frattini subgroup of the groups A_n, S_n and $\mathrm{Dih}(2p)$ where p is an odd prime.

(17) Use (10.2.4) to show that a non-solvable group of order at most 100 must have order 60. [Hint: note that by (10.2.4) the only orders requiring attention are 72, 84 and 90.]

(18) Prove that A_5 is the only non-solvable group with order ≤ 100. [Hint: it is enough to show that a simple group of order 60 must have a subgroup of index 5 and hence is A_5. Consider the number of Sylow 2-subgroups.]

10.3 Theorems on finite solvable groups

The final section of the chapter will take us deeper into the theory of finite solvable groups and several famous theorems will be proved.

Schur's splitting and conjugacy theorem

Suppose that N is a normal subgroup of a group G. A subgroup X such that $G = NX$ and $N \cap X = 1$ is called a *complement* of N in G. In this case G is said to *split over N* and G is the semidirect product of N and X. A *splitting theorem* is theorem asserting that a group splits over a normal subgroup. One can think of such a theorem as resolving a group into a product of potentially simpler groups. The most celebrated splitting theorem in group theory is undoubtedly Schur's theorem.

(10.3.1) (I. Schur). *Let A be an abelian normal subgroup of a finite group G such that $|A|$ and $|G : A|$ are relatively prime. Then G splits over A and all complements of A are conjugate in G.*

Proof. (i) *Existence of a complement.* To start the proof choose an arbitrary transversal to A in G, say $\{t_x \mid x \in Q = G/N\}$ where $x = At_x$. Most likely this transversal will not be a subgroup. The idea behind the proof is to transform the transversal into one which is a subgroup. Let $x, y \in Q$: then $x = At_x$ and $y = At_y$, and in addition $At_{xy} = xy = At_xAt_y = At_xt_y$. Thus it is possible to write

$$t_xt_y = a(x,y)t_{xy} \tag{10.1}$$

for some $a(x,y) \in A$.

The associative law $(t_xt_y)t_z = t_x(t_yt_z)$ imposes conditions on the elements $a(x,y)$. For, on applying the formula (10.1) for the product of two transversal elements, we obtain

$$(t_xt_y)t_z = a(x,y)a(xy,z)t_{xyz},$$

and similarly

$$t_x(t_yt_z) = t_xa(y,z)t_{yz} = (t_xa(y,z)t_x^{-1})t_xt_{yz} = (t_xa(y,z)t_x^{-1})a(x,yz)t_{xyz}.$$

Now conjugation of elements of A by t_x induces an automorphism of A which depends only on x: for, if $u, v \in A$, then $(vt_x)u(vt_x)^{-1} = t_xut_x^{-1}$ since A is abelian. Let us write ^xu for $t_xut_x^{-1}$. Then on equating $(t_xt_y)t_z$ and $t_x(t_yt_z)$ and cancelling t_{xyz}, we arrive at the equation

$$a(x,y)a(xy,z) = {}^x(a(y,z))a(x,yz), \tag{10.2}$$

which is valid for all $x, y, z, \in Q$. A function $a : Q \times Q \to A$ that satisfies the condition (10.2) is called a *factor set* or a *2-cocycle*.

Next define

$$b_x = \prod_{y \in Q} a(x, y),$$

noting that the order of the factors in the product is immaterial since A is abelian. On forming the product of the equations (10.2) above over all z in Q with x and y fixed, we obtain the equation

$$a(x, y)^m b_{xy} = {}^x b_y b_x, \tag{10.3}$$

where $m = |Q| = |G : A|$. Note here that the product over z of all the ${}^x a(y, z)$ is ${}^x b_y$ and the product of all the $a(x, yz)$ is b_x.

Since m is relatively prime to $|A|$, the mapping $u \mapsto u^m$, $u \in A$, is an automorphism of A. Thus we can write b_x as an mth power, say $b_x = (c_x^{-1})^m$ where $c_x \in A$. Substituting for b_x in equation (10.3), we get $(a(x, y)c_{xy}^{-1})^m = (({}^x c_y c_x)^{-1})^m$, from which it follows that

$$c_{xy} = c_x({}^x c_y)a(x, y).$$

We are now ready to form the new transversal. Write $s_x = c_x t_x$ and observe that the s_x, $(x \in Q)$, form a transversal to A. Indeed

$$s_x s_y = c_x t_x c_y t_y = c_x({}^x c_y)t_x t_y = c_x({}^x c_y)a(x, y)t_{xy} = c_{xy}t_{xy} = s_{xy}.$$

This demonstrates that the transversal $H = \{s_x \mid x \in Q\}$ is a subgroup. Since $G = AH$ and $A \cap H = 1$, it follows that H is a complement of A in G and G splits over A.

(ii) *Conjugacy of complements.* Let $H = \{s_x \mid x \in Q\}$ and $H^* = \{s_x^* \mid x \in Q\}$ be two complements of A in G. If $x \in Q$, we can write $x = As_x = As_x^*$ where s_x and s_x^* belong to H and H^* respectively. Thus s_x and s_x^* are related by an equation of the form

$$s_x^* = d(x)s_x$$

where $d(x) \in A$. Since $As_{xy} = xy = As_x As_y = As_x s_y$, we have $s_x s_y = s_{xy}$, and similarly $s_x^* s_y^* = s_{xy}^*$. In the last equation make the substitutions $s_x^* = d(x)s_x$, $s_y^* = d(y)s_y$, $s_{xy}^* = d(xy)s_{xy}$ to get $d(x)s_x d(y)s_y = d(xy)s_{xy}$ and hence

$$d(xy) = d(x)({}^x d(y)) \tag{10.4}$$

for all $x, y \in Q$. Such a function $d : Q \to A$ is called a *derivation* or *1-cocycle.*

Put $u = \prod_{x \in Q} d(x)$ and take the product of all the equations (10.4) over $y \in Q$ with x fixed. This leads to $u = d(x)^m({}^x u)$. Writing $u = v^m$ with $v \in A$, we obtain $v^m = d(x)^m({}^x v)^m$ and hence $v = d(x)({}^x v)$. Thus $d(x) = v({}^x v)^{-1}$. Since ${}^x v = s_x v s_x^{-1}$, we have

$$s_x^* = d(x)s_x = v(^xv)^{-1}s_x = v(s_xv^{-1}s_x^{-1})s_x = vs_xv^{-1}.$$

Therefore $H^* = vHv^{-1}$, so H and H^* are conjugate. □

In fact (10.3.1) is true even when A is non-abelian, a result which is known as the *Schur-Zassenhaus Theorem*. The proof of conjugacy of complements requires use of the Odd Order Theorem: see for example [15].

Hall's theorems on finite solvable groups

To illustrate the usefulness of Schur's splitting theorem we will make a foray into the theory of finite solvable groups by proving the following celebrated result.

(10.3.2) (P. Hall[6]). *Let G be a finite solvable group and write $|G| = mn$ where the positive integers m, n are relatively prime. Then G has a subgroup of order m and all subgroups of this order are conjugate.*

Proof. (i) *Existence.* We argue by induction on $|G| > 1$. The group G has a non-trivial abelian normal subgroup A, for example the smallest non-trivial term of the derived series. Since A is the direct product of its primary components, we can assume that A is a p-group, with $|A| = p^k$, say. There are two cases to consider.

Suppose first that p does not divide m. Then $p^k \mid n$ because m and n are relatively prime. Since $|G/A| = m \cdot (n/p^k)$, the induction hypothesis may be applied to the group G/A to show that it has a subgroup of order m, say K/A. Now $|A|$ is relatively prime to $m = |K : A|$, so (10.3.1) may be applied to K. Hence there is a complement of A in K: this complement is a subgroup order m, as required.

Now assume that p divides m; then $p^k \mid m$ since p cannot divide n. Since $|G/A| = (m/p^k) \cdot n$, induction shows that G/A has a subgroup of order m/p^k, say H/A. Then $|H| = |A| \cdot |H/A| = p^k(m/p^k) = m$, as required.

(ii) *Conjugacy.* Let H and H^* be two subgroups of order m and choose A to be a p-group as in (i). If p does not divide m, then $A \cap H = 1 = A \cap H^*$, and AH/A and AH^*/A are subgroups of G/A with order m. By induction on $|G|$ these subgroups are conjugate and thus $AH = g(AH^*)g^{-1} = A(gH^*g^{-1})$ for some $g \in G$. By replacing H^* by gH^*g^{-1}, we can assume that $AH = AH^*$. But now H and H^* are two complements of A in HA, so (10.3.1) guarantees that they are conjugate.

Finally, assume that p divides m. Then p does not divide $n = |G : H| = |G : H^*|$. Since $|AH : H|$ is a power of p and it also divides n, we conclude that $AH = H$ and $A \le H$. Similarly $A \le H^*$. By induction H/A and H^*/A are conjugate in G/A, as must H and H^* be in G. □

---

6 Philip Hall (1904–1982).

Hall π-subgroups

We will now illustrate the significance of Hall's theorem. Let π denote a non-empty set of primes and let π' be the complementary set of primes. A positive integer is called a *π-number* if it is a product of powers of primes from the set π. A finite group is said to be a *π-group* if its order is a π-number.

Let G be a finite solvable group and write $|G| = mn$ where m is a π-number and n is a π'-number. Then (10.3.2) tells us that G has a subgroup H of order m and index n. Thus H is a π-group and $|G : H|$ is a π'-number: such a subgroup is called a *Hall π-subgroup* of G. Thus (10.3.2) actually asserts that Hall π-subgroups always exist in a finite solvable group for any set of primes π, and that any two Hall π-subgroups are conjugate.

Hall's theorem can be regarded as an extension of Sylow's Theorem since if $\pi = \{p\}$, a Hall π-subgroup is simply a Sylow p-subgroup. However, Sylow's Theorem is valid for any finite group, whereas Hall subgroups need not exist in an insolvable group. For example, A_5 has order $60 = 3 \cdot 20$, but it has no subgroups of order 20, as the reader should verify.

This example is no coincidence since there is in fact a strong converse of Hall's theorem: the mere existence Hall p'-subgroups for all primes p dividing the group order is enough to imply solvability of the group. Here p' is the set of all primes different from p. The proof of this result uses the Burnside pq-Theorem: a group of order $p^m q^n$ is solvable if p and q are primes: this is proved using character theory as (14.4.3).

(10.3.3) (P. Hall). *Let G be a finite group and suppose that for every prime p dividing $|G|$ there is a Hall p'-subgroup. Then G is solvable.*

Proof. Assume the theorem is false and let G be a counterexample of smallest order. We look for a contradiction. Suppose that N is proper non-trivial normal subgroup of G. If H is a Hall p'-subgroup of G, then by consideration of order and index we see that $H \cap N$ and HN/N are Hall p'-subgroups of N and G/N respectively. Therefore N and G/N are solvable by minimality of $|G|$, and thus G is solvable. By this contradiction G must be a simple group.

Write $|G| = p_1^{e_1} p_2^{e_2} \cdots p_k^{e_k}$ where $e_i > 0$ and the p_i are distinct primes. The Burnside pq-Theorem shows that $k > 2$. Let G_i be a Hall p_i'-subgroup of G; thus $|G : G_i| = p_i^{e_i}$. Put $H = G_3 \cap \cdots \cap G_k$ and observe that

$$|G : H| = \prod_{i=3}^{k} |G : G_i| = \prod_{i=3}^{k} p_i^{e_i}$$

by (4.1.13), since the $|G : G_i|$ are all relatively prime. Therefore $|H| = |G|/|G : H| = p_1^{e_1} p_2^{e_2}$ and H is solvable by Burnside's Theorem.

Since $H \neq 1$, it contains a minimal normal subgroup M. By Exercise (10.3.2) below, M is an elementary abelian p-group where $p = p_1$ or p_2: without loss of generality let

$p = p_1$. Now

$$|G : H \cap G_2| = |G : H| \cdot |G : G_2| = \prod_{i=2}^{k} p_i^{e_i}$$

by (4.1.13) once again. Thus $|H \cap G_2| = p_1^{e_1}$, i. e., $H \cap G_2$ is a Sylow p_1-subgroup of H. Hence $M(H \cap G_2)$ is a p_1-group, from which it follows that $M \leq H \cap G_2$. Also $|H \cap G_1| = p_2^{e_2}$ by a previous argument and therefore

$$|(H \cap G_1)G_2| = |H \cap G_1| \cdot |G_2| = p_2^{e_2} \frac{|G|}{p_2^{e_2}} = |G|.$$

Consequently $G = (H \cap G_1)G_2$. Next consider the normal closure of M in G. This is

$$\langle M^G \rangle = \langle M^{(H \cap G_1)G_2} \rangle = \langle M^{G_2} \rangle \leq G_2 < G,$$

since $M \triangleleft H$. It follows that $\langle M^G \rangle$ is a proper non-trivial normal subgroup of G, so G is not simple, a contradiction. □

Hall's theorems are the starting point for a rich theory of finite solvable groups which has been developed over the last ninety years; the standard reference for this is [4].

Exercises (10.3).

(1) Give an example of a finite group G with an abelian normal subgroup A such that G does not split over A.

(2) If G is a finite solvable group with a minimal (non-trivial) normal subgroup N, prove that N is an elementary abelian p-group for some p dividing $|G|$. [Hint: note that $N' \triangleleft G$.]

(3) If M is a maximal subgroup of a finite solvable group G, prove that $|G : M|$ is equal to a prime power. [Hint: use induction on $|G|$ to reduce to the case where M contains no non-trivial normal subgroups of G. Let A be a minimal normal subgroup of G. Show that $G = MA$ and $M \cap A = 1$.]

(4) For which sets of primes π does the group A_5 contain a Hall π-subgroup?

(5) Let G be a finite solvable group and p a prime dividing the order of G. Prove that G has a maximal subgroup with index equal to a power of p. [Hint: apply (10.3.2).]

(6) Let G be a finite group and π a set of primes. Let L be a solvable normal subgroup of G and assume that H is a Hall π-subgroup of L. Prove that $G = LN_G(H)$. [Hint: if $g \in G$, then gHg^{-1} is conjugate to H in L.]

(7) Let G be a finite group with a normal subgroup N. Assume that $|N|$ and $|G : N|$ are relatively prime and also that N is *solvable*. Prove that G splits over N. [Hint: assume that $N \neq 1$ and find a non-trivial abelian subgroup A of G which is contained in N. By induction on the group order the result is true for G/A.]

(8) Let G be a finite group and let p be a prime dividing the order of G. Prove that p divides $|G : \phi(G)|$, where $\phi(G)$ denotes the Frattini subgroup of G. [Hint: assume this is false, so $G/\phi(G)$ is a p'-group. Since $\phi(G)$ is nilpotent, there exists $P \lhd G$ such that $P \leq \phi(G)$, P is a p-group and G/P a p'-group. Now apply Exercise (10.3.7).]

11 The Theory of Fields

Field theory is one of the most attractive parts of algebra. It contains many powerful results on the structure of fields, for example, the Fundamental Theorem of Galois Theory, which establishes a correspondence between subfields of a field and subgroups of the Galois group. In addition field theory can be applied to a wide variety of problems, some of which date from classical antiquity. Among the applications to be described here and in subsequent chapters are: ruler and compass constructions, solution of equations by radicals, orthogonal latin squares and Steiner systems. In short field theory is algebra at its best – deep theorems with convincing applications to problems which might otherwise be intractible.

11.1 Field extensions

Recall from Section 7.4 that a *subfield* of a field F is a subring containing the identity element which is closed with respect to inversion of its non-zero elements. The following is an immediate consequence of the definition.

(11.1.1). *The intersection of any set of subfields of a field is a subfield.*

Suppose that X is a (non-empty) subset of a field F. By (11.1.1) the intersection of all the subfields of F that contain X is a subfield, which is evidently the smallest subfield containing X. This is called the *subfield of F generated by X* and it is easy to describe the form of its elements.

(11.1.2). *If X is a subset of a field F, the subfield generated by X consists of all elements of the form*

$$f(x_1,\ldots,x_m)g(y_1,\ldots,y_n)^{-1}$$

where $f \in \mathbb{Z}[t_1,\ldots,t_m], g \in \mathbb{Z}[t_1,\ldots,t_n], x_i, y_j \in X$ and $g(y_1,\ldots,y_n) \neq 0$.

To prove this, first observe that the set S of elements with the specified form is a subfield of F containing X. Then note that any subfield of F which contains X must also contain all the elements of S, so that S is the smallest subfield that contains X.

Prime subfields

In any field F one can form the intersection of *all* the subfields. This is the unique smallest subfield of F and it is called the *prime subfield* of F. A field which coincides with its prime subfield is called a *prime field*. It is easy to identify the prime fields.

(11.1.3). *A prime field of characteristic 0 is isomorphic with $\mathbb{Q}$: a prime field of characteristic a prime p is isomorphic with $\mathbb{Z}_p$. Conversely, $\mathbb{Q}$ and $\mathbb{Z}_p$ are prime fields.*

https://doi.org/10.1515/9783110691160-011

Proof. Assume that F is a prime field and put $I = \langle 1_F \rangle = \{n1_F \mid n \in Z\}$. Suppose first that F has characteristic 0, so I is infinite cyclic. Define a surjective mapping $\alpha : \mathbb{Q} \to F$ by the rule $\alpha(\frac{m}{n}) = (m1_F)(n1_F)^{-1}$, where $n \neq 0$. It is easily seen that α is a well defined ring homomorphism and its kernel is therefore an ideal of $\mathbb{Q}$. Now 0 and $\mathbb{Q}$ are the only ideals of $\mathbb{Q}$ and $\alpha(1) = 1_F \neq 0_F$, so $\mathrm{Ker}(\alpha) \neq \mathbb{Q}$. It follows that $\mathrm{Ker}(\alpha) = 0$ and $\mathbb{Q} \simeq \mathrm{Im}(\alpha)$. Since F is a prime field and $\mathrm{Im}(\alpha)$ is a subfield, $\mathrm{Im}(\alpha) = F$ and α is an isomorphism. Thus $F \simeq \mathbb{Q}$.

Now suppose that F has characteristic a prime p, so that $|I| = p$. In this situation we define $\alpha : \mathbb{Z} \to F$ by $\alpha(n) = n1_F$. Thus $\alpha(n) = 0_F$ if and only if $n1_F = 0$, i. e., p divides n. Hence $\mathrm{Ker}(\alpha) = p\mathbb{Z}$ and $\mathrm{Im}(\alpha) \simeq \mathbb{Z}/p\mathbb{Z} = \mathbb{Z}_p$. It follows that $\mathbb{Z}_p$ is isomorphic with a subfield of F and, since F is prime, $\mathbb{Z}_p \simeq F$. It is left to the reader to check that $\mathbb{Q}$ and $\mathbb{Z}_p$ are prime fields. □

Field extensions

Consider two fields F and E and suppose there is an injective ring homomorphism $\alpha : F \to E$. Then F is isomorphic with $\mathrm{Im}(\alpha)$, which is a subfield of E: under these circumstances we say that E is an *extension of F*. It is often convenient to assume that F is actually a subfield of E. This is usually a harmless assumption since F could be replaced by the isomorphic field $\mathrm{Im}(\alpha)$. Notice that by (11.1.3) every field is an extension of either $\mathbb{Z}_p$ or $\mathbb{Q}$, according as the characteristic is a prime p or 0.

If E is an extension of F, then E can be regarded as a vector space over F by using the field operations. The vector space axioms are consequences of the field axioms. This simple idea is critical since it allows us to define the *degree of E over F* as

$$(E : F) = \dim_F(E).$$

If this dimension is finite, then E is said to be a *finite extension* of F.

Simple extensions

Let F be a subfield and X a non-empty subset of a field E. The subfield of E generated by $F \cup X$ is denoted by

$$F(X).$$

It follows readily from (11.1.2) that $F(X)$ consists of all elements of the form $f(x_1, \ldots, x_m)g(y_1, \ldots, y_n)^{-1}$ where $f \in F[t_1, \ldots, t_m]$, $g \in F[t_1, \ldots, t_n]$, $x_i, y_j \in X$ and $g(y_1, \ldots, y_n) \neq 0$. If $X = \{x_1, x_2, \ldots, x_l\}$, write

$$F(x_1, x_2, \ldots, x_l)$$

in place of $F\{x_1, x_2, \ldots, x_l\}$. The most interesting case for us is when $X = \{x\}$ and a typical element of $F(x)$ has the form $f(x)g(x)^{-1}$ where $f, g \in F[t]$ and $g(x) \neq 0$. If $E = F(x)$ for some $x \in E$, then E is said to be a *simple extension* of F.

The next result describes the structure of simple extensions.

(11.1.4). *Let $E = F(x)$ be a simple extension of a field F. Then one of the following must hold:*

(i) *$f(x) \neq 0$ for all $0 \neq f \in F[t]$ and $E \simeq F\{t\}$, the field of rational functions in t over F;*

(ii) *$f(x) = 0$ for some monic irreducible polynomial $f \in F[t]$ and $E \simeq F[t]/(f)$.*

Proof. We may assume that $F \subseteq E$. Define a mapping $\theta : F[t] \to E$ by evaluation at x, i. e., $\theta(f) = f(x)$. This is a ring homomorphism whose kernel is an ideal of $F[t]$, say I.

Assume first that $I = 0$, i. e., $f(x) = 0$ implies that $f = 0$. Then θ can be extended to a function $\alpha : F\{t\} \to E$ by the rule $\alpha(\frac{f}{g}) = f(x)g(x)^{-1}$; this function is also a ring homomorphism. Notice that $\alpha(\frac{f}{g}) = 0$ implies that $f(x) = 0$ and hence $f = 0$. Therefore $\mathrm{Ker}(\alpha) = 0$ and $F\{t\}$ is isomorphic with $\mathrm{Im}(\alpha)$, which is a subfield of E. Now $\mathrm{Im}(\alpha)$ contains F and x since $\alpha(a) = a$ if $a \in F$ and $\alpha(t) = x$. Because E is a smallest field containing F and x, it follows that $E = \mathrm{Im}(\alpha) \simeq F\{t\}$.

Now suppose that $I \neq 0$. Then $F[t]/I$ is isomorphic with a subring of the field E, so it is a domain and hence I is a prime ideal of $F[t]$. Since $F[t]$ is a PID, we can apply (7.2.6) to get $I = (f)$ where f is a monic irreducible polynomial in $F[t]$. Thus $F[t]/I$ is a field which is isomorphic with $\mathrm{Im}(\theta)$, a subfield of E containing F and x for reasons given above. Therefore $F[t]/I \simeq \mathrm{Im}(\theta) = E$. $\square$

Algebraic elements

Consider a field extension E of F and let $x \in E$. There are two possible forms for the subfield $F(x)$, as indicated in (11.1.4). If $f(x) \neq 0$ whenever $0 \neq f \in F[t]$, then $F(x) \simeq F\{t\}$ and x is said to be *transcendent over F*.

The other possibility is that x is a root of a monic irreducible polynomial f in $F[t]$. In this case $F(x) \simeq F[t]/(f)$ and x is said to be *algebraic over F*. The polynomial f is the unique monic irreducible polynomial over F which has x as a root: for if g is another such polynomial, then $g \in (f)$ and $f \mid g$, so $f = g$ by irreducibility and monicity. We call f the *irreducible polynomial* of x over F, in symbols

$$\mathrm{Irr}_F(x):$$

thus $F(x) \simeq F[t]/(\mathrm{Irr}_F(x))$.

Now let $f = \mathrm{Irr}_F(x)$ have degree n. For any g in $F[t]$ write $g = fq + r$ where $q, r \in F[t]$ and $\deg(r) < n$, by using the Division Algorithm for $F[t]$, (see (7.1.3)). Then $g + (f) = r + (f)$, which shows that $F(x)$ is generated as an F-vector space by $1, x, x^2, \ldots, x^{n-1}$. In fact these elements are linearly independent over F. For, if $a_0 + a_1 x + \cdots + a_{n-1} x^{n-1} = 0$ with $a_i \in F$, then $g(x) = 0$ where $g = a_0 + a_1 t + \cdots + a_{n-1} t^{n-1}$, and hence $f \mid g$. But

$\deg(g) \le n - 1$, which can only mean that $g = 0$ and all the a_i are zero. It follows that the elements $1, x, x^2, \ldots, x^{n-1}$ form an F-basis of the vector space $F(x)$ and hence $(F(x) : F) = n = \deg(f)$.

These conclusions are summarized in:

(11.1.5). *Let $E = F(x)$ be a simple field extension of F.*
(i) *If x is transcendent over F, then $E \simeq F\{t\}$.*
(ii) *If x is algebraic over F, then $E \simeq F[t]/(\mathrm{Irr}_F(x))$ and $(E : F) = \deg(\mathrm{Irr}_F(x))$.*

Example (11.1.1). Show that $\sqrt{3} - \sqrt{2}$ is algebraic over $\mathbb{Q}$ by finding its irreducible polynomial and hence the degree of $\mathbb{Q}(\sqrt{3} - \sqrt{2})$ over $\mathbb{Q}$.

Put $a = \sqrt{3} - \sqrt{2}$. Our first move is to find a rational polynomial with a as a root. Now $a^2 = 5 - 2\sqrt{6}$, so $(a^2 - 5)^2 = 24$ and $a^4 - 10a^2 + 1 = 0$. Hence a is a root of $f = t^4 - 10t^2 + 1$ and thus is algebraic over $\mathbb{Q}$. If we can show that f is irreducible over $\mathbb{Q}$, it will follow that $\mathrm{Irr}_\mathbb{Q}(a) = f$ and $(\mathbb{Q}(a) : \mathbb{Q}) = 4$.

By Gauss's Lemma (7.3.7) it is enough to show that f is irreducible over $\mathbb{Z}$. Now clearly f has no integer roots, for ± 1 are the only possibilities and neither one is a root. Thus, if f is reducible, there must be a decomposition of the form

$$f = (t^2 + at + b)(t^2 + a_1 t + b_1)$$

where a, b, a_1, b_1 are integers. On equating coefficients of $1, t^3, t^2$ on both sides, we arrive at the equations

$$bb_1 = 1, \quad a + a_1 = 0, \quad aa_1 + b + b_1 = -10.$$

Hence $b = b_1 = \pm 1$ and $a_1 = -a$, so that $-a^2 \pm 2 = -10$. Since this equation has no integer solutions, f is irreducible.

Algebraic extensions

Let E be an extension of a field F. If every element of E is algebraic over F, then E is called an *algebraic extension* of F. Extensions of finite degree are an important source of algebraic extensions.

(11.1.6). *An extension E of a field F with finite degree is algebraic.*

Proof. Let $x \in E$. By hypothesis E has finite dimension as a vector space over F, equal to n say; consequently the set $\{1, x, x^2, \ldots, x^n\}$ is linearly dependent and there are elements $a_0, a_1, \ldots, a_n$ of F, not all zero, such that $a_0 + a_1 x + a_2 x^2 + \cdots + a_n x^n = 0$. Thus x is a root of the non-zero polynomial $a_0 + a_1 t + \cdots + a_n t^n$ and hence is algebraic over F. $\square$

The next result is useful in calculations with degrees.

(11.1.7). *Let $F \subseteq K \subseteq E$ be successive field extensions. If K is finite over F and E is finite over K, then E is finite over F and $(E : F) = (E : K) \cdot (K : F)$.*

Proof. Let $\{x_1, \ldots, x_m\}$ be an F-basis of K and $\{y_1, \ldots, y_n\}$ a K-basis of E. Then each $e \in E$ can be written as $e = \sum_{i=1}^{n} k_i y_i$ where $k_i \in K$. Also each k_i can be written $k_i = \sum_{j=1}^{m} f_{ij} x_j$ with $f_{ij} \in F$. Therefore $e = \sum_{i=1}^{n} \sum_{j=1}^{m} f_{ij} x_j y_i$ and it follows that the elements $x_j y_i$ generate the F-vector space E.

Next assume there is an F-linear relation among the $x_j y_i$, say

$$\sum_{i=1}^{n} \sum_{j=1}^{m} f_{ij} x_j y_i = 0$$

where $f_{ij} \in F$. Then $\sum_{i=1}^{n} (\sum_{j=1}^{m} f_{ij} x_j) y_i = 0$, so that $\sum_{j=1}^{m} f_{ij} x_j = 0$ for all i, since the y_i are K-linearly independent. Finally, $f_{ij} = 0$ for all i and j by linear independence of the x_j over F. Consequently the elements $x_j y_i$ form an F-basis of E and $(E : F) = nm$, that is $(E : F) = (E : K) \cdot (K : F)$. □

Corollary (11.1.8). *Let $F \subseteq K \subseteq E$ be successive field extensions with E algebraic over K and K algebraic over F. Then E is algebraic over F.*

Proof. Let $x \in E$, so that x is algebraic over K; let its irreducible polynomial be $f = a_0 + a_1 t + \cdots + a_{n-1} t^{n-1} + t^n$ where $a_i \in K$. Put $K_i = F(a_0, a_1, \ldots, a_i)$ for $0 \le i < n$ and $K_{-1} = F$. Then a_i is algebraic over F and hence over K_{i-1}. Since $K_i = K_{i-1}(a_i)$, it follows via (11.1.5) that $(K_i : K_{i-1})$ is finite for $i = 0, 1, \ldots, n-1$. Hence $(K_{n-1} : F)$ is finite by (11.1.7). Also x is algebraic over K_{n-1}, so that $(K_{n-1}(x) : K_{n-1})$ is finite and therefore $(K_{n-1}(x) : F)$ is finite. It follows via (11.1.6) that x is algebraic over F. □

Algebraic numbers

Next let us consider the complex field $\mathbb{C}$ as an extension of the rational field $\mathbb{Q}$. If $x \in \mathbb{C}$ is algebraic over $\mathbb{Q}$, then x is called an *algebraic number*: otherwise x is a *transcendental number*. Thus the algebraic numbers are the real and complex numbers which are roots of non-zero rational polynomials.

(11.1.9). *The algebraic numbers form a subfield of $\mathbb{C}$.*

Proof. Let a and b be algebraic numbers. It is sufficient to show that $a \pm b$, ab and ab^{-1} (if $b \ne 0$) are algebraic numbers. To see this note that $(\mathbb{Q}(a) : \mathbb{Q})$ is finite by (11.1.5). Also $\mathbb{Q}(a, b) = (\mathbb{Q}(a))(b)$ is finite over $\mathbb{Q}(a)$ for the same reason. Therefore $(\mathbb{Q}(a, b) : \mathbb{Q})$ is finite by (11.1.7) and hence $\mathbb{Q}(a, b)$ is algebraic over $\mathbb{Q}$ by (11.1.6). The required result now follows. □

The next result shows that not every complex number is an algebraic number.

(11.1.10). *There are countably many algebraic numbers, but uncountably many complex numbers.*

Proof. Of course $\mathbb{C}$ is uncountable by (1.4.7). To see that there are countably many algebraic numbers, observe that $\mathbb{Q}[t]$ is countable since it is a countable union of countable sets – see Exercise (1.4.5). Also each non-zero polynomial in $\mathbb{Q}[t]$ has finitely many roots. It follows that there are only countably many roots of non-zero polynomials in $\mathbb{Q}[t]$: these are precisely the algebraic numbers. □

The existence of transcendental numbers is demonstrated by (11.1.10), but without giving a single example. Indeed it is a good deal harder to find specific examples. The best known transcendental numbers are the numbers π and e. The fact that π is transcendental underlies the impossibility of "squaring the circle" – for this see Section 11.2. A good reference for the transcendence of π, e and many other interesting numbers is [13].

A subfield of $\mathbb{C}$ which is a finite extension of $\mathbb{Q}$ is called an *algebraic number field*: the elements of algebraic number fields constitute all the algebraic numbers. The theory of algebraic number fields is a well developed and active area of algebra; for a detailed account of it see [10].

Algebraic integers

An algebraic number a is called an *algebraic integer* if it is a root of a monic polynomial with integer coefficients. The first thing to notice is that the algebraic integers in an algebraic number field form a subring.

(11.1.11).

(i) *An algebraic number a is an algebraic integer if and only if the subring generated by a and 1 is finitely generated as an abelian group.*

(ii) *The algebraic integers form a subring of the field of algebraic numbers.*

Proof. (i) Assume that a is an algebraic integer. Then there is a polynomial $f(x) = x^n + r_{n-1}x^{n-1} + \cdots + r_1 x + r_0$ with integer coefficients such that $f(a) = 0$. Hence $a^n \in \langle 1, a, \ldots, a^{n-1} \rangle = H$. Thus $a^{n+1} \in \langle a, a^2, \ldots, a^n \rangle \leq H$. One can show by induction on $k \geq n$ that $a^k \in H$. Therefore H is the subring generated by a and 1.

Conversely, assume that the subring generated by a and 1 is finitely generated. Then it must equal $K = \langle 1, a, \ldots, a^{n-1} \rangle$ for some n. Hence $a^n \in K$, which shows that $a^n = r_0 + r_1 a + \cdots + r_{n-1}a^{n-1}$ for certain integers r_i. Hence a is an algebraic integer, which completes the proof of (i).

(ii) Let a_1, a_2 be algebraic integers. By (i) the subrings generated by a_1, a_2 are finitely generated abelian groups. Therefore the ring generated by $\{1, a_1, a_2\}$ is a finitely generated abelian group. Hence the rings generated by $a_1 \pm a_2$ and $a_1 a_2$ are also finitely generated abelian. Therefore $a_1 \pm a_2$ and $a_1 a_2$ are algebraic integers. It follows that the algebraic integers form a subring. □

The next result is very simple, but it will be important when we come to discuss group characters in Section 14.3.

(11.1.12). *An algebraic integer which is a rational number is an integer.*

Proof. Assume that the rational number m/n is an algebraic integer where m, n are relatively prime integers. Then m/n is a root of some monic polynomial $t^r + \ell_{r-1} t^{r-1} + \cdots + \ell_1 t + \ell_0$ where $\ell_i \in \mathbb{Z}$. Hence $m^r + \ell_{r-1} m^{r-1} n + \cdots + \ell_1 mn^{r-1} + \ell_0 n^r = 0$. However, this implies that n divides m^r. Since m and n are relatively prime, it follows that $n = \pm 1$, so the result is proven. □

Algebraically closed fields

It is a very important fact that every field F is contained in a largest algebraic extension called the *algebraic closure*. The construction of such a largest extension is the kind of task for which Zorn's Lemma is well-suited.

Let E be a field extension of F with $F \subseteq E$. Then E is called an *algebraic closure* of F if the following conditions hold:
(i) E is algebraic over F;
(ii) every irreducible polynomial in $E[t]$ has degree 1.

Notice that by the second condition if K is an algebraic extension of E, then $K = E$, so that E is a maximal algebraic extension of F. A field that coincides with its algebraic closure is called an *algebraically closed field*. For example, the complex field $\mathbb{C}$ is algebraically closed by the Fundamental Theorem of Algebra – see (12.3.6) below.

Our objective is to prove the following theorem:

(11.1.13). *Every field has an algebraic closure.*

Proof. Let F be an arbitrary field. The first step is to choose a set that is large enough to accommodate the algebraic closure. What is needed is a set S with cardinal greater than $\aleph_0 \cdot |F|$: for example the set $\mathcal{P}(\mathbb{N} \times F)$ will do – see (1.4.5). In particular $|F| < |S|$, so there is an injection $\alpha : F \to S$. Now use the map α to turn $\mathrm{Im}(\alpha)$ into a field, by defining

$$\alpha(x) + \alpha(y) = \alpha(x + y) \quad \text{and} \quad \alpha(x)\alpha(y) = \alpha(xy)$$

where $x, y \in F$, and $\alpha(0_F)$ and $\alpha(1_F)$ are the zero element and identity element respectively. Clearly $\mathrm{Im}(\alpha)$ is a field isomorphic with F. Thus, replacing F by $\mathrm{Im}(\alpha)$, we may assume that $F \subseteq S$.

To apply Zorn's Lemma we need to introduce a suitable partially ordered set. Let $\mathcal{K}$ denote the set of all subsets E such that $F \subseteq E \subseteq S$ and the field operations of F may be extended to E in such a way that E becomes a field which is algebraic over F. Quite obviously $F \in \mathcal{K}$, so that $\mathcal{K}$ is not empty. A partial order $\preceq$ on $\mathcal{K}$ is defined as follows: if $E_1, E_2 \in \mathcal{K}$, then $E_1 \preceq E_2$ means that $E_1 \subseteq E_2$ and the field operations of E_2 are consistent with those of E_1. Thus E_1 is actually a subfield of E_2. It is quite easy to see that $\preceq$ is a partial order on $\mathcal{K}$. Thus we have our partially ordered set $(\mathcal{K}, \preceq)$.

Next the union U of a chain $\mathcal{C}$ in $\mathcal{K}$ is itself in $\mathcal{K}$. For, by the definition of the partial order $\preceq$, the field operations of all members of $\mathcal{C}$ are consistent, so they may be combined to give the field operations of U. It follows that $U \in \mathcal{K}$ and clearly U is an upper bound for $\mathcal{C}$ in $\mathcal{K}$. Zorn's Lemma may now be applied to yield a maximal element of $\mathcal{K}$, say E.

By definition E is algebraic over F. What needs to be established is that any irreducible polynomial f in $E[t]$ has degree 1. Suppose that in fact $\deg(f) > 1$. Put $E' = E[t]/(f)$, which is an algebraic extension of E and hence of F by (11.1.8). If we write $E_0 = \{a + (f) \mid a \in E\}$, then $E_0 \subseteq E'$ and there is an isomorphism $\beta : E_0 \to E$ given by $\beta(a + (f)) = a$.

It is at this point that the cardinality of the set S is important. One can show without too much trouble that $|E' - E_0| < |S - E|$, by using the inequalities $|E| \leq \aleph_0 \cdot |F|$ and $|E[t]| < |S|$. Accepting this fact, we choose an injective map $\beta_1 : E' - E_0 \mapsto S - E$. Combine β_1 with $\beta : E_0 \to E$ to produce an injection $\gamma : E' \to S$. Thus $\gamma(a + (f)) = a$ for a in E.

Next we use the map γ to make $J = \mathrm{Im}(\gamma)$ into a field, by defining $\gamma(x_1) + \gamma(x_2) = \gamma(x_1 + x_2)$ and $\gamma(x_1)\gamma(x_2) = \gamma(x_1 x_2)$. Then $\gamma : E' \to J$ is an isomorphism of fields and $\gamma(E_0) = E$. Since E' is algebraic over E_0, it follows that J is algebraic over E and therefore $J \in \mathcal{K}$. However, $E \neq J$ since $E_0 \neq E'$, which contradicts the maximality of E and completes the proof. $\square$

While some details in the above proof are tricky, the essential idea is clear: build a largest algebraic extension of F by using Zorn's Lemma. It can be shown, although we shall not do so here, that *every field has a unique algebraic closure up to isomorphism* – see [8] for a proof.

For example, the algebraic closure of $\mathbb{Q}$ is the field of all algebraic numbers. Another example of interest is the algebraic closure of the Galois field $\mathrm{GF}(p)$, which is an algebraically closed field of prime characteristic p.

Exercises (11.1).

(1) Give examples of infinite field extensions of $\mathbb{Q}$ and of $\mathbb{Z}_p$.

(2) Let $a = 2^{\frac{1}{p}}$ where p is a prime. Prove that $(\mathbb{Q}(a) : \mathbb{Q}) = p$ and that $\mathbb{Q}(a)$ has only two subfields.

(3) Let n be an arbitrary positive integer. Construct an algebraic number field of degree n over $\mathbb{Q}$.

(4) Let a be a root of $t^6 - 4t + 2 \in \mathbb{Q}[t]$. Prove that $(\mathbb{Q}(a) : \mathbb{Q}) = 6$.

(5) Let p and q be distinct primes and set $F = \mathbb{Q}(\sqrt{p}, \sqrt{q})$. Establish each of the following statements.
   (i) $(F : \mathbb{Q}) = 4$;
   (ii) $F = \mathbb{Q}(\sqrt{p} + \sqrt{q})$;
   (iii) the irreducible polynomial of $\sqrt{p} + \sqrt{q}$ over $\mathbb{Q}$ is $t^4 - 2(p + q)t^2 + (p - q)^2$.
(6) Let K be a finite extension of a field F and let F_1 be a subfield such that $F \subseteq F_1 \subseteq K$. Prove that F_1 is finite over F and K is finite over F_1.
(7) Prove that every non-constant element of $\mathbb{Q}\{t\}$ is transcendent over $\mathbb{Q}$.
(8) Let $a = 3^{\frac{1}{2}} - 2^{\frac{1}{3}}$. Show that $(\mathbb{Q}(a) : \mathbb{Q}) = 6$ and find $\mathrm{Irr}_{\mathbb{Q}}(a)$.
(9) Let p be a prime and put $a = e^{2\pi i/p}$, a complex primitive pth root of unity. Prove that $\mathrm{Irr}_{\mathbb{Q}}(a) = 1 + t + t^2 + \cdots + t^{p-1}$ and $(\mathbb{Q}(a) : \mathbb{Q}) = p - 1$.

11.2 Constructions with ruler and compass

One of the most striking applications of field theory is to solve certain famous geometric problems dating back to classical Greece. Each problem asks whether it is possible to construct a geometric object by *using ruler and compass only*. Here one has to keep in mind that to the ancient Greeks only mathematical objects constructed by such means had any reality, since Greek mathematics was based on geometry. We will describe four constructional problems and then translate them to field theory.

(i) *Duplication of the cube*. A cube of side one unit is given. The problem is to construct a cube with double the volume using ruler and compass. This problem is said to have arisen when the oracle at Delphi commanded the citizens of Delos to double the size of an altar to the god Apollo which had the shape of a cube.

(ii) *Squaring the circle*. Here the question is whether it is possible to construct, using ruler and compass, a square whose area equals that of a circle with radius one unit? This is perhaps the most notorious of the ruler and compass problems. It is really a question about the nature of the number π.

(iii) *Trisection of an angle*. Another notorious problem asks whether it is always possible to trisect a given angle using ruler and compass.

(iv) *Construction of a regular n-gon*. Here the problem is to construct by ruler and compass a regular n-sided plane polygon with side equal to one unit where $n \geq 3$.

These problems defied the efforts of mathematicians for more than 2000 years despite many ingenious attempts to solve them. It was only with the rise of abstract algebra in the 18th and 19th centuries that it was realized that all four problems had negative solutions.

Constructibility

Our first move must be to formulate precisely what is meant by a ruler and compass construction. Let S be a set of points in the plane containing the points $O(0,0)$ and $I(1,0)$; note that O and I are one unit apart. A point P in the plane is said to be *constructible from S by ruler and compass* if there is a finite sequence of points $P_0, P_1, \ldots, P_n = P$ with P_0 in S where P_{i+1} is obtained from $P_0, P_1, \ldots, P_i$ by a procedure of one of the following types:

(i)   draw a straight line joining two of the points $P_0, P_1, \ldots, P_i$;
(ii) draw a circle with center one of $P_0, P_1, \ldots, P_i$ and radius equal to the distance between two of these points;
(iii) then P_{i+1} is to be a point of intersection of two lines, of a line and a circle or of two circles, where the lines and circles are as described in (i) and (ii).

Finally, a real number r is said to be *constructible from S* if the point $(r,0)$ is constructible from S. The reader will realize that these definitions are designed to express precisely our intuitive idea of a construction by ruler and compass. Each of the four problems asks whether a certain real number is constructible from a given set of points. For example, in the problem of duplicating a cube of side 1, take S to be the set $\{O,I\}$: the question is whether $\sqrt[3]{2}$ is constructible from S.

We begin by showing that the real numbers which are constructible from a given set of points form a field: this explains why field theory is relevant in constructional problems.

(11.2.1). *Let S be a set of points in the plane containing $O(0,0)$ and $I(1,0)$ and let S^* be the set of all real numbers constructible from S. Then S^* is a subfield of $\mathbb{R}$. Also $\sqrt{a} \in S^*$ whenever $a \in S^*$ and $a > 0$.*

Proof. This is entirely elementary plane geometry. Let $a, b \in S^*$; we have first to prove that $a \pm b$, ab and a^{-1} (if $a \neq 0$) belong to S^*. Keep in mind here that by hypothesis a and b are constructible.

To construct $a \pm b$, where say $a \geq b$, draw the circle with center $A(a,0)$ and radius b. This intersects the x-axis at the points $B(a-b,0)$ and $C(a+b,0)$. Hence $a+b$ and $a-b$ are constructible from S and thus belong to S^*.

$$O \quad\bullet\!\!\!\!\!\!\!\!\rule[0.5ex]{2cm}{0.4pt}\!\!\bullet\!\!\rule[0.5ex]{2cm}{0.4pt}\!\!\bullet\!\!\rule[0.5ex]{2cm}{0.4pt}\!\!\bullet\!\!\rule[0.5ex]{2cm}{0.4pt}\!\!\!\!\!\!\rightarrow x$$
$$B(a-b,0) \quad A(a,0) \quad C(a+b,0)$$

It is a little harder to construct ab. Assume that $a \leq 1 \leq b$: in other cases the argument is similar. Let A and B be the points $(a,0)$ and $(b,0)$. Mark the point $B'(0,b)$ on the y-axis; thus $|OB'| = |OB|$. Draw the line IB' and then draw AC' parallel to IB' with C' on the y-axis: elementary geometry tells us how to do this. Mark C on the x-axis so that

$|OC| = |OC'|$.

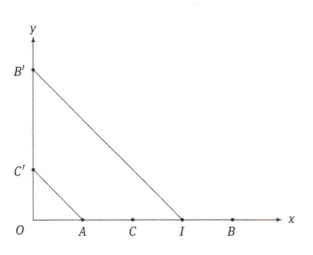

By similar triangles $|OC'|/|OB'| = |OA|/|OI|$; therefore $|OC| = |OC'| = |OA| \cdot |OB'| = ab$. Hence ab is constructible and $ab \in S^*$.

Next we show how to construct a^{-1}: we will assume that $a > 1$, the case $a < 1$ being similar. Let A be the point $(a, 0)$ and mark the point $I'(0, 1)$ on the y-axis. Draw the line IC' parallel to AI' with C' on the y-axis. Mark C on the x-axis so that $|OC| = |OC'|$. Then $|OC'|/|OI'| = |OI|/|OA|$, so $|OC| = |OC'| = a^{-1}$. Hence a^{-1} is constructible and so belongs to S^*.

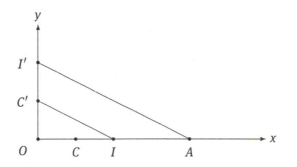

Finally, let $a \in S^*$ where $a > 0$. We have to show how to construct the point $D(\sqrt{a}, 0)$: it will then follow that $\sqrt{a} \in S^*$. We can assume that $a > 1$ – otherwise replace a by a^{-1}. First mark the point $A(a + 1, 0)$. Let C be the mid-point of the line segment OA; thus C is the point $(\frac{a+1}{2}, 0)$ and it is clear how to construct this. Now draw the circle with center C and radius $|OC| = \frac{a+1}{2}$, (not shown). Then draw the perpendicular to the x-axis through the point $I(1, 0)$ and let it meet the upper semicircle at D_1. Mark D on the x-axis so that $|OD| = |ID_1|$.

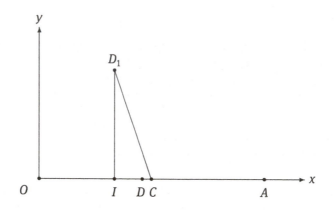

Then

$$|OD|^2 = |ID_1|^2 = |D_1C|^2 - |IC|^2 = \left(\frac{a+1}{2}\right)^2 - \left(\frac{a-1}{2}\right)^2 = a.$$

Hence $|OD| = \sqrt{a}$ and $\sqrt{a}$ is constructible. $\qquad\square$

It is now time to explain the field theoretic aspect of constructibility.

(11.2.2). *Let S be a set of points in the plane containing $O(0,0)$ and $I(1,0)$, and denote by F the subfield of $\mathbb{R}$ generated by the coordinates of the points of S. Let a be any real number. If a is constructible from S, then $(F(a):F)$ is equal to a power of 2.*

Proof. Let P be the point $(a,0)$. Since a is constructible from S, there is by definition a sequence of points $P_0, P_1, \ldots, P_n = P$ with $P_0 \in S$, where P_{i+1} is obtained from $P_0, P_1, \ldots, P_i$ by intersecting lines and circles as explained above. Let P_i be the point (a_i, b_i) and put $E_i = F(a_1, \ldots, a_i, b_1, \ldots, b_i)$ and $E_0 = F$. Then $F(a) \subseteq E_n = E$, say. If P_{i+1} is the point of intersection of two lines whose equations have coefficients in E_i, then a_{i+1} and b_{i+1} are in E_i, as can be seen by solving two linear equations, i. e., $E_i = E_{i+1}$.

If P_{i+1} is a point of intersection of a line and a circle whose equations have coefficients in E_i, then a_{i+1} is a root of a quadratic equation over E_i. If P_{i+1} is a point of intersection of two circles over E_i, subtract the equations of the circles (in standard form) to realize P_{i+1} as a point of intersection of a line and a circle. Hence in either case $(E_i(a_{i+1}) : E_i) \le 2$. Likewise $(E_i(b_{i+1}) : E_i) \le 2$ and therefore $(E_i(a_{i+1})(b_{i+1}) : E_i(a_{i+1})) \le 2$. Note that $E_{i+1} = E_i(a_{i+1})(b_{i+1})$. It follows via (11.1.7) that

$$(E_{i+1} : E_i) = (E_i(a_{i+1})(b_{i+1}) : E_i(a_{i+1})) \cdot (E_i(a_{i+1}) : E_i),$$

which divides 4.

By (11.1.7) again

$$(E : F) = \prod_{i=0}^{n-1} (E_{i+1} : E_i).$$

Therefore $(E : F)$ equals a power of 2, as does $(F(a) : F)$, because $(E : F) = (E : F(a))(F(a) : F))$. $\qquad\square$

The first two ruler and compass problems can now be resolved.

(11.2.3). *It is impossible to duplicate a cube of side* 1 *or to square a circle of radius* 1 *by ruler and compass.*

Proof. Let S consist of the points $O(0,0)$ and $I(1,0)$. In the case of the cube, constructibility would imply that $(\mathbb{Q}(\sqrt[3]{2}) : \mathbb{Q})$ is a power of 2 by (11.2.2). But $(\mathbb{Q}(\sqrt[3]{2}) : \mathbb{Q}) = 3$ since $\mathrm{Irr}_{\mathbb{Q}}(\sqrt[3]{2}) = t^3 - 2$, a contradiction.

If it were possible to square the circle, $\sqrt{\pi}$ would be constructible from S. By (11.2.2) this implies that $(\mathbb{Q}(\sqrt{\pi}) : \mathbb{Q})$ is a power of 2, as is $(\mathbb{Q}(\pi) : \mathbb{Q})$, since $\mathbb{Q}(\pi) \subseteq \mathbb{Q}(\sqrt{\pi})$. But in fact π is transcendental over $\mathbb{Q}$ by a famous result of Lindemann,[1] so $(\mathbb{Q}(\pi) : \mathbb{Q})$ is actually infinite. Therefore it is impossible to square the circle. $\qquad\square$

With a greater effort we can determine which angles can be trisected.

(11.2.4). *An angle α can be trisected by ruler and compass if and only if the polynomial $4t^3 - 3t - \cos\alpha$ is reducible over the field $\mathbb{Q}(\cos\alpha)$.*

Proof. In this problem the angle α is given, so we can construct its cosine by drawing a right angled triangle with angle α and hypotenuse 1. Let S consist of the points O, I and $(\cos\alpha, 0)$. Let $F = \mathbb{Q}(\cos\alpha)$ and put $\theta = \frac{1}{3}\alpha$. The problem is to decide if θ, or equivalently $\cos\theta$, is constructible from S. If this is the case, $(F(\cos\theta) : F)$ must be a power of 2.

Recall the well-known trigonometric identity $\cos 3\theta = 4\cos^3\theta - 3\cos\theta$. Hence $4\cos^3\theta - 3\cos\theta - \cos\alpha = 0$, so that the polynomial $f = 4t^3 - 3t - \cos\alpha \in F[t]$ has $\cos\theta$ as a root. Thus $\mathrm{Irr}_F(\cos\theta)$ has degree at most 3. If θ is constructible, $\mathrm{Irr}_F(\cos\alpha)$ has degree a power of 2 and therefore f is reducible.

Conversely, suppose that f is reducible, so that $\cos\theta$ is a root of a linear or quadratic polynomial over F; thus $\cos\theta$ has the form $u + v\sqrt{w}$ where $u, v, w \in F$ and $w \geq 0$. Let S^* be the set of real numbers constructible from S. Since $F \subseteq S^*$, it follows from (11.2.1) that $\sqrt{w} \in S^*$. Hence $\cos\theta = u + v\sqrt{w} \in S^*$ and $\cos\theta$ is constructible from S, as required. $\qquad\square$

---

[1] Carl Ferdinand von Lindemann (1852–1939).

Example (11.2.1). The angle $\frac{\pi}{4}$ is trisectible by ruler and compass.

Since $\cos\frac{\pi}{4} = \frac{1}{\sqrt{2}}$, the polynomial f in (11.2.4) equals $4t^3 - 3t - \frac{1}{\sqrt{2}}$, which has the root $-\frac{1}{\sqrt{2}}$ in $\mathbb{Q}(\cos(\pi/4)) = \mathbb{Q}(\sqrt{2})$. Hence f is reducible. Now apply (11.2.4) to get the result.

Example (11.2.2). The angle $\frac{\pi}{3}$ is not trisectible by ruler and compass.

In this case $\cos\frac{\pi}{3} = \frac{1}{2}$ and $f = 4t^3 - 3t - \frac{1}{2}$. This polynomial is irreducible over $\mathbb{Q}(\frac{1}{2}) = \mathbb{Q}$ since it has no rational roots. Hence $\frac{\pi}{3}$ is not trisectible.

A complete discussion of the problem of constructing a regular n-gon calls for some Galois theory and will be deferred until (12.3.7).

Exercises (11.2).
(1) Complete the proof that $ab \in S^*$ in (11.2.1) by dealing with the cases $1 \le a \le b$, and $a \le b \le 1$.
(2) Show that a cube of side a can be duplicated if and only if $2a$ is the cube of a rational number.
(3) Consider the problem of doubling the surface area of a cube of side 1. Can a cube with double the surface area be constructed by ruler and compass?
(4) Determine which of the following angles are trisectible: (i) $\frac{\pi}{2}$; (ii) $\frac{\pi}{6}$; (ii) $\frac{\pi}{12}$.
(5) Let p be a prime and suppose that $a = e^{2\pi i/p}$ is constructible from $O(0,0)$ and $I(1,0)$. Show that p must have the form $2^{2^c} + 1$ for some integer $c \ge 0$, i.e., p is a *Fermat prime*. (The known Fermat primes occur for $0 \le c \le 4$.)

11.3 Finite fields

It was shown in (8.2.17) that the order of a finite field is always a power of a prime. More precisely, if F is a finite field of prime characteristic p and $(F : \mathbb{Z}_p) = n$, then $|F| = p^n$. Our main purpose in this section is to show that there are fields with arbitrary prime power order and that fields with the same order are isomorphic.

We begin by identifying finite fields with the splitting fields of certain polynomials. Let F be a field of order $q = p^n$ where p is a prime, namely the characteristic of F. The multiplicative group $U(F)$ has order $q - 1$ and Lagrange's Theorem shows that the order of every element of $U(F)$ divides $q - 1$. This means that $a^{q-1} = 1$ for every $a \ne 0$ in F, so $a^q - a = 0$. Since the zero element also satisfies the last equation, every element of F is a root of the polynomial $t^q - t \in \mathbb{Z}_p[t]$. But $t^q - t$ cannot have more than q roots, so we conclude that the elements of F constitute all the roots of $t^q - t$: thus F is a splitting field of $t^q - t$.

The foregoing discussion suggests that the existence of finite fields can be established by using splitting fields, a hope that is borne out by the next result.

(11.3.1). *Let $q = p^n$ where p is a prime and $n > 0$. Then:*
(i) *a splitting field of the polynomial $t^q - t \in \mathbb{Z}_p[t]$ has order q;*
(ii) *if F is any field of order q, then F is a splitting field of $t^q - t$ over $\mathbb{Z}_p$.*

Proof. We have already proved (ii), so consider the assertion (i) and write F for a splitting field of $t^q - t$. Define $S = \{a \in F \mid a^q = a\}$, i. e., the set of roots of $t^q - t$ in F. First we show that S is a subfield of F. For this purpose let $a, b \in S$. Recall that p divides $\binom{p}{i}$ if $1 \le i < p$ by (2.3.3); therefore the Binomial Theorem for the field F takes the form $(a \pm b)^p = a^p \pm b^p$, (see Exercise (6.1.6)). By repeatedly taking the pth power of both sides we conclude that:

$$(a \pm b)^q = a^q \pm b^q = a \pm b,$$

which shows that $a \pm b \in S$. Also $(ab)^q = a^q b^q = ab$ and $(a^{-1})^q = (a^q)^{-1} = a^{-1}$ if $a \ne 0$. It follows that $ab \in S$ and $a^{-1} \in S$. Therefore S is a subfield of F.

Next the roots of the polynomial $t^q - t$ are all different. For $(t^q - t)' = qt^{q-1} - 1 = -1$, so that $t^q - t$ and its derivative $(t^q - t)'$ are relatively prime; therefore by (7.4.7) the polynomial $t^q - t$ has no repeated roots. It follows that $|S| = q$. Finally, since F is a splitting field of $t^q - t$, it is generated by $\mathbb{Z}_p$ and the roots of $t^q - t$, that is by $\mathbb{Z}_p$ and S. Therefore $F = S$ and $|F| = q$. □

Our next objective is to show that fields with the same finite order are isomorphic. Since every finite field has been identified as a splitting field, our strategy is to prove the general result that any two splitting fields of a given polynomial are isomorphic, plainly a result of independent interest. In proving this we employ a useful lemma which shows how to extend an isomorphism between two given fields to extensions of these fields.

(11.3.2). *Let $E = F(x)$ and $E^* = F^*(x^*)$ be simple algebraic extensions of fields F and F^*. Further assume there is an isomorphism $\alpha : F \to F^*$ such that $\alpha(\mathrm{Irr}_F(x)) = \mathrm{Irr}_{F^*}(x^*)$. Then there is an isomorphism $\theta : E \to E^*$ such that $\theta|_F = \alpha$ and $\theta(x) = x^*$.*

In the statement of this result α has been extended in the natural way to a ring isomorphism $\alpha : F[t] \to F^*[t]$ by the rule $\alpha(\sum_{i=1}^m a_i t^i) = \sum_{i=1}^m \alpha(a_i) t^i$ where $a_i \in F$.

Proof of (11.3.2). Put $f = \mathrm{Irr}_F(x)$ and $f^* = \mathrm{Irr}_{F^*}(x^*)$; then by hypothesis $\alpha(f) = f^*$. This fact permits us to define a mapping

$$\theta_0 : F[t]/(f) \to F^*[t]/(f^*)$$

by the rule $\theta_0(g + (f)) = \alpha(g) + (f^*)$; a simple check shows this to be a well defined isomorphism. Next by (11.1.4) we have $F(x) \simeq F[t]/(f)$ and $F^*(x^*) \simeq F^*[t]/(f^*)$ via the respective assignments $g(x) \mapsto g + (f)$ and $g^*(x^*) \mapsto g^* + (f^*)$. Composition with θ_0

yields an isomorphism $\theta : F(x) \to F^*(x^*)$ where $\theta(g(x)) = \alpha(g(x^*))$, as indicated in the sequence of maps

$$F(x) \to F[t]/(f) \overset{\theta_0}{\to} F^*[t]/(f^*) \to F^*(x^*). \qquad \square$$

The uniqueness of splitting fields is a special case of the next result.

(11.3.3). *Let $\alpha : F \to F^*$ be an isomorphism of fields, and let $f \in F[t]$ and $f^* = \alpha(f) \in F^*[t]$. If E and E^* are splitting fields of f and f^* respectively, there is an isomorphism $\theta : E \to E^*$ such that $\theta|_F = \alpha$.*

Proof. Argue by induction on $n = \deg(f)$. If $n = 1$, then $E = F$, $E^* = F^*$ and $\theta = \alpha$. Assume that $n > 1$. Let a be a root of f in E and put $g = \mathrm{Irr}_F(a)$. Choose any root a^* of $g^* = \alpha(g) \in F^*[t]$. Then $g^* = \mathrm{Irr}_{F^*}(a^*)$. By (11.3.2) we can extend α to an isomorphism $\theta_1 : F(a) \to F^*(a^*)$ such that $\theta_1|_F = \alpha$ and $\theta_1(a) = a^*$.

Now regard E and E^* as splitting fields of the polynomials $f/(t-a)$ and $f^*/(t-a^*)$ over $F(a)$ and $F^*(a^*)$ respectively. By induction on n we can extend θ_1 to an isomorphism $\theta : E \to E^*$; furthermore $\theta|_F = \theta_1|_F = \alpha$, as required. $\qquad \square$

Corollary (11.3.4). *Let f be a non-constant polynomial over a field F. Then up to isomorphism f has a unique splitting field.*

This follows from (11.3.3) on taking $F = F^*$ and α to be the identity map. Since any finite field of order q is a splitting field of $t^q - t$, we deduce from (11.3.4) the fundamental theorem:

(11.3.5) (E. H. Moore[2]). *Finite fields of the same order are isomorphic.*

It is customary to write

$$\mathrm{GF}(q)$$

for the essentially unique field of order q: here "GF" stands for *Galois field*.

It is an important property of finite fields that their multiplicative groups are cyclic. More generally we prove:

(11.3.6). *If F is any field, every finite subgroup of its multiplicative group $U(F)$ is cyclic. Thus if F has finite order q, then $U(F)$ is a cyclic group of order $q - 1$.*

Proof. Let X be a finite subgroup of $U(F)$. Then X is a finite abelian group, so by the Primary Decomposition Theorem (9.2.3) we can write $X = P_1 \times P_2 \times \cdots \times P_k$ where P_i is a finite p_i-group and $p_1, p_2, \ldots, p_k$ are different primes. If each P_i is cyclic, then X is cyclic by Example (4.2.5). Therefore we may assume that X is a finite p-group with p a prime: let $|X| = p^m$.

---

2 Eliakim Hastings Moore (1862–1932).

Choose an element y with maximum order in X, say p^n. Thus $n \le m$. Now every element $x \in X$ satisfies $x^{p^n} = 1$ and must therefore be a root of $t^{p^n} - 1$. Hence $p^m = |X| \le p^n$. It follows that $n = m$ and X has order p^n, which implies that $X = \langle y \rangle$.   □

This result provides another way to represent the elements of a field F of order q. If $U(F) = \langle a \rangle$, then $F = \{0, 1, a, a^2, \dots, a^{q-2}\}$ where $a^{q-1} = 1$. This representation is useful for computational purposes.

Corollary (11.3.7). *Every finite field F is a simple extension of its prime subfield.*

For if $U(F) = \langle a \rangle$, then clearly $F = \mathbb{Z}_p(a)$ where p is the characteristic of F.

Example (11.3.1). Let $F = \mathrm{GF}(27)$ be the Galois field of order 27. Exhibit F as a simple extension of $\mathrm{GF}(3)$ and find a generator of $U(F)$.

The field F may be realized as the splitting field of the polynomial $t^{27} - t$, but it is simpler to choose an irreducible polynomial of degree 3 over $\mathrm{GF}(3)$, for example $f = t^3 - t + 1$. For then $F = (\mathrm{GF}(3)[t])/(f)$ is a field of order 3^3, which by (11.3.5) must be $\mathrm{GF}(27)$. Put $x = t + (f)$. Then, because f has degree 3, each element b of F has the unique form $b = a_0 + a_1 t + a_2 t^2 + (f)$, i. e., $b = a_0 + a_1 x + a_2 x^2$. Thus $F = \mathrm{GF}(3)(x)$ and $\mathrm{Irr}_{\mathrm{GF}(3)}(x) = f = t^3 - t + 1$.

Next we argue that $U(F) = \langle x \rangle$. Since $|U(F)| = 26$, it is enough to prove that $|x| = 26$. Certainly $|x|$ divides 26, so it suffices to show that $x^2 \ne 1$ and $x^{13} \ne 1$. The first statement is true because $f \nmid t^2 - 1$. To show that $x^{13} \ne 1$, use the relation $x^3 = x - 1$ to compute $x^{12} = (x-1)^4 = x^2 + 2$; thus $x^{13} = -1$.

Exercises (11.3).

(1) Let F be a field of order p^m where p is a prime and let K a subfield of F. Prove that $|K| = p^d$ where d divides m.

(2) If F is a field of order p^m and d is a positive divisor of m, show that F has exactly one subfield of order p^d.

(3) Find an element of order 7 in the multiplicative group of $\mathbb{Z}_2[t]/(t^3 + t + 1) \simeq \mathrm{GF}(8)$.

(4) Find elements of order 3, 5 and 15 in the multiplicative group of $\mathbb{Z}_2[t]/(t^4 + t + 1) \simeq \mathrm{GF}(16)$.

(5) Prove that $t^{p^n} - t \in \mathrm{GF}(p)[t]$ is the product of the distinct monic irreducible polynomials with degree dividing n.

(6) Let $\psi(n)$ denote the number of monic irreducible polynomials of degree n in $\mathrm{GF}(p)[t]$ where p is a fixed prime.

   (i) Prove that $p^n = \sum_{d|n} d\psi(d)$ where the sum is over all positive divisors d of n.

   (ii) Deduce that $\psi(n) = \frac{1}{n} \sum_{d|n} \mu(d) p^{n/d}$ where μ is the Möbius function,[3] which is defined as follows: $\mu(1) = 1$, $\mu(n)$ equals $(-1)^r$ where r is the number of distinct prime divisors of n if n is square-free, and $\mu(n) = 0$ otherwise. (You will need

---

[3] August Ferdinand Möbius (1790–1868).

the *Möbius Inversion Formula*: if $f(n) = \sum_{d|n} g(d)$, then $g(n) = \sum_{d|n} \mu(d)f(n/d)$. For an account of the Möbius function see Section 12.2 below.

(7) Find all monic irreducible polynomials over GF(2) with degrees 2, 3, 4 and 5, using Exercise (11.3.6) to check your answer.

12 Galois Theory

In this chapter the Galois group of a field extension is introduced. This will establish the critical link between field theory and group theory in which subfields correspond to subgroups of the Galois group. A major application is to the classical problem of solving polynomial equations by radicals, which is an excellent illustration of the rich rewards that can be reaped when connections are made between different mathematical theories.

12.1 Normal and separable extensions

We begin by introducing two special types of field extension, leading up to the concept of a Galois extension. Let E be an extension of a field F with $F \subseteq E$. Then E is said to be *normal* over F if it is algebraic over F and if every irreducible polynomial in $F[t]$ having a root in E has all its roots in E; thus the polynomial is a product of linear factors over E.

Example (12.1.1). Consider the field $E = \mathbb{Q}(a)$ where $a = 2^{1/3}$. Then E is algebraic over $\mathbb{Q}$ since $(E : \mathbb{Q})$ is finite, but it is not normal over $\mathbb{Q}$. This is because $t^3 - 2$ has one root a in E but not the complex roots $a\omega$, $a\omega^2$ where $\omega = e^{2\pi i/3}$.

Example (12.1.2). Let E be an extension of a field F with $(E : F) = 2$. Then E is normal over F.

In the first place E is algebraic over F. Suppose that $x \in E$ is a root of some monic irreducible polynomial $f \in F[t]$. Then $f = \mathrm{Irr}_F(x)$ and $\deg(f) = (F(x) : F) \leq (E : F) = 2$, which means that $\deg(f) = 1$ or 2. In the first case x is the only root of f. Suppose that $\deg(f) = 2$ with say $f = t^2 + at + b$ and $a, b \in F$; if x' is another root of f in its splitting field, then $xx' = b \in F$, so that $x' \in E$. Therefore E is normal over F.

That there is a close connection between normal extensions and splitting fields of polynomials is demonstrated by the following fundamental result.

(12.1.1). *Let E be a finite extension of a field F. Then E is normal over F if and only if E is the splitting field of some polynomial in $F[t]$.*

Proof. First of all assume that E is normal over F. Since $(E : F)$ is finite, we can write $E = F(x_1, x_2, \ldots, x_k)$. Let $f_i = \mathrm{Irr}_F(x_i)$. Now f_i has the root x_i in E, so by normality of the extension all roots of f_i are in E. Put $f = f_1 f_2 \cdots f_k \in F[t]$. Then f has all its roots in E and these roots together with F generate the field E. Hence E is the splitting field of f.

The converse is harder to prove. Suppose that E is the splitting field of some $f \in F[t]$, and denote the roots of f by $a_1, a_2, \ldots, a_r$, so that $E = F(a_1, a_2, \ldots, a_r)$. Let g be an irreducible polynomial over F with a root b in E. Furthermore let K be the splitting field of g over E. Then $F \subseteq E \subseteq K$. Let $b^* \in K$ be another root of g. Our task is to show that $b^* \in E$.

https://doi.org/10.1515/9783110691160-012

Since $g = \mathrm{Irr}_F(b) = \mathrm{Irr}_F(b^*)$, there is an isomorphism $\theta_0 : F(b) \to F(b^*)$ such that $\theta_0(b) = b^*$ and the restriction of θ_0 to F is the identity map: here we have applied (11.3.2). Put $g_1 = \mathrm{Irr}_{F(b)}(a_1)$ and note that g_1 divides f over $F(b)$ since $f(a_1) = 0$. Now consider $g_1^* = \theta_0(g_1) \in F(b^*)[t]$. Then g_1^* divides $\theta_0(f) = f$ over $F(b^*)$. Hence the roots of g_1^* are among $a_1, a_2, \ldots, a_r$.

Let a_{i_1} be any root of g_1^*. By (11.3.2) once again there is an isomorphism $\theta_1 : F(b, a_1) \to F(b^*, a_{i_1})$ such that $\theta_1(a_1) = a_{i_1}$ and $(\theta_1)_{|F(b_1)} = \theta_0$. Next write $g_2 = \mathrm{Irr}_{F(b,a_1)}(a_2)$ and $g_2^* = \theta_1(g_2)$. The roots of g_2^* are among $a_1, a_2, \ldots, a_r$, by the argument used above. Let a_{i_2} be any root of g_2^*. Now extend θ_1 to an isomorphism $\theta_2 : F(b, a_1, a_2) \to F(b^*, a_{i_1}, a_{i_2})$ such that $\theta_2(a_2) = a_{i_2}$ and $(\theta_2)_{|F(b,a_1)} = \theta_1$.

After r applications of this argument we will have an isomorphism

$$\theta : F(b, a_1, a_2, \ldots, a_r) \to F(b^*, a_{i_1}, a_{i_2}, \ldots, a_{i_r})$$

such that $\theta(a_j) = a_{i_j}$, $\theta(b) = b^*$ and $\theta|_F$ is the identity map. But $b \in E = F(a_1, a_2, \ldots, a_r)$ by hypothesis, so $b^* = \theta(b) \in F(a_{i_1}, a_{i_2}, \ldots, a_{i_r}) \subseteq E$, as required. □

Separable polynomials

Contrary to what one might think, it is possible for an irreducible polynomial to have repeated roots. This phenomenon is called *inseparability*.

Example (12.1.3). Let p be a prime and let f denote the polynomial $t^p - x$ in $\mathbb{Z}_p\{x\}[t]$: here x and t are distinct indeterminates and $\mathbb{Z}_p\{x\}$ is the field of rational functions in x over $\mathbb{Z}_p$. Then f is irreducible over $\mathbb{Z}_p[x]$ by (7.4.9) since x is clearly an irreducible element of $\mathbb{Z}_p[x]$. Gauss's Lemma (7.3.7) shows that f is irreducible over $\mathbb{Z}_p\{x\}$. Let a be a root of f in its splitting field. Then $f = t^p - a^p = (t - a)^p$ since $\binom{p}{i} \equiv 0 \pmod{p}$ if $0 < i < p$. It follows that f has all its roots equal to a.

An irreducible polynomial f over a field F is said to be *separable* if all its roots are different, i. e., f is a product of distinct linear factors over its splitting field. The example above shows that $t^p - x$ is inseparable over $\mathbb{Z}_p\{x\}$, a field with prime characteristic. The criterion which follows shows that the phenomenon of inseparability can only occur for fields of prime characteristic.

(12.1.2). *Let f be an irreducible polynomial over a field F.*
(i) *If $\mathrm{char}(F) = 0$, then f is separable.*
(ii) *If $\mathrm{char}(F) = p > 0$, then f is inseparable if and only if $f = g(t^p)$ for some irreducible polynomial g over F.*

Proof. There is no loss in supposing f to be monic. Assume first that $\mathrm{char}(F) = 0$ and let a be a root of f in its splitting field. If a has multiplicity greater than 1, then (7.4.7) shows that $t - a \mid f'$ where f' is the derivative of f. Thus $f'(a) = 0$. Writing $f = a_0 + a_1 t + \cdots + a_n t^n$, we have $f' = a_1 + 2a_2 t + \cdots + na_n t^{n-1}$. But $f = \mathrm{Irr}_F(a)$, so f divides f'.

Since $\deg(f') < \deg(f)$, this can only mean that $f' = 0$, i. e., $ia_i = 0$ for all $i > 0$ and so $a_i = 0$. Thus f is constant, which is impossible. Therefore a is not a repeated root and f is separable.

Now assume that $\operatorname{char}(F) = p > 0$ and again let a be a multiple root of f. Arguing as before, we have $ia_i = 0$ for all $i > 0$. In this case all we can conclude is that $a_i = 0$ if p does not divide i. Hence

$$f = a_0 + a_p t^p + a_{2p} t^{2p} + \cdots + a_{rp} t^{rp}$$

where rp is the largest positive multiple of p not exceeding n. It follows that $f = g(t^p)$ where $g = a_0 + a_p t + \cdots + a_{rp} t^r$. Notice that g is irreducible since if it were reducible, so would f be.

Conversely, assume that $f = g(t^p)$ where $g = \sum_{i=0}^{r} a_i t^i \in F[t]$. We claim that f is inseparable. Let b_i be a root of $t^p - a_i$ in the splitting field E of the polynomial $(t^p - a_1)(t^p - a_2) \cdots (t^p - a_r)$. Then $a_i = b_i^p$ and hence

$$f = \sum_{i=0}^{r} a_i t^{ip} = \sum_{i=0}^{r} b_i^p t^{ip} = \left(\sum_{i=0}^{r} b_i t^i \right)^p,$$

from which it follows that every root of f has multiplicity at least p. Hence f is inseparable. $\qquad\square$

Separable extensions

Let E be an extension of a field F. An element x of E is said to be *separable over F* if x is algebraic and its multiplicity as a root of $\operatorname{Irr}_F(x)$ is 1. If x is algebraic but inseparable, the final argument of the proof of (12.1.2) shows that its irreducible polynomial is a prime power of a polynomial, so that *all* its roots have multiplicity greater then 1. Therefore $x \in E$ is *separable over F if and only if* $\operatorname{Irr}_F(x)$ *is a separable polynomial.*

If every element of E is separable over F, then E is called a *separable extension* of F. Finally, a field F is said to be *perfect* if every algebraic extension of F is separable. Since any irreducible polynomial over a field of characteristic 0 is separable, *all fields of characteristic 0 are perfect*. There is a simple criterion for a field of prime characteristic to be perfect.

(12.1.3). *Let F be a field of prime characteristic p. Then F is perfect if and only if $F = F^p$ where F^p is the subfield $\{a^p \mid a \in F\}$.*

Proof. In the first place F^p is a subfield of F since $(a \pm b)^p = a^p \pm b^p$, $(a^{-1})^p = (a^p)^{-1}$ and $(ab)^p = a^p b^p$ for $a, b \in F$. Now assume that $F = F^p$. If $f \in F[t]$ is irreducible but inseparable, then $f = g(t^p)$ for some $g \in F[t]$ by (12.1.2). Let $g = \sum_{i=0}^{r} a_i t^i$; then $a_i = b_i^p$

for some $b_i \in F$ since $F = F^p$. Therefore

$$f = \sum_{i=0}^{r} a_i t^{pi} = \sum_{i=0}^{r} b_i^p t^{pi} = \left(\sum_{i=0}^{r} b_i t^i \right)^p,$$

which is impossible since f is irreducible. Thus f is separable. This shows that if E is an algebraic extension of F, then it is separable. Hence F is a perfect field.

Conversely, assume that $F \neq F^p$ and choose $a \in F - F^p$. Consider the polynomial $f = t^p - a$. First we claim that f is irreducible over F. Suppose this is false, so that $f = gh$ where g and h in $F[t]$ are monic with smaller degrees than f. Now $f = t^p - a = (t - b)^p$ where b is a root of f in its splitting field, so it follows that $g = (t - b)^i$ and $h = (t - b)^j$ where $i + j = p$ and $0 < i, j < p$. Since $\gcd\{i, p\} = 1$, we can write $1 = iu + pv$ for suitable integers u, v. Therefore $b = (b^i)^u (b^p)^v = (b^i)^u a^v \in F$ since $b^i \in F$, and hence $a = b^p \in F^p$, a contradiction. Thus f is irreducible and by (12.1.2) it is inseparable. It follows that F cannot be a perfect field. □

(12.1.4). *Every finite field is perfect.*

Proof. Let F be a field of order p^m with p a prime. Every element f of F satisfies the equation $t^{p^m} - t = 0$ by (11.3.1). Hence $F = F^p$ and F is perfect. □

On the other hand, the field $F = \mathbb{Z}_p\{t\}$ is not perfect because $F^p = \mathbb{Z}_p\{t^p\}$ is a proper subfield of F.

It is desirable to have a criterion for a finite extension of a field of prime characteristic to be separable.

(12.1.5). *Let E be a finite extension of a field F with prime characteristic p. Then E is separable over F if and only if $E = F(E^p)$.*

Proof. Assume that E is separable over F and let $a \in E$. Writing $f = \mathrm{Irr}_{F(a^p)}(a)$, we observe that f divides $t^p - a^p = (t - a)^p$. Since f is a separable polynomial, it follows that $f = t - a$ and thus $a \in F(a^p) \subseteq F(E^p)$.

Conversely, assume that $E = F(E^p)$ and let $x \in E$; we need to prove that $f = \mathrm{Irr}_F(x)$ is separable over F. If this is false, then $f = g(t^p)$ for some $g = \sum_{i=0}^{k} a_i t^i \in F[t]$. Since $0 = g(x^p) = a_0 + a_1 x^p + \cdots + a_k x^{kp}$, the field elements $1, x^p, \ldots, x^{kp}$ are linearly dependent over F. On the other hand, $k < kp = \deg(f) = (F(x) : F)$, so that $1, x, \ldots, x^k$ must be linearly independent over F. Extend $\{1, x, \ldots, x^k\}$ to an F-basis of E, say $\{y_1, y_2, \ldots, y_n\}$, using (8.2.6). Thus $n = (E : F)$.

We have $E = Fy_1 + Fy_2 + \cdots + Fy_n$ and thus $E^p \subseteq Fy_1^p + Fy_2^p + \cdots + Fy_n^p$. Therefore $E = F(E^p) = Fy_1^p + Fy_2^p + \cdots + Fy_n^p$. It follows that $y_1^p, y_2^p, \ldots, y_n^p$ are F-linearly independent since $n = (E : F)$. This shows that $1, x^p, \ldots, x^{kp}$ are F-linearly independent, a contradiction. □

Corollary (12.1.6). *Let $E = F(a_1, a_2, \ldots, a_k)$ be an extension of a field F such that each a_i is separable over F. Then E is separable over F.*

Proof. We may assume that char$(F) = p > 0$. Since a_i is separable over F, we have $a_i \in F(a_i^p)$, as in the first paragraph of the preceding proof. Hence $a_i \in F(E^p)$ and $E = F(E^p)$. Therefore E is separable over F by (12.1.5). □

Notice the consequence of the last result: *the splitting field of a separable polynomial is a separable extension.*

We conclude this section by addressing a question which may already have occurred to the reader: when is a finite extension E of F a simple extension, i. e., when is $E = F(x)$ for some x? An important result on this problem is:

(12.1.7) (The Theorem of the Primitive Element). *Let E be a finite separable extension of a field F. Then there is an element a such that $E = F(a)$.*

Proof. The proof is easy when E is finite. For then $E - \{0\}$ is a cyclic group by (11.3.6), generated by a, say. Hence $E = \{0, 1, a, \ldots, a^{q-1}\}$ where $q = |E|$, and thus $E = F(a)$.

From now on assume E is infinite. Since $(E : F)$ is finite, it follows that F is infinite. Next $E = F(u_1, u_2, \ldots, u_n)$ for some u_i in E. The proof proceeds by induction on n. If $n > 2$, then $F(u_1, u_2, \ldots, u_{n-1}) = F(v)$ for some v by induction hypothesis, and hence $E = F(v, u_n) = F(a)$ for some a by the case $n = 2$. Therefore it is enough to deal with the case $n = 2$. From now on write

$$E = F(u, v).$$

We introduce the polynomials $f = \mathrm{Irr}_F(u)$ and $g = \mathrm{Irr}_F(v)$; these are separable polynomials since E is separable over F. Let the roots of f and g be $u = x_1, x_2, \ldots, x_m$ and $v = y_1, y_2, \ldots, y_n$ respectively, in the splitting field of fg over F. Here all the x_i are different, as are all the y_j. From this we conclude that for $j \neq 1$ there is at most one element z_{ij} in F such that

$$u + z_{ij}v = x_i + z_{ij}y_j,$$

namely $z_{ij} = (x_i - u)(v - y_j)^{-1}$. Since F is infinite, it is possible to choose an element z in F which is different from each of these finitely many z_{ij}. Then $u + zv \neq x_i + zy_j$ if $(i, j) \neq (1, 1)$.

With this choice of z, put $a = u + zv \in E$. We will show that $E = F(a)$. Since $g(v) = 0 = f(u) = f(a - zv)$, the element v is a common root of the polynomials g and $f(a - zt) \in F(a)[t]$. Now these polynomials have no other common roots. For if y_j were one, then $a - zy_j = x_i$ for some i, which implies that $u + zv = a = x_i + zy_j$; this is contrary to the choice of z. It follows that $t - v$ is the unique (monic) gcd of g and $f(a - zt)$ in $E[t]$. The gcd of these polynomials actually lies in the subring $F(a)[t]$: for the gcd can be computed by using the Euclidean Algorithm, which is valid for $F(a)[t]$ since it depends only on the Division Algorithm. Therefore $v \in F(a)$ and $u = a - zv \in F(a)$. Finally $E = F(u, v) = F(a)$. □

Since an algebraic number field is by definition a finite extension of $\mathbb{Q}$, we deduce:

Corollary (12.1.8). *If E is an algebraic number field, then $E = \mathbb{Q}(a)$ for some a in E.*

Exercises (12.1).

(1) Which of the following field extensions are normal? (i) $\mathbb{Q}(3^{1/3})$ of $\mathbb{Q}$; (ii) $\mathbb{Q}(3^{1/3}, e^{2\pi i/3})$ of $\mathbb{Q}$; (iii) $\mathbb{R}$ of $\mathbb{Q}$; (iv) $\mathbb{C}$ of $\mathbb{R}$.

(2) Let $F \subseteq K \subseteq E$ be field extensions with all degrees finite. If E is normal over F, show that it is normal over K, but K need not be normal over F.

(3) Let $f \in F[t]$ where char$(F) = p > 0$, and assume that f is monic with degree p^n. If all roots of f are equal in its splitting field, prove that $f = t^{p^n} - a$ for some $a \in F$.

(4) Let E be a finite extension of a field F of characteristic $p > 0$ and assume that $(E : F)$ is not divisible by p. Prove that E is separable over F.

(5) Let $F \subseteq K \subseteq E$ be field extensions with all degrees finite and E separable over F. Prove that E is separable over K.

(6) Let $F \subseteq K \subseteq E$ be field extensions with all degrees finite. If E is separable over K and K is separable over F, show that E is separable over F.

(7) Let E be a finite separable extension of a field F. Prove that there is a finite extension K of E such that K is separable and normal over F. [Hint: use (12.1.7).]

12.2 Automorphisms of fields

Fields, like groups, possess automorphisms and these play a crucial role in field theory. An *automorphism of a field F* is defined to be a bijective ring homomorphism $\alpha : F \to F$; thus $\alpha(x+y) = \alpha(x)+\alpha(y)$ and $\alpha(xy) = \alpha(x)\alpha(y)$. The automorphisms of a field are easily seen to form a group with respect to functional composition. If E is a field extension of F, we interested in *automorphisms of E over F*, i. e., automorphisms of E whose restriction to F is the identity function. For example, complex conjugation is an automorphism of $\mathbb{C}$ over $\mathbb{R}$. The set of all automorphisms of E over F is a subgroup of the group of automorphisms of E and is denoted by

$$\mathrm{Gal}(E/F) :$$

this is the *Galois[1] group* of E over F.

Suppose that $E = F(a)$ is a simple algebraic extension of F with degree n. Then every element of E has the form $x = \sum_{i=0}^{n-1} c_i a^i$ with $c_i \in F$ and thus $\alpha(x) = \sum_{i=0}^{n-1} c_i \alpha(a)^i$ where $\alpha \in \mathrm{Gal}(E/F)$. If b is any root of the polynomial $f = \mathrm{Irr}_F(a)$, then $0 = \alpha(f(b)) = f(\alpha(b))$, so that $\alpha(b)$ is also a root of f in E. Thus each α in $\mathrm{Gal}(E/F)$ gives rise to a permutation $\pi(\alpha)$ of X, the set of distinct roots of f in E: this is given by $\pi(\alpha)(x) = \alpha(x)$

---

[1] Évariste Galois (1811–1831).

where $x \in X$. What is more, the mapping

$$\pi : \mathrm{Gal}(E/F) \to \mathrm{Sym}(X)$$

is evidently a group homomorphism; thus π is a permutation representation of $\mathrm{Gal}(E/F)$ on X.

In fact π a faithful permutation representation of $\mathrm{Gal}(E/F)$ on X. For, if $\pi(\alpha)$ is the identity permutation, then $\alpha(a) = a$ and hence α is the identity automorphism of E. For this reason it is often useful to think of the elements of $\mathrm{Gal}(E/F)$ as permutations of the set of distinct roots X.

Next let b be any element of X. Then $F \subseteq F(b) \subseteq E = F(a)$, and also $(F(b) : F) = \deg(f) = (F(a) : F)$ by (11.1.4) since $f = \mathrm{Irr}_F(b)$. It follows that $F(b) = F(a) = E$ by (11.1.7). Since $\mathrm{Irr}_F(a) = f = \mathrm{Irr}_F(b)$, we may apply (11.3.2) to produce an automorphism α of E over F such that $\alpha(a) = b$. Therefore the group $\mathrm{Gal}(E/F)$ acts *transitively* on the set X. Finally, if α in $\mathrm{Gal}(E/F)$ fixes some b in X, then α must equal the identity since $E = F(b)$. This shows that $\mathrm{Gal}(E/F)$ acts *regularly* on X and it follows from (5.2.2) that $|X| = |\mathrm{Gal}(E/F)|$.

These conclusions are summed up in the following fundamental result.

(12.2.1). *Let $E = F(a)$ be a simple algebraic extension of a field F. Then $\mathrm{Gal}(E/F)$ acts regularly on X, the set of distinct roots of $\mathrm{Irr}_F(a)$ in E. Therefore*

$$\left| \mathrm{Gal}(E/F) \right| = |X| \le (E : F).$$

An extension of a field F which is finite, separable and normal is said to be *Galois over F*. For such extensions we have:

Corollary (12.2.2). *If E is a Galois extension of a field F with degree n, then $\mathrm{Gal}(E/F)$ is isomorphic with a regular subgroup of S_n and*

$$\left| \mathrm{Gal}(E/F) \right| = n = (E : F).$$

For (12.1.7) shows that $E = F(a)$ for some $a \in E$. Also $\mathrm{Irr}_F(a)$ has n distinct roots in E by normality and separability.

The Galois group of a polynomial

Suppose that f is a non-constant polynomial over a field F and let E be the splitting field of f: recall from (11.3.4) that this field is unique up to isomorphism. Then the *Galois group of the polynomial f* is

$$\mathrm{Gal}(f) = \mathrm{Gal}(E/F).$$

This is always a finite group by (12.2.1). The basic properties of the Galois group are given in the next result.

(12.2.3). *Let f be a non-constant polynomial of degree n over a field F. Then:*

(i) *Gal(f) is isomorphic with a permutation group on the set of distinct roots of f; thus |Gal(f)| divides n!;*

(ii) *if all the roots of f are distinct, then f is irreducible if and only if Gal(f) acts transitively on the set of roots of f.*

Proof. Let E denote the splitting field of f, so that $\mathrm{Gal}(f) = \mathrm{Gal}(E/F)$. Let $\alpha \in \mathrm{Gal}(f)$. If a is a root of f in E, then $f(\alpha(a)) = \alpha(f(a)) = 0$, so that $\alpha(a)$ is also a root of f. If α fixes every root of f, then α is the identity automorphism since E is generated by F and the roots of f. Hence $\mathrm{Gal}(f)$ is isomorphic with a permutation group on the set of distinct roots of f. If there are r such roots, then $r \le n$ and $|\mathrm{Gal}(f)| \mid r! \mid n!$, so that $|\mathrm{Gal}(f)| \mid n!$.

Next assume that all the roots of f are different. Let f be irreducible. If a and b are roots of f, then $\mathrm{Irr}_F(a) = f = \mathrm{Irr}_F(b)$, and by (11.3.2) there exists $\alpha \in \mathrm{Gal}(f)$ such that $\alpha(a) = b$. It follows that $\mathrm{Gal}(f)$ acts transitively on the roots of f.

Conversely, suppose that $\mathrm{Gal}(f)$ acts transitively on the roots of f, but f is reducible; write $f = g_1 g_2 \cdots g_k$ where $g_i \in F[t]$ is irreducible and $k \ge 2$. Let a_1 and a_2 be roots of g_1 and g_2 respectively. By transitivity there exists $\alpha \in \mathrm{Gal}(f)$ such that $\alpha(a_1) = a_2$. But $0 = \alpha(g_1(a_1)) = g_1(\alpha(a_1)) = g_1(a_2)$. Hence $g_2 = \mathrm{Irr}_F(a_2)$ divides g_1. Therefore g_2^2 divides f and the roots of f cannot all be different, a contradiction which shows that f is irreducible. ☐

Corollary (12.2.4). *Let f be a separable polynomial of degree n over a field F and let E be its splitting field. Then |Gal(f)| = (E : F) and |Gal(f)| is divisible by n.*

Proof. Note that E is separable and hence Galois over F by (12.1.6). Hence $|\mathrm{Gal}(f)| = |\mathrm{Gal}(E/F)| = (E : F)$ by (12.2.2). Further f is irreducible by definition, so $\mathrm{Gal}(f)$ acts transitively on the n roots of f; therefore n divides $|\mathrm{Gal}(f)|$ by (5.2.2). ☐

Let us consider some polynomials whose Galois groups can be readily computed.

Example (12.2.1). Let $f = t^3 - 2 \in \mathbb{Q}[t]$. Then $\mathrm{Gal}(f) \simeq S_3$.

To see this let E denote the splitting field of f; thus E is Galois over $\mathbb{Q}$. Then $E = \mathbb{Q}(2^{1/3}, e^{2\pi i/3})$ and one can easily check that $(E : \mathbb{Q}) = 6$, so that $|\mathrm{Gal}(f)| = 6$. Since $\mathrm{Gal}(f)$ is isomorphic with a subgroup of S_3, it follows that $\mathrm{Gal}(f) \simeq S_3$.

In fact it is not difficult to write down the six elements of the group $\mathrm{Gal}(f)$. Put $a = 2^{1/3}$ and $\omega = e^{2\pi i/3}$; thus $E = \mathbb{Q}(a, \omega)$. Since $E = \mathbb{Q}(a)(\omega)$ and $t^3 - 2$ is the irreducible polynomial of both a and $a\omega$ over $\mathbb{Q}(\omega)$, there is an automorphism α of E over $\mathbb{Q}$ such that $\alpha(a) = a\omega$, $\alpha(\omega) = \omega$. Clearly α has order 3. Also $\alpha^2(a) = a\omega^2$ and $\alpha^2(\omega) = \omega$. It is easy to identify an automorphism β such that $\beta(a) = a$ and $\beta(\omega) = \omega^2$; indeed β is just complex conjugation. Two more automorphisms of order 2 are formed by composition: $\gamma = \alpha\beta$ and $\delta = \alpha^2\beta$. It is quickly seen that γ maps ω to ω^2 and a to $a\omega$, while δ maps ω to ω^2 and a to $a\omega^2$. Thus the elements of the Galois group $\mathrm{Gal}(f)$ are $1, \alpha, \alpha^2, \beta, \gamma, \delta$.

Example (12.2.2). Let p be a prime and put $f = t^p - 1 \in \mathbb{Q}[t]$. Then $\mathrm{Gal}(f) \simeq U(\mathbb{Z}_p)$, a cyclic group of order $p - 1$.

To see this put $a = e^{2\pi i/p}$, a primitive pth root of unity; the roots of f are $1, a, a^2, \ldots,$ a^{p-1} and its splitting field is $E = \mathbb{Q}(a)$. Now $f = (t - 1)(1 + t + t^2 + \cdots + t^{p-1})$ and the second factor is $\mathbb{Q}$-irreducible by Example (7.4.6). Hence the irreducible polynomial of a is $1 + t + t^2 + \cdots + t^{p-1}$ and $|\mathrm{Gal}(f)| = (E : \mathbb{Q}) = p - 1$.

To show that $\mathrm{Gal}(f)$ is cyclic, we construct a group isomorphism

$$\theta : U(\mathbb{Z}_p) \to \mathrm{Gal}(f).$$

If $1 \le j < p$, define $\theta(j + p\mathbb{Z})$ to be θ_j where $\theta_j(a) = a^j$ and θ_j is trivial on $\mathbb{Q}$; this is an automorphism by (11.3.2). Obviously θ_j is the identity only if $j = 1$, so θ is injective. Since $U(\mathbb{Z}_p)$ and $\mathrm{Gal}(f)$ both have order $p - 1$, they are isomorphic.

Conjugacy in field extensions

Let E be an extension of a field F. Two elements a and b of E are said to be *conjugate over F* if $\alpha(a) = b$ for some $\alpha \in \mathrm{Gal}(E/F)$. In normal extensions conjugacy amounts to the elements having the same irreducible polynomial, as the next result shows.

(12.2.5). *Let E be a finite normal extension of a field F. Then two elements a and b of E are conjugate over F if and only if they have the same irreducible polynomial.*

Proof. If a and b have the same irreducible polynomial, (11.3.2) shows that there is a field isomorphism $\theta : F(a) \to F(b)$ such that $\theta(a) = b$ and θ is the identity map on F. By (12.1.1) E is the splitting field of some polynomial over F and hence over $F(a)$. Consequently, (11.3.3) can be applied to extend θ to an isomorphism $\alpha : E \to E$ such that θ is the restriction of α to $F(a)$. Hence $\alpha \in \mathrm{Gal}(E/F)$ and $\alpha(a) = b$, which shows that a and b are conjugate over F.

To prove the converse, suppose that $b = \alpha(a)$ where $a, b \in E$ and $\alpha \in \mathrm{Gal}(E/F)$. Put $f = \mathrm{Irr}_F(a)$ and $g = \mathrm{Irr}_F(b)$. Then $0 = \alpha(f(a)) = f(\alpha(a)) = f(b)$. Therefore g divides f and it follows that $f = g$ since f and g are monic and irreducible. $\qquad\square$

The next result is of critical importance in Galois theory: it asserts that the only elements of a Galois extension that are fixed by every automorphism are the elements of the base field.

(12.2.6). *Let E be a Galois extension of a field F and let $a \in E$. Then $\alpha(a) = a$ for all automorphisms α of E over F if and only if $a \in F$.*

Proof. Assume that $\alpha(a) = a$ for all $\alpha \in \mathrm{Gal}(E/F)$ and put $f = \mathrm{Irr}_F(a)$. Since E is normal over F, all the roots of f are in E. If b is any such root, it is conjugate to a by (12.2.5), so there exists α in $\mathrm{Gal}(E/F)$ such that $\alpha(a) = b$. Hence $b = a$ and the roots of f are all equal. But f is separable since E is separable over F. Therefore $f = t - a$ and a belongs to F. The converse is obvious. $\qquad\square$

Roots of unity
We will postpone further development of the theory of Galois extensions until the next section and concentrate on roots of unity. Let F be a field and n a positive integer. A root a of the polynomial $t^n - 1 \in F[t]$ is called an *nth root of unity over F*; thus $a^n = 1$. If $a^m \neq 1$ for all proper divisors m of n, then a is said to be a *primitive nth root of unity*. Now if char$(F) = p$ divides n, there are no primitive nth roots of unity over F: for then $t^n - 1 = (t^{n/p} - 1)^p$ and every nth root of unity has order at most n/p. However, if char(F) does not divide n, primitive nth roots of unity over F always exist, as will now be proved.

(12.2.7). *Let F be a field whose characteristic does not divide the positive integer n and let E be the splitting field of $t^n - 1$ over F. Then:*
(i) *primitive nth roots of unity exist in E; furthermore these generate a cyclic subgroup of order n.*
(ii) *Gal(E/F) is isomorphic with a subgroup of $U(\mathbb{Z}_n)$ and is therefore abelian with order dividing $\phi(n)$.*

Proof. (i) Set $f = t^n - 1$, so that $f' = nt^{n-1}$. Since char(F) does not divide n, the polynomials f and f' are relatively prime. It follows via (7.4.7) that f has n distinct roots in its splitting field E, namely the nth roots of unity. Clearly these roots form a subgroup H of $U(E)$ with order n, and by (11.3.6) it is cyclic, say $H = \langle x \rangle$. Here x has order n, so it is a primitive nth root of unity.

(ii) Let a be a primitive nth root of unity in E. Then the roots of $t^n - 1$ are a^i, $i = 0, 1, \ldots, n - 1$, and $E = F(a)$. If $\alpha \in$ Gal(E/F), then α is effectively determined by $\alpha(a) = a^{i(\alpha)}$ where $1 \leq i(\alpha) < n$ and $i(\alpha)$ is relatively prime to n. Furthermore, the assignment $\alpha \mapsto i(\alpha) + n\mathbb{Z}$ yields an injective homomorphism from the Galois group into $U(\mathbb{Z}_n)$. By Lagrange's Theorem $|$Gal$(E/F)|$ divides $|U(\mathbb{Z}_n)| = \phi(n)$. $\square$

Corollary (12.2.8). *The number of primitive nth roots of unity over a field whose characteristic does not divide n is $\phi(n)$, where ϕ is Euler's function.*

For, if a is a fixed primitive nth root of unity, the primitive nth roots of unity are just the powers a^i where $1 \leq i < n$ and i is relatively prime to n.

Cyclotomic polynomials
Assume that F is a field whose characteristic does not divide the positive integer n and denote the primitive nth roots of unity over F by $a_1, a_2, \ldots, a_{\phi(n)}$. The *cyclotomic polynomial of order n over F* is defined to be

$$\Phi_n = \prod_{i=1}^{\phi(n)} (t - a_i),$$

which is a monic polynomial of degree $\phi(n)$. Since every nth root of unity is a primitive dth root of unity for some divisor d of n, we have immediately that

$$t^n - 1 = \prod_{d \mid n} \Phi_d.$$

This leads to the formula

$$\Phi_n = \frac{t^n - 1}{\prod_{d \| n} \Phi_d},$$

where $d \| n$ means that d is a proper divisor of n. By using this formula it is possible to compute Φ_n recursively, i. e., if we know Φ_d for all proper divisors d of n, then we can calculate Φ_n. The formula also shows that $\Phi_n \in F[t]$. For $\Phi_1 = t - 1 \in F[t]$ and if $\Phi_d \in F[t]$ for all proper divisors d of n, then $\Phi_n \in F[t]$, as long division shows.

Example (12.2.3). Since $\Phi_1 = t - 1$,

$$\Phi_2 = \frac{t^2 - 1}{t - 1} = t + 1, \quad \Phi_3 = \frac{t^3 - 1}{t - 1} = t^2 + t + 1,$$

and

$$\Phi_4 = \frac{t^4 - 1}{(t - 1)(t + 1)} = t^2 + 1.$$

There is in fact an explicit formula for Φ_n. This involves the *Möbius function* μ, which is well-known from number theory. It is defined by the rules:

$$\mu(1) = 1, \quad \mu(p_1 p_2 \cdots p_k) = (-1)^k,$$

if $p_1, p_2, \ldots, p_k$ are distinct primes, and

$$\mu(n) = 0$$

if n is divisible by the square of a prime.

(12.2.9). *The cyclotomic polynomial of order n over any field whose characteristic does not divide n is given by*

$$\Phi_n = \prod_{d \mid n} (t^d - 1)^{\mu(n/d)}.$$

Proof. First we note an auxiliary property of the Möbius function,

$$\sum_{d|n} \mu(d) = \begin{cases} 1 & \text{if } n = 1 \\ 0 & \text{if } n > 1 \end{cases}$$

This is obvious if $n = 1$, so assume that $n > 1$ and write $n = p_1^{e_1} p_2^{e_2} \cdots p_k^{e_k}$ where the p_i are distinct primes. If d is a square-free divisor of n, then d has the form $p_{i_1} p_{i_2} \cdots p_{i_r}$ where $1 \le i_1 < i_2 < \cdots < i_r \le n$, which corresponds to the term $(-1)^r t_{i_1} t_{i_2} \cdots t_{i_r}$ in the product $(1 - t_1)(1 - t_2) \cdots (1 - t_n)$; note also that $\mu(d) = (-1)^r$. Therefore we obtain the polynomial identity

$$(1 - t_1)(1 - t_2) \cdots (1 - t_n) = \sum \mu(p_{i_1} p_{i_2} \cdots p_{i_r}) t_{i_1} t_{i_2} \cdots t_{i_r},$$

where the sum is over all i_j satisfying $1 \le i_1 < i_2 < \cdots < i_r \le n$. Set all $t_i = 1$ to get $\sum \mu(p_{i_1}, p_{i_2}, \ldots p_{i_r}) = 0$. Since $\mu(d) = 0$ if d is not square-free, we can rewrite the last equation as $\sum_{d|n} \mu(d) = 0$.

We are now in a position to establish the formula for Φ_n. Define

$$\Psi_n = \prod_{e|n} (t^e - 1)^{\mu(n/e)},$$

so that $\Psi_1 = t - 1 = \Phi_1$. Assume that $\Psi_d = \Phi_d$ for all $d < n$. Then by definition of Ψ_d, we have

$$\prod_{d|n} \Psi_d = \prod_{d|n} \prod_{e|d} (t^e - 1)^{\mu(d/e)} = \prod_{f|n} (t^f - 1)^{\sum_{f|d|n} \mu(d/f)}.$$

Next for a fixed f dividing d we have

$$\sum_{f|d|n} \mu(d/f) = \sum_{\frac{d}{f}|\frac{n}{f}} \mu(d/f),$$

which equals 1 or 0 according as $f = n$ or $f < n$. It therefore follows that

$$\prod_{d|n} \Psi_d = t^n - 1 = \prod_{d|n} \Phi_d.$$

Since $\Psi_d = \Phi_d$ if $d < n$, cancellation yields $\Psi_n = \Phi_n$ and the proof is complete. $\qquad\square$

Example (12.2.4). Use the formula in (12.2.9) to compute the cyclotomic polynomial of order 12 over $\mathbb{Q}$.

The formula yields

$$\Phi_{12} = (t - 1)^{\mu(12)} (t^2 - 1)^{\mu(6)} (t^3 - 1)^{\mu(4)} (t^4 - 1)^{\mu(3)} (t^6 - 1)^{\mu(2)} (t^{12} - 1)^{\mu(1)},$$

which reduces to

$$(t^2 - 1)(t^4 - 1)^{-1}(t^6 - 1)^{-1}(t^{12} - 1) = t^4 - t^2 + 1,$$

since $\mu(12) = \mu(4) = 0$, $\mu(2) = \mu(3) = -1$ and $\mu(6) = \mu(1) = 1$.

Example (12.2.5). If p is a prime, $\Phi_p = 1 + t + t^2 + \cdots + t^{p-1}$.

For $\Phi_p = (t - 1)^{\mu(p)}(t^p - 1)^{\mu(1)} = \frac{t^p - 1}{t - 1} = 1 + t + t^2 + \cdots + t^{p-1}$, since $\mu(p) = -1$.

Since we are interested in computing the Galois group of a cyclotomic polynomial over $\mathbb{Q}$, it is important to know if Φ_n is irreducible. This is certainly true when n is prime – see Example (7.4.6). The general result is:

(12.2.10). *The cyclotomic polynomial Φ_n is irreducible over $\mathbb{Q}$ for all integers n.*

Proof. Assume that Φ_n is reducible over $\mathbb{Q}$; then Gauss's Lemma (7.3.7) tells us that it must be reducible over $\mathbb{Z}$. Since Φ_n is monic, it follows that it is a product of monic irreducible polynomials in $\mathbb{Z}[t]$. Let f be one such polynomial and choose a root a of f; then $f = \mathrm{Irr}_{\mathbb{Q}}(a)$. Now a is a primitive nth root of unity, so, if p is any prime not dividing n, then a^p is also a primitive nth root of unity and thus is a root of Φ_n. Hence a^p is a root of some monic $\mathbb{Q}$-irreducible divisor g of Φ_n in $\mathbb{Z}[t]$. Of course $g = \mathrm{Irr}_{\mathbb{Q}}(a^p)$.

Suppose first that $f \neq g$. Thus $t^n - 1 = fgh$ for some $h \in \mathbb{Z}[t]$ since f and g are distinct $\mathbb{Q}$-irreducible divisors of $t^n - 1$. Also $g(a^p) = 0$ implies that f divides $g(t^p)$ and thus $g(t^p) = fk$ where $k \in \mathbb{Z}[t]$. The canonical homomorphism from $\mathbb{Z}$ to $\mathbb{Z}_p$ induces a homomorphism from $\mathbb{Z}[t]$ to $\mathbb{Z}_p[t]$; let $\bar{f}, \bar{g}, \bar{h}, \bar{k}$, denote the images of f, g, h, k under this homomorphism. Then $\bar{f}\bar{k} = \bar{g}(t^p) = (\bar{g}(t))^p$ since $x^p \equiv x \pmod{p}$ for any integer x. Now $\mathbb{Z}_p[t]$ is a PID and hence a UFD. Since $\bar{f}\bar{k} = \bar{g}^p$, the polynomials $\bar{f}$ and $\bar{g}$ have a common irreducible divisor in $\mathbb{Z}_p[t]$. This means that $\bar{f}\bar{g}\bar{h} \in \mathbb{Z}_p[t]$ is divisible by the square of this irreducible factor and hence $t^n - 1 \in \mathbb{Z}_p[t]$ has a multiple root in its splitting field. However, $(t^n - 1)' = nt^{n-1}$ is relatively prime to $t^n - 1$ in $\mathbb{Z}_p[t]$ since p does not divide n. This is a contradiction by (7.4.7). It follows that $f = g$.

We have proved that a^p is a root of f for all primes p not dividing n. It follows that a^m is a root of f whenever $1 \leq m < n$ and $\gcd\{m, n\} = 1$. Therefore $\deg(f) \geq \phi(n) = \deg(\Phi_n)$. Since f divides Φ_n, we conclude that $f = \Phi_n$ and Φ_n is irreducible. □

We can now compute the Galois group of a cyclotomic polynomial.

(12.2.11). *If n is a positive integer, the Galois group of Φ_n over $\mathbb{Q}$ is isomorphic with $U(\mathbb{Z}_n)$, an abelian group of order $\phi(n)$.*

Proof. Let E denote the splitting field of Φ_n over $\mathbb{Q}$ and let a be a primitive nth root of unity in E. The roots of Φ_n are a^i where $i = 1, 2, \ldots, n - 1$ and $\gcd\{i, n\} = 1$. Hence $E = \mathbb{Q}(a)$ and Φ_n is the irreducible polynomial of a by (12.2.10). Thus $|\mathrm{Gal}(E/F)| = \deg(\Phi_n) = \phi_n$. If $1 \leq i < n$ and i is relatively prime to n, there is an automorphism α_i of E over $\mathbb{Q}$ such that $\alpha_i(a) = a^i$, since a and a^i have the same irreducible polynomial.

Moreover, the map $i + n\mathbb{Z} \mapsto \alpha_i$ is easily seen to be an injective group homomorphism from $U(\mathbb{Z})$ to $\mathrm{Gal}(E/F)$. Since both these groups have order $\phi(n)$, they are isomorphic.
□

The splitting field of $\Phi_n \in \mathbb{Q}[t]$ is called a *cyclotomic number field*. Thus the Galois group of a cyclotomic number field is abelian.

Exercises (12.2).

(1) Give an example of a finite simple extension E of a field F such that $|\mathrm{Gal}(E/F)| = 1$, but $E \neq F$.

(2) If $E = \mathbb{Q}(\sqrt{5})$, find $\mathrm{Gal}(E/F)$.

(3) If $E = \mathbb{Q}(\sqrt{2}, \sqrt{3})$, find $\mathrm{Gal}(E/F)$.

(4) Find the Galois groups of the following polynomials in $\mathbb{Q}[t]$: (i) $t^2 + 1$; (ii) $t^3 - 4$; (iii) $t^3 - 2t + 4$.

(5) Let $f \in F[t]$ and suppose that $f = f_1 f_2 \cdots f_k$ where the f_i are polynomials over the field F. Prove that $\mathrm{Gal}(f)$ is isomorphic with a subgroup of the direct product $\mathrm{Gal}(f_1) \times \mathrm{Gal}(f_2) \times \cdots \times \mathrm{Gal}(f_k)$.

(6) Prove that the Galois group of $\mathrm{GF}(p^m)$ over $\mathrm{GF}(p)$ is a cyclic group of order m and it is generated by the automorphism in which $a \mapsto a^p$.

(7) Give an example to show that $\mathrm{Gal}(\Phi_n)$ need not be cyclic.

(8) Let p be a prime not dividing the positive integer n. Prove that Φ_n is irreducible over $\mathrm{GF}(p)$ if and only if $\phi(n)$ is the smallest positive integer m such that $p^m \equiv 1 \pmod{n}$.

(9) Show that Φ_5 is reducible over $\mathrm{GF}(11)$ and find an explicit factorization of it in terms of irreducibles.

12.3 The Fundamental Theorem of Galois theory

Armed with the techniques of the last two sections, we can now approach the celebrated theorem of the title. First some terminology: let E be an extension of a field F. By an *intermediate field* is meant a subfield S such that $F \subseteq S \subseteq E$. If H is a subgroup of $\mathrm{Gal}(E/F)$, the *fixed field* of H

$$\mathrm{Fix}(H)$$

is the set of elements of E which are fixed by every element of H. It is quickly verified that $\mathrm{Fix}(H)$ is a subfield and $F \subseteq \mathrm{Fix}(H) \subseteq E$, so that $\mathrm{Fix}(H)$ is an intermediate field.

(12.3.1). *Let E be a Galois extension of a field F. Let S be an intermediate field and let H be a subgroup of the Galois group $G = \mathrm{Gal}(E/F)$. Then:*

(i) *the assignments $H \mapsto \mathrm{Fix}(H)$ and $S \to \mathrm{Gal}(E/S)$ are mutually inverse, inclusion reversing bijections;*

(ii) $(E : \text{Fix}(H)) = |H|$ *and* $(\text{Fix}(H) : F) = |G : H|$;
(iii) $(E : S) = |\text{Gal}(E/S)|$ *and* $(S : F) = |G : \text{Gal}(E/S)|$.

Thus the theorem asserts the existence of a bijection from the set of subfields between E and F to the set of subgroups of the Galois group G; furthermore the bijection reverses set inclusions. Such a bijection is called a *Galois correspondence*. The Fundamental Theorem allows us to translate a problem about subfields into a problem about subgroups, which might be easier to solve.

Proof of (12.3.1). (i) In the first place $\text{Fix}(\text{Gal}(E/S)) = S$ by (12.2.6). To show that we have mutually inverse bijections it is necessary to prove that $\text{Gal}(E/\text{Fix}(H)) = H$. By the Theorem of the Primitive Element (12.1.7) we have $E = F(a)$ for some a in E. Define a polynomial f in $E[t]$ by

$$f = \prod_{\alpha \in H}(t - \alpha(a)).$$

Note that all the roots of f are distinct: for $\alpha_1(a) = \alpha_2(a)$, $\alpha_i \in H$, implies that $\alpha_1 = \alpha_2$ since $E = F(a)$. Hence $\deg(f) = |H|$. Also the elements of H permute the roots of f, so that $\alpha(f) = f$ for all $\alpha \in H$. Therefore the coefficients of f lie in $K = \text{Fix}(H)$. In addition $f(a) = 0$, so $\text{Irr}_K(a)$ divides f, and since $E = K(a)$, it follows that

$$(E : K) = \deg(\text{Irr}_K(a)) \le \deg(f) = |H|.$$

Hence $|\text{Gal}(E/K)| \le |H|$. But clearly $H \le \text{Gal}(E/K)$, so that $H = \text{Gal}(E/K)$, as required.

(ii) Since E is Galois over $\text{Fix}(H)$, we have $(E : \text{Fix}(H)) = |\text{Gal}(E/\text{Fix}(H))| = |H|$ by (12.3.1)(i). The second statement follows from

$$(E : \text{Fix}(H)) \cdot (\text{Fix}(H) : F) = (E : F) = |G| = |H| \cdot |G : H|.$$

(iii) The first statement is obvious. For the second statement we have $(E : S) \cdot (S : F) = (E : F)$ and $(E : S) = |\text{Gal}(E/S)|$, while $(E : F) = |G|$. The result now follows. $\square$

Normal extensions and normal subgroups
If E is a Galois extension of a field F, intermediate subfields which are normal over F surely correspond to subgroups of $\text{Gal}(E/F)$ which are in some way special. In fact these are exactly the normal subgroups of $\text{Gal}(E/F)$. To prove this a simple lemma about Galois groups of conjugate subfields is called for. If $\alpha \in \text{Gal}(E/F)$ and $F \subseteq S \subseteq E$, write $\alpha(S) = \{\alpha(a) \mid a \in S\}$. Clearly $\alpha(S)$ is a subfield and $F \subseteq \alpha(S) \subseteq E$: the subfield $\alpha(S)$ is called a *conjugate* of S.

(12.3.2). *Let E be an extension of a field F and let S be an intermediate field. If $\alpha \in \text{Gal}(E/F)$, then $\text{Gal}(E/\alpha(S)) = \alpha\text{Gal}(E/S)\alpha^{-1}$.*

Proof. Let $\beta \in \mathrm{Gal}(E/F)$. Then $\beta \in \mathrm{Gal}(E/\alpha(S))$ if and only if $\beta(\alpha(a)) = \alpha(a)$, i.e., $\alpha^{-1}\beta\alpha(a) = a$, for all $a \in S$, or equivalently $\alpha^{-1}\beta\alpha \in \mathrm{Gal}(E/S)$. Hence $\beta \in \mathrm{Gal}(E/\alpha(S))$ if and only if $\beta \in \alpha\mathrm{Gal}(E/S)\alpha^{-1}$. $\qquad\square$

The connection between normal extensions and normal subgroups can now be made.

(12.3.3). *Let E be a Galois extension of a field F and let S be an intermediate field. Then the following statements about S are equivalent:*
(i) *S is normal over F;*
(ii) $\alpha(S) = S$ *for all* $\alpha \in \mathrm{Gal}(E/F)$;
(iii) $\mathrm{Gal}(E/S) \lhd \mathrm{Gal}(E/F)$.

Proof. (i) *implies* (ii). Let $a \in S$ and write $f = \mathrm{Irr}_F(a)$. Since S is normal over F and f has a root in S, all the roots of f are in S. If $\alpha \in \mathrm{Gal}(E/F)$, then $\alpha(a)$ is also a root of f since $f(\alpha(a)) = \alpha(f(a)) = 0$. Therefore $\alpha(a) \in S$ and $\alpha(S) \subseteq S$. By the same argument $\alpha^{-1}(S) \subseteq S$, so that $S \subseteq \alpha(S)$ and $\alpha(S) = S$.

(ii) *implies* (iii). Suppose that $\alpha \in \mathrm{Gal}(E/F)$. By (12.3.2)

$$\alpha\mathrm{Gal}(E/S)\alpha^{-1} = \mathrm{Gal}(E/\alpha(S)) = \mathrm{Gal}(E/S),$$

which shows that $\mathrm{Gal}(E/S) \lhd \mathrm{Gal}(E/F)$.

(iii) *implies* (i). Starting with $\mathrm{Gal}(E/S) \lhd \mathrm{Gal}(E/F)$, we have for any $\alpha \in \mathrm{Gal}(E/F)$ that $\mathrm{Gal}(E/S) = \alpha\mathrm{Gal}(E/S)\alpha^{-1} = \mathrm{Gal}(E/\alpha(S))$ by (12.3.2). Apply the function Fix to $\mathrm{Gal}(E/S) = \mathrm{Gal}(E/\alpha(S))$ to obtain $S = \alpha(S)$ by the Fundamental Theorem of Galois Theory. Next let f in $F[t]$ be irreducible with a root a in S and suppose b is another root of f. Then $b \in E$ since E is normal over F. Because $\mathrm{Irr}_F(a) = f = \mathrm{Irr}_F(b)$, there exists $\alpha \in \mathrm{Gal}(E/F)$ such that $\alpha(a) = b$. Therefore $b \in \alpha(S) = S$, from which it follows that S is normal over F. $\qquad\square$

(12.3.4). *If E is a Galois extension of a field F and S is an intermediate field which is normal over F, then*

$$\mathrm{Gal}(S/F) \simeq \mathrm{Gal}(E/F)/\mathrm{Gal}(E/S).$$

Proof. Let $\alpha \in \mathrm{Gal}(E/F)$; then $\alpha(S) = S$ by (12.3.3) and thus $\alpha|_S \in \mathrm{Gal}(S/F)$. What is more, the restriction map $\alpha \mapsto \alpha|_S$ is a homomorphism from $\mathrm{Gal}(E/F)$ to $\mathrm{Gal}(S/F)$ with kernel equal to $\mathrm{Gal}(E/S)$. The First Isomorphism Theorem then tells us that $\mathrm{Gal}(E/F)/\mathrm{Gal}(E/S)$ is isomorphic with a subgroup of $\mathrm{Gal}(S/F)$. In addition

$$|\mathrm{Gal}(E/F)/\mathrm{Gal}(E/S)| = (E : F)/(E : S) = (S : F) = |\mathrm{Gal}(S/F)|$$

since S is Galois over F. Therefore $\mathrm{Gal}(E/F)/\mathrm{Gal}(E/S) \simeq \mathrm{Gal}(S/F)$. $\qquad\square$

Example (12.3.1). Let E denote the splitting field of $t^3 - 2 \in \mathbb{Q}[t]$. Thus $E = \mathbb{Q}(a, \omega)$ where $a = 2^{1/3}$ and $\omega = e^{2\pi i/3}$. By Example (12.2.1) $(E : \mathbb{Q}) = 6$ and $G = \mathrm{Gal}(E/F) \simeq S_3$.

Now G has exactly six subgroups, which are displayed in the Hasse diagram below.

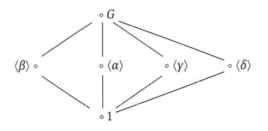

Here $\alpha(a) = a\omega$ and $\alpha(\omega) = \omega$; $\beta(a) = a$ and $\beta(\omega) = \omega^2$; $\gamma(a) = a\omega$ and $\gamma(\omega) = \omega^2$; $\delta(a) = a\omega^2$ and $\delta(\omega) = \omega^2$. Each subgroup H corresponds to its fixed field $\mathrm{Fix}(H)$ under the Galois correspondence. For example, $\mathrm{Fix}(\langle\alpha\rangle) = \mathbb{Q}(\omega)$ and $\mathrm{Fix}(\langle\beta\rangle) = \mathbb{Q}(a)$. The normal subgroups of G are 1, $\langle\alpha\rangle$ and G; the three corresponding normal extensions are E, $\mathbb{Q}(\omega)$ and $\mathbb{Q}$.

The six subfields of E are displayed in the Hasse diagram below.

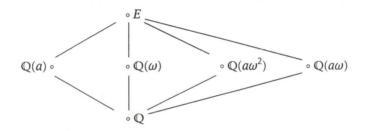

Since every subgroup of an abelian group is normal, we deduce at once from (12.3.3):

Corollary (12.3.5). *If E is a Galois extension of a field F and $\mathrm{Gal}(E/F)$ is abelian, then every intermediate field is normal over F.*

For example, by (12.2.11) the Galois group of the cyclotomic polynomial $\Phi_n \in \mathbb{Q}[t]$ is abelian. Therefore *every subfield of a cyclotomic number field is normal over $\mathbb{Q}$.*

As a demonstration of the power of Galois theory, let us next establish the *Fundamental Theorem of Algebra*, which was stated without proof as (7.4.4). All known proofs of this theorem employ some analysis. Here only the *Intermediate Value Theorem* is used: if f is a continuous function of a real variable which assumes the values a and b, then f assumes all values between a and b. In fact this result is only required for polynomial functions.

(12.3.6). *Let f be a non-constant polynomial over $\mathbb{C}$. Then f is a product of linear factors over $\mathbb{C}$.*

Proof. First note that the polynomial $f\bar{f}$ has real coefficients. Since we can replace f by this polynomial, there is no loss in assuming that f has real coefficients. It may also be assumed that $\deg(f) > 1$. Let E be the splitting field of f over $\mathbb{C}$. Then E is the splitting field of $(t^2 + 1)f$ over $\mathbb{R}$. Hence E is Galois over $\mathbb{R}$, since the characteristic is 0. Put $G = \mathrm{Gal}(E/\mathbb{R})$. Then $|G| = (E : \mathbb{R}) = (E : \mathbb{C}) \cdot (\mathbb{C} : \mathbb{R}) = 2(E : \mathbb{C})$, and G has even order.

Let H be a Sylow 2-subgroup of G and put $F = \mathrm{Fix}(H)$. Then $\mathbb{R} \subseteq F \subseteq E$ and $(F : \mathbb{R}) = |G : H|$ is odd. Let $a \in F$ and set $g = \mathrm{Irr}_{\mathbb{R}}(a)$. Since $\deg(g) = (\mathbb{R}(a) : \mathbb{R})$, which divides $(F : \mathbb{R})$, we conclude that $\deg(g)$ is odd. Also g is monic, so $g(x) > 0$ for large positive x and $g(x) < 0$ for large negative x. This is our opportunity to apply the Intermediate Value Theorem, the conclusion being that $g(x) = 0$ for some real number x. But g is irreducible over $\mathbb{R}$, so $\deg(g) = 1$; hence $a \in \mathbb{R}$ and $F = \mathbb{R}$. This implies that $H = G$ and G is a 2-group.

Let $G_0 = \mathrm{Gal}(E/\mathbb{C}) \leq G$; thus G_0 is a 2-group. Now $G_0 = 1$ implies that $E = \mathbb{C}$ and f is a product on linear factors over $\mathbb{C}$. So assume that $G_0 \neq 1$. Hence there is a maximal (proper) subgroup M of G_0. Now G_0 is nilpotent, so $M \triangleleft G_0$ and $|G_0 : M| = 2$ by (10.2.7). Now put $S = \mathrm{Fix}(M)$. By (12.3.1) we have

$$(S : \mathbb{C}) = |\mathrm{Gal}(E/\mathbb{C}) : \mathrm{Gal}(E/S)| = \frac{|G_0|}{|M|} = 2.$$

Hence any s in $S - \mathbb{C}$ has irreducible polynomial over $\mathbb{C}$ of degree 2, say $t^2 + at + b$. By the quadratic formula $s = -\frac{1}{2}(a \pm \sqrt{a^2 - 4b}) \in \mathbb{C}$ and it follows that $S = \mathbb{C}$, a final contradiction. $\qquad\square$

Constructing regular n-gons

We return to the last of the ruler and compass problems discussed in Section 11.2. The problem is to construct a regular n-gon of side 1 unit using ruler and compass only. We are now in a position to resolve this by using (12.3.6).

Consider a regular n-gon with vertices $A_1, A_2, \ldots, A_n$ and centroid C. Let θ_n be the angle between lines joining the centroid C to two neighboring vertices; thus $\theta_n = \frac{2\pi}{n}$. By elementary geometry, if d is the distance from the centroid C to a vertex, then $d \sin \frac{1}{2}\theta_n = \frac{1}{2}$ and hence

$$d = \frac{1}{2\sin(\frac{1}{2}\theta_n)} = \frac{1}{\sqrt{2(1 - \cos\theta_n)}}.$$

It follows from the discussion of constructibility in Section 11.2 that the regular n-gon is constructible by ruler and compass if and only if $\cos\theta_n$ is constructible from the set $\{(0,0), (1,0)\}$.

The definitive result is next.

(12.3.7). *A regular n-gon of side 1 can be constructed by ruler and compass if and only if n has the form $2^k p_1 p_2 \cdots p_k$ where $k \geq 0$ and the p_j are distinct Fermat primes, i.e., of the form $2^{2^{e_j}} + 1$.*

Proof. Assume that the regular n-gon is constructible, so that $\cos \theta_n$ is constructible. Then $(\mathbb{Q}(\cos \theta_n) : \mathbb{Q})$ must be a power of 2 by (11.2.2). Put $c = e^{2\pi i/n}$, a primitive nth root of unity. Then $\cos \theta_n = \frac{1}{2}(c + c^{-1})$, so that $\mathbb{Q}(\cos \theta_n) \subseteq \mathbb{Q}(c)$. Since $c + c^{-1} = 2 \cos \theta_n$, we have $c^2 - 2c \cos \theta_n + 1 = 0$. Hence $(\mathbb{Q}(c) : \mathbb{Q}(\cos \theta_n)) = 2$ and $(\mathbb{Q}(c) : \mathbb{Q}) = 2^d$ for some d. Recall from (12.2.10) that $\mathrm{Irr}_{\mathbb{Q}}(c) = \Phi_n$, which has degree $\phi(n)$. Writing $n = 2^k p_1^{e_1} \cdots p_r^{e_r}$ with distinct odd primes p_j and $e_j > 0$, we have

$$\phi(n) = 2^{k-1}(p_1^{e_1} - p_1^{e_1 - 1}) \cdots (p_r^{e_r} - p_r^{e_r - 1})$$

by (2.3.8). This must equal 2^d. Hence $e_j = 1$ and $p_j - 1$ is a power of 2 for all j. Since $2^s + 1$ cannot be a prime if s is not a power of 2 (see Exercise (2.2.13)), it follows that p_j is a Fermat prime.

Conversely, assume that n has the form indicated. Since $\mathbb{Q}(c)$ is Galois over $\mathbb{Q}$, we have $(\mathbb{Q}(c) : \mathbb{Q}) = \phi(n)$, which is a power of 2 by the formula for $\phi(n)$. Hence $\mathrm{Gal}(\mathbb{Q}(c)/\mathbb{Q})$ is a finite abelian 2-group. Since $G = \mathrm{Gal}(\mathbb{Q}(\cos \theta)/\mathbb{Q})$ is isomorphic with a quotient of $\mathrm{Gal}(\mathbb{Q}(c) : \mathbb{Q})$, it is also a finite abelian 2-group. Therefore all the factors in a composition series of G have order 2 and by the Fundamental Theorem of Galois Theory there is a chain of subfields

$$\mathbb{Q} = F_0 \subset F_1 \subset \cdots \subset F_\ell = \mathbb{Q}(\cos \theta)$$

such that F_{j+1} is Galois over F_j and $(F_{j+1} : F_j) = 2$.

We argue by induction on j that every element of F_j is constructible. Let $x \in F_{j+1} - F_j$. Then $\mathrm{Irr}_{F_j}(x) = t^2 + at + b$ where $a, b \in F_j$ and thus $x^2 + ax + b = 0$. Hence $(x + \frac{1}{2}a)^2 = \frac{1}{4}a^2 - b > 0$ since x is real. Writing $x' = x + \frac{1}{2}a$, we have $x'^2 \in F_j$. By induction hypothesis x'^2 is constructible and therefore x' is constructible by (11.2.1). Hence x is constructible and so is $\cos \theta$. Thus the proof is complete. $\square$

Example (12.3.2). A regular n-gon is constructible for $n = 3, 4, 5, 6$, but not for $n = 7$.

The only known Fermat primes are 3, 5, 17, $257 = 2^{2^3} + 1$ and $65,537 = 2^{2^4} + 1$. Since 7 is not a Fermat prime, it is impossible to construct a regular 7-gon using ruler and compass.

Exercises (12.3).

(1) For each of the following polynomials over $\mathbb{Q}$ display the lattice of subgroups of the Galois group and the corresponding lattice of subfields of the splitting field: (i) $t^2 - 5$; (ii) $t^4 - 5$; (iii) $(t^2 + 1)(t^2 + 3)$.

(2) Determine the normal subfields of the splitting fields in Exercise (12.3.1).

(3) Use the Fundamental Theorem of Galois Theory and Exercise (12.2.6) to prove that $GF(p^m)$ has exactly one subfield of order p^d for each divisor d of m and no subfields of other orders – see also Exercise (11.3.2).

(4) Let $E = \mathbb{Q}(\sqrt{2}, \sqrt{3})$. Find all the subgroups of $\mathrm{Gal}(E/\mathbb{Q})$ and hence all subfields of E.

(5) Find all finite fields with exactly two subfields and also those with exactly three subfields. [Hint: see Exercise (4.1.5).]

(6) Let E be a Galois extension of a field F and let p^k be the largest power of a prime p dividing $(E : F)$. Prove that there is an intermediate field S such that $(E : S) = p^k$. If $\mathrm{Gal}(E : F)$ is solvable, prove that there is an intermediate field T such that $(T : F) = p^k$. [Hint: see (5.3.8) and (10.3.2).]

(7) If E is a Galois extension of a field F and there is exactly one proper intermediate field, what can be said about $\mathrm{Gal}(E/F)$?

(8) If E is a Galois extension of F and $(E : F)$ is the square of a prime, show that each intermediate field is normal over F.

(9) Prove that a regular 2^k-gon of side 1 is constructible if $k \geq 2$.

(10) For which values of n in the range 10 to 20 can a regular n-gon of side 1 be constructed?

(11) Show that if a is a real number such that $(\mathbb{Q}(a) : \mathbb{Q})$ is a power of 2 and $\mathbb{Q}(a)$ is normal over $\mathbb{Q}$, then a is constructible from the points $(0,0)$ and $(1,0)$. [Hint: let $G = \mathrm{Gal}(\mathbb{Q}(a) : \mathbb{Q})$. Then G is a finite 2-group. Form a composition series in G and then the corresponding chain of subfields E_i, $0 \leq i \leq m$; use induction on i to show that each element of E_i is constructible.]

(12) Let p be a prime and let $f = t^p - t - a \in F[t]$ where $F = GF(p)$. Denote by E the splitting field of f over F.
  (i) If x is a root of f in E, show that the set of all roots of f is $\{x + b \mid b \in F\}$ and that $E = F(x)$.
  (ii) Prove that f is irreducible over F if and only if $a \neq 0$.
  (iii) Prove that $|\mathrm{Gal}(f)| = p$ unless $a = 0$, when $\mathrm{Gal}(f) = 1$.

12.4 Solvability of equations by radicals

One of the oldest parts of algebra is concerned with the problem of solving equations of the form $f(t) = 0$ where f is a non-constant polynomial over $\mathbb{Q}$ or $\mathbb{R}$. The object is to find a formula for the solutions of the equation which involves the coefficients of f, square roots, cube roots, etc. The easiest cases are when $\deg(f) \leq 2$; if the degree is 1, we are solving a single linear equation. If the degree is 2, there is the familiar formula for the solutions of a quadratic equation. For equations of degree 3 and 4 the problem is harder, but solutions had been found by the 16th Century. Thus for $\deg(f) \leq 4$ there are explicit formulas for the roots of $f(t) = 0$, which in fact involve radicals of the form $\sqrt[k]{\ }$ for $k \leq 4$.

The problem of finding formulas for the solutions of equations of degree 5 and higher is one that fascinated mathematicians for hundreds of years. An enormous amount of ingenuity was expended in attempts to solve the general equation of the fifth degree. It was only with the work of Abel, Galois and Ruffini[2] in the early 19th Century that it became clear that all such efforts had been in vain. It is a fact that solvability of a polynomial equation is inextricably linked to the solvability of the Galois group of the polynomial. The symmetric group S_n is solvable for $n < 5$, but is insolvable for $n \geq 5$. This explains why early researchers were able to solve the general equation of degree n only for $n \leq 4$. Without the aid of group theory it is difficult to comprehend the reason for this failure. Our aim here is explain why the solvability of the Galois group governs the solvability of a polynomial equation.

Radical extensions

Let E be an extension of a field F. Then E is called a *radical extension* of F if there is a chain of subfields

$$F = E_0 \subseteq E_1 \subseteq E_2 \subseteq \cdots \subseteq E_m = E$$

such that $E_{i+1} = E_i(a_{i+1})$ where a_{i+1} has its irreducible polynomial over E_i of the form $t^{n_{i+1}} - b_i$ where $0 \leq i < m$. It is natural to refer to a_{i+1} as a *radical* and write $a_{i+1} = \sqrt[n_{i+1}]{b_i}$, but here one has to keep in mind that a_{i+1} may not be uniquely determined by b_i. Since

$$E = F(\sqrt[n_1]{b_0}, \sqrt[n_2]{b_1}, \ldots, \sqrt[n_m]{b_{m-1}}),$$

elements of E are expressible as polynomial functions of the radicals $\sqrt[n_i]{b_i}$.

Let f be a non-constant polynomial over F with splitting field K. Then f, or the equation $f = 0$, is said to be *solvable by radicals* if K is contained in some radical extension of F. This means that the roots of f are obtained by forming a finite sequence of successive radicals, starting with elements of F. The definition gives a precise expression for the intuitive idea of what it means for a polynomial equation to be solvable by radicals.

To make progress with the problem of describing the radical extensions it is necessary to have a better understanding of polynomials of the form $t^n - a$.

---

2 Paolo Ruffini (1765–1822).

(12.4.1). *Let F be a field and n a positive integer. Assume that F contains a primitive nth root of unity. Then for any a in F the group* $\mathrm{Gal}(t^n - a)$ *is cyclic with order dividing n.*

Proof. Let z be a primitive nth root of unity in F and denote by b a root of $f = t^n - a$ in its splitting field E. Then the roots of f are bz^j, $j = 0, 1, \ldots, n - 1$. If $\alpha \in \mathrm{Gal}(f) = \mathrm{Gal}(E/F)$, then $\alpha(b) = bz^{j(\alpha)}$ for some $j(\alpha)$ and α is completely determined by $j(\alpha)$: this is because $\alpha|_F$ is the identity map and $E = F(b)$ since $z \in F$. The assignment $\alpha \mapsto j(\alpha) + n\mathbb{Z}$ is an injective homomorphism from $\mathrm{Gal}(f)$ to $\mathbb{Z}_n$: for $\alpha\beta(b) = \alpha(bz^{j(\beta)}) = \alpha(b)z^{j(\beta)} = bz^{j(\alpha)+j(\beta)}$ and thus $j(\alpha\beta) \equiv j(\alpha) + j(\beta) \pmod{n}$. It follows that $\mathrm{Gal}(f)$ is isomorphic with a subgroup of $\mathbb{Z}_n$, so it is a cyclic group with order dividing n. □

We will also need the following simple result.

(12.4.2). *Let E be a Galois extension of a field F and let K_1 and K_2 be subfields intermediate between F and E. If $H_i = \mathrm{Gal}(E/K_i)$, then $\mathrm{Gal}(E/K_1 \cap K_2) = \langle H_1, H_2 \rangle$.*

Proof. Clearly H_1 and H_2 are contained in $\mathrm{Gal}(E/K_1 \cap K_2)$ and hence $J = \langle H_1, H_2 \rangle \leq \mathrm{Gal}(E/K_1 \cap K_2)$. Next suppose that $x \in E - K_i$. Then there exists $\alpha \in H_i$ such that $\alpha(x) \neq x$. Hence $x \notin \mathrm{Fix}(J)$ and consequently $\mathrm{Fix}(J) \subseteq K_1 \cap K_2$. Taking the Galois group of E over each side and applying (12.3.1), we obtain $J \geq \mathrm{Gal}(E/K_1 \cap K_2)$. □

The principal theorem is now within reach.

(12.4.3). *Let f be a non-constant polynomial over a field F of characteristic 0. If f is solvable by radicals, then $\mathrm{Gal}(f)$ is a solvable group.*

Proof. Let E denote the splitting field of f over F. By hypothesis $E \subseteq R$ where R is a radical extension of F. Hence there are subfields R_i such that

$$F = R_0 \subseteq R_1 \subseteq \cdots \subseteq R_m = R$$

where $R_{i+1} = R_i(a_{i+1})$ and $\mathrm{Irr}_{R_i}(a_{i+1}) = t^{n_{i+1}} - b_i$ with $b_i \in R_i$. It follows that $(R_{i+1} : R_i) = n_{i+1}$ and hence $(R : F) = n_1 n_2 \cdots n_m = n$, say.

Let K and L be the splitting fields of the polynomial $t^n - 1$ over F and R respectively. Note that L may not be normal over F. Let N be the splitting field over F of the product of $t^n - 1$ and all the polynomials $\mathrm{Irr}_F(a_i)$, $i = 1, 2, \ldots, m$. Then $L \subseteq N$ and N is normal over F. Clearly $(N : F)$ is finite and N is separable since the characteristic is zero. Thus N is Galois over F. Put $L_i = K(R_i)$, so there is the chain of subfields

$$K = L_0 \subseteq L_1 \subseteq \cdots \subseteq L_m = L \subseteq N.$$

The relevant subfields are displayed in the Hasse diagram below.

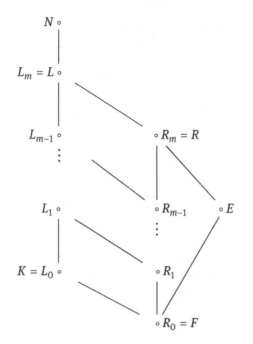

Note that L_{i+1} is the splitting field of $t^{n_{i+1}} - b_i$ over L_i since K contains all n_{i+1}th roots of unity. Thus L_{i+1} is normal and hence Galois, over L_i. Now set $G = \mathrm{Gal}(N/F)$ and $G_i = \mathrm{Gal}(N/L_i)$; hence $G_{i+1} \lhd G_i$ by (12.3.3). Also write $V = \mathrm{Gal}(N/K)$ and $U = \mathrm{Gal}(N/E)$, noting that $U \lhd G$ and $V \lhd G$ since E and K are normal over F. Thus we have the truncated series of subgroups

$$\mathrm{Gal}(N/L) = G_m \lhd G_{m-1} \lhd \cdots \lhd G_1 \lhd G_0 = V.$$

Notice that $G_i/G_{i+1} \simeq \mathrm{Gal}(L_{i+1}/L_i) = \mathrm{Gal}(t^{n_i+1} - b_i)$, and the latter is cyclic by (12.4.1).

Since $G_m \leq U$, there is a series

$$1 = G_m U/U \lhd G_{m-1}U/U \lhd \cdots \lhd G_1 U/U \lhd G_0 U/U = UV/U,$$

and the factors of this series are cyclic since the G_i/G_{i+1} are cyclic. Therefore UV/U is a solvable group. Now from (12.4.2) we have $\mathrm{Gal}(N/K \cap E) = UV \lhd G$ and $K \cap E$ is normal over F. Moreover,

$$G/UV = \mathrm{Gal}(N/F)/\mathrm{Gal}(N/K \cap E) \simeq \mathrm{Gal}(K \cap E/F)$$

and therefore $G/UV \simeq \text{Gal}(K/F)/\text{Gal}(K/K \cap E)$. Since $\text{Gal}(K/F)$ is abelian by (12.2.7), it follows that G/UV is abelian. Therefore G/U is solvable. Finally,

$$\text{Gal}(E/F) \simeq \text{Gal}(N/F)/\text{Gal}(N/E) = G/U$$

by (12.3.4), so that $\text{Gal}(f) = \text{Gal}(E/F)$ is solvable, as required. $\qquad\square$

It can be shown – although we will not do so here – that the converse of (12.4.3) is valid: see [1] or [18] for a proof. As a consequence there is the following definitive result.

(12.4.4). *Let f be a non-constant polynomial over a field of characteristic 0. Then f is solvable by radicals if and only if* $\text{Gal}(f)$ *is a solvable group.*

Let $n = \deg(f)$. Then $\text{Gal}(f)$ is isomorphic with a subgroup of the symmetric group S_n by (12.2.3). If $n \le 4$, then S_n, and hence $\text{Gal}(f)$, is solvable. Therefore by (12.4.4) *every polynomial with degree 4 or less is solvable by radicals.*

On the other hand, when $n \ge 5$, the symmetric group S_n is not solvable since A_n is simple by (10.1.7). Thus we are led to suspect that not every polynomial equation of degree 5 is solvable by radicals. Actual examples of polynomials that are not solvable by radicals are furnished by the next result.

(12.4.5). *Let* $f \in \mathbb{Q}[t]$ *be an irreducible polynomial of prime degree p and assume that f has exactly two complex roots. Then* $\text{Gal}(f) \simeq S_p$ *and hence f is not solvable by radicals if* $p \ge 5$.

Proof. Label the roots of f in its splitting field $a_1, a_2, \ldots, a_p$; these are all different since f is separable. Two of these roots are complex conjugates, say $\bar{a}_1 = a_2$, while the other roots are all real. We can think of $\text{Gal}(f)$ as a group of permutations of the set of roots $\{a_1, a_2, \ldots, a_p\}$ and indeed $\text{Gal}(f)$ acts transitively since f is irreducible. Therefore p divides $|\text{Gal}(f)|$ by (5.2.2), and Cauchy's Theorem (5.3.9) shows that there is an element of order p in $\text{Gal}(f)$. Hence $\text{Gal}(f)$ contains a p-cycle, say $\pi = (a_1 a_{i_2} \ldots a_{i_p})$. Replacing π by a suitable power, we may assume that $i_2 = 2$. Now relabel the remaining roots a_3, $a_4, \ldots, a_p$, so that $\pi = (a_1 a_2 a_3 \ldots a_p)$.

Complex conjugation, i. e., $\sigma = (a_1 a_2)$, is an element of $\text{Gal}(f)$ with order 2. Conjugation of σ by powers of π shows that $\text{Gal}(f)$ contains all the adjacent transpositions $(a_i a_{i+1})$, for $i = 1, 2, \ldots, p - 1$. But any permutation is expressible as a product of adjacent transpositions – see Exercise (3.1.4) – and therefore $\text{Gal}(f) = S_n$. $\qquad\square$

Example (12.4.1). The polynomial $f = t^5 - 6t + 3 \in \mathbb{Q}[t]$ is not solvable by radicals.

In the first place f is irreducible over $\mathbb{Q}$ by Eisenstein's Criterion and Gauss's Lemma. In addition calculus tells us that the curve $f(t) = 0$ crosses the t-axis exactly three times, so there are three real roots and two complex ones. Thus $\text{Gal}(t^5 - 6t + 3) \simeq S_5$ and the result follows via (12.4.3).

Example (12.4.2). The polynomial $f = t^5 + 8t^3 - t^2 + 12t - 2$ is solvable by radicals.

Here the situation is different since f factorizes as $(t^2 + 2)(t^3 + 6t - 1)$. Therefore $\mathrm{Gal}(f)$ is isomorphic with a subgroup of $\mathrm{Gal}(t^2 + 2) \times \mathrm{Gal}(t^3 + 6t - 1)$ by Exercise (12.2.5). The latter is a solvable group. Hence by (10.2.2) the group $\mathrm{Gal}(f)$ is solvable and f is solvable by radicals.

Symmetric functions

As the final topic of the chapter, we present an account of the elementary theory of symmetric functions and explore its relationship with Galois theory. Let F be an arbitrary field and put $E = F\{x_1, x_2, \ldots, x_n\}$, the field of rational functions over F in distinct indeterminates $x_1, x_2, \ldots, x_n$. A *symmetric function* in $x_1, x_2, \ldots, x_n$ over F is an element $g \in E$ such that

$$g(x_{\pi(1)}, x_{\pi(2)}, \ldots, x_{\pi(n)}) = g(x_1, x_2, \ldots, x_n)$$

for all $\pi \in S_n$. Thus g is unaffected by permutations of the indeterminates $x_1, x_2, \ldots, x_n$. It is easy to verify that the symmetric functions form a subfield of E. Next consider the polynomial

$$f = (t - x_1)(t - x_2) \cdots (t - x_n) \in E[t]$$

where t is another indeterminate. Then expansion shows that

$$f = t^n - s_1 t^{n-1} + s_2 t^{n-2} - \cdots + (-1)^n s_n$$

where $s_1 = \sum_{i=1}^{n} x_i$, $s_2 = \sum_{i<j=1}^{n} x_i x_j$, and in general

$$s_j = \sum_{i_1 < i_2 < \cdots < i_j = 1}^{n} x_{i_1} x_{i_2} \cdots x_{i_j},$$

the last sum being over all j-tuples $(i_1, i_2, \ldots, i_j)$ such that $1 \le i_1 < i_2 < \cdots < i_j \le n$. It is evident that the s_j are symmetric functions: they are known as the *elementary symmetric functions* in $x_1, x_2, \ldots, x_n$. For example, when $n = 3$, there are three elementary symmetric functions,

$$s_1 = x_1 + x_2 + x_3, \quad s_2 = x_1 x_2 + x_2 x_3 + x_1 x_3, \quad s_3 = x_1 x_2 x_3.$$

Put $S = F(s_1, s_2, \ldots, s_n)$, which is a subfield of E. Then $f \in S[t]$ and E is generated by S and the roots of f, i. e., $x_1, x_2, \ldots, x_n$. Hence E is the splitting field of f over S. Since all the x_i are distinct, they are separable over S and (12.1.6) shows that E is separable and hence Galois over S. Therefore $\mathrm{Gal}(f) = \mathrm{Gal}(E/S)$ has order $(E : S)$. We now proceed to determine the Galois group of f over S.

With the same notation the definitive result is:

(12.4.6). $\mathrm{Gal}(f) \cong S_n$.

Proof. Since $\mathrm{Gal}(f)$ permutes the roots $x_1, x_2, \ldots, x_n$ faithfully, we may identify it with a subgroup of S_n. Let $\pi \in S_n$ and define $\alpha_\pi : E \to E$ by the rule

$$\alpha_\pi(g(x_1, x_2, \ldots, x_n)) = g(x_{\pi(1)}, x_{\pi(2)}, \ldots, x_{\pi(n)});$$

then α_π is evidently an automorphism of E. Since α_π fixes all the elementary symmetric functions, it fixes every element of $S = F(s_1, s_2, \ldots, s_n)$ and therefore $\alpha_\pi \in \mathrm{Gal}(E/S) = \mathrm{Gal}(f)$. Finally, all the α_π are different, so $\mathrm{Gal}(f) = S_n$. □

From this we deduce a famous theorem.

Corollary (12.4.7) (The Symmetric Function Theorem). *If F is an arbitrary field and s_1, $s_2, \ldots, s_n$ are the elementary symmetric functions in indeterminates $x_1, x_2, \ldots, x_n$, then $F(s_1, s_2, \ldots, s_n)$ is the field of all symmetric functions in $x_1, x_2, \ldots, x_n$. Also the symmetric polynomials form a subring which is generated by F and the $s_1, s_2, \ldots, s_n$.*

Proof. Let $S = F(s_1, s_2, \ldots, s_n) \subseteq E = F\{x_1, x_2, \ldots, x_n\}$. By (12.4.6) $\mathrm{Gal}(E/S)$ effectively consists of all permutations of $\{x_1, x_2, \ldots, x_n\}$. Hence $\mathrm{Fix}(\mathrm{Gal}(E/S))$ is the subfield of all symmetric functions. But by (12.3.1) this is also equal to S. The statement about polynomials is left as an exercise. □

Generic polynomials

Let F be an arbitrary field and write K for the rational function field in indeterminates $x_1, x_2, \ldots, x_n$ over F. The polynomial

$$f = t^n - x_1 t^{n-1} + x_2 t^{n-2} - \cdots + (-1)^n x_n \in K[t]$$

is called a *generic polynomial*. The point to note here is that we can obtain from f any monic polynomial of degree n in $F[t]$ by replacing $x_1, x_2, \ldots, x_n$ by suitable elements of F. It is therefore not surprising that the Galois group of f over K is as large as it could be.

(12.4.8). *With the above notation, $\mathrm{Gal}(f) \cong S_n$.*

Proof. Let $u_1, u_2, \ldots, u_n$ be the roots of f in its splitting field E over K. Thus we have $f = (t - u_1)(t - u_2) \cdots (t - u_n)$ and so $x_i = s_i(u_1, u_2, \ldots, u_n)$ where s_i is the ith elementary symmetric function in n indeterminates $y_1, y_2, \ldots, y_n$, all of which are different from $x_1, x_2, \ldots, x_n, t$.

The assignment $x_i \mapsto s_i$ determines a ring homomorphism

$$\phi_0 : F\{x_1, x_2, \ldots, x_n\} \to F\{y_1, y_2, \ldots, y_n\};$$

observe here that $g(s_1, \ldots, s_n) = 0$ implies that $g(x_1, \ldots, x_n) = 0$, as $x_i = s_i(u_1, \ldots, u_n)$. So ϕ_0 is actually an isomorphism from $K = F\{x_1, x_2, \ldots, x_n\}$ to $L = F(s_1, s_2, \ldots, s_n) \subseteq F\{y_1, y_2, \ldots, y_n\}$. Set $f^* = \phi_0(f)$ with f as above; thus

$$f^* = t^n - s_1 t^{n-1} + s_2 t^{n-2} - \cdots + (-1)^n s_n = (t - y_1)(t - y_2) \cdots (t - y_n),$$

by definition of the elementary symmetric functions s_i.

By (11.3.3) we can extend ϕ_0 to an isomorphism ϕ from E, the splitting field of f over K, to the splitting field of f^* over L. Therefore ϕ induces a group isomorphism from $\mathrm{Gal}(f)$ to $\mathrm{Gal}(f^*)$. But we know that $\mathrm{Gal}(f^*) \simeq S_n$ by (12.4.6). Hence $\mathrm{Gal}(f) \simeq S_n$.

$\square$

Corollary (12.4.9) (Abel, Ruffini). *If F is a field of characteristic 0, the generic polynomial $t^n - x_1 t^{n-1} + x_2 t^{n-2} - \cdots + (-1)^n x_n$ is insolvable by radicals over $F(x_1, x_2, \ldots, x_n)$ if $n \geq 5$.*

Thus, as one would expect, there is no *general formula* for the roots of a polynomial of degree $n \geq 5$ in terms of its coefficients.

Exercises (12.4).

(1) Let $F \subseteq K \subseteq E$ be field extensions with K radical over F and E radical over K. Prove that E is radical over F.

(2) Let $F \subseteq K \subseteq E$ be field extensions with E radical and Galois over F. Prove that E is radical over K.

(3) Show that the polynomial $t^5 - 3t + 2$ in $\mathbb{Q}[t]$ is solvable by radicals. [Hint: the polynomial is reducible, so the Galois group is solvable.]

(4) If p is a prime larger than 11, show that $t^5 - pt + p$ in $\mathbb{Q}[t]$ is not solvable by radicals.

(5) If $f \in F[t]$ is solvable by radicals and $g \mid f$ in $F[t]$, prove that g is solvable by radicals.

(6) Let $f = f_1 f_2$ where $f_1, f_2 \in F[t]$ and F has characteristic 0. If f_1 and f_2 are solvable by radicals, show that f is too. Deduce that every non-constant reducible polynomial of degree less than 6 over $\mathbb{Q}$ is solvable by radicals.

(7) Let F be a field of characteristic 0 and let $f \in F[t]$ be non-constant with splitting field E. Prove that there is a unique smallest intermediate field S such that S is normal over F and f is solvable by radicals over S. [Hint: show first that there is a unique maximum solvable normal subgroup in any finite group.]

(8) For each integer $n \geq 5$ exhibit a polynomial of degree n over $\mathbb{Q}$ which is insolvable by radicals.

(9) Let G be any finite group. Prove that there is a Galois extension E of some algebraic number field F such that $\mathrm{Gal}(E/F) \simeq G$. [You may assume there is an algebraic number field whose Galois group over $\mathbb{Q}$ is isomorphic with S_n.] (Remark: the general problem of whether every finite group is the Galois group of some algebraic number field over $\mathbb{Q}$ is still open; it is known to be true for solvable groups.)

(10) Write each of the following symmetric polynomials as a polynomial in the elementary symmetric functions s_1, s_2, s_3 in x_1, x_2, x_3.

(i) $x_1^2 + x_2^2 + x_3^2$;

(ii) $x_1^2 x_2 + x_1 x_2^2 + x_2^2 x_3 + x_2 x_3^2 + x_1^2 x_3 + x_1 x_3^2$;

(iii) $x_1^3 + x_2^3 + x_3^3$.

12.5 Roots of Polynomials and Discriminants

In this section the concept of the *discriminant* of a polynomial is introduced and it is applied it to the Galois groups of polynomials of degree ≤ 4.

The discriminant of a polynomial

Let f be a non-constant monic polynomial in t over a field F and let $n = \deg(f)$. Let the roots of f in its splitting field E be $a_1, a_2, \ldots, a_n$ and define

$$\Delta = \prod_{i<j=1}^{n} (a_i - a_j),$$

which is an element of E. Note that Δ depends on the order in which the roots are written, so it is only determined up to sign. Also all the roots of f are distinct if and only if $\Delta \neq 0$: let us assume this to be the case. Thus E is Galois over F.

If $\alpha \in \text{Gal}(f) = \text{Gal}(E/F)$, then α permutes the roots $a_1, a_2, \ldots, a_n$, and $\alpha(\Delta) = \pm\Delta$. Indeed $\alpha(\Delta) = \Delta$ precisely when α produces an *even* permutation of the a_i's. Thus in any event α fixes

$$D = \Delta^2.$$

The element D is called the *discriminant* of f: it is independent of the order of the roots of f. Since D is fixed by every automorphism α and E is Galois over F, it follows from (12.2.6) that $D \in F$. The question arises: how is D related to the coefficients of the original polynomial f?

(12.5.1). *Let f be a non-constant polynomial over a field F. Then the discriminant D of f is expressible as a polynomial in the coefficients of f.*

Proof. It can be assumed that f is monic and that it has distinct roots $a_1, a_2, \ldots, a_n$ since otherwise $D = 0$. Then $f = (t - a_1)(t - a_2) \cdots (t - a_n)$, so that

$$f = t^n - s_1 t^{n-1} + s_2 t^{n-2} - \cdots + (-1)^n s_n$$

where $s_1, s_2, \ldots, s_n$ are the elementary symmetric functions of degree $1, 2, \ldots, n$ in a_1, $a_2, \ldots, a_n$. Now $D = \prod_{i<j=1}^{n} (a_i - a_j)^2$ is a symmetric function in $a_1, a_2, \ldots, a_n$. It follows

from the Symmetric Function Theorem (12.4.7) that D is expressible as a polynomial in $s_1, s_2, \ldots, s_n$, i. e., in the coefficients of f. ☐

Next we examine the discriminants of polynomials of degrees 2 and 3 over a field.

Example (12.5.1). Let $f = t^2 + ut + v$. If the roots of f are a_1 and a_2, then $\Delta = a_1 - a_2$ and $D = (a_1 - a_2)^2$. This can be rewritten in the form $D = (a_1 + a_2)^2 - 4a_1a_2$. Now clearly $u = -(a_1 + a_2)$ and $v = a_1a_2$, so we arrive at the familiar formula for the discriminant of the quadratic $t^2 + ut + v$,

$$D = u^2 - 4v.$$

Example (12.5.2). Consider a cubic polynomial $f = t^3 + ut^2 + vt + w$ and let a_1, a_2, a_3 be its roots. Then $D = (a_1 - a_2)^2(a_2 - a_3)^2(a_1 - a_3)^2$. Also $u = -(a_1 + a_2 + a_3)$, $v = a_1a_2 + a_2a_3 + a_1a_3$ and $w = -a_1a_2a_3$. By a rather laborious calculation we can expand D and write it in terms of the elements u, v, w. What emerges is the formula

$$D = u^2v^2 - 4v^3 - 4u^3w - 27w^2 + 18uvw.$$

This expression can be simplified by a judicious change of variable. Put $t' = t + \frac{1}{3}u$, so that $t = t' - \frac{1}{3}u$. On substituting for t in $f = t^3 + ut^2 + vt + w$, we find that $f = t'^3 + pt' + q$ where $p = v - \frac{1}{3}u^2$ and $q = w + \frac{2}{27}u^3 - \frac{1}{3}uv$. Hence no generality is lost in assuming that f does not have a term in t^2 and

$$f = t^3 + pt + q.$$

Now the formula for the discriminant reduces to the more manageable expression

$$D = -4p^3 - 27q^2.$$

Next we relate the discriminant to the Galois group of a polynomial.

(12.5.2). *Let F be a field whose characteristic is not 2 and let f be a monic polynomial in $F[t]$ with distinct roots $a_1, a_2, \ldots, a_n$. Write $\Delta = \prod_{i<j=1}^{n}(a_i - a_j)$. If $G = \mathrm{Gal}(f)$ is identified with a subgroup of S_n, then $\mathrm{Fix}(G \cap A_n) = F(\Delta)$.*

Proof. Let $H = G \cap A_n$ and note that $H \lhd G$ and $|G : H| \leq 2$. If E is the splitting field of f, then $F \subseteq F(\Delta) \subseteq \mathrm{Fix}(H) \subseteq E$ since elements of H, being even permutations, fix Δ. Now E is Galois over F, so we have

$$(F(\Delta) : F) \leq (\mathrm{Fix}(H) : F) = |G : H| \leq 2.$$

If $H = G$, it follows that $F = F(\Delta) = \mathrm{Fix}(H)$ and $\Delta \in F$. The statement is therefore true in this case.

Now suppose that $|G : H| = 2$ and let $\alpha \in G - H$. Then $\alpha(\Delta) = -\Delta$ as α is odd. Since char(F) $\neq 2$, we have $\Delta \neq -\Delta$ and hence $\Delta \notin F$. Therefore $(F(\Delta) : F) = 2$ and Fix$(H) = F(\Delta)$. □

Corollary (12.5.3). *With the above notation,* Gal$(f) \leq A_n$ *if and only if* $\Delta \in F$.

These ideas will now be applied to investigate the Galois groups of polynomials of low degree.

Polynomials of degree at most 4
Let F be a field such that char(F) $\neq 2$.

(i) Consider a quadratic $f = t^2 + ut + v \in F[t]$. Then $|$Gal$(f)| = 1$ or 2. By (12.5.3) $|$Gal$(f)| = 1$ precisely when $\Delta \in F$, i.e., $\sqrt{u^2 - 4v} \in F$. This is the familiar condition for f to have both its roots in F. Of course $|$Gal$(f)| = 2$ if and only if $\Delta \notin F$, which is the irreducible case.

(ii) Next let f be the cubic $t^3 + pt + q \in F[t]$. We saw that

$$\Delta = \sqrt{D} = \sqrt{-4p^3 - 27q^2}.$$

If f is reducible over F, it must have a quadratic factor f_1 and clearly Gal$(f) =$ Gal(f_1), which has order 1 or 2. Thus we can assume f is irreducible. We know from (12.2.3) that Gal$(f) \leq S_3$, and that $|$Gal$(f)|$ is divisible by 3 since it acts transitively on the roots of f. Hence Gal$(f) = A_3$ or S_3. By (12.5.3) Gal$(f) = A_3$ if and only if $\Delta \in F$; otherwise Gal$(f) = S_3$.

(iii) Finally, let f be a monic polynomial of degree 4 in $F[t]$. If f is reducible and $f = f_1 f_2$ with deg$(f_i) \leq 3$, then Gal(f) is isomorphic with a subgroup of Gal$(f_1) \times$ Gal(f_2), (see Exercise (12.2.5)). The structure of Gal(f_i) is known from (i) and (ii), so assume that f is irreducible. Then Gal$(f) \leq S_4$ and 4 divides $|$Gal$(f)|$. The subgroups of S_4 whose orders are divisible by 4 are $\mathbb{Z}_4$, V (the Klein 4-group), Dih(8), A_4 and S_4; thus Gal(f) must be one of these. In fact all five cases can occur, but we will not prove this here.

Explicit formulas for the roots of cubic and quartic equations over $\mathbb{R}$ were found in the early 16th century by Scipione del Ferro (1465–1526), Gerolamo Cardano (1501–1576), Niccolo Tartaglia (1499–1526) and Lodovico Ferrari (1522–1565). An interesting account of their discoveries and of the mathematical life of the times can be found in [21].

Exercises (12.5).
(1) Find the Galois groups of the following quadratic polynomials over $\mathbb{Q}$: (i) $t^2 - 5t + 6$, (ii) $t^2 + 5t + 1$, (iii) $(t + 1)^2$.
(2) Find the Galois group of the following cubic polynomials over $\mathbb{Q}$: (i) $t^3 + 4t^2 + 2t - 7$; (ii) $t^3 - t - 1$; (iii) $t^3 - 3t + 1$; (iv) $t^3 + 6t^2 + 11t + 5$.

(3) Let f be a cubic polynomial over $\mathbb{Q}$ with discriminant D. Show that f has three real roots if and only if $D \geq 0$. Apply this to the polynomial $t^3 + pt + q$.

(4) Let f be an irreducible quartic polynomial over $\mathbb{Q}$ with exactly two real roots. Show that $\mathrm{Gal}(f) \simeq \mathrm{Dih}(8)$ or S_4.

(5) (*How to solve cubic equations*). Let $f = t^3 + pt + q \in \mathbb{R}[t]$. The following procedure, due essentially to Scipione del Ferro, will produce a root of f.

    (i) If $t = u - v$ is a root of f, show that $(u^3 - v^3) + (p - 3uv)(u - v) = -q$.

    (ii) Find a root of the form $u - v$ by solving the equations $u^3 - v^3 = -q$ and $uv = \frac{p}{3}$ to obtain a quadratic equation for u^3.

(6) The procedure of Exercise (12.5.5) yields one root $x_1 = u - v$ of $f = t^3 + pt + q$. Show that two further two roots of f are $x_2 = \omega u - \omega^2 v$ and $x_3 = \omega^2 u - \omega v$ where $\omega = e^{2\pi i/3}$. (These are known as *Cardano's formulas*.) Keep in mind that every cubic equation has at least one real root.

(7) Use the methods of the last two exercises to find the roots of the polynomial $t^3 + 3t + 1$.

(8) Solve the cubic equation $t^3 + 3t^2 + 6t + 3 = 0$ by first transforming it to one of the form $t'^3 + pt' + q = 0$.

13 Tensor Products

The tensor product is a very widely used construction in algebra which can be applied to modules, linear operators and matrices. We will begin by describing the tensor product of modules: here the distinction between left and right modules is essential.

13.1 Definition of the tensor product

Let R be an arbitrary ring and let M_R and $_RN$ be right and left R-modules as indicated. Denote by F the free abelian group whose basis is the set product

$$M \times N = \{(a, b) \mid a \in M, b \in N\}.$$

Thus each element f of F can be uniquely written in the form $f = \sum_{i=1}^{k} \ell_i(a_i, b_i)$ where $\ell_i \in \mathbb{Z}$, $a_i \in M$, $b_i \in N$. Define S to be the (additive) subgroup of F generated by all elements of the forms
(i) $(a_1 + a_2, b) - (a_1, b) - (a_2, b)$,
(ii) $(a, b_1 + b_2) - (a, b_1) - (a, b_2)$,
(iii $(a \cdot r, b) - (a, r \cdot b)$,

where $a, a_i \in M$, $b, b_i \in N$ and $r \in R$. Then the *tensor product* of M and N is defined to be the quotient group

$$M \otimes_R N = F/S.$$

Thus $M \otimes_R N$ is an abelian group generated by all elements of the form

$$a \otimes b = (a, b) + S, \quad (a \in M, \ b \in N):$$

here the elements $a \otimes b$ are called *tensors*.

The reason why one passes to the quotient group is that in this quotient the cosets arising from elements with the forms (i), (ii), (iii) indicated above will vanish. Thus the effect of passing to the quotient group is to enforce linearity on both arguments. This is formalized in the next result, which is an immediate consequence of the definition of tensors and the tensor product. It demonstrates the bilinear nature of tensor products.

(13.1.1). *Let M_R and $_RN$ be modules over a ring R. In the tensor product $M \otimes_R N$ the following rules are valid:*
(i) $(a_1 + a_2) \otimes b = a_1 \otimes b + a_2 \otimes b$;
(ii) $a \otimes (b_1 + b_2) = a \otimes b_1 + a \otimes b_2$;
(iii) $(a \cdot r) \otimes b = a \otimes (r \cdot b)$

where $a, a_i \in M$, $b, b_i \in N$, $r \in R$.

https://doi.org/10.1515/9783110691160-013

We record two simple consequences of (13.1.1):

$$0_M \otimes b = 0_{M \otimes_R N} = a \otimes 0_N, \quad (a \in M, \ b \in N).$$

These follow from (i) and (ii) on setting $a_1 = 0_M = a_2$ and $b_1 = 0_N = b_2$, respectively. It should be stressed that at this point the tensor product is only an abelian group: later we will see how it can be given a module structure.

When $R = \mathbb{Z}$, which is a very common case, it is usual to write

$$M \otimes N$$

instead of $M \otimes_{\mathbb{Z}} N$.

The mapping property of tensor products

We continue the previous notation with modules M_R and $_R N$ over a ring R. A critical property of the tensor product $M \otimes_R N$ is a certain *mapping property*; this involves the concept of a middle linear map, which will now be explained.

Let A be an (additively written) abelian group: a mapping $\alpha : M \times N \to A$ is said to be *R-middle linear* if it has the three properties listed below for all $a, a_i \in M$, $b, b_i \in N$, $r \in R$:

(i) $\alpha((a_1 + a_2, b)) = \alpha((a_1, b)) + \alpha((a_2, b))$;
(ii) $\alpha((a, b_1 + b_2)) = \alpha((a, b_1)) + \alpha((a, b_2))$;
(iii) $\alpha((a \cdot r, b)) = \alpha((a, r \cdot b))$.

For example, the *canonical mapping* $v : M \times N \to M \otimes N$ in which $v((a, b)) = a \otimes b$ is middle linear by virtue of the properties listed in (13.1.1). The crucial mapping property of the tensor product is as follows.

(13.1.2). *Let M_R and $_R N$ be modules over a ring R.*

(i)   *Given a middle linear map $\alpha : M \times N \to A$ with A an abelian group, there is a unique group homomorphism $\beta : M \otimes_R N \to A$ such that $\alpha = \beta v$ where $v : M \times N \to M \otimes_R N$ is the canonical middle linear map in which $(a, b) \mapsto a \otimes b$.*

(ii) *Conversely, if T is an abelian group and $\phi : M \times N \to T$ is a middle linear map such that the pair (T, ϕ) has the mapping property in (i), then $T \simeq M \otimes_R N$.*

The assertion of (13.1.2)(i) is most easily remembered from the triangle diagram below:

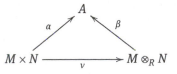

Indeed the relation $\alpha = \beta v$ expresses the *commutativity of the diagram*, in the sense that if we start with an element $x \in M \times N$ and follow it in both directions around the triangle, applying the maps indicated by the arrows, we end up with the same element of A, namely $\alpha(x) = \beta v(x)$. Commutative diagrams are widely used in algebra. For the generalization to categories see Section 16.1.

When (i) and (ii) of (13.1.2) are combined, they demonstrate that the tensor product $M \otimes_R N$, together with the canonical middle linear mapping v, is characterized by the mapping property. Another way of looking at the mapping property is that if a function α from $M \times N$ to an abelian group A is middle linear, then it can be "extended" to a homomorphism from $M \otimes_R N$ to A. It is this form of the mapping property that makes it an indispensable tool in working with tensor products.

Proof of (13.1.2). (i) Let F be the free abelian group with basis $M \times N$. By (9.1.17) there is a homomorphism $\beta' : F \to A$ such that $\beta'((a, b)) = \alpha((a, b))$ for all $a \in M$, $b \in N$. By definition $M \otimes_R N = F/S$ where S is generated by all elements of F of the three types in the definition of the tensor product. Now β' maps each of the listed generators of S to 0 since α is middle linear, and hence $\beta'(s) = 0$ for all $s \in S$. This observation allows us to define *in a unique manner* a function

$$\beta : M \otimes_R N \to A$$

by the rule $\beta(f + S) = \beta'(f)$, $(f \in F.)$ Notice that β is a homomorphism since β' is one. Furthermore

$$\beta v((a, b)) = \beta((a, b) + S) = \beta'((a, b)) = \alpha((a, b))$$

for all $a \in M$, $b \in N$. Therefore $\beta v = \alpha$.

The uniqueness of β remains to be established. Suppose that $\bar{\beta} : M \otimes_R N \to A$ is another homomorphism with the property $\bar{\beta} v = \alpha$. Then $\beta v = \bar{\beta} v$, so that β and $\bar{\beta}$ agree on $\mathrm{Im}(v)$, i. e., on the set of all tensors. But the tensors generate $M \otimes_R N$, so it follows that $\beta = \bar{\beta}$, which completes the proof of (i).

(ii) By the mapping property for the pair $(M \otimes_R N, v)$ there is a homomorphism $\beta : M \otimes_R N \to T$ such that $\phi = \beta v$, and by the mapping property for (T, ϕ) there is a homomorphism $\bar{\beta} : T \to M \otimes_R N$ such that $v = \bar{\beta} \phi$. Thus we have the two commutative triangles that follow:

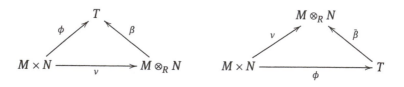

Therefore $\beta\bar{\beta}\phi = \beta v = \phi$ and $\bar{\beta}\beta v = \bar{\beta}\phi = v$: these equations express the commutativity of the two triangles below

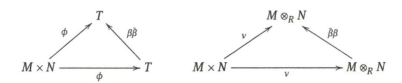

But clearly these triangles will also commute if $\bar{\beta}\beta$ and $\beta\bar{\beta}$ are replaced by the appropriate identity maps. At this point the uniqueness clause in the mapping property is invoked to show that $\bar{\beta}\beta$ and $\beta\bar{\beta}$ are identity maps on $M \otimes_R N$ and T respectively. Hence $\bar{\beta}$ is an isomorphism and $T \overset{R}{\simeq} M \otimes_R N$. $\qquad\square$

Tensor products and homomorphisms

When homomorphisms between pairs of modules are given, there is an induced homomorphism between the tensor products of these modules.

(13.1.3). *Let there be given modules M_R, M'_R and $_RN$, $_RN'$ over a ring R, together with R-module homomorphisms $\alpha : M \to M'$ and $\beta : N \to N'$. Then there is a homomorphism of groups $\alpha \otimes \beta : M \otimes_R N \to M' \otimes_R N'$ such that*

$$\alpha \otimes \beta\left(\sum_{i=1}^{k} \ell_i(a_i \otimes b_i) \right) = \sum_{i=1}^{k} \ell_i(\alpha(a_i) \otimes \beta(b_i)),$$

where $a_i \in M$, $b_i \in N$, $\ell_i \in \mathbb{Z}$.

Proof. The first point to note is that one cannot simply use the formula in the statement of the theorem as the definition of $\alpha \otimes \beta$, the reason being that there is no unique expressibility for elements of $M \otimes_R N$ as linear combinations of tensors. However, an indirect approach using the mapping property succeeds.

To exploit this property we first introduce a function $\theta : M \times N \to M' \otimes_R N'$ by defining $\theta((a, b)) = \alpha(a) \otimes \beta(b)$. Now check the middle linearity of θ, which is easy. By the mapping property there is a group homomorphism $\phi : M \otimes_R N \to M' \otimes_R N'$ such that $\phi v = \theta$ where $v : M \times N \to M \otimes_R N$ is the canonical middle linear map in which $(a, b) \mapsto a \otimes b$. Thus the triangle below commutes:

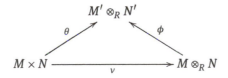

Now define $\alpha \otimes \beta$ to be ϕ and check that this has the required property:

$$\phi\left(\sum_{i=1}^{k} \ell_i(a_i \otimes b_i)\right) = \sum_{i=1}^{k} \ell_i\phi(a_i \otimes b_i) = \sum_{i=1}^{k} \ell_i\phi v((a_i, b_i)) = \sum_{i=1}^{k} \ell_i\theta((a_i, b_i)),$$

which equals $\sum_{i=1}^{k} \ell_i(\alpha(a_i) \otimes \beta(b_i))$ by definition of θ. $\qquad\square$

This use of the mapping property is typical in situations where a mapping from a tensor product is to be defined by its effect on tensors, but the problem of non-uniqueness of expression in terms of tensors has to be addressed.

Important special cases of (13.1.3) arise when α or β is an identity map. Specifically, given module homomorphisms $\alpha : M_R \rightarrow M'_R$ and $\beta : {}_R N \rightarrow {}_R N'$, we can form the *induced homomorphisms*

$$\alpha_* = \alpha \otimes \mathrm{id}_N \quad \text{and} \quad \beta_* = \mathrm{id}_M \otimes \beta.$$

Thus α_* and β_* are homomorphisms from $M \otimes_R N$ to $M' \otimes_R N$ and $M \otimes_R N$ to $M \otimes_R N'$ respectively. Moreover, $\alpha_*(a \otimes b) = \alpha(a) \otimes b$ and $\beta_*(a \otimes b) = a \otimes \beta(b)$ where $a \in M, b \in N$.

Tensor products as modules

As has been observed, in general a tensor product is an abelian group with no module structure other than over $\mathbb{Z}$. However, when the modules in a tensor product have additional module structures, this is inherited by the tensor product.

(13.1.4). *Let ${}_S M_R$ and ${}_R N_T$ be bimodules over rings R, S, T as indicated. Then $M \otimes_R N$ is an (S, T)-bimodule with respect to the ring actions $s \cdot (a \otimes b) = (s \cdot a) \otimes b$ and $(a \otimes b) \cdot t = a \otimes (b \cdot t)$ where $a \in M, b \in N, s \in S, t \in T$.*

Proof. Fix s in S and consider the mapping $\alpha^{(s)} : M_R \rightarrow M_R$ in which $\alpha^{(s)}(a) = s \cdot a$. This is a homomorphism of right R-modules, as an easy check reveals. By (13.1.3) we can form the induced homomorphism $(\alpha^{(s)})_* : M \otimes_R N \rightarrow M \otimes_R N$. This enables us to define a left action of S on $M \otimes_R N$ by $s \cdot x = (\alpha^{(s)})_*(x)$ for $x \in M \otimes_R N$. This is certainly well defined, but we still need to verify the module axioms. First note that $s \cdot (a \otimes b) = \alpha^{(s)}_*(a \otimes b) = (\alpha^{(s)}(a)) \otimes b = (s \cdot a) \otimes b$ where $a \in M, b \in N$.

Turning to the module axioms, we have $s \cdot (x_1 + x_2) = (\alpha^{(s)})_*(x_1 + x_2) = (\alpha^{(s)})_*(x_1) + (\alpha^{(s)})_*(x_2) = s \cdot x_1 + s \cdot x_2$, where $s \in S$ and $x_i \in M \otimes_R N$, since $\alpha^{(s)}$ is a homomorphism. Next let $s_i \in S$; then $(s_1 + s_2) \cdot (a \otimes b) = ((s_1 + s_2) \cdot a) \otimes b = (s_1 \cdot a + s_2 \cdot a) \otimes b = s_1 \cdot (a \otimes b) + s_2 \cdot (a \otimes b)$. This implies that $(s_1 + s_2) \cdot x = s_1 \cdot x + s_2 \cdot x$ for all x in $M \otimes_R N$, since the latter is generated by the tensors $a \otimes b$ and $(\alpha^{(s)})_*$ is a homomorphism. As for the last module axiom, $s_1 \cdot (s_2 \cdot (a \otimes b)) = s_1 \cdot ((s_2 \cdot a) \otimes b) = (s_1 \cdot (s_2 \cdot a)) \otimes b = ((s_1 s_2) \cdot a) \otimes b = (s_1 s_2) \cdot (a \otimes b)$. It follows that $s_1 \cdot (s_2 \cdot x) = (s_1 s_2) \cdot x$ for all $x \in M \otimes_R N$, since $M \otimes_R N$ is generated by the tensors $a \otimes b$.

The right action of T arises in a similar fashion from the map $\beta^{(t)} :_R N \to_R N$, $(t \in T)$, in which $\beta^{(t)}(b) = b \cdot t$ for $b \in N$. Thus $x \cdot t$ is defined to be $(\beta^{(t)})_*(x)$. To complete the proof the reader should check the module axioms and also verify the bimodule condition, $s \cdot (x \cdot t) = (s \cdot x) \cdot t$ for $s \in S, t \in T, x \in M \otimes_R N$, noting that it is enough to do this when x is a tensor. □

We remark that there are versions of (13.1.4) applicable to the module situations $_S M_R$, $_R N$ and M_R, $_R N_T$: in these instances $M \otimes_R N$ is only either a left S-module or a right T-module respectively.

In the case of a commutative ring there is no difference between left and right modules, so the tensor product is always a bimodule. Indeed we have the following obvious result.

(13.1.5). *Let M and N be modules over a commutative ring R. Then $M \otimes_R N$ is an (R, R)-bimodule. Furthermore,*

$$r \cdot (a \otimes b) = (r \cdot a) \otimes b = a \otimes (b \cdot r) = (a \otimes b) \cdot r$$

where $a \in M, b \in N, r \in R$.

Exercises (13.1).

(1) Let R, S, T be rings and $_S M_R$, $_R N_T$ modules. State what module structures the following tensor products possess and write down the module action in each case: $R \otimes_R N$ and $M \otimes_R R$.

(2) Let M and N be R-modules where R is a commutative ring. Prove that $M \otimes_R N \overset{R}{\simeq} N \otimes_R M$.

(3) Let A be an abelian torsion group, i. e., each element of A has finite order. Prove that $A \otimes \mathbb{Q} = 0$.

(4) Let A and B be abelian torsion groups such that elements from A and B have relatively prime orders. Prove that $A \otimes B = 0$.

(5) Let R be a ring and let M_R and $_R N$ be modules. Prove that $M \otimes_R N \simeq N \otimes_{R^{op}} M$. (Here R^{op} is the opposite ring of R – see Section 9.1.)

(6) Let $\alpha : A \to A_1, \beta : B \to B_1, \gamma : A_1 \to A_2, \delta : B_1 \to B_2$ be module homomorphisms. Prove that $(\gamma \otimes \delta)(\alpha \otimes \beta) = (\gamma\alpha) \otimes (\delta\beta)$. [Hint: use (13.1.3).]

(7) Let A be the multiplicative group of all complex p-power roots of unity where p is a prime. Prove that $A \otimes A = 0$.

(8) Let R be a ring and M a right R-module. Also let $\alpha : A \to B$ and $\beta : B \to C$ be homomorphisms of left R-modules. Form the induced homomorphism $\alpha_* = \mathrm{id}_M \otimes \alpha$ and similarly form β_* and $(\beta\alpha)_*$. Prove that $(\beta\alpha)_* = \beta_*\alpha_*$. [Hint: use Exercise (13.1.6).]

13.2 Properties of tensor products

In this section we identify the fundamental properties of tensor products, which aid in their calculation.

(13.2.1). *Let R be a ring with identity and let M_R and $_RN$ be modules. Then*

(i) $M \otimes_R R \overset{R}{\simeq} M$,

(ii) $R \otimes_R N \overset{R}{\simeq} N$,

via the respective isomorphisms in which $a \otimes r \mapsto a \cdot r$ and $r \otimes b \mapsto r \cdot b$, $(a \in M, b \in N, r \in R)$.

Proof. First observe that $M \otimes_R R$ and $R \otimes_R N$ are respectively a right R-module and a left R-module by (13.1.4) since R is a bimodule. Only the first isomorphism will be proved. Consider the map from $M \times R$ to M defined by $(a, r) \mapsto a \cdot r$. This is clearly middle linear, so by the mapping property there is a group homomorphism $\alpha : M \otimes_R R \to M$ such that $\alpha(a \otimes r) = a \cdot r$. In fact α is a homomorphism of right R-modules because

$$\alpha\left(\left(\sum_{i=1}^{k} \ell_i(a_i \otimes r_i) \right) \cdot r \right) = \alpha\left(\sum_{i=1}^{k} \ell_i(a_i \otimes (r_i r)) \right) = \sum_{i=1}^{k} \ell_i \alpha(a_i \otimes (r_i r))$$

$$= \sum_{i=1}^{k} \ell_i(a_i \cdot (r_i r)) = \left(\sum_{i=1}^{k} \ell_i(a_i \cdot r_i) \right) \cdot r = \left(\alpha\left(\sum_{i=1}^{k} \ell_i(a_i \otimes r_i) \right) \right) \cdot r,$$

where $a_i \in M$, $r, r_i \in R$, $\ell_i \in \mathbb{Z}$.

To show that α is an isomorphism we produce an inverse function. Define $\beta : M \to M \otimes_R R$ by $\beta(a) = a \otimes 1_R$. This is certainly well defined and a simple check reveals that $\alpha\beta$ and $\beta\alpha$ are identity functions. The reader is urged to supply the details. Hence $\beta = \alpha^{-1}$. $\qquad\square$

(13.2.2) (Associativity of tensor products). *Let R and S be rings and L_R, $_RM_S$, $_SN$ modules as indicated. Then there is an isomorphism of groups*

$$\alpha : (L \otimes_R M) \otimes_S N \to L \otimes_R (M \otimes_S N)$$

such that $\alpha((a \otimes b) \otimes c) = a \otimes (b \otimes c)$ where $a \in L$, $b \in M$, $c \in N$.

Proof. First note that all of these tensor products exist. Choose and fix $c \in N$; then observe that the assignment $(a, b) \mapsto a \otimes (b \otimes c)$, where $a \in L$, $b \in M$, is an R-middle linear map from $L \times M$ to $L \otimes_R (M \otimes_S N)$. By the mapping property there is a group homomorphism $\beta_c : L \otimes_R M \to L \otimes_R (M \otimes_S N)$ such that $\beta_c(a \otimes b) = a \otimes (b \otimes c)$.

Next the assignment $(x, c) \mapsto \beta_c(x)$ determines an S-middle linear map from $(L \otimes_R M) \times N$ to $L \otimes_R (M \otimes_S N)$ – notice that $\beta_{c_1 + c_2} = \beta_{c_1} + \beta_{c_2}$. Hence there is a homomorphism $\alpha : (L \otimes_R M) \otimes_S N \to L \otimes_R (M \otimes_S N)$ such that $\alpha((a \otimes b) \otimes c) = \beta_c(a \otimes b) = a \otimes (b \otimes c)$. By a similar argument there is a homomorphism $\gamma : L \otimes_R (M \otimes_S N) \to (L \otimes_R M) \otimes_S N$ such that $\gamma(a \otimes (b \otimes c)) = (a \otimes b) \otimes c$. Since α and γ are inverse functions, α is an isomorphism. $\qquad\square$

Here it should be noted that if there is additional module structure in (13.2.2), the map α may be a module isomorphism. Specifically, if we have $_QL_R$ or $_SN_T$ with rings Q and T, then α is a isomorphism of left Q-modules or right T-modules respectively. For example, in the first case, if $a \in L$, $b \in M$, $c \in N$, $q \in Q$, then $\alpha(q \cdot ((a \otimes b) \otimes c)) = \alpha(((q \cdot a) \otimes b) \otimes c) = (q \cdot a) \otimes (b \otimes c) = q \cdot (a \otimes (b \otimes c)) = q \cdot \alpha((a \otimes b) \otimes c)$, which implies that α is a Q-module homomorphism.

(13.2.3) (Distributivity of tensor products). *Let R be a ring and let L_R, $_RM$, $_RN$ be modules. Then there is an isomorphism of groups*

$$\alpha : L \otimes_R (M \oplus N) \to (L \otimes_R M) \oplus (L \otimes_R N)$$

such that $\alpha(a \otimes (b \oplus c)) = (a \otimes b) \oplus (a \otimes c)$, where $a \in L$, $b \in M$, $c \in N$.

Here we are writing $b \oplus c$ for $(b, c) \in M \oplus N$, etc, as it makes the notation more transparent.

Proof. Let $a \in L$, $b \in M$, $c \in N$. Then the assignment $(a, b \oplus c) \mapsto (a \otimes b) \oplus (a \otimes c)$ determines a middle linear map from $L \times (M \oplus N)$ to $(L \otimes_R M) \oplus (L \otimes_R N)$, so there is a group homomorphism $\alpha : L \otimes_R (M \oplus N) \to (L \otimes_R M) \oplus (L \otimes_R N)$ such that $\alpha(a \otimes (b \oplus c)) = (a \otimes b) \oplus (a \otimes c)$.

Next the canonical injections $\iota_M : M \to M \oplus N$ and $\iota_N : N \to M \oplus N$ lead to induced homomorphisms $(\iota_M)_* : L \otimes_R M \to L \otimes_R (M \oplus N)$ and $(\iota_N)_* : L \otimes_R N \to L \otimes_R (M \oplus N)$. Combine $(\iota_M)_*$ and $(\iota_N)_*$ to produce a homomorphism $\beta : (L \otimes_R M) \oplus (L \otimes_R N) \to L \otimes_R (M \oplus N)$ which sends $(a \otimes b) \oplus 0$ to $a \otimes (b \oplus 0)$ and $0 \oplus (a \otimes c)$ to $a \otimes (0 \oplus c)$. Hence $\beta((a \otimes b) \oplus (a \otimes c)) = a \otimes (b \oplus c)$. Since α and β are inverse maps, α is an isomorphism. $\square$

Once again, given the extra module structure $_SL_R$ or $_RM_T$ and $_RN_T$, it is easy to verify that α is a left S- or a right T-module isomorphism respectively.

Tensor products of quotients
There is a useful technique for computing the tensor product of two quotient modules. Let R be a ring and let M_R, $_RN$ be modules with respective submodules M_0 and N_0. Define

$$S = \langle a \otimes b \mid a \in M_0 \text{ or } b \in N_0 \rangle,$$

which is a subgroup of $M \otimes_R N$. With this notation we have the fundamental result that follows.

(13.2.4). *There is an isomorphism of groups*

$$\alpha : (M/M_0) \otimes_R (N/N_0) \to (M \otimes_R N)/S$$

such that $\alpha((a + M_0) \otimes (b + N_0)) = a \otimes b + S$.

Proof. In the first place the assignment $(a + M_0, b + N_0) \mapsto a \otimes b + S$ gives rise to a well defined middle linear mapping from $M/M_0 \times N/N_0$ to $(M \otimes_R N)/S$, by definition of S. Hence there is a homomorphism $\alpha : (M/M_0) \otimes_R (N/N_0) \to (M \otimes_R N)/S$ such that $\alpha((a + M_0) \otimes (b + N_0)) = a \otimes b + S$. Next let $\pi : M \to M/M_0$ and $\sigma : N \to N/N_0$ denote the canonical homomorphisms. Now form the homomorphism $\bar{\beta} = \pi \otimes \sigma$; thus $\bar{\beta}$ sends $a \otimes b$ to $(a + M_0) \otimes (b + N_0)$. Observe that $\bar{\beta}$ maps each generator of S to 0, so that $\bar{\beta}|_S = 0$. Therefore we can define unambiguously a mapping

$$\beta : (M \otimes_R N)/S \to (M/M_0) \otimes_R (N/N_0)$$

by $\beta(x + S) = \bar{\beta}(x)$. Note that $\beta(a \otimes b + S) = \bar{\beta}(a \otimes b) = (a + M_0) \otimes (b + N_0)$. Finally, α and β are inverse maps, so α is an isomorphism. □

As usual when additional module structure in M or N is present, α is a module isomorphism. A first application of (13.2.4) is to compute tensor products in which one factor is a cyclic module. But first recall from (9.1.8) that if R is a ring with identity, a cyclic left R-module is isomorphic with a module $_RR/I$ where I is a left ideal of R, and there is a corresponding statement for cyclic right modules.

(13.2.5). *Let R be a ring with identity and let I, J be left and right ideals of R respectively. Let M_R and $_RN$ be modules. Then*
(i) $M \otimes_R (_RR/I) \simeq M/(M \cdot I)$;
(ii) $(R_R/J) \otimes_R N \simeq N/(J \cdot N)$.

In the statement of this result $M \cdot I$ denotes the subgroup generated by all elements of the form $a \cdot i$ where $a \in M$ and $i \in I$, with a similar explanation for $J \cdot N$.

Proof. Only (i) will be proved. Apply (13.2.4) with $M_0 = 0$ and $N_0 = I$. Then $M \otimes_R (_RR/I) \simeq (M \otimes_R R)/S$ and it is just a matter of identifying the subgroup $S = \langle a \otimes i \mid a \in M, i \in I \rangle$. By (13.2.1) the assignment $a \otimes r \mapsto a \cdot r$ determines an isomorphism $\alpha : M \otimes_R {}_RR \to M$. The image of S under α is generated by the elements $a \cdot i$, where $a \in M, i \in I$; therefore $\alpha(S) = M \cdot I$ and $M \otimes_R (_RR/I) \simeq M/(M \cdot I)$. □

Corollary (13.2.6). *If I and J are respectively left and right ideals of a ring R with identity, the mapping $(r_1 + J) \otimes (r_2 + I) \mapsto r_1 r_2 + (I + J)$ yields an isomorphism*

$$(R_R/J) \otimes_R (_RR/I) \simeq R/(I + J).$$

Moreover, if I and J are two sided ideals, the isomorphism is of (R, R)-bimodules.

Proof. From (13.2.5)(i) we have

$$(R_R/J) \otimes_R (_RR/I) \simeq (R/J)/((R/J) \cdot I) = (R/J)/(I + J/J),$$

which by (9.1.7) is isomorphic with $R/(I + J)$. Composition of the isomorphisms yields the map stated. If I and J are two sided ideals, each module is an (R, R)-bimodule and clearly the isomorphism is of R-modules. □

For example, if m, n are positive integers with $d = \gcd\{m, n\}$, then $d\mathbb{Z} = m\mathbb{Z} + n\mathbb{Z} = (m) + (n)$ and it follows from (13.2.6) that

$$\mathbb{Z}_m \otimes \mathbb{Z}_n = \mathbb{Z}/(m) \otimes \mathbb{Z}/(n) \simeq \mathbb{Z}/((m) + (n)) = \mathbb{Z}/(d) = \mathbb{Z}_d. \tag{13.1}$$

Example (13.2.1). Let $A = \mathbb{Z} \oplus \mathbb{Z}_3 \oplus \mathbb{Z}_{5^2}$ and $B = \mathbb{Z} \oplus \mathbb{Z}_{3^2} \oplus \mathbb{Z}_{5^2} \oplus \mathbb{Z}_7$. Applying the distributive property together with (13.2.1) and the isomorphism (13.1), we obtain

$$A \otimes B \simeq \mathbb{Z} \oplus \mathbb{Z}_3 \oplus \mathbb{Z}_3 \oplus \mathbb{Z}_{3^2} \oplus \mathbb{Z}_{5^2} \oplus \mathbb{Z}_{5^2} \oplus \mathbb{Z}_{5^2} \oplus \mathbb{Z}_7.$$

Tensor products of free modules

A tensor product of free modules over a commutative ring with identity is in fact always a free module. For simplicity of presentation we will discuss only the case where the free modules are finitely generated.

(13.2.7). *Let R be a commutative ring with identity and let M and N be finitely generated free R-modules with respective bases $\{x_1, x_2, \ldots, x_m\}$ and $\{y_1, y_2, \ldots, y_n\}$. Then $M \otimes_R N$ is a free R-module with basis $\{x_i \otimes y_j \mid i = 1, 2, \ldots, m, j = 1, 2, \ldots, n\}$.*

Proof. We have $N = R \cdot y_1 \oplus R \cdot y_2 \oplus \cdots \oplus R \cdot y_n$ and hence by the distributive law

$$M \otimes_R N \overset{R}{\simeq} (M \otimes_R (R \cdot y_1)) \oplus (M \otimes_R (R \cdot y_2)) \oplus \cdots \oplus (M \otimes_R (R \cdot y_n)).$$

Now $R \cdot y_j \overset{R}{\simeq} R$: for $r \cdot y_j = 0$ implies that $r = 0$, since the y_j form a basis. Thus $M \otimes_R (R \cdot y_j) \overset{R}{\simeq} M \otimes_R R \overset{R}{\simeq} M$, and for a fixed j the image of $x_i \otimes y_j$ under the composite of these isomorphisms is x_i. Therefore the $x_i \otimes y_j$, $i = 1, 2, \ldots, m$, are R-linearly independent, so they form a basis of $M \otimes_R (R \cdot y_j)$, which implies the result. □

Corollary (13.2.8). *Let M and N be free modules of finite rank over R, a commutative ring with identity. Then $\operatorname{rank}(M \otimes_R N) = \operatorname{rank}(M) \cdot \operatorname{rank}(N)$.*

For the of rank of a free module see (9.1.19) and its sequel. Note that for a vector space rank equals dimension; thus if V and W are finite dimensional vector spaces over a field F, then $V \otimes_F W$ is a finite dimensional F-space and $\dim(V \otimes_F W) = \dim(V) \cdot \dim(W)$.

Tensor products of matrices

We have seen how to form the tensor product of module homomorphisms in (13.1.3). The close connection between matrices and linear mappings suggests that there should be a corresponding way to form tensor products of matrices.

Let A and B be $m \times n$ and $p \times q$ matrices respectively over a field F. Then there are corresponding linear transformations $\alpha : F^n \to F^m$ and $\beta : F^q \to F^p$ defined by equations $\alpha(X) = AX$ and $\beta(Y) = BY$. Let $E_i^{(n)}$ denote the ith column of the $n \times n$ identity matrix I_n. Thus $\{E_i^{(n)} \mid i = 1, \ldots, n\}$ is the standard basis of F^n. The linear transformation α is represented with respect to the bases $\{E_i^{(n)}\}$ and $\{E_j^{(m)}\}$ by the matrix A. There is a similar statement for β and B.

By definition of the linear mapping $\alpha \otimes \beta : F^n \otimes_F F^q \to F^m \otimes_F F^p$,

$$\alpha \otimes \beta(E_i^{(n)} \otimes E_j^{(q)}) = \alpha(E_i^{(n)}) \otimes \beta(E_j^{(q)}),$$

which equals

$$\sum_{k=1}^{m} a_{ki} E_k^{(m)} \otimes \sum_{\ell=1}^{p} b_{\ell j} E_\ell^{(p)} = \sum_{k=1}^{m} \sum_{\ell=1}^{p} a_{ki} b_{\ell j} (E_k^{(m)} \otimes E_\ell^{(p)}).$$

By (13.2.7) the $E_i^{(n)} \otimes E_j^{(q)}$ form a basis for $F^n \otimes_F F^q$, as do the $E_k^{(m)} \otimes E_\ell^{(p)}$ for $F^m \otimes_F F^p$. Let these bases be ordered lexicographically, i. e., by first subscript, then second subscript. With this choice of ordered bases we can read off the $mp \times nq$ matrix M which represents the linear mapping $\alpha \otimes \beta$. The rows of M are labelled by the pairs (k, ℓ), $1 \leq k \leq m$, $1 \leq \ell \leq p$, and the columns by the pairs (i, j), $1 \leq i \leq n$, $1 \leq j \leq q$. Therefore the $((k, \ell), (i, j))$ entry of M is

$$a_{ki} b_{\ell j}.$$

The foregoing discussion suggests that we define the tensor product $A \otimes B$ of A and B to be the $mp \times nq$ matrix M. In essence the entries of $A \otimes B$ are formed by taking all possible products of an entry of A and an entry of B. Writing the matrix in block form, we obtain the more easily remembered formula

$$M = A \otimes B = \begin{bmatrix} a_{11}B & a_{12}B & \ldots & a_{1n}B \\ a_{21}B & a_{22}B & \ldots & a_{2n}B \\ \cdot & \cdot & \ldots & \cdot \\ a_{m1}B & a_{m2}B & \ldots & a_{mn}B \end{bmatrix}.$$

The tensor product of matrices is sometimes called the *Kronecker product*.[1]

Example (13.2.2). Consider the matrices

$$A = \begin{bmatrix} a_{11} & a_{12} \\ a_{21} & a_{22} \end{bmatrix} \quad \text{and} \quad B = \begin{bmatrix} b_{11} & b_{12} \\ b_{21} & b_{22} \end{bmatrix}.$$

---

1 Leopold Kronecker (1823–1891).

The tensor product is

$$A \otimes B = \begin{bmatrix} a_{11}b_{11} & a_{11}b_{12} & a_{12}b_{11} & a_{12}b_{12} \\ a_{11}b_{21} & a_{11}b_{22} & a_{12}b_{21} & a_{12}b_{22} \\ a_{21}b_{11} & a_{21}b_{12} & a_{22}b_{11} & a_{22}b_{12} \\ a_{21}b_{21} & a_{21}b_{22} & a_{22}b_{21} & a_{22}b_{22} \end{bmatrix}.$$

Right exactness of tensor products

The section concludes with a discussion of the right exactness property of tensor products, a fundamental result that is used constantly in advanced work. Exact sequences of modules were defined in Section 9.1.

(13.2.9). *Let M_R and $_RN$ be modules over a ring R.*

(i) *Let $A \xrightarrow{\alpha} B \xrightarrow{\beta} C \to 0$ be an exact sequence of left R-modules. Then there is an exact sequence of abelian groups and induced homomorphisms*

$$M \otimes_R A \xrightarrow{\alpha_*} M \otimes_R B \xrightarrow{\beta_*} M \otimes_R C \to 0.$$

(ii) *Let $A \xrightarrow{\alpha} B \xrightarrow{\beta} C \to 0$ be an exact sequence of right R-modules. Then there is an exact sequence of abelian groups and induced homomorphisms*

$$A \otimes_R N \xrightarrow{\alpha_*} B \otimes_R N \xrightarrow{\beta_*} C \otimes_R N \to 0.$$

Proof. Only (i) will be proved. The first step is to show that β_* is surjective. Let $m \in M$ and $c \in C$. Since β is surjective, $c = \beta(b)$ for some $b \in B$. Hence

$$\beta_*(m \otimes b) = (\mathrm{id}_M \otimes \beta)(m \otimes b) = m \otimes \beta(b) = m \otimes c.$$

Since $M \otimes_R C$ is generated by the tensors $m \otimes c$, it follows that β_* is surjective.

It remains to prove that $\mathrm{Im}(\alpha_*) = \mathrm{Ker}(\beta_*)$, which is harder. In the first place, $\beta_* \alpha_* = (\beta\alpha)_* = 0_* = 0$ by Exercise (13.1.8), so that $\mathrm{Im}(\alpha_*) \subseteq \mathrm{Ker}(\beta_*)$. To establish the reverse inclusion form the commutative triangle

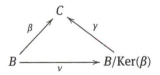

where ν is the canonical homomorphism and γ is the isomorphism in the First Isomorphism Theorem which sends $b + \mathrm{Ker}(\beta)$ to $\beta(b)$. Commutativity of the diagram is easily checked, so $\gamma\nu = \beta$. This implies that $\gamma_* \nu_* = (\gamma\nu)_* = \beta_*$. Since γ is an isomorphism, so is the induced map γ_* and hence $\mathrm{Ker}(\beta_*) = \mathrm{Ker}(\nu_*)$.

Consider the subgroup

$$S = \langle m \otimes k \mid m \in M, \; k \in \text{Ker}(\beta) \rangle.$$

Then $S = \text{Im}(\alpha_*)$ since $\text{Ker}(\beta) = \text{Im}(\alpha)$. Next $S \subseteq \text{Ker}(\nu_*)$; for, if $m \in M$ and $k \in \text{Ker}(\beta)$, we have $\nu_*(m \otimes k) = m \otimes (k + \text{Ker}(\beta)) = m \otimes 0 = 0$. Hence ν_* induces a homomorphism $\lambda : (M \otimes_R B)/S \to M \otimes_R (B/\text{Ker}(\beta))$ such that $\lambda(u + S) = \nu_*(u)$ for $u \in M \otimes_R B$. Thus $\lambda(m \otimes b + S) = \nu_*(m \otimes b) = m \otimes (b + \text{Ker}(\beta))$. By (13.2.4) there is an isomorphism $\theta : M \otimes (B/\text{Ker}(\beta)) \to (M \otimes_R B)/S$ such that $\theta(m \otimes (b + \text{Ker}(\beta))) = m \otimes b + S$. Notice that θ and λ are mutually inverse maps, so $\lambda = \theta^{-1}$ is an isomorphism. If $u \in \text{Ker}(\nu_*)$, then, since ν_* induces λ, we have $u + S \in \text{Ker}(\lambda) = 0_{M \otimes_R B/S}$ and $u \in S$. Hence $\text{Ker}(\nu_*) \subseteq S$ and finally $\text{Ker}(\beta_*) = \text{Ker}(\nu_*) \subseteq S = \text{Im}(\alpha_*)$, so that $\text{Ker}(\beta_*) = \text{Im}(\alpha_*)$, as required. $\qquad\square$

The right exactness property of tensor products should be compared with the left exactness of Hom in Chapter Nine – see (9.1.25). The "duality" between the tensor product and Hom indicated by (13.2.9) and (9.1.25) is just the beginning of a fundamental duality in homological algebra between homology and cohomology.

Exercises (13.2).

(1) Given a module $_RN_S$ where R and S are rings and R has identity, prove that $R \otimes_R N \overset{S}{\cong} N$.

(2) Simplify $(\mathbb{Z} \oplus \mathbb{Q} \oplus \mathbb{Z}_{18}) \otimes (\mathbb{Q} \oplus \mathbb{Z}_5 \oplus \mathbb{Z}_{24})$ as far as possible.

(3) Show by an example that the tensor product does not have the left exactness property, i. e., if M is a right R-module and $0 \to A \overset{\alpha}{\to} B \overset{\beta}{\to} C$ is an exact sequence of left R-modules, the induced sequence $0 \to M \otimes_R A \overset{\alpha_*}{\to} M \otimes_R B \overset{\beta_*}{\to} M \otimes_R C$ is not exact in general. [Hint: apply $\mathbb{Z}_2 \otimes -$ to the sequence $0 \to \mathbb{Z} \to \mathbb{Q} \to \mathbb{Q}/\mathbb{Z}$.]

(4) Let A and B be $m \times m$ and $n \times n$ matrices over a field. Prove that $\det(A \otimes B) = (\det(A))^n(\det(B))^m$. Deduce that the tensor product of non-singular matrices is non-singular. [Hint: define $\bar{A}$ to be the $mn \times mn$ block matrix whose (i, j) block is $a_{ij}I_n$ and let $B^{\sharp}$ be the $mn \times mn$ block matrix with B on the diagonal and 0 elsewhere. Show that $A \otimes B = \bar{A}B^{\sharp}$. Then take the determinant of both sides.]

(5) Let Q, R, S be rings and $_QL_R$, $_RM_S$, $_SN$ modules as indicated. Prove that there is an isomorphism of left Q-modules $\alpha : (L \otimes_R M) \otimes_S N \to L \otimes_R (M \otimes_S N)$.

(6) Let $_RN$ be a module over an arbitrary ring R. Suppose that $A \overset{\alpha}{\to} B \overset{\beta}{\to} C \to 0$ is an exact sequence of right R-modules. Prove that the sequence of abelian groups and induced homomorphisms $A \otimes_R N \overset{\alpha_*}{\to} B \otimes_R N \overset{\beta_*}{\to} C \otimes_R N \to 0$ is exact.

(7) (Adjoint associativity). Let R and S be rings and A_R, $_RB_S$, C_S modules. (i) Explain why $\text{Hom}_S(B, C)$ is a right R-module. (ii) Establish the isomorphism $\text{Hom}_S(A \otimes_R B, C) \simeq \text{Hom}_R(A, \text{Hom}_S(B, C))$.

13.3 Extending the ring of operators

Suppose we have a module over a ring R: is there a way to make it into a module over a different ring S? Of course the question is vague, but one situation in which this is clearly possible is if a ring homomorphism $\phi : S \rightarrow R$ is given. For, if M is a left R-module, a left action of S on M can be defined by the rule $s \cdot a = \phi(s) \cdot a$ for $s \in S$ and $a \in M$. The simple task of verifying the module axioms is left to the reader.

It is a less trivial exercise to go in the opposite direction: let M be a left R-module and let $\phi : R \rightarrow S$ be a ring homomorphism. The question is: how can one make M into a left S-module? At this point tensor products come to our aid. First observe that S is a (S, R)-bimodule where the left action comes from the ring product in S and the right action of R on S is given by $s \cdot r = s\phi(r)$, $(s \in S, r \in R)$. Again verification of the module axioms is easy. Thus we can form the tensor product $S \otimes_R M$, which is a left S-module by (13.1.4), and also a left R-module via ϕ.

One can ask how the new R-module $S \otimes_R M$ is related to the original module M. If the ring S has an identity element, there is an obvious mapping

$$\theta : M \rightarrow S \otimes_R M$$

given by $\theta(a) = 1_S \otimes a$. Observe that $\theta(r \cdot a) = 1_S \otimes (r \cdot a) = (1_S \cdot r) \otimes a = (1_S\phi(r)) \otimes a = (\phi(r)1_S) \otimes a = (r \cdot 1_S) \otimes a = r \cdot (1_S \otimes a) = r \cdot (\theta(a))$, where $r \in R$, $a \in M$. Therefore θ *is a homomorphism of left R-modules.*

A case of particular interest is where ϕ is injective, so that R is essentially a subring of S. In this circumstance we are extending the ring of operators on a module from the subring R to S. The interesting question is whether θ is also injective. A detailed investigation of the problem would take us too far afield, so we will restrict ourselves to the special, but important, case where R is a domain and ϕ is the canonical injection from R into its field of fractions F. Thus we are trying to embed an R-module in an F-vector space.

Tensor products and localizations

Let R be a domain with field of fractions F. Recall from Section 6.3 that each element of F is a fraction over R with the form $\frac{r_1}{r_2}$ where $r_i \in R$ and $r_2 \neq 0$. Also there is an injective ring homomorphism $\phi : R \rightarrow F$ in which $r \mapsto \frac{r}{1}$: this is by (6.3.10).

Assume now that M is a torsion-free R-module. We are interested in the mapping $\psi : M \rightarrow F \otimes_R M$ where $\psi(a) = 1_F \otimes a$, the aim being to prove that it is injective. Before that can be done, a better understanding of the elements of the vector space $F \otimes_R M$ is needed. For this purpose a "model" of $F \otimes_R M$ will be constructed. The construction that follows should be compared with that of the field of fractions of a domain.

We start by forming the set

$$S = \{(a, r) \mid a \in M, \, 0 \neq r \in R\}$$

and then introduce a binary relation $\sim$ on S by

$$(a, r) \sim (a', r') \quad \Leftrightarrow \quad r \cdot a' = r' \cdot a.$$

The motivation here is the rule of equality of two rational numbers. By a simple check $\sim$ is an equivalence relation on S, but notice that for the transitive law to hold it is essential that M be torsion-free. The $\sim$-equivalence class of (a, r) will be written

$$\frac{a}{r}.$$

and referred to as a *fraction* over R. Denote the set of all such fractions by $R^{-1} \cdot M$.

The plan is to turn $R^{-1} \cdot M$ into an F-module by defining

$$\frac{a_1}{r_1} + \frac{a_2}{r_2} = \frac{r_2 \cdot a_1 + r_1 \cdot a_2}{r_1 r_2} \quad \text{and} \quad \left(\frac{r_1}{r_2}\right) \cdot \frac{a}{r} = \frac{r_1 \cdot a}{r_2 r}.$$

Since these are operations on equivalence classes, it is essential to verify that they are well defined, i. e., there is no dependence on the choice of elements (a_i, r_i) from $\frac{a_i}{r_i}$ or (a, r) from $\frac{a}{r}$. This consists of routine calculations, at least some of which the reader should perform.

Then the module axioms must be checked. For example,

$$\left(\frac{r}{r'}\right) \cdot \left(\frac{a_1}{r_1} + \frac{a_2}{r_2}\right) = \left(\frac{r}{r'}\right) \cdot \left(\frac{r_2 \cdot a_1 + r_1 \cdot a_2}{r_1 r_2}\right) = \frac{r r_2 \cdot a_1 + r r_1 \cdot a_2}{r' r_1 r_2}. \tag{13.2}$$

Also

$$\left(\frac{r}{r'}\right) \cdot \left(\frac{a_1}{r_1}\right) + \left(\frac{r}{r'}\right) \cdot \left(\frac{a_2}{r_2}\right) = \frac{r \cdot a_1}{r' r_1} + \frac{r \cdot a_2}{r' r_2} = \frac{r r' r_2 \cdot a_1 + r r' r_1 \cdot a_2}{r'^2 r_1 r_2},$$

which is seen to equal the final expression in (13.2) on cancelling the common factor r' in the numerator and denominator.

The module $R^{-1} \cdot M$ is called the *localization of M*. The result that we are aiming for is as follows.

(13.3.1). *Let R be an integral domain and F its field of fractions. If M is a torsion-free R-module, then*

$$R^{-1} \cdot M \stackrel{F}{\simeq} F \otimes_R M.$$

Proof. The assignment $(\frac{r_1}{r_2}, a) \mapsto \frac{r_1 \cdot a}{r_2}$ yields a well defined R-middle linear mapping from $F \times M$ to $R^{-1} \cdot M$. For example, the map sends $(\frac{r_1}{r_2}, a_1 + a_2) - (\frac{r_1}{r_2}, a_1) - (\frac{r_1}{r_2}, a_2)$ to

$$\frac{r_1 \cdot (a_1 + a_2)}{r_2} - \frac{r_1 \cdot a_1}{r_2} - \frac{r_1 \cdot a_2}{r_2} = 0.$$

The other verifications are at a similar level of difficulty.

It follows that there is a homomorphism $\pi : F \underset{R}{\otimes} M \to R^{-1} \cdot M$ such that $\pi(\frac{r_1}{r_2} \otimes a) = \frac{r_1 \cdot a}{r_2}$. Now check that π is an F-module homomorphism. Let $r, r', r_i \in R$, $a \in M$; then

$$\pi\left(\frac{r}{r'} \cdot \left(\frac{r_1}{r_2} \otimes a\right)\right) = \pi\left(\frac{rr_1}{r'r_2} \otimes a\right) = \frac{rr_1 \cdot a}{r'r_2} = \frac{r}{r'} \cdot \left(\frac{r_1 \cdot a}{r_2}\right) = \frac{r}{r'} \cdot \left(\pi\left(\frac{r_1}{r_2} \otimes a\right)\right),$$

which is sufficient because $F \otimes_R M$ is generated by the tensors $\frac{r_1}{r_2} \otimes a$.

Next define a mapping $\psi : R^{-1} \cdot M \to F \otimes_R M$ by $\psi(\frac{a}{r}) = \frac{1}{r} \otimes a$. To show that ψ is well defined, suppose that $(a, r) \sim (a', r')$. Thus $r \cdot a' = r' \cdot a$ and

$$\frac{1}{r} \otimes a = \frac{r'}{rr'} \otimes a = \frac{1}{rr'} \otimes (r' \cdot a) = \frac{1}{rr'} \otimes (r \cdot a') = \frac{r}{rr'} \otimes a' = \frac{1}{r'} \otimes a',$$

as required. Also it is routine to check that ψ is a homomorphism.

Finally, π and ψ are mutually inverse maps: for $\pi\psi(\frac{a}{r}) = \pi(\frac{1}{r} \otimes a) = \frac{1 \cdot a}{r} = \frac{a}{r}$, while

$$\psi\pi\left(\frac{r_1}{r_2} \otimes a\right) = \psi\left(\frac{r_1 \cdot a}{r_2}\right) = \frac{1}{r_2} \otimes (r_1 \cdot a) = \frac{r_1}{r_2} \otimes a.$$

This is implies that $\psi\pi$ is the identity on $F \otimes_R M$ since the latter is generated by the tensors $\frac{r_1}{r_2} \otimes a$. Therefore ψ is an F-isomorphism. $\square$

Corollary (13.3.2). *Let R be an integral domain and F its field of fractions. If M is a torsion-free R-module, the assignment $a \mapsto 1 \otimes a$ determines an injective R-module homomorphism $\theta : M \to F \otimes_R M$.*

Proof. Assume that $\theta(a) = 0$ for some $a \in M$, so that $1 \otimes a = 0$. Apply the isomorphism π in the proof of (13.3.1) to both sides of this equation to get $\frac{a}{1} = \pi(1 \otimes a) = \frac{0}{1}$, which implies that $a = 0$. $\square$

This corollary provides some insight into the nature of torsion-free abelian groups, but first some terminology. A torsion-free abelian group A is said to have *finite rank* if it has no infinite linearly independent subsets. In this event A must possess a finite maximal linearly independent subset $\{a_1, a_2, \ldots, a_r\}$, for otherwise there would exist infinite linearly independent subsets.

(13.3.3). *If A is a torsion-free abelian group, then A is isomorphic with a subgroup of the rational vector space $V = \mathbb{Q} \otimes A$. If A has finite rank, then V has finite dimension.*

Proof. Let $\theta : A \to \mathbb{Q} \otimes A$ be the mapping in which $a \mapsto 1 \otimes a$. By (13.3.2) this is an injective homomorphism, which proves the first statement. Now assume that A has finite rank and $S = \{a_1, \ldots, a_r\}$ is a maximal linearly independent subset of A. Then $T = \theta(S)$ is linearly independent since θ is injective. If $\frac{r_1}{r_2} \otimes a$ is a typical tensor in V with $r_i \in \mathbb{Z}$, $a \in A$, then $r_2(\frac{r_1}{r_2} \otimes a) = r_1(1 \otimes a) \in \langle T \rangle$, which shows that every element of $\mathbb{Q} \otimes A$ is a $\mathbb{Q}$-linear combination of elements of T and consequently that T is a $\mathbb{Q}$-basis for V. Hence $\dim_{\mathbb{Q}}(V) = r$ is finite. $\qquad\square$

The proof also shows that all maximal linearly independent subsets of A have the same number of elements, namely $\dim_{\mathbb{Q}}(\mathbb{Q} \otimes A)$.

While (13.3.3) provides a familiar setting for torsion-free abelian groups of finite rank in the sense that they exist inside finite dimensional rational vector spaces, this placement does not materially advance the classification of the groups. In fact torsion-free abelian groups of finite rank can have extremely complex structure, far beyond that of finitely generated abelian groups. A standard reference for infinite abelian groups is [5].

Exercises (13.3).

(1) Let F be a subfield of a field K and let V be an n-dimensional vector space over F. Prove that $V \otimes_F K$ is an n-dimensional vector space over K.

(In the exercises that follow R is a domain with field of fractions F with $R \subseteq F$, and M is an R-module).

(2) Prove that the module operations specified for $R^{-1} \cdot M$ are well defined.

(3) Prove that *every* element of $F \otimes_R M$ has the form $\frac{1}{r} \otimes a$ where $r \in R$, $a \in M$.

(4) Let T denote the torsion submodule of M.
   (i) Prove that $F \otimes_R T = 0$.
   (ii) Prove that $F \otimes_R M \overset{R}{\cong} F \otimes_R (M/T)$. [Hint: start with the exact sequence $0 \to T \to M \to M/T \to 0$ and apply the right exactness property of tensor products.]

(5) (The flatness property of F). Let $\alpha : A \to B$ be an injective R-module homomorphism. Prove that the induced map $\alpha_* : F \otimes_R A \to F \otimes_R B$ is also injective. [Hint: by Exercise (13.3.4) A and B may be assumed to be torsion-free. Form the commutative square with horizontal sides $A \overset{\alpha}{\to} B$ and $F \otimes_R A \overset{\alpha_*}{\to} F \otimes_R B$, and vertical sides the canonical maps $A \to F \otimes_R A$ and $B \to F \otimes_R B$. The vertical maps are injective by (13.3.2). Argue that the lower horizontal map is also injective by using Exercise (13.3.3).]

(6) If $0 \to A \overset{\alpha}{\to} B \overset{\beta}{\to} C \to 0$ is an exact sequence of R-modules, show that the induced sequence $0 \to F \otimes_R A \overset{\alpha_*}{\to} F \otimes_R B \overset{\beta_*}{\to} F \otimes_R C \to 0$ is also exact.

(7) Prove that $F \otimes_R F \overset{R}{\cong} F$. [Hint: apply $F \otimes_R -$ to the exact sequence $0 \to R \to F \to F/R \to 0$.]

14 Representations of groups

The topic of this chapter is the representation of finite groups by a groups of linear operators on a finite dimensional vector space. Aside from its importance within the body of algebra, and module theory in particular, this is an area which has many applications to the structure of finite groups. Our aim is first to provide coverage of the basic concepts, then to apply the theory to prove the Burnside $p - q$ theorem. We will concentrate on the so called non-modular case, when the characteristic of the field does not divide the group order. In addition the field will usually be algebraically closed and may have characteristic 0. Thus the theory applies to the complex number field in particular, which was the case studied when representation theory was first developed.

14.1 Representations and group rings

Let G be a group, F a field and V a non-zero finite dimensional vector space over F. Recall that GL(V) denotes the group of all invertible, linear operators on V. A homomorphism

$$\rho : G \rightarrow \mathrm{GL}(V)$$

is called a *(linear) representation* of G over F. If $\mathrm{Ker}(\rho) = 1$, so that G is isomorphic with a subgroup of $GL(V)$, the representation is said to be *faithful*. The *degree* of the representation is the dimension of the vector space V.

There is an immediate connection with matrices. Assume that the representation ρ has degree n, so the vector space V has an ordered basis $v_1, v_2, \ldots, v_n$. Recall from Section 8.3 that a linear operator on V can be represented by an $n \times n$ matrix over F which depends on the choice of basis. This correspondence shows that $GL(V)$ is isomorphic with $\mathrm{GL}_n(F)$, the general linear group of degree n over the field F. By composing ρ with this isomorphism we obtain a homomorphism

$$\rho^* : G \rightarrow \mathrm{GL}_n(F),$$

the *associated matrix representation* of G over F. For $g \in G$ let the (i, j) entry of the matrix $\rho^*(g)$ be written $\rho_{ij}(g)$. Then by the discussion in Section 8.3

$$\rho(g)(v_i) = \sum_{j=1}^{n} \rho_{ji}(g)v_j, \quad \text{for } i = 1, 2, \ldots, n.$$

Example (14.1.1). The most obvious example of a representation of a group G is the *trivial representation*, in which every element of G is mapped to the identity linear transformation. More interesting examples arise from permutation representations.

https://doi.org/10.1515/9783110691160-014

Let $\pi : G \to \mathrm{Sym}(X)$ be a permutation representation of a group G on a finite non-empty set X. Let V be a vector space over a field F with basis $\{v_x \mid x \in X\}$. Then we obtain a linear representation $\rho : G \to \mathrm{GL}(V)$ of degree $|X|$ by defining $\rho(g)(v_x) = v_{\pi(x)}$. Note that in the matrix representation corresponding to ρ the matrices are permutation matrices.

Representations and modules

Let G be a group and R a ring with identity. Recall from Section 6.1 that the group ring

$$RG$$

consists of all finite formal sums $\sum_{x \in G} r_x x$ where $r_x \in R$. The ring operations are formal addition and term by term multiplication of such sums. When the ring is a field F, the group ring FG is an algebra over F known as the *group algebra*.

There is a close connection between representations of G over a field F and FG-modules. In the first place, if $\rho : G \to \mathrm{GL}(V)$ is a representation of G on a vector space V of finite dimension $n > 0$ over a field F, then V has the structure of a left FG-module if we define

$$\left(\sum_{x \in G} r_x x \right) \cdot v = \sum_{x \in G} r_x \rho(x)(v), \quad \text{where } v \in V, r_x \in R.$$

It is a simple matter to check the validity of the module axioms.

Conversely, if M is an FG-module with F-dimension $n > 0$, there is a corresponding representation $\rho : G \to \mathrm{GL}(M)$ of degree n defined by $\rho(x)(a) = x \cdot a$ where $x \in G, a \in M$. This observation makes it clear that there is a bijection between F-representations of G with degree n and left FG-modules of F-dimension n, so the theories of these algebraic objects are essentially identical. In this chapter it is tacitly assumed that all modules are left modules.

Equivalent representations

Two F-representations ρ and σ of a group G are said to be *equivalent* if they arise from isomorphic FG-modules. In particular equivalent representations have the same degree.

(14.1.1). *Let ρ and σ be representations of a group G over a field F arising from FG-modules M and N. Then the following hold.*
(i) *ρ and σ are equivalent if and only if there is an F-isomorphism $\alpha : M \to N$ such that $\alpha\rho(g) = \sigma(g)\alpha$ for all $g \in G$.*
(ii) *If ρ and σ are equivalent, there exists $A \in \mathrm{GL}_n(F)$ such that $\sigma^*(g) = A\rho^*(g)A^{-1}$ for all $g \in G$, where n is the degree of ρ and σ.*

Proof. (i) Assume that ρ and σ are equivalent, so there is an FG-isomorphism $\alpha : M \to N$. Thus $\alpha(g \cdot a) = g \cdot \alpha(a)$ for $a \in M$, $g \in G$. Therefore $\alpha\rho(g)(a) = \sigma(g)(\alpha(a))$ for all $a \in M$ and hence $\alpha\rho(g) = \sigma(g)\alpha$. The argument is reversible, so the converse statement is true.

(ii) Assume that ρ and σ are equivalent; then for some F-isomorphism $\alpha : M \to N$ we have $\alpha\rho(g) = \sigma(g)\alpha$ for $g \in G$ by (i). Choose bases for the modules M and N and pass to the associated matrix representations ρ^* and σ^*. Let α be represented by $A \in \mathrm{GL}_n(F)$ with respect to these bases. Taking the matrix form of the equation $\alpha\rho(g) = \sigma(g)\alpha$, we obtain $A\rho^*(g) = \sigma^*(g)A$. Since A is non-singular, the result follows. $\quad\square$

As a consequence equivalent representations represent the group G by conjugate subgroups of $\mathrm{GL}_n(F)$.

Irreducible representation

An F-representation of a group G is said to be *irreducible* if the corresponding FG-module is a simple module, i. e., it has no proper non-zero submodules. Otherwise ρ is *reducible*. Recall that by (9.1.9) a simple FG-module is isomorphic with a module of the form FG/L where L is a maximal left ideal of FG. This fact suggests that knowledge of the ideal structure of the group algebra will be useful in studying the irreducible representations of a group.

Let M be an FG-module with finite F-dimension $n > 0$. Recall that M is semisimple if there is a direct decomposition

$$M = M_1 \oplus M_2 \oplus \cdots \oplus M_k$$

where the M_i are simple submodules. The corresponding representation afforded by M is called *completely reducible*. Let ρ_i denote the irreducible F-representation of G afforded by M_i.

In a real sense the representation ρ is determined by the irreducible components ρ_i. There is basis of M for which each representing matrix $\rho^*(g)$ is a block matrix with the matrices $\rho_i^*(g)$ placed on the principal diagonal. To see this just choose an F-basis for each submodule M_i and form the union to get a basis of M. The matrix representing g with respect to this basis will have the form specified. Thus it is reasonable to regard ρ as a direct sum of the ρ_i and write $\rho = \rho_1 \oplus \rho_2 \oplus \cdots \oplus \rho_k$.

Maschke's Theorem[1]

The most famous condition for a representation to be completely reducible is the following result of H. Maschke.

---

1 Heinrich Maschke (1853–1908).

(14.1.2). *Let G be a group of finite order and let F be a field whose characteristic does not divide the order of G. Then every F-representation of G is completely irreducible.*

Proof. Let M be an FG-module; we need to show that M is semisimple and by (9.1.12) it is enough to prove that any submodule N is a direct summand of M. Now N is an F-subspace of the F-vector space M, so there is a subspace L such that $M = N \oplus L$: to see this take a basis of N, extend it to a basis of M and let L be the subspace generated by the additional basis elements.

Now L may not be a FG-submodule, so we modify it by using an averaging technique. Let $\pi : M \to N$ denote the canonical projection in which $n \oplus \ell \mapsto n$, $n \in N$, $\ell \in L$. Here π is only an F-linear transformation, but it can be transformed into an FG-module homomorphism. Define a function $\theta : M \to M$ by

$$\theta(a) = \frac{1}{m} \sum_{x \in G} x^{-1} \cdot \pi(x \cdot a),$$

where $m = |G|$ and $a \in M$. This definition makes sense as division by m is well defined in M: for m is indivisible by the characteristic of F. Note that $\theta(a) \in N$ for all $a \in M$. Clearly θ is an F-linear transformation; we will show that it is an FG-homomorphism. Let $a \in M$ and $g \in G$; it follows from the definition that

$$g^{-1} \cdot \theta(g \cdot a) = \frac{1}{m} g^{-1} \cdot \sum_{x \in G} x^{-1} \cdot \pi((xg) \cdot a) = \frac{1}{m} \sum_{x \in G} (xg)^{-1} \cdot \pi((xg) \cdot a).$$

In the last sum replace x by $y = xg$ and sum over y to get

$$\frac{1}{m} \sum_{y \in G} y^{-1} \cdot \pi(y \cdot a) = \theta(a).$$

Therefore $\theta(g \cdot a) = g \cdot \theta(a)$, which establishes the claim that θ is an FG-module homomorphism.

We conclude the proof by showing that $M = N \oplus K$ where $K = \mathrm{Ker}(\theta)$, which is of course an FG-submodule. Note that $\pi|_N$ is the identity function by definition of π. Let $a \in N$; thus $\pi(x \cdot a) = x \cdot a$, so that $\theta(a) = \frac{1}{m} \sum_{x \in G} a = a$ and θ is the identity map on N. Let $b \in M$; then $\theta(b) \in N$, so that $\theta^2(b) = \theta(\theta(b)) = \theta(b)$. Hence $\theta(\theta(b) - b) = 0$ and $\theta(b) - b \in K$. Therefore $b \in \mathrm{Im}(\theta) + K \subseteq N + K$ and $M = N + K$. Finally, $N \cap K = 0$, since $\theta|_N$ is the identity and hence $M = N \oplus K$ as required. □

Maschke's theorem makes it clear that one should study the irreducible representations of a group, at least if the characteristic of the field does not divide the group order. The key to understanding these is knowledge of the simple modules over the group algebra. Recall from Schur's Lemma (9.1.22) that $\mathrm{End}_{FG}(M)$ is a division ring if M is a simple module. In fact when F is an algebraically closed field, it is even a field.

(14.1.3). *Let G be a finite group, F an algebraically closed field and M a simple FG-module. Then M has finite F-dimension and $\mathrm{End}_{FG}(M)$ consists of scalar multiplications by elements of F, so that $\mathrm{End}_{FG}(M) \simeq F$.*

Proof. In the first place M has finite F-dimension since G is finite and $M = (FG)a$ for any $0 \neq a \in M$ by simplicity of M.

Let $0 \neq \alpha \in \mathrm{End}_{FG}(M)$. Since F is algebraically closed and α is a linear operator on the finite dimensional F-space M, we can be sure that α has an eigenvalue in F, say f with associated eigenvector $m \in M$, by (8.4.1). Let N denote the set of all $\ell \in M$ such that $\alpha(\ell) = f\ell$ and observe that N is an F-subspace containing m. If $g \in G$, $a \in N$, then $\alpha(g \cdot a) = g \cdot (\alpha(a)) = g \cdot fa = f(g \cdot a)$, which shows that $g \cdot a \in N$. Therefore N is an FG-submodule of M. But M is a simple module and $N \neq 0$ since $0 \neq m \in N$. Consequently $N = M$ and thus α is multiplication in M by f. $\qquad\square$

Next we apply (14.1.3) to determine the irreducible representations of a finite abelian group over an algebraically closed field.

(14.1.4). *Let G be a finite abelian group and let F be an algebraically closed field whose characteristic does not divide $|G|$. Then the irreducible F-representations of G all have degree 1 and they correspond to the elements of $\mathrm{Hom}(G, U(F))$.*

Proof. Let M be a simple FG-module. Then $\mathrm{End}_{FG}(M)$ consists of F-scalar multiplications on M by (14.1.3). Let $g \in G$; the assignment $a \mapsto g \cdot a$ determines an F-endomorphism β of M. In fact it is an FG-endomorphism: for we have $\beta(x \cdot a) = g \cdot (x \cdot a) = (gx) \cdot a = (xg) \cdot a = x \cdot (g \cdot a) = x \cdot \beta(a)$ for all $x \in G$, since G is abelian. Therefore $\beta \in \mathrm{End}_{FG}(M)$ and β is a scalar multiplication. This means that every F-subspace of M is a submodule. Since M is a simple module, it follows that M has F-dimension 1. Thus the irreducible F-representations of G correspond to the elements of the $\mathrm{Hom}(G, U(F))$. $\qquad\square$

Exercises (14.1).

(1) Let G be a group and F a field. Let M, N be FG-modules affording representations ρ, σ respectively. If there is an F-isomorphism $\alpha : M \simeq N$ such that $\alpha\rho(g) = \sigma(g)\alpha$ for all $g \in G$, prove that ρ and σ are equivalent.

(2) Show that every permutation representation of degree > 1 is reducible.

(3) Find all the inequivalent irreducible complex representations of a cyclic group of order 6.

(4) Let G be a finite p-group and F a finite field of prime characteristic p. Prove that every irreducible F-representation of G has degree 1. [Hint: let M be a simple G-module. Note that the semidirect product of M and G is a finite p-group and hence is nilpotent. Apply Exercise (10.2.12).]

(5) Use Exercise (14.1.4) to show that Maschke's Theorem is not valid if the characteristic of the field divides the group order.

(6) Let the degrees of the irreducible representations of a finite group G over an arbitrary field F be $n_1, n_2, \dots$. Prove that $\sum_i n_i \leq |G|$, so that there are only finitely many inequivalent irreducible F-representations of G. [Hint: form a series of left ideals in FG of maximum length. The factors in the series are simple FG-modules.]

(7) Prove that a cyclic group of order n has a faithful irreducible $\mathbb{Q}$-representation of degree $\phi(n)$ where ϕ is Euler's function. [Hint: map a generator of the group to a root of the cyclotomic polynomial Φ_n.]

(8) Let G be a finite group which has a *unique* minimal normal subgroup $N \neq 1$ and let F be a field whose characteristic does not divide $|G|$. Prove that G has at least one faithful irreducible F-representation. [Hint: assume there are no faithful irreducible representations. Write FG as the direct sum of simple FG-modules M_i. Then N acts trivially on each M_i, and hence on FG, by left multiplication. Conclude that $N = 1$.]

(9) Let G be a perfect group, i.e., $G = G'$, and let F be any field. Prove that every F-representation of G with degree 1 is trivial.

14.2 The structure of group algebras

Complete information about the irreducible representations of a finite group is contained within the group algebra. The reason for this is that the simple modules that determine the irreducible representations arise as quotients of the group algebra. Thus group algebras are the key to understanding the irreducible representations of a finite group.

The most complete results are obtained in the case of an algebraically closed field whose characteristic does not divide the group order. The fundamental theorem that follows gives valuable information regarding the number and degrees of the irreducible representations.

(14.2.1). *Let G be a finite group and F an algebraically closed field whose characteristic does not divide the order of G. Then:*

(i) *$FG = I_1 \oplus I_2 \oplus \cdots \oplus I_h$ where I_i is an ideal of FG which is isomorphic with $M_{n_i}(F)$, the ring of all $n_i \times n_i$ matrices over F;*

(ii) *$|G| = n_1^2 + n_2^2 + \cdots + n_h^2$;*

(iii) *each simple FG-module is isomorphic with a minimal left ideal of some I_i and has dimension n_i. Thus the n_i are the degrees of the irreducible representations.*

(iv) *the number h of inequivalent irreducible F-representations of G is equal to the class number of G.*

The proof of (14.2.1) is one of the longer ones in this book. In order to prove this major result two preparatory lemmas are needed.

First a definition. Let R and S be rings and let $\alpha : R \to S$ be a homomorphism of abelian groups. Then α is called an *anti-homomorphism* of rings if

$$\alpha(r_1 r_2) = \alpha(r_2)\alpha(r_1)$$

for all $r_i \in R$. An anti-homomorphism is called an *anti-isomorphism* if it is bijective. For example, the transpose mapping in which $A \mapsto A^T$ is an anti-isomorphism from $M_n(F)$ to itself because of the matrix identity $(AB)^T = B^T A^T$.

The next result is quite elementary: it makes the first transition, from a ring with identity to the ring of endomorphisms of a module.

(14.2.2). *Let S be a ring with identity and define a function $\theta : S \to \text{End}_S({}_S S)$ by $\theta(s) : x \mapsto xs$ where $s, x \in S$. Then θ is an anti-isomorphism.*

Proof. First we show that θ is an anti-homomorphism. By definition

$$\theta(s_1 s_2)(x) = x(s_1 s_2) = (xs_1)s_2 = \theta(s_2)(xs_1) = \theta(s_2)\theta(s_1)(x)$$

where $x, s_i \in S$. Hence $\theta(s_1 s_2) = \theta(s_2)\theta(s_1)$. By an even simpler computation $\theta(s_1 + s_2) = \theta(s_1) + \theta(s_2)$.

It remains to prove that θ is bijective. Suppose that $\theta(s) = 0$ for some $s \in S$. Then $xs = 0$ for all $x \in S$ and taking x to be 1_S, we deduce that $s = 1_S s = 0$. Hence θ is injective. Finally, let $\xi \in \text{End}_S({}_S S)$ and put $t = \xi(1_S)$. Then for any $s \in S$

$$\xi(s) = \xi(s 1_S) = s\xi(1_S) = st = \theta(t)(s),$$

since ξ is an homomorphism of S-modules. It follows that $\xi = \theta(t)$ and θ is surjective. $\square$

Endomorphisms of direct sums

The next move is to transition from the ring of endomorphisms of a semisimple module to a ring of matrices. This calls for precise information about the endomorphism ring of a direct sum of modules.

Let R be a ring with identity and let M be an R-module which has a direct decomposition into submodules

$$M = M_1 \oplus M_2 \oplus \cdots \oplus M_k.$$

Let ξ be an R-endomorphism of M. The idea here is to represent ξ by a matrix; this is very similar to the matrix representation of a linear operator on a vector space described in Section 8.3.

If $a \in M_i$, define $\xi_{ji}(a)$ to be the M_j-component of $\xi(a)$. Since $\xi \in \text{End}_R(M)$, it follows that $\xi_{ji} : M_i \to M_j$ is an R-module homomorphism. Writing $a = \sum_{i=1}^{k} a_i$ where

$a_i \in M_i$, we have

$$\xi(a) = \sum_{i=1}^{k} \xi(a_i) = \sum_{i=1}^{k}\sum_{j=1}^{k} \xi_{ji}(a_i) = \sum_{j=1}^{k}\sum_{i=1}^{k} \xi_{ji}(a_i). \tag{14.1}$$

Next denote by ξ^* the $k \times k$ matrix whose (i,j)th entry is ξ_{ji}. Define $[a]$ to be the column of components of $a \in M$. Then equation (14.1) above takes the matrix form

$$[\xi(a)] = \xi^*[a].$$

Thus the endomorphism ξ corresponds to left multiplication by the matrix ξ^*.

The information required in the proof of (14.2.1) is embodied in the following result – here the notation of the foregoing discussion is maintained.

(14.2.3). *The ring* $\mathrm{End}_R(M)$ *is isomorphic with the ring of* $k \times k$ *matrices* ξ^* *with* (i,j)-*entry* ξ_{ji}, *where the matrix operations are the usual operations of matrix algebra.*

To prove this it suffices to show that the assignment $\xi \mapsto \xi^*$ is a ring isomorphism, which is a straightforward calculation along the lines of (8.3.12). We are now in a position to undertake the proof of the main theorem.

Proof of (14.2.1). Write $R = FG$. By (14.1.2) the left R-module $_RR$ is semisimple, so it is a direct sum of finitely many simple submodules, that is to say, of minimal left ideals. Every simple submodule of R is isomorphic with a quotient of R and hence with one of the simple direct summands by (9.1.10). Let $S_1, S_2, \ldots, S_h$ be a complete set of non-isomorphic simple submodules of R and let I_i denote the sum of all the simple submodules that are isomorphic with S_i. Notice that

$$I_i \cap \sum_{j \neq i} I_j = 0.$$

For if the intersection were non-zero, it would contain a simple submodule which would have to be isomorphic with S_i *and* some S_j where $j \neq i$. Thus

$$_RR = I_1 \oplus I_2 \oplus \cdots \oplus I_h$$

and I_i is the direct sum of say n_i minimal left ideals, all of them isomorphic with S_i.

Next we show that each I_i is an ideal of R: certainly it is a left ideal. Let $r \in R$ and observe that the assignment $\xi : x \mapsto xr$ determines a homomorphism ξ of (left) R-modules. Let U be a simple submodule of I_i. Then $\xi(U)$ is a quotient of U, so either it is 0 or it is isomorphic with S_i. Either way $\xi(U) \leq I_i$ and therefore $\xi(I_i) \leq I_i$. Hence I_i is a right ideal and therefore an ideal. Consequently $I_j I_i \leq I_j \cap I_i = 0$ for $i \neq j$.

Now I_i is certainly a subring of R; we claim that it has an identity element. Write $1_R = e_1 + e_2 + \cdots + e_h$ with $e_i \in I_i$. Suppose that some $e_i = 0$. Then $I_i = 1_R I_i = 0$ since $I_j I_i = 0$ if $i \neq j$. This is impossible, so $e_i \neq 0$ for all i. Next let $r \in I_i$. Then we have $r = r1_R = re_i$.

In the same way $r = e_i r$. Thus e_i is the identity element of I_i. It follows from (14.2.2) that there is an anti-isomorphism from I_i to $\text{End}_{I_i}(I_i)$. Furthermore, $\text{End}_R(I_i) = \text{End}_{I_i}(I_i)$ since $I_j I_i = 0$ if $i \neq j$.

We still have to investigate the structure of $E_i = \text{End}_R(I_i)$. Recall that I_i is a direct sum of n_i simple submodules, each isomorphic with S_i. By (14.1.3) $\text{End}_R(S_i) \simeq F$. Thus (14.2.3) can be applied to show that $E_i \simeq M_{n_i}(F)$. This means that there are mappings

$$I_i \to E_i \simeq M_{n_i}(F)$$

whose composite is an anti-isomorphism $I_i \to M_{n_i}(F)$. Follow this up by applying the transpose map $A \mapsto A^T$ to get a ring ismorphism $I_i \to M_{n_i}(F)$. This completes the proof of (i).

To establish (ii) take the F-dimension of each side in the equation in (i), noting that $R = FG$ has dimension $|G|$, while I_i has dimension n_i^2 since $I_i \simeq M_{n_i}(F)$. Hence $|G| = \sum_{i=1}^h n_i^2$.

By (9.1.9) every simple R-module is an image of R and hence is isomorphic with a minimal left ideal of R; the latter is contained in some I_i. Hence the n_i are the degrees of the irreducible representations. If X is a left ideal of I_i, it is automatically a left ideal of R, since $I_j X \leq I_i \cap I_j = 0$ if $i \neq j$. Thus a minimal left ideal of R which is contained in I_i is minimal in I_i.

To establish (iii) it remains to show that a minimal left ideal of $M_n(F)$ has dimension n as an F-space. This can be proved by a short matrix calculation – see Exercise (14.2.6).

The final step in the proof is to establish (iv), that h equals the class number of G. Let C denote the center of the ring R, i. e., the set of all $r \in R$ such that $rx = xr$ for all $x \in R$. Recall from Exercise (6.1.10) that C is a commutative subring of R and hence is an F-algebra. Observe also that C coincides with the sum of the centers of the I_i. But the center of $M_{n_i}(F)$ is the subring of scalar matrices, which has F-dimension 1. It follows that C has F-dimension h.

On the other hand, the center of FG can be computed directly. Let $C_1, C_2, \ldots, C_\ell$ denote the conjugacy classes of G and write $s_i = \sum_{x \in C_i} x$. Clearly $s_i \in C$. Notice that the s_i are linearly independent over F, so $\ell \leq h$. To complete the proof write $S = Fs_1 + Fs_2 + \cdots + Fs_\ell$. Let $c \in R$ and write $c = \sum_{x \in G} f_x x$ with $f_x \in F$. Then for all $g \in G$

$$c = g\left(\sum_{x \in G} f_x x\right) g^{-1} = \sum_{x \in G} f_x (gxg^{-1}) = \sum_{y \in G} f_{g^{-1}yg}\, y,$$

where in the last sum $y = gxg^{-1}$. Consequently $f_y = f_{g^{-1}yg}$ for all g, y, showing that the function f is constant on each conjugacy class C_i. Let f_i be the value of f on C_i. Then $c = \sum_{i=1}^\ell f_i s_i \in S$, so the s_i form a basis of C. Hence $C = S$ and the proof of (14.2.1) is complete. $\qquad\square$

As an illustration of the usefulness of (14.2.1) in determining irreducible representations we present an explicit example.

Example (14.2.1). Find all the irreducible complex representations of the dihedral group of order 8.

Let $G = \mathrm{Dih}(8)$ and write $G = \langle x \rangle \ltimes \langle a \rangle$, where $a^4 = x^2 = 1$, $xax^{-1} = a^{-1} = a^3$. This group has five conjugacy classes

$$\{1\}, \{a, a^3\}, a^2, \{x, xa^2\}, \{xa, xa^3\},$$

so we expect to find five irreducible representations.

Observe that G^{ab} is a Klein 4-group, so G has four irreducible representations of degree 1 by (14.1.4). We can assume that $n_1 = n_2 = n_3 = n_4 = 1$. Since $\sum_{i=1}^5 n_i^2 = 8$, it follows that $n_5 = 2$, and there is an irreducible representation of degree 2. A little experimentation reveals the representation to be

$$a \mapsto \begin{bmatrix} 0 & 1 \\ -1 & 0 \end{bmatrix}, \quad x \mapsto \begin{bmatrix} 1 & 0 \\ 0 & -1 \end{bmatrix}.$$

Exercises (14.2).

(1) Find all the irreducible complex representations of S_3, the symmetric group of degree 3.

(2) The same question for the alternating group A_4.

(3) Show that the number of irreducible representations of a finite group over $\mathbb{Q}$ can be less than the class number.

(4) Find the number of inequivalent irreducible complex representations of the symmetric group S_n.

(5) Find the degrees of the irreducible complex representations of S_4.

(6) Prove that the ring $R = M_n(F)$ is the direct sum of n minimal left ideals of F-dimension n. (This fact was used in the proof of (14.2.1).) [Hint: let E_{ij} denote the $n \times n$ matrix whose (i, j) entry is 1 and other entries are 0. Put $S_i = \sum_{j=1}^n FE_{ji}$. Show that S_i is a minimal left ideal of R and $R = \bigoplus_{i=1}^n S_i$.]

(7) Let G be a non-abelian group of order $p(p-1)$ where p is a prime. Find the degrees of the irreducible complex representations of G. [Hint: G has a normal subgroup N of order p and G/N is cyclic of order $p-1$; also the class number is p and there are $p-1$ representations of degree 1.]

14.3 Group characters

Associated with a representation of a finite group is a function called the character, a concept of great significance in representation theory. Let F be a field and G a finite group. An FG-module M gives rise to a F-representation $\rho : G \rightarrow \mathrm{GL}(M)$ of G. By choosing an F-basis for M we obtain the associated matrix representation $\rho^* : G \rightarrow \mathrm{GL}_n(F)$. Choice of a different basis will result in an equivalent representation σ. By (14.1.1) the matrices $\rho^*(x)$ and $\sigma^*(x)$ are similar for each $x \in G$; therefore these matrices have the same trace. If A is a square matrix, write $\mathrm{tr}(A)$ for its trace: recall that this is the sum of the entries on the principal diagonal and is also the sum of the eigenvalues.

Define the *character* of ρ to be the function $\chi : G \rightarrow F$ where

$$\chi(x) = \mathrm{tr}(\rho^*(x)), \quad x \in G.$$

The function χ, unlike ρ, is independent of the choice of basis for M. The significance of the character of a representation is underscored by the next result.

(14.3.1). *Equivalent representations have the same character.*

Proof. Let ρ and σ be equivalent representations of a finite group G over a field F. Then by (14.1.1) the matrices $\rho^*(x)$ and $\sigma^*(x)$ are similar for any $x \in G$, so they have the same trace. It follows that ρ and σ have the same character. $\qquad\square$

Class functions

Let G be a finite group and F a field. Then $\mathrm{Fun}(G, F)$ denotes the set of all functions from G to F. This set acquires the structure of a vector space over F if we define the sum and scalar multiple in the obvious way:

$$\alpha + \beta(x) = \alpha(x) + \beta(x) \quad \text{and} \quad f\alpha(x) = f(\alpha(x)),$$

where $x \in G$, $f \in F$, $\alpha, \beta \in \mathrm{Fun}(G, F)$.

A *class function* from G to F is a function which is constant on each conjugacy class of G. A simple check reveals that the class functions form a subspace of $\mathrm{Fun}(G, F)$: this will be denoted by

$$\mathrm{Cl}(G, F).$$

Our interest in class functions stems from the following observation.

(14.3.2). *Group characters are class functions.*

Proof. Let χ be the character of a representation ρ of a finite group G over a field F. Then $\rho^*(gxg^{-1}) = \rho^*(g)\rho^*(x)(\rho^*(g))^{-1}$ where $x, g \in G$. Then, since similar matrices have the same trace,

$$\chi(gxg^{-1}) = \mathrm{tr}(\rho^*(g)\rho^*(x)(\rho^*(g)^{-1})) = \mathrm{tr}(\rho^*(x)) = \chi(x). \qquad\square$$

(14.3.3). *Let G be a finite group and F a field. Then the vector spaces $\mathrm{Fun}(G, F)$ and $\mathrm{Cl}(G, F)$ have F-dimensions equal to $|G|$ and the class number of G respectively.*

Proof. For any $x \in G$ define $\delta_x \in \mathrm{Fun}(G, F)$ by $\delta_x(x) = 1$ and $\delta_x(y) = 0$ if $y \neq x$. The functions δ_x are linearly independent over F. For, if $\sum_{x \in G} f_x \delta_x = 0$ where $f_x \in F$, then $0 = \sum_{x \in G} f_x \delta_x(y) = f_y$ for all $y \in G$. Also for any $\alpha \in \mathrm{Fun}(G, F)$ we have $\alpha = \sum_{x \in G} \alpha(x) \delta_x$ by a check of functional values. Hence $\{\delta_x | x \in G\}$ is a basis of $\mathrm{Fun}(G, F)$, which therefore has dimension $|G|$.

Next let $C_1, C_2, \ldots, C_h$ denote the conjugacy classes of G, so that h is the class number. Define $y_i \in \mathrm{Cl}(G, F)$, $(1 \leq i \leq n)$, by mapping elements of C_i to 1_F and elements in other conjugacy classes to 0_F. The y_i form a basis of $\mathrm{Cl}(G, F)$ by an argument similar to that used in the previous paragraph. Hence $\mathrm{Cl}(G, F)$ has dimension h. $\qquad\square$

The character of an irreducible representation is called an *irreducible character*. The next result shows that it is the irreducible characters that matter.

(14.3.4). *In any finite group each character is a sum of irreducible characters.*

Proof. Let G be a finite group and F a field. Let ρ be the F-representation of G arising from an FG-module M and denote its character by χ. Since the F-dimension of M is finite, there is a series of FG-submodules of maximum length $0 = M_0 < M_1 < \cdots < M_k = M$. Then by the Correspondence Theorem – see Section 9.1 – the module M_{i+1}/M_i is simple, and therefore affords an irreducible F-representation ρ_i of G: denote the character of ρ_i by χ_i.

Choose an F-basis of M_1, extend it to a basis of M_2, then to one of M_3 and so on. This procedure yields a basis of M which is a union of bases of the M_i. For each $g \in G$ form the $k \times k$ matrix that represents the FG-endomorphism $\rho(g)$. Our choice of basis ensures that the matrix has the block form

$$
\rho^*(g) = \begin{bmatrix} A_1 & * & \cdots & * \\ 0 & A_2 & * & \cdots & * \\ \cdot & \cdot & \cdot & \cdots & \cdot \\ 0 & 0 & 0 & \cdots & A_k \end{bmatrix},
$$

where $\rho_i^*(g) = A_i$. Since the trace of $\rho^*(g)$ is the sum of the traces of the matrices A_i, it follows that $\chi = \chi_1 + \chi_2 + \cdots + \chi_k$. Here χ_i is an irreducible character, since it is derived from the simple module M_{i+1}/M_i. $\qquad\square$

Orthogonality relations

Let G be a finite group and F a field. Let ρ be an F-representation of a finite group G arising from an FG-module M. Choose an F-basis of M and as usual write ρ^* for the corresponding matrix representation of G. Let

$$
\rho_{ij}(g)
$$

denote the (i, j) entry of the matrix $\rho^*(g)$ where $g \in G$. Thus ρ_{ij} is a function from G to F.

The irreducible characters satisfy certain fundamental orthogonality relations, the basis for which is given in the next result. Here δ_{ij} denotes the *Kronecker delta*: its value is 1 if $i = j$ and 0 otherwise.

(14.3.5). *Let G be a finite group and F a field. Let ρ, σ be two irreducible F-representations with respective degrees m, n and write ρ_{ij} and σ_{ij} for the (i, j) entries of the associated matrix functions with respect to fixed bases. Let $1 \le r, s \le m$ and $1 \le i, j \le n$.*
(i) *If ρ and σ are inequivalent, then $\sum_{x \in G} \sigma_{ij}(x^{-1})\rho_{rs}(x) = 0$.*
(ii) *If F is algebraically closed with characteristic not dividing $|G|$, then*

$$\sum_{x \in G} \rho_{ij}(x^{-1})\rho_{rs}(x) = \frac{|G|}{n}\delta_{is}\delta_{jr}.$$

Proof. Assume that ρ and σ arise from respective FG-modules M, N. If $\eta \in \mathrm{Hom}_F(M, N)$, define $\bar{\eta} : M \to N$ by

$$\bar{\eta} = \sum_{x \in G} \sigma(x^{-1})\eta\rho(x).$$

Clearly $\bar{\eta}$ is an F-linear mapping.

In fact $\bar{\eta}$ is a map of FG-modules. Indeed for any $g \in G$,

$$\sigma(g)\bar{\eta} = \sum_{x \in G} \sigma(g)\sigma(x^{-1})\eta\rho(x) = \sum_{x \in G} \sigma(gx^{-1})\eta\rho(x).$$

Set $y = xg^{-1}$ and replace x in the sum by yg. Thus

$$\sigma(g)\bar{\eta} = \sum_{y \in G} \sigma(y^{-1})\eta\rho(yg) = \sum_{y \in G} \sigma(y^{-1})\eta\rho(y)\rho(g) = \bar{\eta}\rho(g). \tag{14.2}$$

Keep in mind that if $g \in G$, $a \in M$, $b \in N$, then $\rho(g)(a) = g \cdot a$ and $\sigma(g)(b) = g \cdot b$. On applying the function in (14.2) to $a \in M$, we obtain the module form $g \cdot (\bar{\eta}(a)) = \bar{\eta}(g \cdot a)$ for any $g \in G$, $a \in M$. Hence $\bar{\eta}$ is an FG-module homomorphism.

In order to prove (i) and (ii) we make a special choice of η. Let η be the F-linear map that sends the jth basis element of M to the rth basis element of N and sends all other basis elements to 0. Then the matrix representing η has its (k, ℓ) entry equal to $\delta_{jk}\delta_{r\ell}$. The matrix form of the equation defining $\bar{\eta}$ is

$$\bar{\eta}_{is} = \sum_{x \in G} \sum_{k=1}^{n} \sum_{\ell=1}^{m} \sigma_{ik}(x^{-1})\delta_{jk}\delta_{r\ell}\rho_{\ell s}(x) = \sum_{x \in G} \sigma_{ij}(x^{-1})\rho_{rs}(x). \tag{14.3}$$

If ρ and σ are inequivalent, the simple modules M and N are not isomorphic, and hence $\mathrm{Hom}_{FG}(M, N) = 0$. Therefore $\bar{\eta} = 0$, which establishes (i).

In the next part of the proof we assume that $\sigma = \rho$ and $M = N$, using a single basis for M. The function η is as defined above and $\bar{\eta} = \sum_{x \in G} \rho(x^{-1})\eta\rho(x)$. Since F is algebraically closed, $\mathrm{Hom}_{FG}(M, M) = \mathrm{End}_{FG}(M)$ consists of scalar multiplications by (14.1.3). Therefore $\bar{\eta}$ is multiplication by some $f_{jr} \in F$. In the matrix representation of $\bar{\eta}$, we have $\bar{\eta}_{is} = f_{jr}\delta_{is}$. Hence by equation (14.3)

$$f_{jr}\delta_{is} = \sum_{x \in G} \rho_{ij}(x^{-1})\rho_{rs}(x) = \sum_{y \in G} \rho_{rs}(y^{-1})\rho_{ij}(y) = f_{si}\delta_{rj},$$

where we have replaced x in the sum by $y = x^{-1}$. Therefore we can assume that $i = s$ and $j = r$, since otherwise the sum in (ii) equals 0; thus

$$f_{jj} = \sum_{x \in G} \rho_{ij}(x^{-1})\rho_{ji}(x) = \sum_{y \in G} \rho_{ji}(y^{-1})\rho_{ij}(y) = f_{ii}, \qquad (14.4)$$

which shows that $f = f_{ii}$ is independent of i. Form the sum of the equations (14.4) for $j = 1, 2, \ldots, n$ to get

$$nf = \sum_{j=1}^{n} \sum_{x \in G} \rho_{ij}(x^{-1})\rho_{ji}(x) = \sum_{x \in G} \sum_{j=1}^{n} \rho_{ij}(x^{-1})\rho_{ji}(x).$$

But $\sum_{j=1}^{n} \rho_{ij}(x^{-1})\rho_{ji}(x)$ is the (i, i)th entry of the matrix $\rho^*(x^{-1})\rho^*(x) = \rho^*(1) = I_n$, that is, the identity $n \times n$ matrix. Therefore $nf = \sum_{x \in G} 1 = |G|$, so $f = |G|/n$. The result now follows from (14.4). □

On the basis of the last result we are able to establish the fundamental orthogonality relations for irreducible characters.

(14.3.6). *Let G be a finite group and F a field. Let χ and ψ be distinct irreducible F-characters of G. Then the following hold.*
(i) $\sum_{x \in G} \chi(x^{-1})\psi(x) = 0$.
(ii) *If F is algebraically closed with characteristic not dividing $|G|$, then*
$\sum_{x \in G} \chi(x^{-1})\chi(x) = |G|$.

Proof. Let ρ, σ be irreducible F-representations with respective characters χ, ψ and degrees n, m. By (14.3.1) ρ and σ are inequivalent. As before we write ρ_{ij} and σ_{ij} for the (i, j) entries of the associated matrix representations with respect to fixed bases. Thus $\chi = \sum_{i=1}^{n} \rho_{ii}$ and $\psi = \sum_{j=1}^{m} \sigma_{jj}$. Therefore by (14.3.5)

$$\sum_{x \in G} \chi(x^{-1})\psi(x) = \sum_{x \in G} \sum_{i=1}^{n} \sum_{j=1}^{m} \rho_{ii}(x^{-1})\sigma_{jj}(x) = \sum_{i=1}^{n} \sum_{j=1}^{m} \sum_{x \in G} \rho_{ii}(x^{-1})\sigma_{jj}(x) = 0,$$

which establishes (i).

Now assume we are in the situation of (ii), so $\rho = \sigma$ and $\chi = \psi$. By (14.3.5) again

$$\sum_{x \in G} \chi(x^{-1})\chi(x) = \sum_{i=1}^{n} \sum_{j=1}^{n} \sum_{x \in G} \rho_{ii}(x^{-1})\rho_{jj}(x) = \sum_{i=1}^{n} \sum_{j=1}^{n} \frac{|G|}{n} \delta_{ij}^2,$$

which equals $|G|$. □

Inner products of characters

Let G be a finite group and F a field whose characteristic does not divide $|G|$. For any $\alpha, \beta \in \operatorname{Fun}(G, F)$ define

$$\langle \alpha, \beta \rangle_G = \frac{1}{|G|} \sum_{x \in G} \alpha(x^{-1})\beta(x),$$

which is an element of F. Then it is easy to see that $\langle \ , \ \rangle_G$ is a bilinear form on the F-space $\operatorname{Fun}(G, F)$. Moreover, the bilinear form is symmetric; for

$$\langle \beta, \alpha \rangle_G = \frac{1}{|G|} \sum_{x \in G} \beta(x^{-1})\alpha(x) = \frac{1}{|G|} \sum_{y \in G} \alpha(y^{-1})\beta(y) = \langle \alpha, \beta \rangle_G,$$

where we have put $y = x^{-1}$. (For bilinear forms see for example [16].)

Suppose that $\langle \alpha, \beta \rangle_G = 0$ for all α; then $\beta = 0$. For, if $\beta(x) \neq 0$, we could choose α so that $\alpha(x^{-1}) = \beta(x)^{-1}$ and $\alpha(y) = 0$ for $y \neq x^{-1}$; then it follows that

$$\langle \alpha, \beta \rangle_G = \frac{1}{|G|} \sum_{y \in G} \alpha(y^{-1})\beta(y) = \frac{1}{|G|} \alpha(x^{-1})\beta(x) = \frac{1}{|G|} \neq 0.$$

Therefore $\beta = 0$ and $\langle \ , \ \rangle_G$ is a non-degenerate bilinear form.

It follows that $\langle \ , \ \rangle_G$, which may be written just $\langle \ , \ \rangle$, is an inner product on the vector space $\operatorname{Fun}(G, F)$ – for an account of inner products see, for example, [16]. This observation allows us to reinterpret the properties of irreducible characters listed in (14.3.6).

(14.3.7). *Let G be a finite group and F an algebraically closed field whose characteristic does not divide $|G|$. Then the distinct irreducible F-characters of G form an orthonormal basis of $\operatorname{Cl}(G, F)$, the vector space of class functions on G over F.*

Proof. Let $\chi_1, \chi_2, \ldots, \chi_h$ be the distinct irreducible characters of G. By (14.3.6) and the definition of the inner product

$$\langle \chi_i, \chi_j \rangle_G = \frac{1}{|G|} \sum_{x \in G} \chi_i(x^{-1})\chi_j(x) = \frac{1}{|G|} |G|\delta_{ij} = \delta_{ij}.$$

Hence $\{\chi_1, \chi_2, \ldots, \chi_h\}$ an orthonormal set. Now by (14.2.1) the number of irreducible characters equals h, the class number of G, which by (14.3.3) is the dimension of the

F-space $\mathrm{Cl}(G,F)$. Therefore the χ_i generate $\mathrm{Cl}(G,F)$ and they form an orthonormal basis of it. $\qquad\square$

The *permutation character* of a group is defined to be the character of the (left) regular permutation representation. Next we apply (14.3.7) to express the permutation character in terms of the irreducible characters.

(14.3.8). *Let G be a finite group and F an algebraically closed field whose characteristic does not divide $|G|$. Denote the distinct irreducible F-characters of G by $\chi_1, \chi_2, \ldots, \chi_h$ where h is the class number of G. Then the permutation character is equal to*

$$\psi = \sum_{j=1}^{h} \ell_j \chi_j$$

where ℓ_j is the degree of χ_j.

Proof. Since the χ_j form an F-basis for $\mathrm{Cl}(G,F)$, there is an expression $\psi = \sum_{j=1}^{h} f_j \chi_j$ with $f_j \in F$. Then $\langle \psi, \chi_i \rangle = f_i$ since $\langle \chi_j, \chi_i \rangle_G = \delta_{ij}$. Now left multiplication on G by a non-identity element has no fixed points. Therefore the permutation matrix representing any $x \neq 1$ in G has zeros on the principal diagonal and thus $\psi(x) = 0$. Hence

$$f_i = \langle \psi, \chi_i \rangle = \frac{1}{|G|} \sum_{x \in G} \psi(x^{-1}) \chi_i(x) = \frac{1}{|G|} \psi(1) \chi_i(1).$$

Since $\psi(1) = |G|$ and $\chi_i(1) = \ell_i$, it follows that $f_i = \frac{1}{|G|}|G| \cdot \ell_i = \ell_i$. $\qquad\square$

An important property of the degrees of the irreducible characters is that they divide the group order when the field is algebraically closed and has characteristic 0. In order to establish this we will need a further result. First recall that an algebraic integer is a complex number that is a root of some monic polynomial in $\mathbb{Z}[t]$.

(14.3.9). *Let G be a finite group and F an algebraically closed field of characteristic 0. Let χ be an irreducible F-character of G of degree n. If $g \in G$ has exactly ℓ conjugates, then $\frac{\ell \chi(g)}{n}$ is an algebraic integer.*

Proof. Let $C_1, C_2, \ldots, C_h$ be the conjugacy classes of G and write $d_i = \sum_{x \in C_i} x$. It was observed at the end of the proof of (14.2.1) that the d_i form an F-basis of the center C of the ring FG. Since C is a subring, $d_i d_j \in C$ for $1 \leq i, j \leq h$. Consider the product

$$d_i d_j = \sum_{x \in C_i} \sum_{y \in C_j} xy.$$

Let $x \in C_i$, $y \in C_j$; then $z = xy \in C_r$ for some r and z must occur in the product $d_i d_j$. If $g \in G$, then gzg^{-1} also occurs in $d_i d_j$ since the latter element lies in C, the center of FG. Define $m_{ij}^{(r)}$ to be the number of pairs (x,y) such that $x \in C_i$, $y \in C_j$ and xy equals a fixed element z_r of C_r. Notice that $m_{ij}^{(r)}$ does not depend on the choice of z_r. These considerations show that

$$d_i d_j = \sum_{r=1}^{h} m_{ij}^{(r)} d_r. \tag{14.5}$$

Let ρ be an irreducible representation of G with character χ. Extend the functions ρ and χ from G to FG in the obvious way. Now ρ arises from some simple FG-module M. Let i be fixed; since $d_i \in C$, we have $\rho(d_i) \in \operatorname{End}_{FG}(M)$. By (14.1.3) the latter consists of scalars: hence $\rho^*(d_i) = f_i I_n$ where $f_i \in F$. (Here as usual ρ^* denotes the corresponding matrix representation with respect to some fixed basis of M). Hence $\chi(d_i) = \operatorname{tr}(\rho^*(d_i)) = n f_i$. Also

$$\chi(d_i) = \sum_{x \in C_i} \chi(x) = \ell_i \chi^{(i)}$$

where $\ell_i = |C_i|$ and $\chi^{(i)}$ is the value of χ on the conjugacy class C_i. Hence $n f_i = \ell_i \chi^{(i)}$ and $f_i = \ell_i \chi^{(i)}/n = \ell_i \chi(g)/n$ for all $g \in C_i$.

Next apply the function ρ to the equation (14.5) to get $f_i f_j = \sum_{r=1}^{h} m_{ij}^{(r)} f_r$, since $\rho^*(d_i) = f_i I_n$. This may be rewritten as

$$\sum_{r=1}^{h} (f_i \delta_{jr} - m_{ij}^{(r)}) f_r = 0, \quad j = 1, 2, \dots, r. \tag{14.6}$$

With i fixed, let A denote the $h \times h$ matrix whose (j, r) entry is $f_i \delta_{jr} - m_{ij}^{(r)}$. Then in matrix form the system of equations (14.6) becomes

$$A[f_1 f_2 \dots f_h]^T = 0. \tag{14.7}$$

Suppose that $\det(A) \neq 0$, so A^{-1} exists. Multiply the equation (14.7) on the left by A^{-1} and conclude that $f_1 = f_2 = \cdots = f_h = 0$. Therefore $0 = f_i = \ell_i \chi(g)/n$ for all $g \in C_i$, $1 \leq i \leq h$. Hence χ is zero on C. But this is impossible because $\chi(1_G) = n \neq 0$ in F.

Thus we are forced to the conclusion that $\det(A) = 0$. On expanding $\det(A) = \det(f_i \delta_{jr} - m_{ij}^{(r)})$, we find that f_i is a root of a monic polynomial over $\mathbb{Z}$. Hence $f_i = \ell_i \chi(g)/n$ is an algebraic integer for $g \in C_i$ and $i = 1, 2, \dots, h$. $\qquad\square$

It is now possible to establish the divisibility property of the character degrees.

(14.3.10). *Let G be a finite group and F an algebraically closed field of characteristic 0. Then the degrees of the irreducible F-characters of G divide the order of the group.*

Proof. Let ρ be an irreducible F-representation of G with degree n and denote its character by χ. Set $m = |G|$ and let $C_i, i = 1, 2, \dots, h$ be the conjugacy classes of G. By (14.3.6) we have

$$\frac{m}{n} = \frac{1}{n} \sum_{x \in G} \chi(x^{-1})\chi(x).$$

Put $\ell_i = |C_i|$ and write $C_{i^*} = (C_i)^{-1}$, noting that $(C_i)^{-1}$ is a conjugacy class. Changing the summation in the last equation to one over conjugacy classes, we find that

$$\frac{m}{n} = \frac{1}{n} \sum_{i=1}^{h} \ell_i \chi^{(i)} \chi^{(i^*)} = \sum_{i=1}^{h} f_i \chi^{(i^*)}, \tag{14.8}$$

where $f_i = \ell_i \chi^{(i)}/n$. Now f_i is an algebraic integer by (14.3.9). Also $\chi^{(i^*)}$ is a sum of eigenvalues of eigenvalues of $\rho(g)$ for $g \in G$. The latter are roots of unity, since $g^m = 1$ implies that $f^m = 1$ with f is an eigenvalue of $\rho(g)$. Thus $\chi^{(i^*)}$ an algebraic integer and it follows that $\sum_{i=1}^{h} f_i \chi^{(i^*)}$ is also an algebraic integer. Finally, a rational number which is an algebraic integer is an integer by (11.1.12). Therefore by equation (14.8) n divides m. $\square$

Character tables

Let G be a finite group and F an algebraically closed field whose characteristic does not divide $|G|$. Let the class number of G be h. Write $C_1, C_2, \ldots, C_h$ for the conjugacy classes of G and $\chi_1, \chi_2, \ldots, \chi_h$ for the irreducible characters. The value of χ_i on the conjugacy class C_j will be written

$$\chi_i^{(j)}.$$

The character values can be displayed in a convenient tabular form called the *character table* of G.

	C_1	C_2	...	C_h
χ_1	$\chi_1^{(1)}$	$\chi_1^{(2)}$	...	$\chi_1^{(h)}$
χ_2	$\chi_2^{(1)}$	$\chi_2^{(2)}$	...	$\chi_2^{(h)}$
.	.	.	...	.
χ_h	$\chi_h^{(1)}$	$\chi_h^{(2)}$	...	$\chi_h^{(h)}$

Usually χ_1 is taken to be the trivial character and C_1 the conjugacy class $\{1\}$. With this convention, $\chi_1^{(i)} = 1$ for all i. Also $\chi_i^{(1)} = \chi_i(1) = n_i$, the degree of χ_i, since the trace of the $n_i \times n_i$ identity matrix is n_i. Thus the first column of the character table displays the degrees of the irreducible characters. The properties of the characters recorded in (14.3.6) can be translated into orthogonality properties of the rows and columns of the character table. These can be helpful in computing the characters of particular groups.

The orthogonality properties of the character table are listed in the following result. Here the χ_i are the distinct irreducible characters, ℓ_i is the degree of χ_i and the C_i are the conjugacy classes, while $C_{i^*} = (C_i)^{-1}$.

(14.3.11). *Let G be a finite group and F an algebraically closed field whose characteristic does not divide $m = |G|$. Then the following hold:*
(i) $\sum_{r=1}^{h} \ell_r \chi_i^{(r^*)} \chi_j^{(r)} = m\delta_{ij}$, *for all i, j, (orthogonality of rows);*
(ii) $\sum_{i=1}^{h} \chi_i^{(r)} \chi_i^{(s^*)} = \frac{m}{\ell_s} \delta_{rs}$, *for all r, s, (orthogonality of columns).*

Proof. From (14.3.6) we have $\sum_{x \in G} \chi_i(x^{-1}) \chi_j(x) = |G|\delta_{ij}$. Forming the sum over conjugacy classes yields

$$\sum_{r=1}^{h} \ell_r \chi_i^{(r^*)} \chi_j^{(r)} = m\delta_{ij} \tag{14.9}$$

since $|C_r| = \ell_r$. Thus (i) is established.

Let X and Y be the $h \times h$ matrices whose (i, r) and (r, j) entries are given by $X_{ir} = \ell_r \chi_i^{(r^*)}$ and $Y_{rj} = \chi_j^{(r)}$ respectively. Then equation (14.9) may be expressed in matrix form as $XY = mI_h$ where I_h is the identity $h \times h$ matrix. Now $\det(Y) \neq 0$ since $\det(XY) = \det(X)\det(Y) = m^h$; hence Y^{-1} exists. Therefore $YX = Y(XY)Y^{-1} = mI_h$. Taking the (r, s) entry on each side of the equation $YX = mI_h$, we derive

$$\sum_{i=1}^{h} Y_{ri} X_{is} = \sum_{i=1}^{h} \chi_i^{(r)} \ell_s \chi_i^{(s^*)} = m\delta_{rs}.$$

Observe that ℓ_s divides m, so ℓ_s^{-1} exists in F. Therefore we can divide by ℓ_s and the validity of (ii) follows. $\qquad\square$

The case of the complex field

The obvious field to which (14.3.11) is applicable is the complex field. It is worthwhile recording the simpler form that the orthogonality relations take in this case. Let ρ be a $\mathbb{C}$-representation of a finite group G and let f be an eigenvalue of $\rho(x)$ where $x \in G$. Then $x^m = 1$ for $m > 0$ implies that $f^m = 1$, so f is a complex root of unity. Therefore $f^{-1} = \bar{f}$, the complex conjugate of f. Thus $\chi(x^{-1})$ is the complex conjugate of $\chi(x)$, which we write as

$$\chi(x^{-1}) = \bar{\chi}(x).$$

Now the formulas in (14.3.11) take the following form:

$$\sum_{r=1}^{h} \ell_r \bar{\chi}_i^{(r)} \chi_j^{(r)} = m\delta_{ij} \quad \text{and} \quad \sum_{i=1}^{h} \chi_i^{(r)} \bar{\chi}_i^{(s)} = \frac{m}{\ell_s} \delta_{rs}.$$

Here is a simple example of a character table.

Example (14.3.1). Construct the character table of the symmetric group S_3 over $\mathbb{C}$.

The first step is to find the conjugacy classes of the group $G = S_3$. Write $G = \langle x, a \rangle$ where $x^2 = 1 = a^3$ and $xax^{-1} = a^2$. The conjugacy classes are $C_1 = \{1\}$, $C_2 = \{a, a^2\}$, $C_3 = \{x, xa, xa^2\}$ and the class number is $h = 3$.

By (14.2.1) the degrees of the three irreducible representations satisfy $n_1^2 + n_2^2 + n_3^2 = |G| = 6$. Thus $n_1 = n_2 = 1$, $n_3 = 2$. Let χ_1 be the trivial character and χ_2 the character arising from $G \to G^{ab} \to \langle -1 \rangle$. Hence the values of χ_2 on the three conjugacy classes are 1, 1, −1. So far the extent of our knowledge of the character table is:

	C_1	C_2	C_3
χ_1	1	1	1
χ_2	1	1	−1
χ_3	2	s	t

Orthogonality properties of columns allow us to determine s and t. Note that in this example $\chi_i(g^{-1}) = \chi_i(g)$. From columns 1 and 2 we obtain $1 + 1 + 2s = 0$ and $s = -1$. Also from columns 1 and 3 there follows $1 - 1 + 2t = 0$, so that $t = 0$. Hence the character table of G is

	C_1	C_2	C_3
χ_1	1	1	1
χ_2	1	1	−1
χ_3	2	−1	0

It is not hard to identify the irreducible representation whose character is χ_3. A little experimentation reveals that it is

$$a \mapsto \begin{bmatrix} \omega & 0 \\ 0 & \omega^2 \end{bmatrix}, \quad x \mapsto \begin{bmatrix} 0 & 1 \\ 1 & 0 \end{bmatrix},$$

where $\omega = e^{\frac{2\pi i}{3}}$, a primitive cubic root of unity. The traces of these matrices are −1 and 0 respectively since $\omega + \omega^2 = -1$, which is in agreement with character table.

Exercises (14.3).

(1) Construct the character table of A_4 over $\mathbb{C}$.

(2) The same problem for Dih(8).

(3) The same problem for the quaternion group Q_8. (For this group see Exercise (6.3.13).)

(4) Show that non-isomorphic groups can have the same character table.

(5) Show that the degree of an irreducible $\mathbb{Q}$- character need not divide the group order.

(6) Let A be the $h{\times}h$ matrix whose (i, r) entry is $\sqrt{\ell_r/m}/\chi_i^{(r)}$ with the notation of (14.3.11) and $F = \mathbb{C}$. Prove that A is a unitary matrix, i. e. $A\bar{A}^T = 1$.

(7) Complete the proof of (14.3.3) by showing that the dimension of $\mathrm{Cl}(G, F)$ equals the class number of G. [Hint: let the conjugacy classes be C_i, $i = 1, 2, \ldots, h$ and define $\beta_i \in \mathrm{Cl}(G, F)$ by $\beta_i(x) = 1$ if $x \in C_i$ and $\beta_i(x) = 0$ if $x \notin C_i$. Show that the β_i form a basis of $\mathrm{Cl}(G, F)$.]

(8) Let G be a finite group and F an algebraically closed field whose characteristic does not divide $|G|$. If χ an F-character of G, prove that χ is irreducible if and only if $\langle \chi, \chi \rangle = 1$. [Hint: for necessity see (14.3.7). To prove sufficiency write $\chi = \sum_{i=1}^{h} m_i \chi_i$ where the χ_i are the irreducible characters and $m_i > 0$. Compute $\langle \chi, \chi \rangle$.]

(9) Let G be a perfect group. Use (14.2.1) and (14.3.10) to show that there are integers $r_i > 1$ dividing $|G|$ such that $|G| - 1 = \sum_{i=1}^{k} r_i^2$.

14.4 The Burnside $p - q$ Theorem

In this section our aim is to prove that *every group whose order is divisible by at most two primes p, q is solvable*. This is the famous Burnside $p - q$ theorem. It was one of the first great successes of representation theory. Indeed for many years no proof of it was known that did not involve group characters, although some low order cases can be handled using only Sylow's theorem – see (10.2.4). The theorem will be established via two preliminary results.

(14.4.1). *Let ρ be an irreducible representation of a finite group G over the complex field $\mathbb{C}$ with character χ. Suppose that g is an element of G with exactly ℓ conjugates where ℓ is relatively prime to the degree of ρ. Then either $\chi(g) = 0$ or else $\rho(g)$ is scalar.*

Proof. First recall from (14.3.9) that if n is the degree of χ, then $\frac{\ell\chi(g)}{n}$ is an algebraic integer. Since ℓ and n are relatively prime, there are integers r and s such that $1 = r\ell + sn$. Write

$$t = \frac{\chi(g)}{n} = \frac{r\ell\chi(g)}{n} + s\chi(g).$$

Since $\chi(g)$ is a sum of roots of unity, t is an algebraic integer.

Let $f_1, f_2, \ldots, f_n$ denote the eigenvalues of $\rho(g)$; then $\chi(g) = \sum_{i=1}^{n} f_i$. Since the f_i are complex roots of unity, $|f_i| = 1$. By definition

$$t = \frac{1}{n} \sum_{i=1}^{n} f_i,$$

which shows that $|t| \leq \frac{1}{n} \sum_{i=1}^{n} |f_i| \leq 1$. Suppose that all the f_i are equal; then $\rho(g)$ is scalar, as can be seen by applying Maschke's theorem to $\rho|_{\langle g \rangle}$. Therefore we can assume that this is not the case. Now it is an elementary exercise to prove that if

$z_1, z_2, \ldots, z_n$ are complex numbers such that $|z_i| = 1$ for all i and $|\sum_{i=1}^n z_i| = n$, then $z_1 = z_2 = \cdots = z_n$. Applying this to our situation with $z_i = f_i$, we deduce that $|t| < 1$. Write K for the field $\mathbb{Q}(f_1, f_2, \ldots, f_n)$ and let $\alpha \in \mathrm{Gal}(K/\mathbb{Q})$. Now the $\alpha(f_i)$ cannot all be equal, so $|\alpha(t)| < 1$. It follows that the element

$$u = \prod_{\alpha \in \mathrm{Gal}(K/\mathbb{Q})} \alpha(t)$$

satisfies $|u| < 1$. Next $\alpha(u) = u$ for all α, so the Fundamental Theorem of Galois Theory (12.3.1) guarantees that $u \in \mathbb{Q}$. But u is an algebraic integer because it is the product of algebraic integers $\alpha(t)$. Therefore u is an integer by (11.1.12). Since $|u| < 1$, it follows that $u = 0$ and thus $t = 0$. Therefore $f_i = 0$ for all i, and hence $\chi(g) = 0$. $\qquad\square$

(14.4.2). *Let G be a finite group which has a conjugacy class with exactly $p^m > 1$ elements where p is a prime. Then G is not a simple group.*

Proof. Assume to the contrary that G is simple: of course it cannot be abelian. By hypothesis there exists an element $g \in G$ with p^m conjugates. Let ρ be a non-trivial irreducible $\mathbb{C}$-representation of G and let χ be its character. Assume that $\chi(g) \neq 0$ and p does not divide the degree of χ. Then by (14.4.1) the linear operator $\rho(g)$ is scalar, so it lies in the center of $\rho(G)$. Now $\mathrm{Ker}(\rho) \neq G$, since ρ is not the trivial representation. But G is simple, so $\mathrm{Ker}(\rho) = 1$. Therefore ρ maps G isomorphically onto $\rho(G)$, which implies that $g \in Z(G)$ and hence $p^m = 1$. By this contradiction every non-trivial irreducible character χ of G either has degree divisible by p or satisfies $\chi(g) = 0$.

Next let σ denote the regular permutation representation of G and write ψ for its character. By (14.3.8) there is an expression $\psi = \sum_{i=1}^h \ell_i \chi_i$ where ℓ_i is the degree of the irreducible character χ_i and h is the class number of G. Now $\ell_1 = 1$ and $\chi_1(g) = 1$ since χ_1 is the trivial character. Consider the expression $\psi(g) = \sum_{i=1}^h \ell_i \chi_i(g)$. If $i > 1$, then either $\chi_i(g) = 0$ or else ℓ_i is divisible by p. Therefore $\psi(g) \equiv 1 \mod p$. On the other hand, $\sigma(g)$ arises from left multiplication in G by $g \neq 1$, so that $\sigma(g)$ has no fixed points. Therefore $\psi(g) = \mathrm{tr}(\sigma^*(g)) = 0$ and a contradiction ensues. $\qquad\square$

We are now in a position to establish the principal result of the section.

(14.4.3) (The Burnside $p - q$ Theorem). *If p and q are primes, then any group of order $p^m q^n$ is solvable.*

Proof. Assume the statement is false and let G be a counterexample of smallest order. If N is a proper non-trivial normal subgroup, then N and G/N are both solvable, being groups with smaller order than G. But this implies that G is solvable. Therefore G must be a simple group.

Let Q be a Sylow q-subgroup of G. Certainly $Q \neq 1$ since otherwise G would be a p-group and hence solvable. Therefore Q has non-trivial center by (5.3.6). Let $1 \neq g \in Z(Q)$; thus $Q \leq C_G(g) < G$. It follows that $|G : C_G(g)|$ divides $|G : Q|$ and hence is a power of p: moreover this is the number of conjugates of g in G. By (14.4.2) the group G is not simple, a final contradiction. $\qquad\qquad\qquad\qquad\qquad\qquad\qquad\qquad\qquad\qquad$ $\square$

Exercises (14.4).

(1) Show that there are insolvable groups of each order $2^{\ell}3^m5^n$ where $\ell \geq 2$, $m, n \geq 1$. [Hint: A_5 has order $2^2 \cdot 3 \cdot 5$.]

(2) Let G be a group of order $p^m qr$ where p, q, r are distinct primes. Assume that G has a subgroup of order qr where q does not divide $r - 1$ and r does not divide $q - 1$. Prove that G is solvable. [For groups of order qr see Exercise (5.3.2).]

(3) Suppose that G is a finite group with a nilpotent subgroup of prime power index. Prove that G is a solvable group. [Hint: let G be a minimal counterexample and argue that G must be a simple group. Now apply (14.4.2).]

Needless to say we have only scratched the surface of representation theory; for a deeper account see [9] or [20].

15 Presentations of groups

15.1 Free groups

When groups entered the mathematical arena towards the close of the 18th century, they were exclusively permutation groups and were often studied in connection with the theory of equations. A hundred years later groups arose from a different source, geometry, and these groups were most naturally specified by listing a set of *generators* and a set of *defining relations* which the generators had to satisfy. A very simple example is where there is a single generator x subject to one defining relation $x^n = 1$ where n is a positive integer. Intuitively one would expect these to determine a cyclic group of order n.

As another example, suppose that a group has two generators x and y subject to the three relations $x^2 = 1 = y^2$ and $xy = yx$. Now the Klein 4-group fits this description, with $x = (12)(34)$ and $y = (13)(24)$. Thus it seems reasonable that a group with these generators and relations should be a Klein 4-group.

Of course these claims cannot be substantiated until we have explained what is meant by a group with a given set of generators subject to a set of defining relations. Even when the generators are subject to no defining relations at all, a precise definition is lacking: this is the important case of a *free group*. Thus our first task must be to define a free group.

Free groups

A free group is defined by a certain *mapping property*. Let F be a group, X a non-empty set and $\sigma : X \to F$ a function. Then F, or more precisely the pair (F, σ), is said to be *free on X* if, for each function α from X to a group G there is a *unique* homomorphism $\beta : F \to G$ such that $\beta\sigma = \alpha$, i. e., the triangle below commutes:

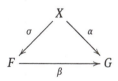

First a comment on the definition. *The function $\sigma : X \to F$ is necessarily injective.* For suppose that $\sigma(x_1) = \sigma(x_2)$ where $x_1 \neq x_2$. Let G be any group with two or more elements and choose a function $\alpha : X \to G$ such that $\alpha(x_1) \neq \alpha(x_2)$ in G. From $\sigma(x_1) = \sigma(x_2)$ it follows that $\beta\sigma(x_1) = \beta\sigma(x_2)$ and hence $\alpha(x_1) = \alpha(x_2)$, a contradiction.

This observation indicates that we can replace X by the set $\text{Im}(\alpha)$, which has the same cardinality, and take X to be a subset of F with σ the inclusion map. What the mapping property then asserts is that every mapping from the subset X to a group G can be extended to a unique homomorphism from F to G.

https://doi.org/10.1515/9783110691160-015

Free groups, and also free modules, which were defined in Section 9.1, are special cases of free objects in a category: these are discussed at length in Section 16.3 below.

At first sight the definition of a free group may seem abstract and certainly it offers no clue as to the nature or even the existence of free groups. Soon concrete examples of free groups will be given. In the meantime the first order of business is to show that free groups exist. Unlike the case of free modules, where these are easily defined as direct sums, free groups have to be constructed from scratch.

(15.1.1). *Let X be any non-empty set. Then there exist a group F and a function $\sigma : X \to F$ such that (F, σ) is free on X. Moreover, F is generated by $\mathrm{Im}(\sigma)$.*

Proof. Roughly speaking, the idea of the proof is to construct F by forming "words" in X which are combined in a formal manner by juxtaposition, while at the same time allowing for cancellation of word segments like xx^{-1} or $x^{-1}x$ where $x \in X$.

The first step is to choose a set disjoint from X with the same cardinality. Since the purpose of this move is to accommodate inverses of elements of X, it is appropriate to denote the set of inverses by $X^{-1} = \{x^{-1} \mid x \in X\}$. But keep in mind that x^{-1} is merely a symbol at this point and does not denote an inverse. By *a word in X* is meant any finite sequence w of elements of the set $X \cup X^{-1}$, written for convenience in the form

$$w = x_1^{q_1} x_2^{q_2} \cdots x_r^{q_r},$$

where $q_i = \pm 1$, $x_i^1 = x_i \in X$ and $r \geq 0$. The case $r = 0$, when the sequence is empty, is the *empty word*, which is written 1. Two words are said to be *equal* if they have the same entries in each position, i. e., they look exactly alike.

The *product* of words $w = x_1^{q_1} \cdots x_r^{q_r}$ and $v = y_1^{p_1} \cdots y_s^{p_s}$ is formed in the obvious way by juxtaposition, i. e.,

$$wv = x_1^{q_1} \cdots x_r^{q_r} y_1^{p_1} \cdots y_s^{p_s},$$

with the convention that $w1 = w = 1w$. This is clearly an associative binary operation on the set S of all words in X. The *inverse* of the word w is defined to be

$$w^{-1} = x_r^{-q_r} \cdots x_1^{-q_1},$$

with the convention that $1^{-1} = 1$. Thus far S, together with the product operation, is a semigroup with an identity element, i. e., a monoid. Next a device will be introduced which permits the cancellation of segments of a word with the form xx^{-1} or $x^{-1}x$. Once this is done, instead of a monoid, we will have a group.

A relation $\sim$ on the set S is defined in the following way: $w \sim v$ means that it is possible to pass from w to v by means of a finite sequence of operations of the following types:

(i) insertion of xx^{-1} or $x^{-1}x$ as consecutive symbols in a word where $x \in X$;

(ii) deletion of any such sequence from a word.

For example, $xyy^{-1}z \sim t^{-1}txz$ where $x, y, z, t \in X$. It is easy to check that $\sim$ is an equivalence relation on S. Let F denote the set of all equivalence classes of words $[w]$, $w \in S$. Our aim is to make F into a group: this will turn out to be a free group on the set X.

If $w \sim w'$ and $v \sim v'$, then it is readily seen that $wv \sim w'v'$. It is therefore meaningful to define the *product* of the equivalence classes $[w]$ and $[v]$ by the rule

$$[w][v] = [wv].$$

It follows from this that $[w][1] = [w]$ and $[1][v] = [v]$ for all words w, v. Also $[w][w^{-1}] = [1] = [w^{-1}][w]$, since ww^{-1} and $w^{-1}w$ are plainly equivalent to the empty word 1. Finally, we verify the associative law:

$$([u][v])[w] = [uv][w] = [(uv)w] = [u(vw)] = [u][vw] = [u]([v][w]).$$

Consequently, F is a group in which [1] is the identity element and $[w^{-1}]$ is the inverse of $[w]$. Furthermore, F is generated by the subset $\bar{X} = \{[x] \mid x \in X\}$: for, if $w = x_1^{q_1} x_2^{q_2} \cdots x_r^{q_r}$ with $x_i \in X$, $q_i = \pm 1$, then

$$[w] = [x_1]^{q_1} [x_2]^{q_2} \cdots [x_r]^{q_r} \in \langle \bar{X} \rangle.$$

It remains to prove that F is a free group on X. To this end define a function $\sigma : X \to F$ by the rule $\sigma(x) = [x]$; thus the image of σ is $\bar{X} = \{[x] \mid x \in X\}$ and this subset generates F. Next let $\alpha : X \to G$ be a map from X into some group G. To show that (F, σ) is free on X we need to produce a unique homomorphism $\beta : F \to G$ such that $\beta\sigma = \alpha$. There is only one reasonable candidate here: define β by the rule

$$\beta([x_1^{q_1} x_2^{q_2} \cdots x_r^{q_r}]) = \alpha(x_1)^{q_1} \alpha(x_2)^{q_2} \cdots \alpha(x_r)^{q_r},$$

where $x_i \in X$, $q_i = \pm 1$. The first thing to observe is that β is well-defined: for any other element in the equivalence class $[x_1^{q_1} x_2^{q_2} \cdots x_r^{q_r}]$ differs from $x_1^{q_1} x_2^{q_2} \cdots x_r^{q_r}$ only by segments of the form xx^{-1} or $x^{-1}x$, $(x \in X)$, and these will contribute to the image under β merely $\alpha(x)\alpha(x)^{-1}$ or $\alpha(x)^{-1}\alpha(x)$, i.e., the identity. It is a simple direct check that β is a homomorphism. Notice also that $\beta\sigma(x) = \beta([x]) = \alpha(x)$, so that $\beta\sigma = \alpha$.

Finally, we have to establish the uniqueness of β. If $\beta' : F \to G$ is another homomorphism for which $\beta'\sigma = \alpha$, then $\beta\sigma = \beta'\sigma$ and thus β and β' agree on $\operatorname{Im}(\sigma)$. But $\operatorname{Im}(\sigma)$ generates the group F, so $\beta = \beta'$. Therefore (F, σ) is free on X. □

Reduced words

Now that free groups are known to exist, it is necessary to find a convenient form for their elements. Let F be the free group on the set X that has just been constructed.

A word in X is called *reduced* if it contains no pairs of consecutive symbols xx^{-1} or $x^{-1}x$ with $x \in X$. The empty word is considered to be reduced. Now if w is any word, we can delete subsequences xx^{-1} and $x^{-1}x$ from w until a reduced word is obtained. Thus each equivalence class $[w]$ contains at least one reduced word. The important point to establish is that there is a *unique* reduced word in each equivalence class.

(15.1.2). *Each equivalence class of words on X contains a unique reduced word.*

Proof. There are likely to be multiple ways to cancel segments xx^{-1} or $x^{-1}x$ from a word. For this reason a direct approach to proving uniqueness would be complicated. An indirect argument will be used which avoids this difficulty.

Let R denote the set of all reduced words in X. The idea behind the proof is to introduce a permutation representation of the free group F on the set R. Let $u \in X \cup X^{-1}$: then a function $u' : R \to R$ is determined by the following rule

$$u'\left(x_1^{q_1}x_2^{q_2}\cdots x_r^{q_r}\right) = \begin{cases} ux_1^{q_1}x_2^{q_2}\cdots x_r^{q_r} & \text{if } u \neq x_1^{-q_1} \\ x_2^{q_2}\cdots x_r^{q_r} & \text{if } u = x_1^{-q_1}. \end{cases}$$

Here $x_1^{q_1}x_2^{q_2}\cdots x_r^{q_r}$ denotes a reduced word; observe that after applying the function u' we still have a reduced word. Next u' is a permutation of R since its inverse is the function $(u^{-1})'$. Now let $\alpha : X \to \mathrm{Sym}(R)$ be defined by the assignment $u \mapsto u'$.

By the mapping property of the free group F there is a homomorphism $\beta : F \to \mathrm{Sym}(R)$ such that $\beta\sigma = \alpha$: hence $\alpha(x) = \beta\sigma(x) = \beta([x])$ for $x \in X$: here $\sigma : X \to F$ is the function used in the construction of F, so $\sigma(x) = [x]$. Thus the diagram below is commutative.

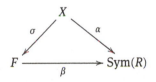

Now suppose that v and w are two equivalent reduced words; we will show that $v = w$. Certainly $[v] = [w]$, so $\beta([v]) = \beta([w])$. If $v = x_1^{q_1}x_2^{q_2}\cdots x_r^{q_r}$, then $[v] = [x_1^{q_1}][x_2^{q_2}]\cdots[x_r^{q_r}]$ and we have

$$\beta([v]) = \beta([x_1^{q_1}])\beta([x_2^{q_2}])\cdots\beta([x_r^{q_r}]) = \beta([x_1])^{q_1}\beta([x_2])^{q_2}\cdots\beta([x_r])^{q_r}.$$

Now $\beta([x_i]) = \beta\sigma(x_i) = \alpha(x_i)$. Therefore

$$\beta([v]) = \alpha(x_1)^{q_1}\alpha(x_2)^{q_2}\cdots\alpha(x_r)^{q_r} = (x_1')^{q_1}(x_2')^{q_2}\cdots(x_r')^{q_r}.$$

Applying the function $\beta([v])$ to the empty word 1, which is reduced, we obtain $x_1^{q_1}x_2^{q_2}\cdots x_r^{q_r} = v$ since this word is reduced. Similarly $\beta([w])$ sends the empty word to w. Therefore $v = w$. $\qquad\square$

Normal form in free groups

The proof of (15.1.2) is subtle and it is well worth rereading. The importance of this result is that it provides a unique way of representing the elements of the constructed free group F on the set X. Each element of F has the form $[w]$ where w is a uniquely determined reduced word, say $w = x_1^{q_1} x_2^{q_2} \cdots x_r^{q_r}$ where $q_i = \pm 1$, $r \geq 0$. No consecutive terms xx^{-1} or $x^{-1}x$ occur in w. Now $[w] = [x_1]^{q_1}[x_2]^{q_2} \cdots [x_r]^{q_r}$; on combining consecutive terms of this product which involve the same x_i, we conclude that the element $[w]$ can be uniquely written in the form

$$[w] = [x_1]^{\ell_1}[x_2]^{\ell_2} \cdots [x_s]^{\ell_s},$$

where $x_i \in X$, $s \geq 0$, ℓ_i is a non-zero integer and $x_i \neq x_{i+1}$.

To simplify the notation let us agree to drop the distinction between x and $[x]$, so that now $X \subseteq F$. Then every element w of F has the unique form

$$w = x_1^{\ell_1} x_2^{\ell_2} \cdots x_s^{\ell_s}$$

where $x_i \in X$, $s \geq 0$, $\ell_i \neq 0$ and $x_i \neq x_{i+1}$. This is called the *normal form* of w. For example, if $X = \{x\}$, each element of F has the unique normal form x^ℓ, where $\ell \in \mathbb{Z}$. Thus $F = \langle x \rangle$ is an infinite cyclic group.

In fact the existence of a normal form is characteristic of free groups in the sense of the next result.

(15.1.3). *Let X be a subset of a group G and suppose that each element g of G can be uniquely written in the form $g = x_1^{\ell_1} x_2^{\ell_2} \cdots x_s^{\ell_s}$ where $x_i \in X$, $s \geq 0$, $\ell_i \neq 0$, and $x_i \neq x_{i+1}$. Then G is free on X.*

Proof. Let F be the free group of equivalence classes of words in the set X which was constructed in (15.1.1), and let $\sigma : X \to F$ be the associated injection; thus $\sigma(x) = [x]$. Apply the mapping property with $\alpha : X \to G$ the inclusion map, i. e., $\alpha(x) = x$ for all $x \in X$. Hence there is a homomorphism $\beta : F \to G$ such that $\beta\sigma = \alpha$, so $\mathrm{Im}(\alpha) \subseteq \mathrm{Im}(\beta)$. Since $X = \mathrm{Im}(\alpha)$ generates G, it follows that $\mathrm{Im}(\beta) = G$ and β is surjective. Finally, β is injective. For, if $\beta([x_1]^{\ell_1} \cdots [x_r]^{\ell_r}) = 1$ with $r > 0$, $x_i \neq x_{i+1}$, $\ell_i \neq 0$, then $(\beta\sigma(x_1))^{\ell_1} \cdots (\beta\sigma(x_r))^{\ell_r} = 1$, and hence $x_1^{\ell_1} \cdots x_r^{\ell_r} = 1$: this contradicts the uniqueness of expression in terms of the x_i. Therefore β is an isomorphism and $F \simeq G$, so that G is free on X. $\square$

Up to this point we have worked with a particular free group on a set X, namely the group constructed from equivalence classes of words in X. However, all free groups on the same set are isomorphic, a fact that allows us to deal only with free groups of words. This follows from the next result.

(15.1.4). *Let F_i be a free group on X_i, $i = 1, 2$, where $|X_1| = |X_2|$. Then $F_1 \simeq F_2$.*

Proof. Let $\sigma_1 : X_1 \to F_1$ and $\sigma_2 : X_2 \to F_2$ be the respective injections associated with the free groups F_1 and F_2, and let $\alpha : X_1 \to X_2$ be a bijection, which exists since $|X_1| = |X_2|$. By the mapping property there are commutative diagrams

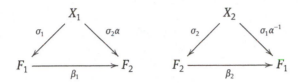

in which β_1 and β_2 are homomorphisms. Thus $\beta_1\sigma_1 = \sigma_2\alpha$ and $\beta_2\sigma_2 = \sigma_1\alpha^{-1}$. Hence $\beta_2\beta_1\sigma_1 = \beta_2\sigma_2\alpha = \sigma_1\alpha^{-1}\alpha = \sigma_1$. Similarly $\beta_1\beta_2\sigma_2 = \beta_1\sigma_1\alpha^{-1} = \sigma_2\alpha\alpha^{-1} = \sigma_2$. Therefore the diagrams below commute,

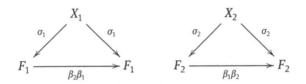

But the identity maps on F_1 and F_2 will also make these diagrams commute. Hence $\beta_2\beta_1$ and $\beta_1\beta_2$ must equal these identity maps by the uniqueness clause in the mapping property. Therefore $\beta_1 : F_1 \to F_2$ is an isomorphism. $\qquad\square$

In fact the converse of (15.1.4) is true: *if $F_1 \simeq F_2$, then $|X_1| = |X_2|$*. For a hint on how to prove this see Exercise (15.1.8). It follows that a free group is determined up to isomorphism by the cardinality of the set on which it is free. This cardinality is called the *rank* of the free group.

Examples of free groups

At this point free groups may still appear to the reader to be mysterious abstract objects, despite our success in constructing them. It is time to remedy this by exhibiting some real life examples.

Example (15.1.1). Consider the functions α and β on the set $\mathbb{C} \cup \{\infty\}$ which are defined by the rules

$$\alpha(x) = x + 2 \quad \text{and} \quad \beta(x) = \frac{1}{2 + \frac{1}{x}}.$$

Here the symbol ∞ is required to satisfy the formal rules $\frac{1}{\infty} = 0$, $\frac{1}{0} = \infty$, $2 + \infty = \infty$. Thus $\alpha(\infty) = \infty$, $\beta(0) = 0$ and $\beta(\infty) = \frac{1}{2}$. The first thing to notice is that α and β are bijections since they have inverses. These are given by $\alpha^{-1}(x) = x - 2$ and $\beta^{-1}(x) = \frac{1}{\frac{1}{x} - 2}$. This can be checked by computing the composites $\alpha\alpha^{-1}$, $\alpha^{-1}\alpha$, $\beta\beta^{-1}$, $\beta^{-1}\beta$.

Define F to be the subgroup $\langle \alpha, \beta \rangle$ of the symmetric group on the set $\mathbb{C} \cup \{\infty\}$. We are going to prove that F is a free group on $\{\alpha, \beta\}$. To accomplish this it is enough to show that no non-trivial reduced word in α and β can equal 1: for then each element of F has a unique normal form and (15.1.3) can be invoked to show that F is free.

Since direct calculations with the functions α and β would be tedious, a geometric approach is adopted. Observe that each non-trivial power of α maps the interior of the unit circle in the complex plane to its exterior. Also a non-trivial power of β maps the exterior of the unit circle to its interior with $(0, 0)$ removed: the truth of the last statement is seen from the equation $\beta(\frac{1}{x}) = \frac{1}{x+2}$. It follows from this observation that no mapping of the form $\alpha^{\ell_1} \beta^{m_1} \cdots \alpha^{\ell_r} \beta^{m_r}$ can be trivial unless all the ℓ_i and m_i are 0.

Example (15.1.2). An even more concrete example of a free group is provided by the matrices

$$A = \begin{bmatrix} 1 & 2 \\ 0 & 1 \end{bmatrix} \quad \text{and} \quad B = \begin{bmatrix} 1 & 0 \\ 2 & 1 \end{bmatrix};$$

for these generate a subgroup F_1 of $GL_2(\mathbb{Z})$ which is free on $\{A, B\}$.

To see why this is true, first consider a matrix

$$U = \begin{bmatrix} a & b \\ c & d \end{bmatrix} \in GL_2(\mathbb{C}).$$

Thus $ad - bc \neq 0$. There is a corresponding permutation $\theta(U)$ of $\mathbb{C} \cup \{\infty\}$ defined by

$$\theta(U) : x \mapsto \frac{ax + b}{cx + d} = \frac{a + \frac{b}{x}}{c + \frac{d}{x}}.$$

Note that $\theta(U)(\infty) = \frac{a}{c}$ if $c \neq 0$ and ∞ otherwise. This is called a *linear fractional transformation*. It is easy to verify that $\theta(UV) = \theta(U)\theta(V)$, so that $\theta : GL_2(\mathbb{C}) \rightarrow \text{Sym}(\mathbb{C} \cup \{\infty\})$ is a homomorphism. The linear fractional transformations form a subgroup $\text{Im}(\theta)$ of $\text{Sym}(\mathbb{C} \cup \{\infty\})$. Now $\theta(A) = \alpha$ and $\theta(B) = \beta$. Hence, if some non-trivial reduced word in A and B were to equal the identity matrix, the corresponding word in α and β would equal the identity permutation, which is impossible by Example (15.1.1). Therefore F_1 is free on $\{A, B\}$ by (15.1.3).

Next we will use normal form to obtain some structural information about free groups.

(15.1.5). *Let F be a free group on a set X. Then:*
(i) *each non-trivial element of F has infinite order; hence free groups are torsion-free;*
(ii) *if F is not infinite cyclic, i. e. $|X| > 1$, then $Z(F) = 1$.*

Proof. (i) Let $1 \neq f \in F$ and suppose that $f = x_1^{\ell_1} x_2^{\ell_2} \cdots x_s^{\ell_s}$ is the normal form. If $x_1 = x_s$, we can replace f by the conjugate $x_s^{\ell_s} f x_s^{-\ell_s} = x_1^{\ell_1 + \ell_s} x_2^{\ell_2} \cdots x_{s-1}^{\ell_{s-1}}$, which has the same order as f. For this reason there is nothing to be lost in assuming that $x_1 \neq x_s$. Let m be a positive integer; then

$$f^m = (x_1^{\ell_1} \cdots x_s^{\ell_s})(x_1^{\ell_1} \cdots x_s^{\ell_s}) \cdots (x_1^{\ell_1} \cdots x_s^{\ell_s}),$$

with m factors in parentheses, which is in normal form since $x_1 \neq x_s$. It follows that $f^m \neq 1$ and f has infinite order.

(ii) Assume that $1 \neq f \in Z(F)$ and let $f = x_1^{\ell_1} x_2^{\ell_2} \cdots x_s^{\ell_s}$ be the normal form of f. If $s = 1$, we can choose $x \neq x_1$ from X, since $|X| > 1$. But then $xf \neq fx$; therefore $s > 1$. By conjugating f as in (i), we may assume that $x_1 \neq x_s$. Then $fx_1 = x_1^{\ell_1} x_2^{\ell_2} \cdots x_s^{\ell_s} x_1$ and $x_1 f = x_1^{\ell_1 + 1} x_2^{\ell_2} \cdots x_s^{\ell_s}$ are both in normal form, except that $x_1^{\ell_1 + 1}$ is trivial if $\ell_1 = -1$; but in any event $fx_1 \neq x_1 f$ and so $f \notin Z(G)$. $\qquad\square$

Subgroups of free groups

During the discussion of free modules in Chapter Nine it was shown that a submodule of a free module over a principal ideal domain is always free – see (9.2.4). In particular, a subgroup of a free abelian group is free abelian. It is natural to ask if a subgroup of a free group is necessarily a free group. An affirmative answer is given by a well known theorem.

(15.1.6) (The Nielsen[1]-Schreier Theorem). *Let F be a free group of rank n and let H be a subgroup of F. Then H is a free group. Moreover, if H has finite index m in F, the rank of H is exactly $mn + 1 - m$.*

This theorem is much less obvious than the corresponding result for modules. There are many different approaches to proving it. Since the algebraic proof is somewhat technical, it will not be presented here. For an account of it see [15].

Exercises (15.1).

(1) Let (F, σ) be free on a set X where $\sigma : X \to F$. Prove that $F = \langle \text{Im}(\sigma) \rangle$. [Hint: use (15.1.4) and the construction of a free group on X.]

(2) Let F be the free group on a set X. Prove that an element f of F belongs to the derived subgroup F' if and only if the sum of the exponents of x in the normal form of f is 0 for every x in X.

(3) If F is a free group, prove that F/F' is a direct product of infinite cyclic groups, i. e., it is a free abelian group.

(4) Let G be the subgroup of $GL_2(\mathbb{R})$ generated by the matrices $\left[\begin{smallmatrix} 1 & a \\ 0 & 1 \end{smallmatrix}\right]$ and $\left[\begin{smallmatrix} 1 & 0 \\ a & 1 \end{smallmatrix}\right]$ where a is real and $a \geq 2$. Prove that G is a free group.

---

1 Jacob Nielsen 1890–1959.

(5) (*The projective property of free groups*). Let there be given groups and homomorphisms $\alpha : F \to H$ and $\beta : G \to H$ where F is a free group and β is surjective. Show that there is a homomorphism $\gamma : F \to G$ such that $\beta\gamma = \alpha$, i. e., the triangle below commutes.

(6) Let G be a group with a normal subgroup N such that G/N is a free group. Prove that there is free subgroup H such that $G = HN$ and $H \cap N = 1$.

(7) Let H be a subgroup with finite index in a free group F. If $1 \neq K \leq F$, prove that $H \cap K \neq 1$. [Hint: let C be the normal core of H in F, i. e., the intersection of all the conjugates of H in F. Show that F/C is finite, whence so is $K/C \cap K$.]

(8) Let F_1 and F_2 be free groups on sets X_1 and X_2 respectively. If $F_1 \simeq F_2$, prove that $|X_1| = |X_2|$. Thus a free group is determined up to isomorphism by the cardinality of the set on which it is free. [Hint: consider F_i/F_i^2 as a vector space over GF(2).]

(9) Let $1 \neq f \in F$ where F is a free group. Prove that $C_F(f)$ is cyclic. [Hint: use the Nielsen-Schreier theorem.]

15.2 Generators and relations

In this section it is shown that free groups occupy a key position in group theory in the sense that their quotients account for all groups. The next result should be compared with (9.1.17), which is the corresponding result for free modules.

(15.2.1). *Let G be a group and X a set of generators for G. If F is a free group on the set X, there is a surjective homomorphism $\theta : F \to G$ and hence $G \simeq F/\mathrm{Ker}(\theta)$.*

Proof. Let (F, σ) be free on X. The existence of the homomorphism θ follows on applying the mapping property of the free group F to obtain the commutative diagram

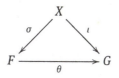

where ι is the inclusion map. Thus $x = \iota(x) = \theta\sigma(x) \in \mathrm{Im}(\theta)$ for all x in X. Hence $G = \mathrm{Im}(\theta) \simeq F/\mathrm{Ker}(\theta)$. $\qquad\square$

We are now ready to explain what it means for a group to be given by a set of generators and defining relations. Let X be a non-empty set and let F be the free group on X with $X \subseteq F$. Let R be a subset of F and define

$$N = \langle R^F \rangle,$$

the normal closure of R in F: thus N is the subgroup generated by all conjugates in F of elements of R – for normal closures see Section 4.2. Let

$$G = F/N.$$

Certainly the group G is generated by the elements xN where $x \in X$; also $r(xN) = r(x)N = N = 1_G$ for all $r \in R$. Hence the relations $r = 1$ hold in G. Here $r(xN)$ is the element of G obtained from r by replacing each x by xN. Then G is called *the group with generators X and defining relations $r = 1$* where $r \in R$: in symbols

$$G = \langle X \mid r = 1, \forall r \in R \rangle.$$

Elements of R are called *defining relators* and the group may also be written

$$G = \langle X \mid R \rangle.$$

The pair $(X|R)$ is said to be a *presentation* of G. An element w in the normal subgroup N is a *relator*; it is expressible as a product of conjugates of defining relators and their inverses. The relator w is said to be a *consequence* of the defining relators in R. Finally, a presentation $(X|R)$ is called *finite* if X and R are both finite.

Our first concern is to show that every group can be defined by a presentation.

(15.2.2). *Every group has a presentation.*

Proof. Let G be an arbitrary group and choose a set of generators for it; for example $X = G$ will do. Let F be a free group on X. Then by (15.2.1) there is a surjective homomorphism $\theta : F \to G$, so $G \simeq F/\mathrm{Ker}(\theta)$. Next choose a subset R of $\mathrm{Ker}(\theta)$ whose normal closure in F is $\mathrm{Ker}(\theta)$ – for example we could take R to be $\mathrm{Ker}(\theta)$. Then $G \simeq F/\mathrm{Ker}(\theta) = G/(R^F) = \langle X \mid R \rangle$, so we have a presentation of G. $\square$

In the proof just given there are many possible choices for X and R, so a group has many presentations. This is one reason why it can be difficult to extract useful information about the structure of a group from a specific presentation. However, despite the difficulties inherent in working with presentations of groups, there is one very useful tool available.

(15.2.3) (Von Dyck's[2] Theorem). *Let G and H be groups given by presentations, $G = \langle X \mid R \rangle$ and $H = \langle Y \mid S \rangle$. Assume that there is given a surjective map $\alpha : X \to Y$ such that*

---

2 Walter von Dyck (1856–1934).

$\alpha(x_1)^{\ell_1}\alpha(x_2)^{\ell_2}\cdots\alpha(x_k)^{\ell_k}$ *is a relator of* H, *i. e., a consequence of the words in* S, *whenever* $x_1^{\ell_1}x_2^{\ell_2}\cdots x_k^{\ell_k}$ *is a defining relator of* G. *Then there is a surjective homomorphism* $\theta : G \to H$ *such that* $\theta|_X = \alpha$.

Proof. Let F be the free group on X; then $G = F/N$ where N is the normal closure of R in F. By the mapping property of free groups there is a homomorphism $\theta_0 : F \to H$ such that $\theta_0|_X = \alpha$. By hypothesis $\theta_0(r) = 1$ for all $r \in R$ and thus $\theta_0(a) = 1$ for all a in $N = \langle R^F \rangle$. Hence θ_0 induces a homomorphism $\theta : G \to H$ such that $\theta(fN) = \theta_0(f)$. Finally, $Y \subseteq \mathrm{Im}(\theta_0)$ since α is surjective, so θ_0, and hence θ, is surjective. $\square$

We will shortly see how Von Dyck's Theorem can be used to extract structural information about a group from a specific presentation.

Finitely presented groups

A presentation of a group is said to be *finite* if it contains finitely many generators and finitely many relators. A group is called *finitely presented* if it has at least one finite presentation. For example, cyclic groups are finitely presented. Likewise free groups of finite rank are finitely presented, since the set of relators is empty. Not surprisingly finite groups are finitely presented.

(15.2.4). *Every finite group is finitely presented.*

Proof. Let $G = \{g_1, g_2, \ldots, g_n\}$ be a finite group of order n. Then $g_i g_j = g_{v(i,j)}$ and $g_i^{-1} = g_{u(i)}$ where $1 \le u(i), v(i,j) \le n$. Now let $\bar{G}$ be the group with generators $\bar{g}_1, \bar{g}_2, \ldots, \bar{g}_n$ and defining relations $\bar{g}_i \bar{g}_j = \bar{g}_{v(i,j)}$, $\bar{g}_i^{-1} = \bar{g}_{u(i)}$, where $i, j = 1, 2, \ldots, n$. This is clearly a finite presentation of $\bar{G}$. Apply Von Dyck's Theorem to $\bar{G}$ and G where α is the assignment $\bar{g}_i \mapsto g_i$, noting that each defining relator of $\bar{G}$ is mapped to a relator of G. It follows that there is a surjective homomorphism $\theta : \bar{G} \to G$ such that $\theta(\bar{g}_i) = g_i$.

Now every element $\bar{g}$ of $\bar{G}$ is expressible as a product of $\bar{g}_i$'s and their inverses. Moreover, repeated use of the defining relations for $\bar{G}$ shows that $\bar{g}$ is equal to some $\bar{g}_k$; it follows that $\bar{G}$ is finite and $|\bar{G}| \le n$. But $G \simeq \bar{G}/\mathrm{Ker}(\theta)$, so $|\mathrm{Ker}(\theta)| = |\bar{G}|/|G| \le 1$. Hence $\mathrm{Ker}(\theta) = 1$ and $G \simeq \bar{G}$. $\square$

The next result shows that the property of being finitely presented does not depend on any particular set of generators.

(15.2.5). *If a group is finitely presented, then it has a finite presentation in any finite set of generators.*

Proof. Let $G = \langle y_1, \ldots, y_m \mid s_1, \ldots, s_\ell \rangle$ be a finitely presented group and suppose that $\{x_1, \ldots, x_n\}$ is some other finite set of generators for G. There are expressions for x_j in terms of the y_i and for y_i in terms of the x_j, say $y_i = w_i(x)$ and $x_j = v_j(y)$. Therefore the following relations in the x_i hold in G:

$$s_k(w_1(x), \ldots, w_m(x)) = 1, \quad x_j = v_j(w_1(x), \ldots, w_m(x)), \tag{15.1}$$

where $k = 1, \ldots, \ell, j = 1, \ldots, n$. Notice that there are finitely many such relations.

Next form the group $\bar{G}$ with a presentation in which the generators are $\bar{x}_1, \ldots, \bar{x}_n$ and the defining relations are those in equation (15.1), but written in the $\bar{x}_j$ instead of the x_j. Certainly $\bar{G}$ is a finitely presented group. By (15.2.3) there is a surjective homomorphism $\theta : \bar{G} \to G$ such that $\theta(\bar{x}_j) = x_j$. Define $\bar{y}_i = w_i(\bar{x})$; then $\bar{G} = \langle \bar{y}_1, \ldots, \bar{y}_m \rangle$ by the relations (15.1). Since $s_k(\bar{y}) = 1$, there is a homomorphism $\phi : G \to \bar{G}$ such that $\phi(y_i) = \bar{y}_i$ by (15.2.3) once again. Next verify that θ and ϕ are mutually inverse functions, so that they are isomorphisms. Indeed we have

$$\theta\phi(y_i) = \theta(\bar{y}_i) = \theta(w_i(\bar{x})) = w_i(\theta(\bar{x})) = w_i(x) = y_i$$

for $i = 1, \ldots, m$. Therefore $\theta\phi$ is the identity function on G since the y_i generate G. A similar argument shows that $\phi\theta$ is the identity function on $\bar{G}$. Hence $G \simeq \bar{G}$, so that G has a finite presentation in the x_i. $\qquad\square$

Further examples of finitely presented groups can be read off from the following result.

(15.2.6). *Let N be a normal subgroup of a group G. Assume that N and G/N are finitely presented. Then G is finitely presented.*

Proof. Let N have generators $x_1, x_2, \ldots, x_m$ and relators $r_1(x), r_2(x), \ldots, r_k(x)$ and let G/N have generators $y_1 N, y_2 N, \ldots, y_n N$ and relators $s_1(yN), s_2(yN), \ldots, s_\ell(yN)$. It is clear that the x_i and y_j together generate G. We need to produce finitely many defining relators in these generators. There are for sure relations of the following types:

$$r_i(x) = 1, \; s_j(y) = t_j(x), \quad (i = 1, \ldots, k, \; j = 1, \ldots, \ell), \tag{15.2}$$

for certain words t_j. There are also relations that express the normality of the subgroup N. These have the form

$$y_j x_i y_j^{-1} = u_{ij}(x), \quad y_j^{-1} x_i y_j = v_{ij}(x), \tag{15.3}$$

for certain words u_{ij}, v_{ij} where $i = 1, \ldots, m, j = 1, \ldots, n$.

The next step is to form a group $\bar{G}$ having a presentation with generators

$$\bar{x}_1, \ldots, \bar{x}_m, \bar{y}_1, \ldots, \bar{y}_n$$

and the relations (15.2) and (15.3) above expressed in terms of the $\bar{x}_i, \bar{y}_j$. By (15.2.3) there is a surjective homomorphism $\alpha : \bar{G} \to G$ such that $\alpha(\bar{x}_i) = x_i$ and $\alpha(\bar{y}_j) = y_j$: set $K = \text{Ker}(\alpha)$. By (15.2.3) again there is a homomorphism from N to $\bar{N} = \langle \bar{x}_1, \ldots, \bar{x}_m \rangle$ in which $x_i \mapsto \bar{x}_i$ since the relators r_i map to the identity of $\bar{N}$; clearly this homomorphism is the inverse of the restriction of α to $\bar{N}$. Hence $K \cap N = 1$.

Next $\bar{N} \lhd \bar{G}$ since the $\bar{y}_j \bar{x}_i \bar{y}_j^{-1}$ and $\bar{y}_j^{-1} \bar{x}_i \bar{y}_j$ belong to $\bar{N}$. Now α induces a surjective homomorphism from $\bar{G}/\bar{N}$ to G/N. There is also a homomorphism from G/N to $\bar{G}/\bar{N}$ in which $y_j N \mapsto \bar{y}_j \bar{N}$; clearly it is the inverse of the previous one. Hence $K \le N$ and thus $K = 1$. It follows that $\bar{G} \simeq G$, so G is finitely presented. $\qquad \square$

As a consequence of the last result, *a group which has a series of finite length with cyclic factors is finitely presented*. Such groups are termed *polycyclic* and evidently they are solvable. On the other hand, it is known that there exist finitely generated solvable groups which are not finitely presented – see [15].

As has been remarked, it can be difficult to extract information about a group from a presentation. There is a deep reason for this difficulty, namely the *insolvability of the word problem* for finitely presented groups. Roughly speaking, this means that there does not exist an algorithm which can decide if a given word in the generators of an arbitrary finitely presented group is equal to the identity element. For a very readable account of the word problem see [17]. One consequence of this failure is the need to exploit special features of a presentation in order to obtain structural information about the group it presents.

To illustrate this here are some examples of groups given by a finite presentation where the structure of the group can be determined.

Example (15.2.1). Let $G = \langle x \mid x^n \rangle$ where n is a positive integer.

The free group F on $\{x\}$ is generated by x: thus $F \simeq \mathbb{Z}$ and $G = F/F^n \simeq \mathbb{Z}/n\mathbb{Z} = \mathbb{Z}_n$, a cyclic group of order n, as expected.

Example (15.2.2). Let $G = \langle x, y \mid xy = yx, x^2 = 1 = y^2 \rangle$.

Since $xy = yx$, the group G is abelian; also every element of G has the form $x^i y^j$ where $i, j \in \{0, 1\}$, because $x^2 = 1 = y^2$; hence $|G| \le 4$. On the other hand, the Klein 4-group V is generated by the permutations $a = (12)(34)$ and $b = (13)(24)$, and the relations $ab = ba$ and $a^2 = 1 = b^2$ hold in V. Hence Von Dyck's Theorem can be applied to yield a surjective homomorphism $\theta : G \to V$ such that $\theta(x) = a$ and $\theta(y) = b$. Thus $G/\mathrm{Ker}(\theta) \simeq V$. Since $|G| \le 4 = |V|$, it follows that $\mathrm{Ker}(\theta) = 1$ and θ is an isomorphism. Therefore G is a Klein 4-group.

For a greater challenge consider the following presentation.

Example (15.2.3). Let $G = \langle x, y \mid x^2 = y^3 = (xy)^2 = 1 \rangle$.

Our first move is to find an upper bound for $|G|$. Let $H = \langle y \rangle$; this is a subgroup of order 1 or 3. Write $\mathcal{S} = \{H, xH\}$; we will argue that $\mathcal{S}$ is the set of *all* left cosets of H in G. To establish this it is sufficient to show that $x\mathcal{S} = \mathcal{S} = y\mathcal{S}$: for then it will follow that $g\mathcal{S} = \mathcal{S}$ for all $g \in G$ and thus $\mathcal{S}$ contains all left cosets of H. Certainly $x\mathcal{S} = \mathcal{S}$ since $x^2 = 1$. Next $xyxy = 1$ and hence $yx = x^{-1}y^{-1} = xy^2$, since $y^{-1} = y^2$. It follows that

$yxH = xy^2H = xH$ and thus $y\mathcal{S} = \mathcal{S}$. Since $|H| \leq 3$ and $|G : H| = |\mathcal{S}| \leq 2$, we deduce that $|G| \leq 6$.

Next observe that the symmetric group S_3 is generated by the permutations $a = (12)(3)$ and $b = (123)$, and that $a^2 = b^3 = (ab)^2 = 1$ since $ab = (1)(23)$. By Von Dyck's theorem there is a surjective homomorphism $\theta : G \to S_3$. Since $|G| \leq 6$, it follows that θ is an isomorphism and $G \simeq S_3$.

The method of the last two examples can be effective when a finite group is given by a presentation. The general procedure is to choose a subgroup for whose order one has an upper bound, and then by coset enumeration to find an upper bound for the index. This gives an upper bound for the order of the group. The challenge is then to identify the group by comparing it with known groups whose order equals the upper bound and in which the defining relations hold.

Exercises (15.2). In the following three exercises identify the groups with the given presentations.

(1) $\langle x, y \mid x^2 = 1 = y^4, xy = yx \rangle$.

(2) $\langle x, y \mid x^3 = (xy)^2 = y^3 = 1 \rangle$.

(3) $\langle x, y \mid x^2 = (xy)^2 = y^5 = 1 \rangle$.

(4) Let G be a group which has a presentation with n generators and r defining relators. If $r < n$, prove that G is infinite. [Hint: consider the abelian group G/G' and use Exercise (15.1.3).]

(5) Establish the converse of Exercise (15.1.5): a group G which has the projective property is free. [Hint: let $\pi : F \to G$ be a surjective homomorphism with F a free group. Apply the projective property to the identity map on G and π and use (15.1.6).]

(6) Let G be a finitely generated group and let $N \lhd G$. Assume that G/N is finitely presented. Prove that there is a finite subset R such that $N = \langle gRg^{-1} \mid g \in G \rangle$. [Hint: choose a surjective homomorphism $\pi : F \to G$ where F is a free group of finite rank. Let S be the preimage of N under π and note that $F/S \simeq G/N$. Then apply (15.2.5).]

(7) Complete the proof of (15.2.5) by showing that $\phi\theta$ is the identity function on $\bar{G}$.

15.3 Free products

In this section we will describe a construction called the free product of a set of groups. Roughly speaking it is the "largest" group that can be generated by isomorphic copies of a given set of groups.

Let $\{G_\lambda \mid \lambda \in \Lambda\}$ be a non-empty set of groups. By a *free product* of the G_λ is meant a set

$$\{G, \alpha_\lambda \mid \lambda \in \Lambda\}$$

where G is a group and $\alpha_\lambda : G_\lambda \to G$ is a homomorphism: it is required that the following mapping property hold. Given a set of homomorphisms $\phi_\lambda : G_\lambda \to H$ for each $\lambda \in \Lambda$ and some group H, there is a *unique* homomorphism $\theta : G \to H$ such that $\theta\alpha_\lambda = \phi_\lambda$ for all $\lambda \in \Lambda$. This amounts to requiring that all the diagrams below commute.

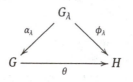

As will be seen in Section 16.3, what this definition asserts is that a free product is a coproduct in the category of groups.

A point to note here is that *the mappings $\alpha_\lambda : G_\lambda \to G$ are necessarily injective.* To see this assume that $1 \neq x \in \mathrm{Ker}(\alpha_\lambda)$. Apply the mapping property with $H = G_\lambda$; also let ϕ_λ be the identity map and ϕ_μ the trivial homomorphism if $\mu \neq \lambda$. By the mapping property there exists a homomorphism $\theta : G \to H$ such that $\theta\alpha_\lambda = \phi_\lambda$. Hence $x = \phi_\lambda(x) = \theta\alpha_\lambda(x) = \theta(1) = 1$, a contradiction.

Our first task is to show that free products actually exist.

(15.3.1). *Every non-empty set of groups $\{G_\lambda \mid \lambda \in \Lambda\}$ has a free product.*

Proof. The construction is similar to that of free groups in Section 15.1. There is no loss of generality in assuming that $G_\lambda \cap G_\mu = \emptyset$ if $\lambda \neq \mu$: for if necessary, G_λ can always be replaced by a suitable isomorphic copy.

Set $U = \bigcup_{\lambda \in \Lambda} G_\lambda$ and form the set S of all *words in U*, i. e., all finite sequences

$$w = g_1 g_2 \ldots g_r$$

where $g_i \in G_{\lambda_i}$ and $\lambda_i \in \Lambda$. The case $r = 0$ is the *empty word*, written 1. The *product vw* of words v and w is defined by simple juxtaposition, v followed by w. By convention $1w = w = w1$. The *inverse* of a word $w = g_1 g_2 \ldots g_r$ is defined to be $w^{-1} = g_r^{-1} \ldots g_2^{-1} g_1^{-1}$, with the convention that $1^{-1} = 1$.

Next we introduce a device by means of which consecutive entries of a word that belong to the same G_λ can be combined. An equivalence relation on the set S is defined as follows. Let $v, w \in S$. Then v is equivalent to w if there is a finite chain of operations leading from v to w of the following types:
(i)   insertion or deletion of the identity element from one of the G_λ;
(ii) contraction, i. e., replacement of two consecutive entries belonging to the same group G_λ by their product;
(iii) expansion, i. e., replacement of an entry belonging to a G_λ by two elements of G_λ whose product it equals.
This is evidently an equivalence relation on S. Denote the equivalence class of w by $[w]$. One readily verifies that if v is equivalent to v' and w to w', then vw is equivalent

to $v'w'$ and w^{-1} is equivalent to $(w')^{-1}$. Then the set of equivalence classes

$$G = \{[w] \mid w \in S\}$$

becomes a group with the well defined group operations

$$[v][w] = [vw] \quad \text{and} \quad [w]^{-1} = [w^{-1}].$$

Next let $\alpha_\lambda : G_\lambda \to G$ be the mapping in which $x \mapsto [x]$ where $x \in G_\lambda$: this is readily seen to be a homomorphism of groups.

We aim to prove that $\{G, \alpha_\lambda \mid \lambda \in \Lambda\}$ is a free product of the G_λ. To this end, let there be given homomorphisms $\phi_\lambda : G_\lambda \to H$ for $\lambda \in \Lambda$ and some group H. It is necessary to produce a unique homomorphism $\theta : G \to H$ such that $\theta\alpha_\lambda = \phi_\lambda$ for all $\lambda \in \Lambda$. Let $w \in S$ and write $w = g_1 g_2 \ldots g_r$ with $g_i \in G_{\lambda_i}$. Define

$$\theta([w]) = \phi_{\lambda_1}(g_1)\phi_{\lambda_2}(g_2) \cdots \phi_{\lambda_r}(g_r).$$

First observe that θ is well defined. The reason is that application of one of the operations (i), (ii), (iii) to the word w has no effect on the element $\phi_{\lambda_1}(g_1)\phi_{\lambda_2}(g_2) \cdots \phi_{\lambda_r}(g_r)$ of H. It is then evident that θ is a homomorphism. If $x \in G_\lambda$, we have $\theta\alpha_\lambda(x) = \theta([x]) = \phi_\lambda(x)$ and hence $\theta\alpha_\lambda = \phi_\lambda$.

Finally, there is the uniqueness requirement to verify. Let $\theta' : G \to H$ be another homomorphism such that $\theta'\alpha_\lambda = \phi_\lambda$ for all λ. Then $\theta\alpha_\lambda = \theta'\alpha_\lambda$, so that θ and θ' agree on each $\mathrm{Im}(\alpha_\lambda)$. But the $\mathrm{Im}(\alpha_\lambda)$ generate G since

$$[g_1 g_2 \ldots g_r] = [g_1][g_2] \cdots [g_r] = \alpha_{\lambda_1}(g_1)\alpha_{\lambda_2}(g_2) \cdots \alpha_{\lambda_r}(g_r),$$

where $g_i \in G_{\lambda_i}$. Therefore $\theta = \theta'$. $\square$

Now that free products have been shown to exist, the question of their uniqueness arises.

(15.3.2). *Suppose that $\{G, \alpha_\lambda | \lambda \in \Lambda\}$ and $\{\bar{G}, \bar{\alpha}_\lambda | \lambda \in \Lambda\}$ are two free products for a set of groups $\{G_\lambda \mid \lambda \in \Lambda\}$. Then $G \simeq \bar{G}$.*

The proof of this result follows the pattern established for free groups in (15.1.4). It is in fact a special case of the uniqueness of coproducts – see (16.3.1) below.

The free product G of $\{G_\lambda, \lambda \in \Lambda\}$ is usually written without specifying the associated homomorphisms, as

$$G = \mathrm{Fr}_{\lambda \in \Lambda} G_\lambda.$$

In the case of a finite set $\Lambda = \{\lambda_1, \lambda_2, \ldots, \lambda_r\}$ the free product is denoted by

$$G = G_{\lambda_1} * G_{\lambda_2} * \cdots * G_{\lambda_r}.$$

Reduced words

Consider a free product $G = \mathrm{Fr}_{\lambda \in \Lambda} G_\lambda$. In order to determine the structural properties of G, we need to find a standard form for its elements. By (15.3.2) we can assume that G consists of equivalence classes of words in $U = \bigcup_{\lambda \in \Lambda} G_\lambda$, as described in the proof of (15.3.1). A word is said to be *reduced* if none of its symbols is an identity element and no two consecutive symbols belong to the same group G_λ. By convention the empty word 1 is reduced.

Let w be a word in U and suppose we want to find a reduced word which is equivalent to w. There is an obvious way to do this. First of all delete all entries in w which are identity elements. By a *segment* of w is meant a subsequence of w with all its entries in the same group G_λ and which is not part of a longer sequence of the same type. Replace each segment by the product of its elements in the given order. Repetition of these procedures will eventually lead to a reduced word which is equivalent to the original word. It follows that every equivalence class of words contains a reduced word. The critical step is to prove that there is a unique reduced word in each equivalence class.

(15.3.3). *Each equivalence class of words in $\bigcup_{\lambda \in \Lambda} G_\lambda$ contains exactly one reduced word.*

Proof. Let v and w be equivalent reduced words; it must be shown that $v = w$. This will be achieved by an indirect argument, as in the case of free groups. Let R denote the set of all reduced words. An action of the free product G on R is defined as follows.

First define an action of G_λ on R. Let $u \in G_\lambda$; then we define a permutation u' of R as follows. If $u = 1_{G_\lambda}$, then u' is the identity; assume this is not the case. Write $v = x_1 x_2 \ldots x_r$ with $x_i \in G_{\lambda_i}$ and define $u'(x_1 x_2 \ldots x_r)$ to be $u x_1 x_2 \ldots x_r$ or $(u x_1) x_2 \ldots x_r$ according as $\lambda \neq \lambda_1$ or $\lambda = \lambda_1$ respectively. There is the additional stipulation that if $u x_1 = 1_{G_1}$, then $u x_1$ is to be deleted from the word. This ensures that the permuted word is also reduced. It is easy to see that $u \mapsto u'$ is a homomorphism from G_λ to $\mathrm{Sym}(R)$. By the mapping property of free products these homomorphisms combine to give a homomorphism $\theta : G \to \mathrm{Sym}(R)$ such that $\theta([x]) = x'$ where $x \in G_\lambda$.

Next $v = x_1 x_2 \ldots x_r$ with $x_i \in G_{\lambda_i}$. Hence $[v] = [x_1][x_2] \cdots [x_r]$ and $\theta([v]) = x_1' x_2' \cdots x_r'$. It follows from the definition of x_i' that $\theta([v])$ sends the empty word 1 to v. Similarly $\theta([w])$ sends 1 to w. Since $[v] = [w]$, it follows that $v = w$, as claimed. □

Normal form in free products

Consider a free product $G = \mathrm{Fr}_{\lambda \in \Lambda} G_\lambda$. We can assume that each element of G has the form $[w]$ where w is a uniquely determined reduced word, say $w = g_1 g_2 \ldots g_r$ where $1 \neq g_i \in G_{\lambda_i}$ and $\lambda_i \neq \lambda_{i+1}$. Let $\bar{G}_\lambda$ be the subgroup of all $[x]$ where $x \in G_\lambda$. Thus $[g_i] \in \bar{G}_{\lambda_i}$ and each $[w] \in G$ has a unique expression as a product of elements of the $\bar{G}_\lambda$, namely

$$[w] = [g_1][g_2] \cdots [g_r].$$

This is called the *normal form* of w.

The notation will be much simplified if we agree to identify $x \in G_\lambda$ with $[x] \in \bar{G}_\lambda$, so that the normal form of w becomes just $w = g_1 g_2 \ldots g_r$. Observe that as a consequence of the uniqueness of normal form

$$G_\lambda \cap \langle G_\mu \mid \mu \neq \lambda, \mu \in \Lambda \rangle = 1 \quad \text{where } \lambda \in \Lambda.$$

The existence of a normal form is typical of free products in the following sense.

(15.3.4). *Let G be a group which is generated by subgroups $G_\lambda, \lambda \in \Lambda$. Assume that each element g of G can be uniquely written in the form $g = g_1 g_2 \cdots g_r$ where $r \geq 0, 1 \neq g_i \in G_{\lambda_i}$, $\lambda_i \neq \lambda_{i+1}$. Then $G \simeq \mathrm{Fr}_{\lambda \in \Lambda} G_\lambda$.*

Proof. Let $\{F, \alpha_\lambda \mid \lambda \in \Lambda\}$ be the free product of the G_λ as constructed in the proof of (15.3.1). By the mapping property of the free product there is a homomorphism $\theta : F \to G$ making all the diagrams below commute

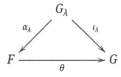

where $\iota_\lambda : G_\lambda \to G$ is the inclusion map. Thus $\theta \alpha_\lambda = \iota_\lambda$. Now θ is surjective since $x = \iota_\lambda(x) = \theta \alpha_\lambda(x) \in \mathrm{Im}(\theta)$ for $x \in G_\lambda$. Therefore $\mathrm{Im}(\theta) = G$, as the G_λ generate G.

Suppose that $1 \neq [f] \in \mathrm{Ker}(\theta)$ where $f = g_1 g_2 \cdots g_r$ is a reduced word with $g_i \in G_{\lambda_i}$, $r > 0$. Thus $[f] = [g_1][g_2] \cdots [g_r]$ is in normal form. Then

$$1 = \theta([f]) = \prod_{i=1}^{r} \theta([g_i]) = \prod_{i=1}^{r} \theta \alpha_{\lambda_i}(g_i) = \prod_{i=1}^{r} \iota_{\lambda_i}(g_i) = \prod_{i=1}^{r} g_i.$$

This contradicts the unique expressibility in G. Hence $\mathrm{Ker}(\theta) = 1$ and θ is an isomorphism, showing that $G \simeq F$. □

Examples of free products

(i) *A free product of free groups is a free group.* To see this let F_λ be a group which is free on a set X_λ for $\lambda \in \Lambda$. Assume that the sets X_λ are disjoint, which is no real restriction since X_λ could be replaced by any set with the same cardinality. Then F is free on the set $\bigcup_{\lambda \in \Lambda} X_\lambda$, which follows on applying (15.1.3). In particular, a free product of infinite cyclic groups is a free group.

(ii) *The free product of two groups of order 2 is an infinite dihedral group.* For let $G = \langle x \rangle * \langle y \rangle$ where $x^2 = 1 = y^2$. Set $a = xy$; then we have $G = \langle x, a \rangle$. Also $xax^{-1} = x^2 yx^{-1} = yx = a^{-1}$. It follows from (15.2.3) that G is a homomorphic image of the infinite dihedral group $\bar{G} = \langle \bar{x}, \bar{a} \mid \bar{x}^2 = 1, \bar{x}\bar{a}\bar{x}^{-1} = \bar{a}^{-1} \rangle$. On the other hand, $a, a^2, a^3, \ldots$ are distinct elements of G by uniqueness of the normal form in the free product. Therefore a has infinite order. Hence G is the semidirect product of $\langle x \rangle$ and $\langle a \rangle$, so it is an infinite dihedral group.

Some elementary properties of free products are recorded in the next two results.

(15.3.5). *Let $G = \mathrm{Fr}_{\lambda \in \Lambda} G_\lambda$ be a free product of groups. Then the following statements are true.*

(i) *Let $g = g_1 g_2 \cdots g_r$ be the normal form of $g \in G$ where $g_i \in G_{\lambda_i}$. If $\lambda_1 \neq \lambda_r$, then g has infinite order.*

(ii) *If at least two of the free factors G_λ are non-trivial, then G has an element of infinite order.*

(iii) *An element of finite order in G is conjugate to an element in one of the G_λ.*

Proof. (i) By hypothesis $r > 1$. If $m > 0$, then g^m has normal form of length $mr > 0$ since the initial and final entries of g belong to different free factors. Therefore $g^m \neq 1$ and g has infinite order.

(ii) This follows at once from (i).

(iii) Assume that $g = g_1 g_2 \cdots g_r$ is in normal form and has finite order. Suppose that g has been chosen with r least such that no conjugate of g belongs to any G_λ. Then g_1 and g_r must belong to the same free factor by (i). On conjugating g by g_r, we obtain $\bar{g} = g_r g g_r^{-1} = (g_r g_1) g_2 \cdots g_{r-1}$, which also has finite order. After deleting a possible identity element from $\bar{g}$ we obtain an element whose normal form has length less than r, which is a contradiction. $\qquad \square$

A consequence of the last theorem is that *a free product of torsion-free groups is torsion-free*. The next result shows that commutativity of elements in a free product is quite limited.

(15.3.6). *Let $G = \mathrm{Fr}_{\lambda \in \Lambda} G_\lambda$ and let $1 \neq g \in G_\lambda$. Then $C_G(g) \leq G_\lambda$.*

Proof. Let $x \in C_G(g)$ have normal form $x = x_1 x_2 \cdots x_r$ where $x_i \in G_{\lambda_i}$. Assume that $x \notin G_\lambda$ and that x has minimal length with this property: thus $r > 0$. If $\lambda_1 \neq \lambda$, then $xg \neq gx$ since these elements have different normal forms. Hence $\lambda_1 = \lambda$ and in a similar way $\lambda_r = \lambda$. Now form the conjugate $x' = x_r x x_r^{-1} = (x_r x_1) x_2 \cdots x_{r-1}$. Then x' has normal form of shorter length than x and $x' \in x_r C_G(g) x_r^{-1} = C_G(x_r g x_r^{-1})$. Also $x_r g x_r^{-1} \in G_\lambda$. Therefore by minimality of r we have $C_G(x_r g x_r^{-1}) \leq G_\lambda$, which implies that $C_G(g) \leq G_\lambda$. $\qquad \square$

One consequence of this result is that *if a free product has at least two non-trivial factors, then its center is trivial*.

Subgroups of free products

There is a deep theory that describes the subgroup structure of free products, culminating in what is known as the *Kuroš*[3] *subgroup theorem*.

(15.3.7). *Let H be a subgroup of the free product $G = \mathrm{Fr}_{\lambda \in \Lambda} G_\lambda$. Then H is a free product of the form*

$$H = F * \mathrm{Fr}_{\lambda, d_\lambda}(H \cap (d_\lambda G_\lambda d_\lambda^{-1})),$$

where F is a free group and the free product is formed over all $\lambda \in \Lambda$ and certain (H, G_λ)-double coset representatives d_λ.

For double cosets see Exercise (4.1.13). The algebraic proof of the Kuroš subgroup theorem is complicated and it will not be presented here – however, see [15] for an account of it.

Free products with amalgamation

We conclude the chapter by mentioning a useful generalization of the free product. Roughly speaking, this is the "largest" group that can be generated by a given set of groups when certain isomorphic subgroups of the groups in the set are identified. Details of the construction follow.

Let $\{G_\lambda \mid \lambda\}$ be a non-empty set of groups and let H be a fixed group. Assume there are given injective homomorphisms

$$\phi_\lambda : H \to G_\lambda, \quad \lambda \in \Lambda.$$

Let $F = \mathrm{Fr}_{\lambda \in \Lambda} G_\lambda$ be the free product of the G_λ and define N to be the normal closure in F of the subset

$$\{(\phi_\lambda(h))^{-1} \phi_\mu(h) \mid \lambda, \mu \in \Lambda, \ h \in H\}.$$

Then the group

$$G = F/N$$

is called the *free product of the groups G_λ with the amalgamated subgroup H*. It is often referred to as a *generalized free product*. The idea here is that the isomorphic subgroups $\phi_\lambda(H)$ are all identified in the group F/N. In general G will depend upon the homomorphims ϕ_λ as well as H. In order to develop the properties of generalized free products it is necessary to introduce a normal form for their elements. For details of

---

3 Aleksandr Gennadievich Kuroš, 1908–1971.

the normal form in a free product with amalgamation see [15]. Here we will be content to analyze one example.

Example (15.3.1). Consider the group with the finite presentation shown.

$$G = \langle x, y \mid x^2 = y^2 \rangle.$$

The first move is to identify G as a generalized free product. Let F be the free group with basis $\{x, y\}$. Then $G = F/N$ where $N = \langle (x^{-2}y^2)^F \rangle$. Next let $H = \langle h \rangle$ be an infinite cyclic group and define injective homomorphisms $\phi : H \to X = \langle x \rangle$ and $\psi : H \to Y = \langle y \rangle$ by $\phi(h) = x^2$ and $\psi(h) = y^2$. Then G is the free product of X and Y with the subgroups $\langle x^2 \rangle$ and $\langle y^2 \rangle$ amalgamated by means of ϕ and ψ.

Let us see what can be said about the structure of G. Since there is no normal form at our disposal, we will have to rely on a bare hands approach. Observe that the cyclic subgroup $Z = \langle x^2 \rangle = \langle y^2 \rangle$ is contained in the center of G; thus $Z \triangleleft G$. Moreover, G/Z is generated by two groups of order 2. Thus by Von Dyck's theorem (15.2.3) the group G/Z is a quotient of an infinite dihedral group. Consequently G is a solvable, and even polycyclic, group. In fact Z is infinite cyclic and G/Z is infinite dihedral by an indirect argument: this is sketched in Exercise (15.3.9).

Exercises (15.3).
(1) Prove (15.3.2).
(2) Let $\{F, \alpha_\lambda \mid \lambda \in \Lambda\}$ be a free product of groups G_λ, $\lambda \in \Lambda$. Prove that the subgroups $\mathrm{Im}(\alpha_\lambda)$ generate F. [Hint: this is true for the free product constructed in (15.3.1). Apply (15.3.2) to deduce the result for F.]
(3) Let $G = \mathrm{Fr}_{\lambda \in \Lambda} G_\lambda$ and $H = \mathrm{Fr}_{\lambda \in \Lambda} H_\lambda$ be free products, and let $\phi_\lambda : G_\lambda \to H_\lambda$ be a homomorphism for each $\lambda \in \Lambda$. Prove that there is a unique homomorphism $\phi : G \to H$ whose restriction to G_λ is ϕ_λ. In addition show that $\mathrm{Ker}(\phi)$ is the normal closure in G of $\bigcup_{\lambda \in \Lambda} \mathrm{Ker}(\phi_\lambda)$.
(4) Prove that $(\mathrm{Fr}_{\lambda \in \Lambda} G_\lambda)^{ab} \simeq \mathrm{Dr}_{\lambda \in \Lambda} (G_\lambda)^{ab}$.
(5) Let $G = \mathrm{Fr}_{\lambda \in \Lambda} G_\lambda$ and let $H_\lambda \leq G_\lambda$. If $H = \langle H_\lambda \mid \lambda \in \Lambda \rangle$, prove that $H \simeq \mathrm{Fr}_{\lambda \in \Lambda} H_\lambda$.
(6) Let $G = \mathrm{Fr}_{\lambda \in \Lambda} G_\lambda$ and let $N_\lambda \triangleleft G_\lambda$. Denote the normal closure of $\bigcup_{\lambda \in \Lambda} N_\lambda$ in G by N. Prove that $G/N \simeq \mathrm{Fr}_{\lambda \in \Lambda} (G_\lambda/N_\lambda)$.
(7) Let $F = A * B$. If F is abelian, show that $A = 1$ or $B = 1$.
(8) Let $F = A * B$ and suppose that H is an abelian subgroup of F. Prove that either H is cyclic or else it is conjugate to a subgroup of A or B. [Hint: use Exercise (15.3.7) and the Kuroš subgroup theorem.]
(9) Let $A = \langle a \rangle \times \langle b \rangle$ be a free abelian group of rank 2 and let $\langle x \rangle$ be an infinite cyclic group. Define an action of x on A by $x \cdot a = a, x \cdot b = b^{-1}$. Let G_0 denote the semidirect product of A and $\langle x \rangle$ determined by this action. Note that a and x^2 both belong to the center of G_0, so that $N = \langle x^{-2} a \rangle \triangleleft G_0$. Define $\bar{G} = G_0/N$ and write $\bar{x} = xN$, $\bar{b} = bN$ and $\bar{y} = \bar{b}\bar{x}$.
  (i) Prove that $\bar{G} = \langle \bar{x}, \bar{y} \rangle$ and $\bar{x}^2 = \bar{y}^2$.

(ii) Use Von Dyck's theorem to construct a surjective homomorphism from the group G of Example (15.3.1) to $\bar{G}$.

(iii) Deduce that Z is infinite cyclic and G/Z is an infinite dihedral group.

For a further account of groups given by presentations see [11].

16 Introduction to category theory

Categories were introduced in the 1940's by S. Eilenberg[1] and S. Mac Lane[2] in order to elucidate relationships between algebraic topology and algebra. Since that time they have become a standard tool in many branches of mathematics. Categories by their nature are highly abstract structures, which is scarcely surprising given their wide applicability. The reader who is able to see beyond the abstraction will recognize many ideas and concepts that have appeared in earlier chapters, on groups, rings and modules in particular. Thus categories emphasize common features of these topics. Recently the utility of categories in theoretical computer science has been recognized as a potent means of expressing concepts involved in programming languages and data structures: see for example [14]. Thus there are many reasons to study category theory. There is a change of emphasis in this chapter, where the concern is to explain and elucidate concepts rather than to prove theorems.

16.1 Categories

A category C consists of three entities:
(i)   a class of *objects* denoted by $\mathrm{obj}(C)$;
(ii) for each pair of objects (A, B) a set of *morphisms*, written $\mathrm{Mor}_C(A, B)$ or $\mathrm{Mor}(A, B)$, which may be empty;
(iii) a *law of composition* of morphisms, that is, a function

$$\mathrm{Mor}(A, B) \times \mathrm{Mor}(B, C) \to \mathrm{Mor}(A, C),$$

denoted by $(\alpha, \beta) \mapsto \beta\alpha$, for every triple of objects (A, B, C), where $\alpha \in \mathrm{Mor}(A, B)$, $\beta \in \mathrm{Mor}(B, C)$.

It will simplify considerations if we agree to use the *arrow notation* for morphisms. But it must be emphasized that morphisms need not be functions nor need objects be sets. Thus, if $\alpha \in \mathrm{Mor}(A, B)$, we will write

$$\alpha : A \to B \quad \text{or} \quad A \xrightarrow{\alpha} B.$$

For these entities to form a category we require that the following properties hold:
(a) the sets $\mathrm{Mor}(A, B)$ are disjoint;
(b) for each object A there is an *identity morphism* ι_A in $\mathrm{Mor}(A, A)$ such that $\alpha\iota_A = \alpha = \iota_B\alpha$ for $\alpha \in \mathrm{Mor}(A, B)$;

---

1 Samuel Eilenberg 1913–1998.
2 Saunders Mac Lane 1909–2005.

https://doi.org/10.1515/9783110691160-016

(c) composition of morphisms is an associative operation, that is, $\gamma(\beta\alpha) = (\gamma\beta)\alpha$ for every triple of morphisms $\alpha : A \rightarrow B, \beta : B \rightarrow C, \gamma : C \rightarrow D$.

Morphisms $\alpha : A \rightarrow B$ and $\beta : B \rightarrow A$ are said to be *mutually inverse* if $\alpha\beta = \iota_B$ and $\beta\alpha = \iota_A$. A morphism which has an inverse is called an *isomorphism* and the objects A and B are then said to be *isomorphic*, in symbols $A \simeq B$. While the concept of a category may appear abstruse at this point, many familiar examples will soon follow.

Before proceeding further we point out that the objects in a category generally form a proper class and not a set, whereas the morphisms between two objects always constitute a set. A category whose objects form a set is called a *small category*: otherwise it is a *large category*.

As has been our custom throughout this work, we shall be fairly relaxed about the exact set theoretic basis for category theory. To put matters on a completely sound footing it is necessary to work within a fixed universe: for further details of this see [12].

Examples of categories

(i) The category of sets and functions is denoted by **Set**. Here the objects are sets and the morphisms are the functions between sets. Composition is just functional composition and the identity morphism for a set X is the identity function on X. The elementary properties of sets guarantee that **Set** a category. **Set** is the example of a category that comes most immediately to mind. (Strictly speaking, we should assign a symbol as the identity morphism of the empty set with the formal properties required in the definition. This is sometimes referred to as the empty function.)

(ii) The categories **Mon, Gp** and **Rg** of all monoids, groups and rings respectively. In each case the morphisms are the homomorphisms of the relevant type for the structure together with functional composition.

(iii) Further natural categories are **Ab,** $_R$**Mod, Mod**$_R$, where R is a ring. These are the categories of abelian groups, left R-modules and right R-modules. In each case the morphisms are the homomorphisms of the relevant type together with functional composition.

(iv) **Top**. This is the category of topological spaces with the continuous functions between spaces as morphisms.

So far in all the categories mentioned the objects have been sets and the morphisms functions between them, with functional composition as the law of composition. But it is not hard to find categories in which this is not the case. Here are some examples.

(v) Let S be a fixed non-empty set. The category S associated with S has as its objects the elements of S, so obj$(S) = S$, and the only morphisms are identity morphisms. This means that every non-empty set can be regarded as a category.

(vi) Let n be a positive integer. A category **n** is defined to have objects $\{0, 1, 2, \ldots, n-1\}$, while its morphisms are identities and all composites of the arrows $0 \rightarrow 1, 1 \rightarrow 2,$ $\ldots n-2 \rightarrow n-1$.

(vii) If M is a fixed monoid, define a category $\mathcal{C}_M$ to have the single object M. The moprhisms are the elements of M, the law of composition of morphisms being the binary operation of M. This is an associative operation and the identity morphism is the identity element of the monoid. Thus $\mathcal{C}_M$ is category. In fact every single object category arises from a monoid in this manner: see Exercise (16.2.7).

Diagrams in categories

A *diagram* in a category $\mathcal{C}$ is a directed graph, with multiple edges and loops allowed, in which the vertices are objects and the edges are morphisms in the category. Diagrams provide a convenient way to visualize morphisms and their relationships, even in abstract categories where the objects are not sets and the morphisms are not functions.

For example, let $\alpha : A \rightarrow B$ and $\beta : B \rightarrow C$ be morphisms in a category. Using these morphisms and their composite $\beta\alpha$, we construct the following triangle diagram:

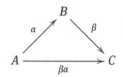

A diagram is said to be *commutative* if for paths between two fixed vertices – always following the directions of the arrows – the composite of the morphisms along a path is the same for all paths between the vertices. For example, the triangle diagram above is commutative since this merely expresses the law of composition of morphisms.

For another example consider the square of objects and morphisms:

This diagram is commutative if and only if $\beta\alpha = \gamma\delta$. Commutative diagrams are powerful tools that are used throughout algebra: indeed they have already been encountered in Chapter 13.

Subcategories

A category $\mathcal{C}$ is said to be a *subcategory* of a category $\mathcal{D}$ if

(i) $\mathrm{obj}(\mathcal{C}) \subseteq \mathrm{obj}(\mathcal{D})$;

(ii) $\mathrm{Mor}_{\mathcal{C}}(A,B) \subseteq \mathrm{Mor}_{\mathcal{D}}(A,B)$ for all $A, B \in \mathrm{obj}(\mathcal{C})$;

(iii) the law of composition of morphisms in $\mathcal{C}$ is consistent with that in $\mathcal{D}$.

Examples of subcategories are easy to find: **Gp** is a subcategory of **Set** and **Ab** is a subcategory of **Gp**.

Let $\mathcal{C}$ be a subcategory of category $\mathcal{D}$; then $\mathcal{C}$ is said to be a *full subcategory* if $\mathrm{Mor}_{\mathcal{C}}(A,B) = \mathrm{Mor}_{\mathcal{D}}(A,B)$ for every pair of objects A, B in $\mathcal{C}$, i. e., the morphism sets are the same in each category for every pair of objects in $\mathcal{C}$. For example, **Ab** is a full subcategory of **Gp**, but **Gp** is not a full subcategory of **Set** since not every map between groups is a homomorphism.

Exercises (16.1).

(1) Prove that the identity morphism associated with an object in a category is unique.

(2) Prove that an isomorphism between two objects in a category has a unique inverse.

(3) How to represent a partially ordered set $(S, \preceq)$ by a category $\mathcal{C}$. Define $\mathrm{obj}(\mathcal{C})$ to be S. Then let $\mathrm{Mor}_{\mathcal{C}}(x,y)$ have a single element, say $\sigma(x,y)$, if $x \preceq y$: otherwise $\mathrm{Mor}_{\mathcal{C}}(x,y)$ is empty. What should the law of composition of morphisms be and what properties should the $\sigma(x,y)$ have if $\mathcal{C}$ is to be a category?

(4) An object A of a category is called *initial* if there is a unique morphism $A \to C$ for every object C. An object Z of a category is called *terminal* if there is a unique morphism $C \to Z$ for every object C. Prove that if an initial object or terminal object exists in a category, then it is unique up to isomorphism.

(5) Do initial and terminal objects exist in the category **Set**? If so, what are they? [Hint: it depends on whether one allows morphisms from the empty set.]

(6) Give an example of a category with no initial or terminal objects.

(7) (Product categories). Let $\mathcal{C}$ and $\mathcal{D}$ be categories. The *product category* $\mathcal{C} \times \mathcal{D}$ has objects (A,B) where $A \in \mathrm{obj}(\mathcal{C})$, $B \in \mathrm{obj}(\mathcal{D})$ and morphisms (α, β) where α, β are morphisms in $\mathcal{C}$, $\mathcal{D}$ respectively. The law of composition of morphisms is $((\alpha, \beta), (\alpha', \beta')) \mapsto (\alpha'\alpha, \beta'\beta)$. Verify that $\mathcal{C} \times \mathcal{D}$ is in fact a category.

(8) An object of a category is called a *zero object* if it is both an initial object and a terminal object. Prove that the zero submodule is the zero object in $_R$**Mod**.

16.2 Functors

A functor between two categories relates the objects and morphisms of one category to objects and morphisms of the other, and also connects their laws of composition. Let $\mathcal{C}$ and $\mathcal{D}$ be categories. A *covariant functor F* from $\mathcal{C}$ to $\mathcal{D}$, in symbols

$$F : \mathcal{C} \to \mathcal{D},$$

is a collection of functions (also denoted by F) of the form

$$F : \mathrm{obj}(\mathcal{C}) \rightarrow \mathrm{obj}(\mathcal{D}) \quad \text{and} \quad F : \mathrm{Mor}_{\mathcal{C}}(A, B) \rightarrow \mathrm{Mor}_{\mathcal{D}}(F(A), F(B))$$

for each pair of objects (A, B) of $\mathcal{C}$. The functions are required to have the following properties:

(i)   if $\alpha \in \mathrm{Mor}_{\mathcal{C}}(A, B)$ and $\beta \in \mathrm{Mor}_{\mathcal{C}}(B, C)$, then $F(\beta\alpha) = F(\beta)F(\alpha)$.

(ii) $F(\iota_A) = \iota_{F(A)}$ for all $A \in \mathrm{obj}(\mathcal{C})$.

The property (i) in the definition may be expressed by saying that the functor F can be applied to the commutative triangle

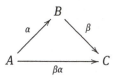

to produce the new commutative triangle

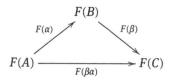

More generally, this observation allows a functor to be applied to any commutative diagram in a category without disturbing commutativity.

Examples of functors

(i)   The *identity functor* $1_{\mathcal{C}} : \mathcal{C} \rightarrow \mathcal{C}$ fixes all objects and morphisms in a category $\mathcal{C}$.

(ii)   The *forgetful functor* from **Gp** to **Set** maps each group to its underlying set and leaves all homomorphisms fixed. Thus the functor simply forgets about the group operation. There are many other forgetful functors, for example from **Rg** to **Set**.

(iii)   A more interesting example is the *abelianizing functor*. Define

$$F^{ab} : \mathbf{Gp} \rightarrow \mathbf{Gp}$$

by $F^{ab}(G) = G^{ab} = G/G'$ for any group G. If $\alpha : G \rightarrow H$ is a homomorphism of groups, define $F^{ab}(\alpha) : G^{ab} \rightarrow H^{ab}$ by $F^{ab}(\alpha) : xG' \mapsto \alpha(x)H'$ for $x \in G$. Here it is essential to observe that $\alpha(G') \leq H'$ to ensure that $F^{ab}(\alpha)$ is well defined.

Contravariant functors

There is a second type of functor which reverses the direction of the arrows when applied to morphisms. Let C and D be categories. A *contravariant functor*

$$F : C \to D$$

is a collection of functions

$$F : \mathrm{obj}(C) \to \mathrm{obj}(D), \quad \text{and} \quad F : \mathrm{Mor}_C(A, B) \to \mathrm{Mor}_D(F(B), F(A))$$

for each pair of objects (A, B) of C. Thus if $\alpha : A \to B$ is a morphism in C, then $F(\alpha) :$ $F(B) \to F(A)$ is a morphism in D. These functions are required to have the following properties:

(i)   if $\alpha \in \mathrm{Mor}_C(A, B)$ and $\beta \in \mathrm{Mor}_D(B, C)$, then $F(\beta\alpha) = F(\alpha)F(\beta)$ in D.

(ii) $F(\iota_A) = \iota_{F(A)}$ for all $A \in \mathrm{obj}(C)$.

As in the case of covariant functors, property (i) can be expressed by saying that the functor F may be applied to a commutative triangle, thereby retaining commutativity, but reversing all the arrows. Thus application of the functor F to the triangle below

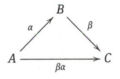

produces the new commutative triangle

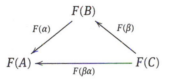

We will sometimes speak of the *variance* of a functor, referring to whether it is covariant or contravariant.

Opposite categories

A useful concept that connects the two types of functors is the opposite of a category. Let C be a category. The *opposite* of C

$$C^{op}$$

is the category whose objects are the objects of $\mathcal{C}$ and whose morphisms have the form $\alpha^{op} : B \to A$, where $\alpha : A \to B$ is a morphism in $\mathcal{C}$. Note the reversal in direction of the arrow. The rule of composition of morphisms in $\mathcal{C}^{op}$ is given by

$$\alpha^{op}\beta^{op} = (\beta\alpha)^{op}$$

where $\alpha \in \text{Mor}_{\mathcal{C}}(A, B)$, $\beta \in \text{Mor}_{\mathcal{C}}(B, C)$. To confirm that this the "correct" rule, consider the commutative diagram expressing the law of composition of morphisms in $\mathcal{C}$: this is

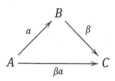

The commutative diagram in $\mathcal{C}^{op}$ obtained from this by reversing all the arrows is:

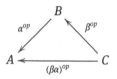

As was to be expected, its commutativity expresses the law of composition in $\mathcal{C}^{op}$. Of course $\iota_A{}^{op}$ is the identity morphism for an object A in $\mathcal{C}^{op}$. It is straightforward to verify that $\mathcal{C}^{op}$ is a category.

Next if $F : \mathcal{C} \to \mathcal{D}$ is a functor, define a new functor

$$F^{op} : \mathcal{C}^{op} \to \mathcal{D}^{op}$$

by $F^{op}(A) = F(A)$ for $A \in \text{obj}(\mathcal{C})$ and $F^{op}(\alpha^{op}) = F(\alpha)^{op}$ where $\alpha \in \text{Mor}_{\mathcal{C}}(A, B)$.

Now suppose that F is covariant. Let us check that F^{op} is also covariant. Let $\alpha \in \text{Mor}_{\mathcal{C}}(A, B)$, $\beta \in \text{Mor}_{\mathcal{C}}(B, C)$. Thus

$$F^{op}(\alpha^{op}\beta^{op}) = F^{op}((\beta\alpha)^{op}) = (F(\beta\alpha))^{op},$$

while

$$F^{op}(\alpha^{op})F^{op}(\beta^{op}) = F(\alpha)^{op}F(\beta)^{op} = (F(\beta)F(\alpha))^{op} = (F(\beta\alpha))^{op}.$$

Therefore $F^{op}(\alpha^{op}\beta^{op}) = F^{op}(\alpha^{op})F^{op}(\beta^{op})$ and F^{op} is a covariant functor.

By a similar argument, if we start with a contravariant functor $F : \mathcal{C} \to \mathcal{D}$, one can verify that $F^{op} : \mathcal{C}^{op} \to \mathcal{D}^{op}$ is also contravariant – see Exercise (16.2.1). These conclusions are summed up in

(16.2.1). *If $F : C \to D$ is functor, then $F^{op} : C^{op} \to D^{op}$ is a functor with the same variance.*

Contravariant functors as covariant functors

As has been seen, there are two types of functors, covariant and contravariant. However, for many purposes it is enough to deal with covariant functors. Let $F : C \to D$ be a functor of either variance. Then define a new functor

$$\bar{F} : C^{op} \to D$$

by $\bar{F}(A) = F(A)$ for $A \in \mathrm{obj}(C)$ and $\bar{F}(\alpha^{op}) = F(\alpha)$ where α is a morphism in C. This is a functor of the opposite variance, as the next result shows.

(16.2.2). *If $F : C \to D$ is a functor, then $\bar{F} : C^{op} \to D$ is a functor of the opposite variance.*

Proof. Let $\alpha : A \to B$ and $\beta : B \to C$ be morphisms in C and suppose that F is a contravariant functor. We will show that $\bar{F}$ is a covariant functor, arguing as follows.

$$\bar{F}(\alpha^{op}\beta^{op}) = \bar{F}((\beta\alpha)^{op}) = F(\beta\alpha) = F(\alpha)F(\beta) = \bar{F}(\alpha^{op})\bar{F}(\beta^{op}).$$

In a similar manner it can be shown that if F is covariant, then $\bar{F}$ is contravariant. $\quad\square$

This result shows that in suitable circumstances it is sufficient to study covariant functors. For, if a property has been proved for covariant functors and F is a contravariant functor, then the property holds for the covariant functor $\bar{F}$. It may then be possible to deduce that the property is valid for F.

Hom functors

Two of the most important functors in algebra are formed by using the additive group of homomorphisms between modules. More generally, take any category C and choose and fix an object A of C. Define a covariant functor

$$F_A : C \to \mathbf{Set}$$

in the following manner. Put $F_A(B) = \mathrm{Mor}(A, B)$ for any $B \in C$, noting that this is a set. Next, if $\beta \in \mathrm{Mor}(B, B')$, define a function $F_A(\beta) : \mathrm{Mor}(A, B) \to \mathrm{Mor}(A, B')$ by $F_A(\beta)(\alpha) = \beta\alpha$, where $\alpha : A \to B$ is a morphism in C. Observe that if $\beta \in \mathrm{Mor}(B, B')$ and $\beta' \in \mathrm{Mor}(B', B'')$, then

$$F_A(\beta'\beta)(\alpha) = \beta'\beta\alpha = F_A(\beta')(\beta\alpha) = F_A(\beta')F_A(\beta)(\alpha),$$

so that $F_A(\beta'\beta) = F_A(\beta')F_A(\beta)$. Also it is clear that $F_A(\iota_B)$ is the identity morphism on $\mathrm{Mor}(A, B)$. Hence F_A is a covariant functor, which will be denoted by

$$\mathrm{Mor}(A, -).$$

There is a dual contravariant functor defined by switching arguments. As before let $\mathcal{C}$ be a category with a fixed object A. Define a functor

$$G^A : \mathcal{C} \to \mathbf{Set}$$

as follows. Let $G^A(B) = \mathrm{Mor}(B, A)$ for any $B \in \mathrm{obj}(\mathcal{C})$. If $\beta \in \mathrm{Mor}(B, B')$, define a function $G^A(\beta) : \mathrm{Mor}(B', A) \to \mathrm{Mor}(B, A)$ by $G^A(\beta)(\alpha) = \alpha\beta$ where $\alpha : B' \to A$ is a morphism in $\mathcal{C}$. We claim that G^A is a contravariant functor. Let $\beta \in \mathrm{Mor}(B, B')$ and $\beta' \in \mathrm{Mor}(B', B'')$; then

$$G^A(\beta'\beta)(\alpha) = \alpha\beta'\beta = G^A(\beta)(\alpha\beta') = G^A(\beta)G^A(\beta')(\alpha).$$

Therefore $G^A(\beta'\beta) = G^A(\beta)G^A(\beta')$. Also $G^A(\iota_B)$ is the identity morphism for the object $\mathrm{Mor}(B, A)$. Thus G^A is a contravariant functor, which is written

$$\mathrm{Mor}(-, A).$$

The most important cases are when $\mathcal{C} = \mathbf{Mod}_R$ or $_R\mathbf{Mod}$ and the morphisms are R-module homomorphisms, R being a ring. In this situation the notations used for the functors F_A and G^A are

$$\mathrm{Hom}_R(A, -) \quad \text{and} \quad \mathrm{Hom}_R(-, A)$$

respectively. These are functors from $_R\mathbf{Mod}$ or $\mathbf{Mod}_R$ to $\mathbf{Ab}$. The Hom functors are ubiquitous in module theory.

It is often convenient to write

$$\beta_* = \mathrm{Hom}_R(A, \beta) \quad \text{and} \quad \beta^* = \mathrm{Hom}_R(\beta, A).$$

We may refer to β_* as an *induced homomorphism* and β^* as a *co-induced homomorphism*. These terminologies were already adopted in Section 9.1.

Tensor product functors

Another important source of functors is tensor products over a ring R; these were introduced in Chapter 13. Let M be a fixed right R-module and define a functor $F : {}_R\mathbf{Mod} \to \mathbf{Ab}$ by $F(A) = M \otimes_R A$ where $A \in {}_R\mathrm{Mod}$. If $\alpha \in \mathrm{Hom}_R(A, B)$, let $F(\alpha) = \alpha_*$ be the induced map $id_M \otimes \alpha$, as defined in Section 13.1. We check that F is a covariant functor.

Let $\alpha \in \text{Hom}_R(A, B)$, $\beta \in \text{Hom}_R(B, C)$; then by (13.1.3) we have

$$(\beta\alpha)_* = id_M \otimes \beta\alpha = (id_M \otimes \beta)(id_M \otimes \alpha) = \beta_* \alpha_*,$$

Therefore F is a covariant functor.

We can form another covariant functor from $\mathbf{Mod}_R$ to $\mathbf{Ab}$ by placing the fixed left module M on the right of the tensor product. A convenient notation for these two functors is

$$M \otimes_R - : {}_R\mathbf{Mod} \longrightarrow \mathbf{Ab} \quad \text{and} \quad - \otimes_R M : \mathbf{Mod}_R \longrightarrow \mathbf{Ab}.$$

Concrete categories

A general category is an abstract entity far removed from the realm of sets and functions. Thus special interest attaches to categories which can be represented in a meaningful way by sets and functions.

We start with the concept of a faithful functor. A functor $T : \mathcal{C} \to \mathcal{D}$ is called *faithful* if it is injective on morphisms, i. e., $T(\alpha_1) = T(\alpha_2)$ implies that $\alpha_1 = \alpha_2$. (Actually, this implies that T is also injective on objects: see Exercise (16.2.9).

A *concrete category* is defined as a pair $(\mathcal{C}, U)$ where $\mathcal{C}$ is a category and $U : \mathcal{C} \to \mathbf{Set}$ is a faithful covariant functor. Thus in $\mathcal{C}$ each object A has an associated set $U(A)$ and each morphism $\alpha : A \to B$ has an associated function $U(\alpha) : U(A) \to U(B)$. Note that $U(\alpha\beta) = U(\alpha)U(\beta)$ since U is a covariant functor.

More generally still, a category $\mathcal{C}$ is said to be *concretizable* if there exists a faithful covariant functor $U : \mathcal{C} \to \mathbf{Set}$ such that $(\mathcal{C}, U)$ is a concrete category. Many of the categories that we have encountered are seen to be concretizable by pairing them with the forgetful functor: for example, $\mathbf{Gp}$ and $\mathbf{Rg}$ are concretizable. On the other hand, non-concretizable categories exist. In fact the category $\mathbf{hTop}$ of topological spaces whose morphisms are homotopy classes of continous functions is known to be non-concretizable.

Additive functors

A category $\mathcal{C}$ is said to be *pre-additive* if each morphism set $\text{Mor}(A, B)$ in $\mathcal{C}$ is endowed with a binary operation, usually written +, which makes the sets of morphisms into abelian groups. Next suppose that $\mathcal{C}$ and $\mathcal{D}$ are pre-additive categories. A functor $F : \mathcal{C} \to \mathcal{D}$ is called an *additive functor* if, for any pair of morphisms $\alpha, \beta : A \to B$ in $\mathcal{C}$, the following rule holds:

$$F(\alpha + \beta) = F(\alpha) + F(\beta).$$

Among the most important additive functors are the Hom functors and the tensor product functors.

(16.2.3). *Let R be a ring and A a left R-module. Then:*

(i) $\mathrm{Hom}_R(A, -)$ *is an additive covariant functor and* $\mathrm{Hom}_R(-, A)$ *is an additive contravariant functor from* $_R\mathbf{Mod}$ *to* **Ab**.

(ii) $A \otimes_R -$ *and* $- \otimes_R A$ *are additive covariant functors from* $_R\mathbf{Mod}$ *to* **Ab** *and from* $\mathbf{Mod}_R$ *to* **Ab** *respectively.*

Proof. Take the case of $\mathrm{Hom}_R(A, -)$. The category $_R\mathbf{Mod}$ is pre-additive since $\mathrm{Hom}_R(A, B)$ is an abelian group, as was seen in Section 9.1. Let $\phi \in \mathrm{Hom}_R(B, B')$. Recall that $\phi_* = \mathrm{Hom}_R(A, \phi)$ was defined by $\phi_*(\theta) = \phi\theta$. Hence

$$\phi_*(\alpha + \beta) = \phi(\alpha + \beta) = \phi\alpha + \phi\beta = \phi_*(\alpha) + \phi_*(\beta).$$

The other cases are left to the reader as exercises. $\qquad\square$

Exact functors

In the remainder of this section only functors from $_R\mathbf{Mod}$ or $\mathbf{Mod}_R$ to **Ab**, where R is a ring, will be considered. Consider the effect of applying a functor to the exact sequence of R-modules with the form

$$0 \to A \xrightarrow{\alpha} B \xrightarrow{\beta} C. \tag{16.1}$$

(For exact sequences see Section 9.1.) Let $F : {}_R\mathbf{Mod} \to \mathbf{Ab}$ be a covariant functor. Then F is called *left exact* if, for every exact sequence of R-modules as in equation (16.1), the sequence arising from it on applying the functor F,

$$0 \to F(A) \xrightarrow{F(\alpha)} F(B) \xrightarrow{F(\beta)} F(C),$$

is also exact.

Similarly F is *right exact* if, given an exact sequence of R-modules

$$A \xrightarrow{\alpha} B \xrightarrow{\beta} C \to 0,$$

the sequence

$$F(A) \xrightarrow{F(\alpha)} F(B) \xrightarrow{F(\beta)} F(C) \to 0$$

is exact.

There are corresponding notions for a contravariant functor $F : {}_R\mathbf{Mod} \to \mathbf{Ab}$. Call F *left exact* if, given the exact sequence $A \xrightarrow{\alpha} B \xrightarrow{\beta} C \to 0$, the sequence

$$0 \to F(C) \xrightarrow{F(\beta)} F(B) \xrightarrow{F(\alpha)} F(A),$$

is exact. On the other hand, F is said to be *right exact* if, given the exact sequence $0 \to A \xrightarrow{\alpha} B \xrightarrow{\beta} C$, the sequence

$$F(C) \xrightarrow{F(\beta)} F(B) \xrightarrow{F(\alpha)} F(A) \to 0$$

is exact. Notice how the arrows are reversed by the contravariant functor F.

The exactness properties of the Hom and tensor product functors are described in the next result.

(16.2.4). *Let R be a ring and M an R-module. Then:*

(i) $\mathrm{Hom}_R(M, -)$ *is a left exact covariant functor and* $\mathrm{Hom}_R(-, M)$ *is a left exact contravariant functor.*

(ii) $M \otimes_R -$ *and* $- \otimes_R M$ *are right exact covariant functors.*

These are simply re-statements of results already proven, namely (9.1.25) and (13.2.9).

Exercises (16.2).

(1) Complete the proof of (16.2.1) by showing that if $F : \mathcal{C} \to \mathcal{D}$ is a contravariant functor, then $F^{op} : \mathcal{C}^{op} \to \mathcal{D}^{op}$ is contravariant.

(2) If $F : {}_R\mathbf{Mod} \to \mathbf{Ab}$ is an additive functor, prove that F fixes the zero module and the zero homomorphism.

(3) Let $\mathcal{C}$ and $\mathcal{D}$ be categories and recall that $\mathcal{C} \times \mathcal{D}$ is the product category (see Exercise (16.1.7)). Define a functor $F : \mathcal{C} \times \mathcal{D} \to \mathcal{C}$ by $F((A, B)) = A$ and $F((\alpha, \beta)) = \alpha$. Verify that F is a covariant functor.

(4) Let $F : \mathcal{C} \to \mathcal{D}$ and $G : \mathcal{D} \to \mathcal{E}$ be functors. Define the *composite functor* $G \circ F : \mathcal{C} \to \mathcal{E}$ by $G \circ F(A) = G(F(A))$ and $G \circ F(\alpha) = G(F(\alpha))$ where A is an object and α a morphism in $\mathcal{C}$. Verify that with these definitions $G \circ F$ is in fact a functor.

(5) Let F, G be functors such that $G \circ F$ exists. Prove that $G \circ F$ is covariant if F and G have the same variance and contravariant if they have opposite variances.

(6) Let $F : \mathcal{C} \to \mathcal{C}'$ and $G : \mathcal{D} \to \mathcal{D}'$ be functors of the same variance. Define the *product* functor $F \times G : \mathcal{C} \times \mathcal{D} \to \mathcal{C}' \times \mathcal{D}'$ by $F \times G((A, B)) = (F(A), G(B))$ and $F \times G((\alpha, \beta)) = (F(\alpha), G(\beta))$. Prove that $F \times G$ is a functor with the same variance as F and G.

(7) Let $\mathcal{C}$ be a category with just one object. Prove that the morphisms of $\mathcal{C}$ form a monoid M whose binary operation is composition of morphisms. Then prove that $\mathcal{C}$ is isomorphic with the monoidal category $\mathcal{C}_M$ defined in Example (vii) of Section 16.1.

(8) Let $\mathcal{C}$ be a category with the property that $\mathrm{Mor}(A, B) = \emptyset$ for every pair of distinct objects A, B of $\mathcal{C}$. Prove that $\mathcal{C}$ is the union of disjoint subcategories of the monoidal type in Exercise (16.2.7).

(9) Let $F : \mathcal{C} \to \mathcal{D}$ be a faithful functor, i. e., one that is injective on morphisms. Prove that F is also injective on objects.

16.3 Categorical constructions

In this section we describe some standard constructions that can be performed in certain important categories. Some of these constructs have appeared in earlier chapters in the context of groups and modules. All the constructs are characterized by mapping properties of diagrams in categories. We begin with coproducts and products, which already appeared in Chapter 9 in the guise of restricted and unrestricted direct sums of modules.

Coproducts

Let C be a category and $\{A_i \mid i \in I\}$ a *set* of objects in C. A *coproduct* of the A_i in C consists of an object

$$\coprod_{j \in I} A_j$$

and a set of morphisms $\alpha_i : A_i \to \coprod_{j \in I} A_j$ in C. These entities are required to have the following mapping property. If there are given morphisms $\phi_i : A_i \to X$ in C, then there is a *unique* morphism $\theta : \coprod_{j \in I} A_j \to X$ such that $\theta \alpha_i = \phi_i$ for all $i \in I$. This condition is expressed by commutativity of the triangles

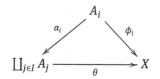

for all $i \in I$. One can think of the condition as requiring that the morphisms ϕ_i can be combined to form the morphism θ.

It is not claimed that coproducts always exist in a category, but if they do exist, they are essentially unique in the sense of the following result.

(16.3.1). *Let $\{A_i \mid i \in I\}$ be a set of objects in a category C. If $\{A, \alpha_i | i \in I\}$ is a coproduct of the A_i in C, then A is unique up to isomorphism.*

Proof. Suppose that $\{B, \beta_i \mid i \in I\}$ is another coproduct of the A_i in C. Apply the mapping property for each coproduct to obtain morphisms θ, τ making the two triangles below commute for each $i \in I$:

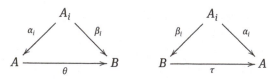

Hence $\theta\alpha_i = \beta_i$ and $\tau\beta_i = \alpha_i$, which combine to yield $(\theta\tau)\beta_i = \theta\alpha_i = \beta_i$ and $(\tau\theta)\alpha_i = \tau\beta_i = \alpha_i$. Thus we have two more commutative triangles

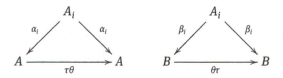

However, the morphisms $\tau\theta$ and $\theta\tau$ in these triangles could be replaced by identity morphisms for A and B respectively without disturbing commutativity. Consequently, by the uniqueness clause in the definition of the coproduct

$$\tau\theta = \iota_A \quad \text{and} \quad \theta\tau = \iota_B.$$

Hence θ and τ are mutually inverse morphisms, so they are isomorphisms. Therefore $A \simeq B$. □

The reader should take note of the form of the last proof, since it is used frequently to establish the uniqueness of constructed objects in a category.

Examples

(i) In the categories of R-modules $_R\mathbf{Mod}$, $\mathbf{Mod}_R$ the coproduct of a set of modules $\{M_i \mid i \in I\}$ is their restricted direct sum $\bigoplus_{i\in I} M_i$. The associated morphisms are the canonical injections $\mu_i : M_i \to \bigoplus_{j\in I} M_j$. (For direct sums of modules see Section 9.1.)

(ii) In the category $\mathbf{Gp}$ the coproduct of a set of groups $\{G_i \mid i \in I\}$ is less obvious. It is in fact the free product $\mathrm{Fr}_{i\in I}G_i$. The associated morphisms are the natural injections $G_i \to \mathrm{Fr}_{j\in I}G_j$. For free products see Section 15.3.

We will prove the statement (i). Let there be given module homomorphisms $\psi_i : M_i \to X$, $i \in I$. Define $\theta : \bigoplus_{i\in I} M_i \to X$ as follows. If $a \in \bigoplus_{i\in I}M_i$, write $a = \sum_{i\in I}\mu_i(m_i)$, where $m_i \in M_i$ and the μ_i are the canonical injections. Note that only finitely many m_i are non-zero, so the sum is actually a finite one. Now define $\theta(a) = \sum_{i\in I}\psi_i(m_i)$. Then $\theta\mu_i(m_i) = \psi_i(m_i)$ since $\mu_i(m_i)$ has only one non-zero component. Thus $\theta\mu_i = \psi_i$.

If θ' is another homomorphism such that $\theta'\mu_i = \psi_i$, then $\theta\mu_i = \theta'\mu_i$ for all $i \in I$. This implies that $\theta = \theta'$, since the submodules $\mathrm{Im}(\mu_i)$ generate $\bigoplus_{i\in I}M_i$. This shows that the direct sum is a coproduct in $_R\mathbf{Mod}$ and by (15.3.1) we can be sure that it is unique up to isomorphism.

Products

The concept of a product is dual to that of a coproduct in the sense that it is obtained by reversing all the arrows in the definition. Let $\{A_i \mid i \in I\}$ be a set of objects in a category C. A *product* of the A_i in C consists of an object

$$\prod_{j\in I} A_j$$

and a set of morphisms $\pi_i : \prod_{j\in I} A_j \to A_i$ such that, given morphisms $\alpha_i : X \to A_i$ in C, there is a *unique* morphism $\theta : X \to \prod_{j\in I} A_j$ such that $\pi_i\theta = \alpha_i$ for $i \in I$. Thus the triangle below commutes for all $i \in I$:

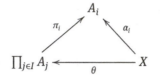

There is no claim that products exist in C, but, as in the case of coproducts, if they do exist, they are unique.

(16.3.2). *Let $\{A_i \mid i \in I\}$ be a set of objects in a category C. If $\{A, \pi_i \mid i \in I\}$ is a product of the A_i in C, then A is unique up to isomorphism.*

The proof is entirely analogous to the proof of (16.3.1) and involves writing down four commuting triangles. Indeed, all one need do is reverse all the arrows in the proof of (16.3.1). Alternatively one could simply observe that a product in a category C is a coproduct in C^{op}, and conversely.

Examples

(i) The product of a set of groups $\{G_i \mid i \in I\}$ in the category **Gp** is the unrestricted direct (or cartesian) product $\text{Cr}_{i\in I} G_i$, together with the associated canonical projections: for this see Section 4.2.

(ii) In the categories of R-modules $_R$**Mod**, **Mod**$_R$ the product of a set of modules $\{M_i \mid i \in I\}$ is their unrestricted direct sum, again with the associated canonical projections.

We describe next two further constructions that occur frequently in algebra and in homological algebra in particular.

Pullbacks

Let $\beta : B \to A$ and $\gamma : C \to A$ be morphisms in a category C. A *pullback* of the diagram

is a triple (X, λ, μ) consisting of an object X and morphisms $\lambda : X \to B$, $\mu : X \to$ in $\mathcal{C}$
such that the square below commutes:

Thus $\beta\lambda = \gamma\mu$.

In addition, the triple (X, λ, μ) must have the additional property that, if (X', λ', μ')
is another such triple making the corresponding square commute, there exists a
unique morphism $\theta : X' \to X$ such that the following diagram commutes:

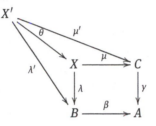

Thus $\lambda\theta = \lambda'$ and $\mu\theta = \mu'$. One can think of the additional requirement as expressing a
kind of "minimality" of the pullback (X, λ, μ) with respect to the commutative square
property. For brevity we may also refer to X as the pullback.

Of course, one cannot expect pullbacks to exist in a general category, but when
they do exist, they are essentially unique.

(16.3.3). *If (X, λ, μ) and (X', λ', μ') are two pullbacks of morphisms $\beta : B \to A$ and
$\gamma : C \to A$ in a category $\mathcal{C}$, then there is an isomorphism $\theta : X' \to X$ such that $\lambda' = \lambda\theta$
and $\mu' = \mu\theta$.*

Proof. This follows the same general pattern as the proof of (16.3.1). There are morphisms $\theta : X' \to X$ and $\phi : X \to X'$ making the triangles below commute, since we
have two pullbacks of $\beta : B \to A$ and $\gamma : C \to A$.

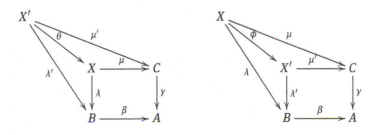

From the commutativity of the two triangles in each diagram we obtain $\lambda\theta = \lambda'$, $\mu\theta = \mu'$ and $\lambda'\phi = \lambda$, $\mu'\phi = \mu$ respectively. Hence

$$\lambda\theta\phi = \lambda'\phi = \lambda, \quad \mu\theta\phi = \mu'\phi = \mu$$

and

$$\lambda'\phi\theta = \lambda\theta = \lambda', \quad \mu'\phi\theta = \mu\theta = \mu'.$$

These equations show that the following diagrams commute:

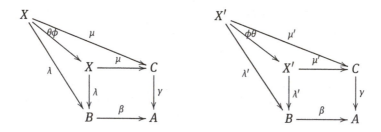

On the other hand, it is obvious that if the identity morphisms ι_X and $\iota_{X'}$ are substituted for $\theta\phi$ and $\phi\theta$ respectively, the triangles will still commute. Therefore by the uniqueness requirement in the definition $\theta\phi = \iota_X$ and $\phi\theta = \iota_{X'}$. Thus θ and ϕ are isomorphisms and hence $\theta : X' \to X$ is an isomorphism. □

It is important to observe that pullbacks exist in the categories of modules over a ring.

(16.3.4). *Pullbacks exist in the categories $_R\mathbf{Mod}$ and $\mathbf{Mod}_R$ for any ring R.*

Proof. Let $\beta : B \to A$ and $\gamma : C \to A$ be homomorphisms of R-modules. Define a submodule X of the R-module $B \oplus C$ by

$$X = \{(b,c) \mid b \in B, c \in C, \beta(b) = \gamma(c)\}.$$

Let $\lambda : X \to B$ and $\mu : X \to C$ be the natural projections $(b,c) \mapsto b$ and $(b,c) \mapsto c$ respectively. It is claimed that (X, λ, μ) is a pullback of β and γ. First of all for $(b,c) \in X$ we have

$$\gamma\mu(b,c) = \gamma(c) = \beta(b) = \beta\lambda(b,c),$$

so that $\beta\lambda = \gamma\mu$. Therefore β, γ, λ, μ form the commutative square in the next diagram.

To complete the proof we need to establish the minimality property. Suppose that (X', λ', μ') is another triple making the square commute. It must be shown that there is a unique homomorphism θ making the diagram below commute.

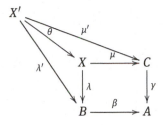

Define the map $\theta : X' \to X$ by $\theta(x') = (\lambda'(x'), \mu'(x'))$ where $x' \in X'$. Note that $\theta(x') \in X$ since $\beta\lambda'(x') = \gamma\mu'(x')$. Now check that the two triangles in the diagram commute. We have $\lambda\theta(x') = \lambda(\lambda'(x'), \mu'(x')) = \lambda'(x')$, so that $\lambda\theta = \lambda'$ and in a similar way $\mu\theta = \mu'$.

The final step is to verify that θ is the only homomorphism that will make the diagram commute. Suppose that $\bar{\theta} : X' \to X$ is another one. Then $\lambda\theta = \lambda\bar{\theta}$ and $\mu\theta = \mu\bar{\theta}$. Let $x' \in X'$; then $\theta(x')$ and $\bar{\theta}(x')$ have the same components in $B \oplus C$, from which it follows that $\theta(x') = \bar{\theta}(x')$ and hence $\theta = \bar{\theta}$. This completes the proof that (X, λ, μ) is a pullback. □

Pushouts

The next construction is the dual of a pullback, which means that effectively it is defined by reversing all the arrows in the definition of a pullback. Let $y : A \to C$ and $\beta : A \to B$ be morphisms in a category $\mathcal{C}$. Thus we have the diagram:

$$
\begin{array}{ccc}
A & \xrightarrow{\gamma} & C \\
\beta \downarrow & & \\
B & &
\end{array}
$$

A *pushout* of this diagram is a triple (X, λ, μ) where $X \in \text{obj}(\mathcal{C})$ and $\lambda : B \to X, \mu : C \to X$ are morphisms in $\mathcal{C}$ such that $\lambda\beta = \mu\gamma$, which is to say the square below commutes.

$$
\begin{array}{ccc}
A & \xrightarrow{\gamma} & C \\
\beta \downarrow & & \downarrow \mu \\
B & \xrightarrow{\lambda} & X
\end{array}
$$

In addition the following property must be satisfied: if (X', λ', μ') is another such triple leading to a commutative square, then there is a unique morphism $\theta : X \to X'$ such that $\theta\lambda = \lambda'$ and $\theta\mu = \mu'$. In short, θ causes the two triangles in the diagram below to commute.

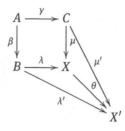

This can be thought of as a minimality property of the pushout. Next we note that pushouts are unique when they exist.

(16.3.5). *If (X, λ, μ) and (X', λ', μ') are two pushouts of morphisms $\beta : A \to B$ and $\gamma : A \to C$ in a category $\mathcal{C}$, then there is an isomorphism $\theta : X \to X'$ such that $\lambda' = \theta\lambda$ and $\mu' = \theta\mu$.*

This is proved in a similar way to (16.3.3). Alternatively, observe that a pushout in $\mathcal{C}$ is a pullback in $\mathcal{C}^{op}$ and use (16.3.3). The relevant point is that the commutativity of a diagram is unaffected by reversal of all the arrows.

The next result is dual to (16.3.4) and is important in module theory.

(16.3.6). *Pushouts exist in the categories $_R\mathbf{Mod}$ and $\mathbf{Mod}_R$ for any ring R.*

Proof. Here is an outline of the proof. Let $\beta : A \to B$ and $\gamma : A \to C$ be homomorphisms of R-modules. The pushout is defined to be a certain quotient of the module $B \oplus C$. Let

$$S = \{(\beta(a), -\gamma(a)) \mid a \in A\},$$

which is a submodule of $B \oplus C$, and set

$$X = (B \oplus C)/S.$$

Define also $\lambda(b) = (b, 0) + S$ and $\mu(c) = (0, c) + S$. It is claimed that the triple (X, λ, μ) is a pushout of the homomorphisms β, γ. For $a \in A$ we have

$$\lambda\beta(a) = (\beta(a), 0) + S = (0, \gamma(a)) + S = \mu\gamma(a),$$

showing that $\lambda\beta = \mu\gamma$. The rest of the proof establishes the minimality property of pushouts. This is Exercise (16.3.7). $\qquad\square$

Free objects in a category

The concept of a free entity is widespread in algebra and has already been encountered in the case of free modules and free groups. Here we will explain how free objects in a category can be defined by means of a certain mapping property.

To begin let $(\mathcal{C}, U)$ denote a concrete category: thus $U : \mathcal{C} \to$ **Set** is a faithful functor which represents objects in $\mathcal{C}$ by sets and morphisms in $\mathcal{C}$ by functions. Let X be a non-empty set. An object F in $\mathcal{C}$ is said to be *free on X* if there is a function $\mu : X \to U(F)$ with the following property. If $A \in \mathrm{obj}(\mathcal{C})$ and $\alpha : X \to U(A)$ is a function, there is a *unique* morphism $\beta : F \to A$ such that $U(\beta)\mu = \alpha$, which is just to say that the triangle below commutes:

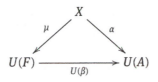

In many cases, but not always, the mapping property implies that the function μ is injective: see Exercise (16.3.12) in this connection.

It has already been shown that free objects exist in the categories of R-modules and groups – see Sections 9.1 and 15.1 respectively. It is not to be expected that free objects will exist in an arbitrary concrete category, but, if a free object on a set exists, then it is unique up to isomorphism.

(16.3.7). *Let F and F' be objects in a concrete category $(\mathcal{C}, U)$ which are both free on a (non-empty) set X. Then there is an isomorphism β in $\mathcal{C}$ such that $\beta(F) = F'$; thus $F \simeq F'$.*

Proof. This takes a form that should by now be familiar, cf. (16.3.1) and (16.3.2). Let $\mu : X \to U(F)$ and $\mu' : X \to U(F')$ be the associated functions for F and F'. Display the commutative triangles that express the mapping properties of F and F' for certain morphisms $\beta : F \to F'$ and $\beta' : F' \to F$.

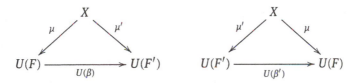

Therefore $U(\beta)\mu = \mu'$ and $U(\beta')\mu' = \mu$. From these it follows that $U(\beta')U(\beta)\mu = \mu$ and $U(\beta)U(\beta')\mu' = \mu'$, so we have two additional commutative triangles:

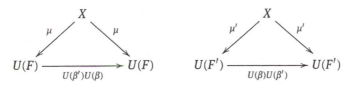

But the roles of $U(\beta')U(\beta)$ and $U(\beta)U(\beta')$ can be played here by the identity functions on $U(F)$ and $U(F')$ respectively, while preserving commutativity of the diagrams. By the uniqueness clause in the definition

$$U(\beta'\beta) = U(\beta')U(\beta) = \text{id}_{U(F)} = U(\iota_F)$$

and similarly $U(\beta\beta') = U(\iota_{F'})$. Since the functor U is faithful, it follows that $\beta'\beta = \iota_F$ and $\beta\beta' = \iota_{F'}$. Hence β and β' are isomorphisms and F and F' are isomorphic.   □

Exercises (16.3).
(1) Prove (16.3.2), which asserts that products are unique up to isomorphism.
(2) Verify that the free product is the coproduct in the category **Gp**.
(3) Prove that the disjoint union is the coproduct in **Set**.
(4) Prove that the set product is the product in **Set**.
(5) Let B, C be subsets of a set A and let $\iota_1 : B \to A$ and $\iota_2 : C \to A$ be inclusion maps. Prove that $B \cap C$ is the pullback of these maps in **Set**.
(6) Let A be a subset of sets B and C. Let $\iota_1 : A \to B$ and $\iota_2 : A \to C$ be inclusion maps. Prove that $B \cup C$ is the pushout of these maps in **Set**.
(7) Complete the proof of (16.3.6) on the existence of pushouts in a module category by establishing the minimality property.
(8) Let $\beta : B \to A$ and $\gamma : C \to A$ be module homomorphisms and let (X, λ, μ) be the pullback of β and γ in a module category. If β is surjective or injective, prove that μ is surjective or injective respectively.
(9) Let $\beta : A \to B$ and $\gamma : A \to C$ be module homomorphisms and let (X, λ, μ) be the pushout of β and γ in a module category. If β is surjective or injective, prove that μ is surjective or injective respectively.
(10) Let F and F' be free objects on sets X and X' in a concrete category $(\mathcal{C}, U)$, with respective associated functions μ and μ'. If X and X' have equal cardinality, prove that there is an isomorphism θ in $\mathcal{C}$ such that $\theta(F) = F'$. [Hint: by hypothesis there is a bijective function $\pi : X \to X'$. Follow the method of proof of (16.3.7), using the functions $\mu'\pi : X \to U(F')$ and $\mu\pi^{-1} : X' \to U(F)$.]
(11) Establish the existence of free objects in the category **Mon** of monoids and monoidal homomorphisms by arguing as follows. Let X be a non-empty set and let $W(X)$ denote the set of all words in X, i. e., finite formal sequences $x_1x_2\ldots x_n$, where $x_i \in X$, including the empty word. This is a monoid in which the binary operation is juxtaposition and the identity element is the empty word. Prove that $W(X)$ is free on the set X in the category **Mon**.
(12) Let (F, μ) be free on a set X in a concrete category $(\mathcal{C}, U)$. Assume that there exists an object A in $\mathcal{C}$ such that the set $U(A)$ has at least two elements. Prove that the function $\mu : X \to U(F)$ is necessarily injective. [Hint: follow the method of proof used for free groups in Section 15.1.]

16.4 Natural transformations

Two functors with the same variance can be compared by means of a collection of morphisms called a natural transformation. In detail let $F, G : \mathcal{C} \to \mathcal{D}$ be two covariant functors. A *natural transformation* τ from F to G is a collection of morphisms $\{\tau_A \mid A \in \mathrm{obj}(\mathcal{C})\}$ in $\mathcal{D}$ where

$$\tau_A : F(A) \to G(A),$$

such that, if $\alpha : A \to A'$ is a morphism in $\mathcal{C}$, the diagram below commutes:

$$
\begin{array}{ccc}
F(A) & \xrightarrow{F(\alpha)} & F(A') \\
\tau_A \downarrow & & \downarrow \tau_{A'} \\
G(A) & \xrightarrow{G(\alpha)} & G(A')
\end{array}
$$

that is to say, $G(\alpha)\tau_A = \tau_{A'}F(\alpha)$. Here one might imagine the natural transformation τ as "sliding" the functorial action in the top row of the square onto the bottom row, thus relating the two functors. The notation

$$\tau : F \xrightarrow{\cdot} G$$

will be used to indicate a natural transformation from F to G. If τ_A is an isomorphism for every $A \in \mathrm{obj}(\mathcal{C})$, then τ is called a *natural isomorphism*.

Examples

(i) Let $F^{ab} : \mathbf{Gp} \to \mathbf{Gp}$ be the abelianizing functor. Thus $F^{ab}(x) = xG'$ for $x \in G$, and if $\alpha : G \to H$ is a homomorphism, $F^{ab}(\alpha) : xG' \to \alpha(x)H'$ – see the third example of a functor in Section 16.2. We have the commutative square

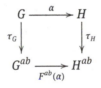

where τ_G denotes the canonical homomorphism from G to G^{ab}, which is a morphism in **Gp**. Let $\tau = \{\tau_G \mid G \in \mathrm{obj}(\mathbf{Gp})\}$. Thus we have a natural transformation $\tau : 1_{\mathbf{Gp}} \xrightarrow{\cdot} F^{ab}$ from the identity functor on **Gp** to the abelianizing functor,

(ii) Let R be a ring with identity. Then the functor

$$\mathrm{Hom}_R(R, -) : {}_R\mathbf{Mod} \to {}_R\mathbf{Mod}$$

is naturally isomorphic with the identity functor $1_{{}_R\mathbf{Mod}}$.

To see this, recall from (9.1.21) the isomorphism

$$\theta_A : \mathrm{Hom}_R(R, A) \rightarrow A$$

defined by $\theta_A(\phi) = \phi(1_R)$ for any left R-module A. Put $\boldsymbol{\theta} = \{\theta_A \mid A \in {}_R\mathbf{Mod}\}$. If $\alpha : A \rightarrow A'$ is a homomorphism of left R-modules, recall that α_* denotes the induced mapping $\mathrm{Hom}_R(R, \alpha)$. Then we have the commutative square:

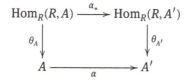

It is routine to check commutativity. Therefore we have a natural transformation $\boldsymbol{\theta} : \mathrm{Hom}_R(R, -) \overset{\cdot}{\rightarrow} 1_{{}_R\mathbf{Mod}}$. Indeed, since each θ_A is an isomorphism, $\boldsymbol{\theta}$ a natural isomorphism.

(iii) Again let R be a ring with identity. This time consider the covariant functor $R \otimes -$: ${}_R\mathbf{Mod} \rightarrow {}_R\mathbf{Mod}$. This functor is naturally isomorphic with the identity functor $1_{{}_R\mathbf{Mod}}$. To establish this, first recall from (13.2.1) that there is an R-module isomorphism

$$\theta_A : R \otimes_R A \rightarrow A$$

for any left R-module A: here $\theta_A(r \otimes a) = r \cdot a$ for $r \in R$, $a \in A$. Let $\alpha : A \rightarrow A'$ be a homomorphism of left R-modules. Then

$$\alpha_* = R \otimes_R \alpha : R \otimes_R A \rightarrow R \otimes_R A'$$

is the induced homomorphism in which $r \otimes a \mapsto r \otimes \alpha(a)$ for $r \in R$, $a \in A$. Hence the square below commutes:

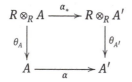

as the reader should verify. Set $\boldsymbol{\theta} = \{\theta_A \mid A \in {}_R\mathbf{Mod}\}$; then

$$\boldsymbol{\theta} : R \otimes_R - \overset{\cdot}{\rightarrow} 1_{{}_R\mathbf{Mod}}$$

is a natural isomorphism.

(iv) (*Dual spaces.*) Another well known example of a natural transformation involves the double dual of a vector space – for this see Section 8.3 above. Let K be a field and

$V \in {}_K\mathbf{Mod}$, so that V is a K-vector space. Recall that $F = \operatorname{Hom}_K(-, K)$ is a contravariant functor: write $F(V) = V^*$, the dual space, and $F(\alpha) = \alpha^*$. One can also form the covariant functor $D = F \circ F : {}_K\mathbf{Mod} \to {}_K\mathbf{Mod}$, the so called *double dual*. Write $D(V) = V^{**}$ and $D(\alpha) = (\alpha^*)^* = \alpha^{**}$.

It was shown in (8.3.11) that there is an injective K-homomorphism $\theta_V : V \to V^{**}$ given by $\theta_V(v)(\phi) = \phi(v)$ where $v \in V$, $\phi \in V^*$. Let $\alpha : V \to V'$ be a K-linear mapping and consider the diagram

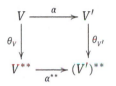

This diagram commutes: for if $v \in V$ and $\beta \in V'^*$, then

$$\theta_{V'}(\alpha(v))(\beta) = \beta(\alpha(v)) = \beta\alpha(v).$$

Also $\alpha^{**}\theta_V(v) = (\alpha^*)^*\theta_V(v) = \theta_V(v)\alpha^*$. Therefore

$$\alpha^{**}\theta_V(v)(\beta) = \theta_V(v)\alpha^*(\beta) = \theta_V(v)(\beta\alpha) = \beta\alpha(v),$$

which shows that $\alpha^{**}\theta_V = \theta_{V'}\alpha$, as required.

Finally, put $\boldsymbol{\theta} = \{\theta_V \mid V \in {}_K\mathbf{Mod}\}$. What has just been established is that $\boldsymbol{\theta} : 1_{{}_K\mathbf{Mod}} \overset{\bullet}{\to} D$ is a natural transformation from the identity functor to the double dual functor. If we restrict to the subcategory of finite dimensional K-spaces, then by (8.3.11) the maps θ_V are isomorphisms, so that $\boldsymbol{\theta}$ is a natural isomorphism from $1_{{}_K\mathbf{Mod}}$ to the double dual functor D.

Composites of natural transformations

Just as for morphisms and functors, there is an obvious way to compose natural transformations. Let $\mathcal{C}$ and $\mathcal{D}$ be categories with covariant functors $F, G, H : \mathcal{C} \to \mathcal{D}$. Let $\boldsymbol{\sigma} : F \overset{\bullet}{\to} G$ and $\boldsymbol{\tau} : G \overset{\bullet}{\to} H$ be natural transformations from F to G and G to H respectively. Then the *composite* of $\boldsymbol{\tau}$ and $\boldsymbol{\sigma}$,

$$\boldsymbol{\tau} \bullet \boldsymbol{\sigma} : F \overset{\bullet}{\to} H,$$

is defined as the collection of morphisms $\{\tau_A \sigma_A \mid A \in \operatorname{obj}(\mathcal{C})\}$.

To verify that $\tau \bullet \sigma$ is a natural transformation, let $\alpha : A \to A'$ be a morphism in $\mathcal{C}$ and form the diagram

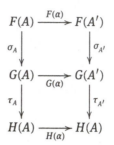

The two small squares in the diagram commute since σ and τ are natural transformations. This implies that the square below commutes:

$$\begin{array}{ccc} F(A) & \xrightarrow{F(\alpha)} & F(A') \\ \tau_A \sigma_A \downarrow & & \downarrow \tau_{A'} \sigma_{A'} \\ H(A) & \xrightarrow{H(\alpha)} & H(A') \end{array}$$

For

$$H(\alpha)(\tau_A \sigma_A) = \tau_{A'} G(\alpha)\sigma_A = (\tau_{A'} \sigma_{A'})F(\alpha).$$

Therefore $\tau \bullet \sigma : F \xrightarrow{\bullet} H$ is a natural transformation. It is routine to prove that *composition of natural transformations is an associative operation.*

Functor categories

We have seen that natural transformations can be composed in an associative manner. This observation suggests the possibility of forming a new category in which the objects are functors and the morphisms are natural transformations between functors.

Let $\mathcal{C}$ and $\mathcal{D}$ be categories with $\mathcal{C}$ a small category: this means that $\mathrm{obj}(\mathcal{C})$ is a set. The *functor category*

$$\mathcal{D}^{\mathcal{C}}$$

has as its objects the functors from $\mathcal{C}$ to $\mathcal{D}$. If $F, G : \mathcal{C} \to D$ are two such functors, the morphisms for $\mathcal{D}^{\mathcal{C}}$ are to be the natural transformations from F to G. The law of composition of morphisms in $\mathcal{D}^{\mathcal{C}}$ is to be the composition of natural transformations defined above. This is an associative operation and the identity natural transformation plays the role of the identity morphism. We conclude the chapter with an example of a functor category.

Example (16.4.1). Any non-empty set C can be made into a category $\mathcal{C}$ by declaring that $\mathrm{obj}(\mathcal{C}) = C$ and all morphisms are identities. Conversely, a category in which every morphism is an identity is just a set, if we ignore the identities. Let C and D be non-empty sets. Then a functor from $\mathcal{C}$ to $\mathcal{D}$ is simply a function from C to D, since all morphisms are identities.

Next suppose that τ is a natural transformation between functors $F, G : \mathcal{C} \to \mathcal{D}$. Since the morphisms in $\mathcal{D}$ are identities, it follows that τ is just a collection of identities. Now consider the functor category $\mathcal{D}^{\mathcal{C}}$. Its objects are the functions from C to D, while its morphisms are natural transformations between functions. Thus, identities aside, the category $\mathcal{D}^{\mathcal{C}}$ is just the set of all functions from C to D, that is, $\mathcal{D}^{\mathcal{C}} = \mathrm{Fun}(C, D)$.

Exercises (16.4).

(1) By checking commutativity of the relevant diagram complete the proof that if R is a ring with identity, then $\mathrm{Hom}_R(R, -) \stackrel{\bullet}{\simeq} 1_{R\mathbf{Mod}}$.

(2) By checking commutativity of the relevant diagram complete the proof that if R is a ring with identity, then $R \otimes_R - \stackrel{\bullet}{\simeq} 1_{R\mathbf{Mod}}$.

(3) Let $F, G : {}_R\mathbf{Mod} \to \mathbf{Ab}$ be additive functors. If $F \stackrel{\bullet}{\simeq} G$ and F is left exact, show that G is left exact.

(4) Let H be a fixed group. Define a functor $F_H = H \times - : \mathbf{Gp} \to \mathbf{Gp}$ by $F_H(G) = H \times G$ and $F_H(\alpha) : (h, g) \to (h, \alpha(g))$ for $h \in H$, $g \in G$. Let $\theta : H \to K$ be a group homomorphism. Define $\tau_G : F_H(G) \to F_K(G)$ by $\tau_G : (h, g) \mapsto (\theta(h), g)$. Let $\tau = \{\tau_G \mid G \in \mathbf{Gp}\}$. Prove that $\tau : F_H \stackrel{\bullet}{\to} F_K$ is a natural transformation.

(5) Prove that composition of natural transformations is an associative operation.

(6) Show that if σ and τ are natural isomorphisms that can be composed, then $\tau \bullet \sigma$ is a natural isomorphism. Also prove that σ^{-1} is a natural isomorphism and that $(\tau \bullet \sigma)^{-1} = \sigma^{-1} \bullet \tau^{-1}$.

(7) Let $\mathbf{Fd}$ denote the category of fields and ring homomorphisms. Let $K \in \mathbf{Fd}$. Define two functors GL_n and U from $\mathbf{Fd}$ to $\mathbf{Gp}$ as follows. (i) $\mathrm{GL}_n : K \to \mathrm{GL}_n(K)$ and $\mathrm{GL}_n(\alpha) : \mathrm{GL}_n(K) \to \mathrm{GL}_n(K')$ where $\alpha : K \to K'$ is a ring homomorphism: (ii) $U : K \to U(K)$ and $U(\alpha) : U(K) \to U(K')$. (Here $U(K)$ denotes the multiplicative group of K.) Define $\det_K$ to be the determinant function on $\mathrm{GL}_n(K)$: this is a morphism in $\mathbf{Gp}$. Put $\mathbf{det} = \{\det_K \mid K \in \mathbf{Fd}\}$. Now verify that the following square commutes:

$$\begin{array}{ccc} \mathrm{GL}_n(K) & \xrightarrow{\mathrm{GL}_n(\alpha)} & \mathrm{GL}_n(K') \\ {\scriptstyle \det_K}\downarrow & & \downarrow{\scriptstyle \det_{K'}} \\ U(K) & \xrightarrow[U(\alpha)]{} & U(K') \end{array}$$

and conclude that $\mathbf{det} : \mathrm{GL}_n \stackrel{\bullet}{\to} U$ is a natural transformation.

17 Applications

This chapter presents applications of algebraic concepts and techniques to various problems that lie largely outside the domain of algebra. It will be seen that groups acting on sets, finite fields and methods from linear algebra are especially useful. Sometimes it is the notational sharpness that algebra provides which leads to fresh insights. Sections 17.1–17.3 contain applications to combinatorial problems, while Section 17.4 is an introduction to error correcting codes; finally Section 17.5 uses algebraic techniques to construct models of accounting systems.

17.1 Set labelling problems

Group actions can often be used effectively to solve certain types of labelling problems. As an example of such a problem, suppose we wish to color the six faces of a cube and five colors are available. How many different coloring schemes are there? At first sight one might answer 5^6 since each of the six faces can be colored in five different ways. However, this answer is incorrect since by merely rotating the cube it is possible to pass from one coloring scheme to another one. Clearly two such coloring schemes are not really different. Thus not all of the 5^6 colorings schemes are distinct.

Let us pursue the idea of rotating the cube. The group of rotations of the cube acts on the set of all possible coloring schemes. If two colorings belong to the same rotational orbit, they should be considered identical since one arises from the other by a suitable rotation. Thus what we really need to do is to count the number of orbits of colorings and for this purpose the Frobenius-Burnside Theorem (5.2.3) is ideally suited.

The problem is really about the labelling of sets. Let X and L be two non-empty finite sets, with L referred to as the set of *labels*. Suppose that a label is to be assigned to each element of the set X: such a labelling is specified by a function

$$\alpha : X \to L :$$

call such a function α a *labelling* of X by L. Thus the set of all labellings of X by L is

$$\text{Fun}(X, L).$$

Now suppose that G is a finite group that acts on the set X on the left. Then G can be made to act on the set of labellings in a natural way by the rule

$$(g \cdot \alpha)(x) = \alpha(g^{-1} \cdot x),$$

https://doi.org/10.1515/9783110691160-017

where $g \in G$, $x \in X$ and $\alpha \in \text{Fun}(X, L)$. Notice that this is equivalent to $(g \cdot \alpha)(g \cdot x) = \alpha(x)$, i. e., the labelling $g \cdot \alpha$ is to assign to the set element $g \cdot x$ the same label as α assigns to x. The example of the cube should convince the reader that this is the correct action.

First we must verify that this really is an action of G on $\text{Fun}(X, L)$. To do this let $g_1, g_2 \in G$, $x \in X$ and $\alpha \in \text{Fun}(X, L)$; then

$$(g_1 \cdot (g_2 \cdot \alpha))(x) = (g_2 \cdot \alpha)(g_1^{-1} \cdot x) = \alpha(g_2^{-1} \cdot (g_1^{-1} \cdot x))$$
$$= \alpha((g_1 g_2)^{-1} \cdot x)$$
$$= ((g_1 g_2) \cdot \alpha)(x).$$

Hence $g_1 \cdot (g_2 \cdot \alpha) = (g_1 g_2) \cdot \alpha$. Also $1_G \cdot \alpha(x) = \alpha(1_G \cdot x) = \alpha(x)$, so that $1_G \cdot \alpha = \alpha$. Therefore we have an action of G on $\text{Fun}(X, L)$.

Our goal is to count the G-orbits in $\text{Fun}(X, L)$, which is achieved in the following fundamental result.

(17.1.1) (Polya[1]). *Let G be a finite group acting on a finite set X, and let L be a finite set of labels. Then the number of G-orbits of labellings of X by L is*

$$\frac{1}{|G|} \left(\sum_{g \in G} \ell^{m(g)} \right)$$

where $\ell = |L|$ and $m(g)$ is the number of disjoint cycles in the permutation of X corresponding to g.

Proof. By (5.2.3) the number of G-orbits of labellings is

$$\frac{1}{|G|} \left(\sum_{g \in G} |\text{Fix}(g)| \right)$$

where $\text{Fix}(g)$ is the set of labellings fixed by g. We have to count these labellings. Now $\alpha \in \text{Fix}(g)$ if and only if $g \cdot \alpha(x) = \alpha(x)$, i. e., $\alpha(g^{-1} \cdot x) = \alpha(x)$ for all $x \in X$. This equation asserts that α is constant on the $\langle g \rangle$-orbit $\langle g \rangle \cdot x$. Now the $\langle g \rangle$-orbits arise from the disjoint cycles involved in the permutation of X corresponding to g. Therefore, to construct a labelling in $\text{Fix}(g)$ all we need to do is to assign a label to each cycle of g. This can be done in $\ell^{m(g)}$ ways where $m(g)$ is the number of cycles; consequently $|\text{Fix}(g)| = \ell^{m(g)}$ and we have our formula. $\square$

Polya's Theorem will now be applied to solve some counting problems.

---

1 George Polya (1887–1985).

Example (17.1.1). How many ways are there to design a necklace of 11 beads if c different colors of beads are available?

Here it is assumed that the beads are identical apart from color. The necklace can be visualized as a regular 11-gon with the beads as vertices, labelled $1, 2, \ldots, 11$. The labels are the c colors and one color has to be assigned to each vertex. Clearly a symmetry of the 11-gon can be applied without changing the design of the necklace. Recall from Section 3.2 that G, the group of symmetries of the 11-gon, is the dihedral group Dih(22). It consists of the identity, rotations through $(\frac{2\pi}{11})i$ for $i = 1, 2, \ldots, 10$, and reflections in a line joining a vertex to the midpoint of the opposite edge.

For each $g \in G$ count the number $m(g)$ of $\langle g \rangle$-orbits in the set of vertices $X = \{1, 2, \ldots, 11\}$, so that Polya's formula can be applied. The results of the count can be conveniently displayed in a tabular form in the following table.

Type of element	Cycle type	Number of elements	m
identity	eleven 1-cycles	1	11
rotation through $\frac{2\pi i}{11}$, $1 \leq i \leq 10$	one 11-cycle	10	1
reflection	one 1-cycle, five 2-cycles	11	6

From the table and Polya's formula we deduce that the number of different designs is

$$\frac{1}{22}(c^{11} + 11c^6 + 10c) = \frac{1}{22}c(c^5 + 1)(c^5 + 10).$$

Next we tackle the cube-coloring problem with which the section began.

Example (17.1.2). How many ways are there to color the faces of a cube using c different colors?

In this problem the relevant group is the rotation group G of the cube since this group acts on the set of colorings. In fact $G \simeq S_4$: the easiest way to see this is to observe that each rotation permutes the four diagonals of the cube.

The labels are the c colors: let X consist of the six faces of the cube. To identify the rotations in G, we examine the various axes of symmetry of the cube. For each rotation record the cycle type and the m-value, i. e., the number of cycles in the corresponding permutation of X. The results are displayed in the table which follows:

Type of element	Cycle type	Number of elements	m
identity	six 1-cycles	1	6
rotation about line through centroids of opposite faces through			
$\pi/2$	two 1-cycles, one 4-cycle	3	3
π	two 1-cycles, two 2-cycles	3	4
$\frac{3\pi}{2}$	two 1-cycles, one 4-cycle	3	3
rotation about a diagonal through			
$\frac{2\pi}{3}$	two 3-cycles	4	2
$\frac{4\pi}{3}$	two 3-cycles	4	2
rotation about line joining midpoints of opposite edges through π	three 2-cycles	6	3

The data in the table confirm that $|G| = 24$, and on applying Polya's formula, we obtain the answer $\frac{1}{24}(c^6 + 3c^3 + 3c^4 + 3c^3 + 4c^2 + 4c^2 + 6c^3)$, which factorizes as

$$\frac{1}{24}c^2(c+1)(c^3 - c^2 + 4c + 8).$$

When $c = 5$, the formula yields 800, so there are 800 different ways to color the faces of a cube using 5 colors.

It is apparent from these examples that Polya's theorem enables one to solve complex combinatorial problems which might otherwise be intractable.

Exercises (17.1).
(1) Show that there are $\frac{1}{10}c(c^2+1)(c^2+4)$ ways to label the vertices of a regular pentagon using c labels.
(2) The same problem for the *edges* of the pentagon.
(3) A baton has n bands of equal width. Show that there are $\frac{1}{2}(c^n + c^{\lceil \frac{n+1}{2} \rceil})$ ways to color it using c colors. (The baton can only be rotated through 180°.)
(4) The faces of a regular tetrahedron are to be painted using c colors. Prove that there are $\frac{1}{12}c^2(c^2+11)$ ways to do it.
(5) A necklace has p beads of identical shape and size where p is an odd prime number. Beads of c colors available. How many necklace designs are possible?
(6) How many ways are there to place eight identical checkers on an 8×8 chessboard of squares if only rotations of the board are allowed?
(7) Prove that the number of ways to design a necklace with n beads of c different colors is

$$\frac{1}{2n}\left(\sum_{\substack{i \geq 1 \\ i|n}} \phi(i)c^{\frac{n}{i}}\right) + \frac{1}{4}(c^{\lceil \frac{n+1}{2} \rceil} + c^{\lceil \frac{n+2}{2} \rceil}),$$

where ϕ is Euler's function. [Hint: the group acting is Dih($2n$). Record the m-values of the rotations and then the reflections; for the latter distinguish between the cases n even and odd.]

17.2 Enumerating graphs

In this section we show how to count the number of graphs with a fixed set of vertices by using Polya's theorem. First a few remarks about graphs.

An (undirected) *graph* Γ consists of a non-empty set V of *vertices* and a relation E on V which is symmetric and *irreflexive*, i. e., $v \not\mathrel{E} v$ for all $v \in V$. If $u \mathrel{E} v$, call the 2-element set $\{u, v\}$ an *edge* of Γ. Since E is symmetric, we can identify E with the set of all edges of Γ.

A graph can be visualized by representing the vertices by points in the plane or in 3-space and the edges by lines joining appropriate vertices. Simple examples of graphs are:

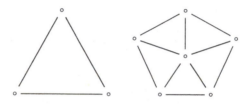

Note that loops and multiple edges are not permitted. Graph theory has many applications outside mathematics, for example to transportation systems, telephone networks and electrical circuits.

Two graphs $\Gamma_i = (V_i, E_i)$, $i = 1, 2$, are said to be *isomorphic* if there is a bijection $\theta : V_1 \to V_2$ such that $\{u, v\} \in E_1$ if and only if $\{\theta(u), \theta(v)\} \in E_2$. Two graphs may appear to be different, yet be isomorphic: for example, the graphs

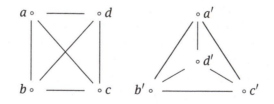

are isomorphic because of the bijection $a \mapsto a', b \mapsto b', c \mapsto c', d \mapsto d'$.

The problem of interest to us here is to compute the number of non-isomorphic graphs on a given set of n vertices. For this purpose it is enough to count isomorphism classes of graphs with a fixed vertex set $V = \{1, 2, \ldots, n\}$. The first step is to observe that a graph $\Gamma = (V, E)$ is determined by its *edge function*

$$\alpha_\Gamma : V^{[2]} \rightarrow \{0, 1\}$$

where $V^{[2]}$ is the set of all sets of vertices $\{u, v\}$, with $u \neq v$ in V, and

$$\alpha_\Gamma(\{u, v\}) = \begin{cases} 0 & \text{if } (u, v) \notin E \\ 1 & \text{if } (u, v) \in E. \end{cases}$$

Thus we can think of a graph as a labelling of $V^{[2]}$ by the set $\{0, 1\}$. The symmetric group S_n acts on the vertex set V in the natural way and this leads to an action of S_n on $V^{[2]}$ in which

$$\pi \cdot \{u, v\} = \{\pi(u), \pi(v)\}$$

where $\pi \in S_n$. Thus S_n acts on the set of all edge functions for V, i. e., on

$$F = \text{Fun}(V^{[2]}, \{0, 1\}).$$

It is a consequence of the definition of an isomorphism that graphs $\Gamma_1 = (V, E_1)$ and $\Gamma_2 = (V, E_2)$ are isomorphic if and only if there exists $\pi \in S_n$ such that $\pi \cdot \alpha_{\Gamma_1} = \alpha_{\Gamma_2}$, i. e., α_{Γ_1} and α_{Γ_2} belong to the same S_n-orbit of F. Thus we have to count the S_n-orbits of F. Now (17.1.1) can be applied to this situation with $G = S_n$, $X = V^{[2]}$ and $L = \{0, 1\}$. This allows us to derive a formula for the number of isomorphism classes of graphs with vertex set V.

(17.2.1). *The number of non-isomorphic graphs with a fixed set of n vertices is given by*

$$g(n) = \frac{1}{n!}\left(\sum_{\pi \in S_n} 2^{m(\pi)}\right)$$

where $m(\pi)$ is the number of disjoint cycles present in the permutation of $V^{[2]}$ induced by π.

To use this result one must able to compute $m(\pi)$, the number of S_n-orbits in $V^{[2]}$. While formulas for $m(\pi)$ are available, we will be content to calculate these numbers directly for small values of n.

Example (17.2.1). Show that there are exactly 11 non-isomorphic graphs with 4 vertices.

We need to compute $m(\pi)$ for π of each cycle type in S_4. Note that $|V^{[2]}| = \binom{4}{2} = 6$. Of course, $m(1) = 6$. If π is a 4-cycle, say (1234), there are two cycles in the permutation of $V^{[2]}$ produced by π, namely $(\{1, 2\}, \{2, 3\}, \{3, 4\}, \{4, 1\})$ and $(\{1, 3\}, \{2, 4\})$; thus $m(\pi) = 2$. Also there are six 4-cycles in S_4.

If π is a 3-cycle, say (123)(4), there are two cycles, ($\{1, 2\}, \{2, 3\}, \{1, 3\}$) and ($\{1, 4\}$, $\{2, 4\}, \{3, 4\}$); thus $m(\pi) = 2$ and there are eight such 3-cycles.

If π has two 2-cycles, say π = (12)(34), there are four cycles ($\{1, 2\}$), ($\{3, 4\}$), ($\{1, 3\}, \{2, 4\}$), ($\{1, 4\}, \{2, 3\}$); hence $m(\pi) = 4$. There are three such π's.

Finally, there are six transpositions π and it is easy to see that for each one $m(\pi) = 4$. The formula in (17.2.1) therefore yields

$$g(4) = \frac{1}{4!}(2^6 + 6 \cdot 2^2 + 8 \cdot 2^2 + 3 \cdot 2^4 + 6 \cdot 2^4) = 11.$$

This result can be verified by actually enumerating the graphs.

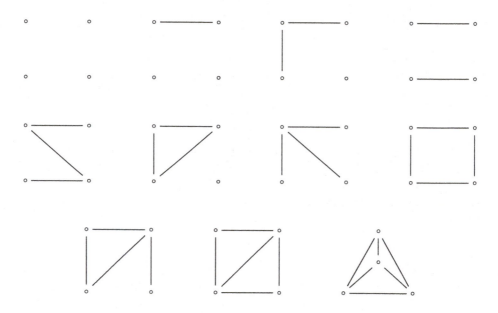

Notice that all these graphs are *planar*, i. e., they can be drawn in the plane in such a way that no edges cross except at vertices.

Exercises (17.2).

(1) Prove that the number of isomorphism types of graphs with n vertices is at most $2^{n(n-1)/2}$.

(2) Show that there are four isomorphism types of graphs with three vertices.

(3) Show that there are 34 isomorphism types of graphs with five vertices.

17.3 Latin squares and Steiner systems

A *latin square of order n* is an $n \times n$ matrix with entries from a set of n symbols such that each symbol occurs exactly once in each row and once in each column. Examples of latin squares are easily found – for convenience we will write Latin squares as matrices.

Example (17.3.1).

(i) The matrices

$$
\begin{bmatrix} a & b \\ b & a \end{bmatrix} \quad \text{and} \quad \begin{bmatrix} a & b & c \\ b & c & a \\ c & a & b \end{bmatrix}
$$

are latin squares of orders 2 and 3 respectively.

(ii) Let $G = \{g_1, g_2, \ldots, g_n\}$ be a (multiplicatively written) group of order n. Then the multiplication table of G is a latin square of order n. For, if the first row is $g_1, g_2, \ldots, g_n$, the entries of the ith row are $g_i g_1, g_i g_2, \ldots, g_i g_n$, which are clearly all different. A similar argument applies to the columns. On the other hand, not every latin square determines a group table since the associative law may fail to hold. In fact a latin square determines a more general type of algebraic structure called a *quasigroup* – for this concept see Exercises (17.3.4) and (17.3.5) below. Latin squares frequently occur in puzzles, but they also have a serious use in the design of statistical experiments. Here is an example to illustrate this use.

Example (17.3.2). Five types of washing powder P_1, P_2, P_3, P_4, P_5 are to be tested in five machines A, B, C, D, E over five days D_1, D_2, D_3, D_4, D_5. Each washing powder is to be used once each day and tested once on each machine. How can this be done?

The intention here is to allow for differences in the machines and in the water supply on different days, while keeping the number of tests to a minimum. A schedule of tests can be given in the form of a latin square of order 5 whose rows correspond to the washing powders and whose columns correspond to the days; the symbols are the machines. For example, we could use the latin square

$$
\begin{bmatrix}
A & B & C & D & E \\
B & C & D & E & A \\
C & D & E & A & B \\
D & E & A & B & C \\
E & A & B & C & D
\end{bmatrix}.
$$

This would mean, for example, that washing powder P_3 will be used on day D_4 in machine A. There are of course many other possible schedules.

The number of latin squares

Let $L(n)$ denote the number of latin squares of order n which can be formed from a given set of n symbols. It is clear that $L(n)$ must increase rapidly with n. A rough upper bound for $L(n)$ can be found by counting derangements.

(17.3.1). *The number $L(n)$ of latin squares of order n that can be formed from n given symbols satisfies the inequality*

$$L(n) \leq (n!)^n \left(1 - \frac{1}{1!} + \frac{1}{2!} - \cdots + \frac{(-1)^n}{n!} \right)^{n-1},$$

and hence $L(n) = O((n!)^n/e^{n-1})$.

Proof. Taking the symbols to be $1, 2, \ldots, n$, we note that each row of a latin square of order n corresponds to a permutation of $\{1, 2, \ldots, n\}$, i. e., to an element of the symmetric group S_n. Thus there are $n!$ choices for the first row. Now rows 2 through n must be derangements of row 1 since no column can have a repeated element. Recall from (3.1.11) that the number of derangements of n symbols is

$$d_n = n! \left(1 - \frac{1}{1!} + \frac{1}{2!} - \cdots + \frac{(-1)^n}{n!} \right).$$

Hence rows 2 through n of the latin square can be chosen in at most $(d_n)^{n-1}$ ways. Therefore $L(n) \leq (n!)(d_n)^{n-1}$ and the result follows. $\quad\square$

It can be shown that $L(n) \geq \frac{(n!)^{2n}}{n^{n^2}}$ – for details of this see [2, 6.5].

Orthogonal latin squares

Suppose that $A = [a_{ij}]$ and $B = [b_{ij}]$ are two latin squares of order n. Then A and B are said to be *mutually orthogonal latin squares* (or MOLS) if the n^2 ordered pairs (a_{ij}, b_{ij}) are all different.

Example (17.3.3). The latin squares

$$\begin{bmatrix} a & b & c \\ b & c & a \\ c & a & b \end{bmatrix} \quad \text{and} \quad \begin{bmatrix} \alpha & \beta & \gamma \\ \gamma & \alpha & \beta \\ \beta & \gamma & \alpha \end{bmatrix}$$

are mutually orthogonal, as can be seen by listing the nine pairs of entries. On the other hand, there are no pairs of MOLS of order 2 since these would have to be of the form

$$\begin{bmatrix} a & b \\ b & a \end{bmatrix}, \quad \begin{bmatrix} a' & b' \\ b' & a' \end{bmatrix}$$

and the pair (a, a') is repeated.

One reason for the interest in mutually orthogonal latin squares is that they have statistical applications, as can be seen from an elaboration of the washing powder example.

Example (17.3.4). Suppose that in Example (17.3.2) there are also five washing machine operators $\alpha, \beta, \gamma, \delta, \varepsilon$. Each operator is to test each powder once and to carry out one test per day. In addition, for reasons of economy, the same combination of machine and operator cannot be repeated for any powder and day.

What is called for here is a pair of MOLS of order 5. A latin square with the schedule of machines was given in Example (17.3.2). By a little experimentation another latin square for the machines can be found such that the two are mutually orthogonal. The pair of MOLS is

$$\begin{bmatrix} A & B & C & D & E \\ B & C & D & E & A \\ C & D & E & A & B \\ D & E & A & B & C \\ E & A & B & C & D \end{bmatrix}, \quad \begin{bmatrix} \alpha & \beta & \gamma & \delta & \varepsilon \\ \gamma & \delta & \varepsilon & \alpha & \beta \\ \varepsilon & \alpha & \beta & \gamma & \delta \\ \beta & \gamma & \delta & \varepsilon & \alpha \\ \delta & \varepsilon & \alpha & \beta & \gamma \end{bmatrix}.$$

In both matrices the rows correspond to washing powders and the columns to days: in the second matrix the entries are the operators. Direct enumeration of the 25 pairs of entries from the two latin squares reveals that all are different. The two latin squares tell us the schedule of operations: thus, for example, powder P_3 is to be tested on day D_4 by operator γ in machine A.

We are interested in determining the maximum number of MOLS of order n, say

$$f(n).$$

In the first place there is an easy upper bound for $f(n)$.

(17.3.2). *If $n \geq 1$, then $f(n) \leq n - 1$.*

Proof. Assume that there exist r MOLS of order n, namely $A_1, A_2, \ldots, A_r$, and let the $(1, 1)$ entry of A_i be a_i. Consider row 2 of A_1. It has an a_1 in the $(2, i_1)$ position for some $i_1 \neq 1$ since there is already an a_1 in the first column. Hence there are $n - 1$ possibilities for i_1. Next in A_2 there is an a_2 in row 2, say as the $(2, i_2)$ entry where $i_2 \neq 1$; also $i_2 \neq i_1$ since the pair (a_1, a_2) has already occurred and cannot be repeated. Therefore there are $n - 2$ possibilities for i_2. Continuing this line of argument until A_r is reached, we conclude that a_r is the $(2, i_r)$ entry of A_r where there are $n - r$ possibilities for i_r. Therefore $n - r > 0$ and $r \leq n - 1$, as required. □

The really interesting question is whether $f(n) > 1$ for $n > 2$; recall that $f(2) = 1$ since there do not exist two MOLS of order 2.

The intervention of field theory

The mere existence of finite fields of every prime power order is enough to make a decisive advance in the construction of MOLS of prime power order.

(17.3.3). *Let p be a prime and m a positive integer. Then $f(p^m) = p^m - 1$.*

Proof. Let F be a field of order p^m, which exists by (11.3.1). For each $a \neq 0$ in F define a $p^m \times p^m$ matrix $A(a)$ over F with rows and columns labelled by the elements of F, written in some fixed order: the (u, v) entry of $A(a)$ is to be computed from the formula

$$[A(a)]_{u,v} = ua + v$$

where $u, v \in F$. In the first place $A(a)$ is a latin square of order p^m. For $ua + v = u'a + v$ implies that $ua = u'a$ and $u = u'$ since $0 \neq a \in F$. Also $ua + v = ua + v'$ implies that $v = v'$.

Next we show that $A(a)$'s are mutually orthogonal. Suppose that $A(a_1)$ and $A(a_2)$ are not orthogonal where $a_1 \neq a_2$: then

$$(ua_1 + v, ua_2 + v) = (u'a_1 + v', u'a_2 + v')$$

for some $u, v, u', v' \in F$. Then $ua_1 + v = u'a_1 + v'$ and $ua_2 + v = u'a_2 + v'$. Subtraction of the second equation from the first leads to $u(a_1 - a_2) = u'(a_1 - a_2)$. Since $a_1 - a_2 \neq 0$ and F is a field, it follows that $u = u'$ and hence $v = v'$. Thus we have constructed $p^m - 1$ MOLS of order p^m, which is the maximum number possible by (17.3.2). Therefore $f(p^m) = p^m - 1$. $\qquad\square$

Example (17.3.5). Construct three MOLS of order 4.

In the first place $f(4) = 3$. To construct three MOLS, start with a field F of order 4, obtained from $t^2 + t + 1$, the unique irreducible polynomial of degree 2 in $\mathbb{Z}_2[t]$. If a is a root of this polynomial, then $F = \{0, 1, a, 1 + a\}$ where $a^2 = a + 1$. Now form the three MOLS $A(1)$, $A(a)$, $A(1 + a)$, using the formula indicated in the proof of (17.3.3):

$$A(1) = \begin{bmatrix} 0 & 1 & a & 1+a \\ 1 & 0 & 1+a & a \\ a & 1+a & 0 & 1 \\ 1+a & a & 1 & 0 \end{bmatrix},$$

$$A(a) = \begin{bmatrix} 0 & 1 & a & 1+a \\ a & 1+a & 0 & 1 \\ 1+a & a & 1 & 0 \\ 1 & 0 & 1+a & a \end{bmatrix},$$

$$A(1 + a) = \begin{bmatrix} 0 & 1 & a & 1+a \\ 1+a & a & 1 & 0 \\ 1 & 0 & 1+a & a \\ a & 1+a & 0 & 1 \end{bmatrix}.$$

To construct MOLS whose order is not a prime power, a direct product construction can be used. Let A and B be latin squares of orders m and n respectively. The *direct product* $A \times B$ is defined to be the $mn \times mn$ matrix whose entries are pairs of elements $(a_{ij}, b_{i'j'})$. The matrix can be visualized in the block form

$$\begin{bmatrix} (a_{11}, B) & (a_{12}, B) & \dots & (a_{1m}, B) \\ (a_{21}, B) & (a_{22}, B) & \dots & (a_{2m}, B) \\ \vdots & \vdots & \vdots & \vdots \\ (a_{m1}, B) & (a_{m2}, B) & \dots & (a_{mm}, B) \end{bmatrix}$$

where (a_{ij}, B) means that a_{ij} is paired with each entry of B in the natural matrix order. It is easy to see that $A \times B$ is a latin square of order mn.

Example (17.3.6). Given latin squares

$$A = \begin{bmatrix} a & b \\ b & a \end{bmatrix} \quad \text{and} \quad B = \begin{bmatrix} \alpha & \beta & \gamma \\ \beta & \gamma & \alpha \\ \gamma & \alpha & \beta \end{bmatrix},$$

we can form

$$A \times B = \begin{bmatrix} (a,\alpha) & (a,\beta) & (a,\gamma) & (b,\alpha) & (b,\beta) & (b,\gamma) \\ (a,\beta) & (a,\gamma) & (a,\alpha) & (b,\beta) & (b,\gamma) & (b,\alpha) \\ (a,\gamma) & (a,\alpha) & (a,\beta) & (b,\gamma) & (b,\alpha) & (b,\beta) \\ (b,\alpha) & (b,\beta) & (b,\gamma) & (a,\alpha) & (a,\beta) & (a,\gamma) \\ (b,\beta) & (b,\gamma) & (b,\alpha) & (a,\beta) & (a,\gamma) & (a,\alpha) \\ (b,\gamma) & (b,\alpha) & (b,\beta) & (a,\gamma) & (a,\alpha) & (a,\beta) \end{bmatrix},$$

which is a latin square of order 6.

Suppose that we have MOLS $A_1, A_2, \dots, A_r$ of order m and $B_1, B_2, \dots, B_s$ of order n where $r \leq s$; then the latin squares $A_1 \times B_1, A_2 \times B_2, \dots, A_r \times B_r$ have order mn and they are mutually orthogonal, as a check of the entry pairs shows. On the basis of this observation we can state:

(17.3.4). *If* $n = n_1 n_2$, *then* $f(n) \geq \min\{f(n_1), f(n_2)\}$.

This result can be used to give further information about the integer $f(n)$. Let $n = p_1^{e_1} p_2^{e_2} \cdots p_k^{e_k}$ be the primary decomposition of n. Then

$$f(n) \geq \min\{p_i^{e_i} - 1 \mid i = 1, 2, \dots, k\}$$

by (17.3.3) and (17.3.4). Therefore $f(n) > 1$ provided that $p_i^{e_i} \neq 2$ for all i. This will be the case if n is either odd or divisible by 4, i. e., $n \not\equiv 2 \pmod 4$. Hence we have:

(17.3.5). *If* $n \not\equiv 2 \pmod 4$, *then* $f(n) > 1$, *so there exist at least two mutually orthogonal latin squares of order* n.

In 1782 Euler conjectured that the converse is true, i. e. if $n \equiv 2 \pmod 4$, there cannot be a pair of $n \times n$ MOLS. As evidence for this, in 1900 Tarry[2] was able to confirm that there does not exist a pair of 6×6 MOLS; thus $f(6) = 1$. However, in the end it turned out that Euler was wrong; for in 1959 in a remarkable work Bose,[3] Shrikhande[4] and Parker[5] were able to prove that there is a pair of $n \times n$ MOLS for all even integers $n \neq 2, 6$.

The case $n = 6$ is Euler's celebrated *Problem of the Thirty Six Officers*. Suppose there are thirty six officers of six ranks and six regiments, with six of each regiment and six of each rank. Euler asked if it is possible for the officers to march in six rows of six, so that in each row and in each column there is exactly one officer of each rank and one of each regiment, with *no combination of rank and regiment being repeated*. He was really asking if there are two mutually orthogonal latin squares of order 6, the symbols of the first latin square being the ranks and those of the second the regiments of the officers. By Tarry's result the answer is negative.

Steiner triple systems

Another striking use of finite fields is to construct certain combinatorial objects known as Steiner[6] triple systems. We begin with a brief explanation of these. A *Steiner triple system of order n* is a pair $(X, \mathcal{T})$ where X is a set with n elements, called *the points*, and $\mathcal{T}$ is a set of 3-element subsets of X, called *the triples*, such that every pair of points occurs in exactly one triple. Steiner triple systems belong to a wide class of combinatorial objects called designs which are frequently used in the design of experiments.

Example (17.3.7). A Steiner triple system of order 7.

Consider the diagram consisting of an equilateral triangle with the three medians drawn. Let X be the set of seven points consisting of the vertices, the midpoints of the sides and the centroid, labelled A, B, C, D, E, F, G.

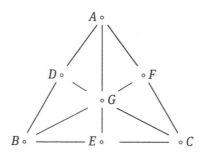

2 Gaston Tarry (1843–1913).
3 Raj Chandra Bose (1901–1987).
4 Sharadchandra Shankar Shrikhande (1917–2020).
5 Ernest Tilden Parker (1926–1991).
6 Jakob Steiner (1796–1863).

The set of triples is to be

$$\mathcal{T} = \{ADB, BEC, CFA, AGE, BGF, CGD, DEF\}.$$

Here, for example, we have written ADB for the triple $\{A, D, B\}$. So the triples are the sets of points lying on each of the six lines in the diagram, together with DEF. This configuration is well known as the projective plane over $\mathbb{Z}_2$ with seven points and seven lines: for the triple DEF corresponds to the line in the projective plane with equation $x + y + z = 0$ if the vertices are suitably labelled. It is clear from the diagram that each pair of points belongs to a unique triple.

We will consider the question: for which positive integers n do there exist Steiner triple systems of order n? It is easy to derive necessary conditions on n from the next result.

(17.3.6). *Suppose that $(X, \mathcal{T})$ is a Steiner triple system of order n. Then:*
(i) *each point belongs to exactly $\frac{n-1}{2}$ triples;*
(ii) *the number of triples is $\frac{n(n-1)}{6}$.*

Proof. (i) Let $x, y \in X$ with x fixed. The idea behind the proof is to count in two different ways the pairs (y, T) such that $y \in T$, $y \neq x$, $T \in \mathcal{T}$. There are $n - 1$ choices for y; then, once y has been chosen, there is a unique $T \in \mathcal{T}$ containing x and y, so the number of such pairs is $n - 1$. On the other hand, let r denote the number of triples in $\mathcal{T}$ to which x belongs. Once a triple $T \in \mathcal{T}$ containing x has been chosen, there are two choices for y in T. Thus the number of pairs is $2r$. Therefore $2r = n - 1$ and $r = \frac{n-1}{2}$.

(ii) In a similar manner count in two different ways the pairs (x, T) such that $x \in T$ and $T \in \mathcal{T}$. If t is the total number of triples, the number of pairs is $3t$ since there are three choices for x in T. On the other hand, we may also choose x in n ways and a triple T containing x in $\frac{n-1}{2}$ ways by (i). Therefore $3t = \frac{n(n-1)}{2}$ and $t = \frac{1}{6}n(n - 1)$. $\qquad\square$

Taking $n = 7$ in (17.3.6), we see that a Steiner triple system of order 7 necessarily has 7 triples. From (17.3.6) we can deduce a necessary condition on n for a Steiner triple system of order n to exist.

Corollary (17.3.7). *If a Steiner triple system of order n exists, then $n \equiv 1$ or $3 \pmod 6$.*

Proof. In the first place $\frac{n-1}{2}$ must be an integer, so n is odd. Thus we can write $n = 6k + \ell$ where $\ell = 1$, 3 or 5. If $\ell = 5$, then $\frac{1}{6}n(n - 1) = \frac{1}{3}(6k + 5)(3k + 2)$, which is not an integer. Hence $l = 1$ or 3 and $n \equiv 1$ or $3 \pmod 6$. $\qquad\square$

The fundamental theorem on Steiner triple systems asserts that the converse of (17.3.7) is true. *If $n \equiv 1$ or $3 \pmod 6$, there is a Steiner triple system of order n.* We will prove a special case of this theorem to illustrate how field theory can be applied.

(17.3.8). *If q is a prime power such that $q \equiv 1 \pmod 6$, there is a Steiner triple system of order q.*

Proof. Let F be a finite field of order q. Recall from (11.3.6) that $U(F)$ is a cyclic group of order $q - 1$. Since $6 \mid q - 1$ by hypothesis, it follows from (4.1.6) that $U(F)$ contains an element z of order 6. Thus $|U(F) : \langle z \rangle| = \frac{q-1}{6}$. Choose a transversal to $\langle z \rangle$ in $U(F)$, say $\{t_1, t_2, \ldots, t_{\frac{q-1}{6}}\}$. Now define subsets

$$T_i = \{0, t_i, t_i z\}$$

for $i = 1, 2, \ldots, \frac{q-1}{6}$.

The points of the Steiner triple system are to be the elements of the field F, while the set of triples is designated as

$$\mathcal{T} = \left\{ a + T_i \,\middle|\, a \in F, \ i = 1, 2, \ldots, \frac{q-1}{6} \right\}.$$

Here $a + T_i$ denotes the set $\{a + x \mid x \in T_i\}$. We claim that $(X, \mathcal{T})$ is a Steiner triple system. First we make an observation. Let D_i denote the set of differences of pairs of elements in T_i; thus

$$D_i = \{0, \pm t_i, \pm t_i z, \pm t_i(1 - z)\}.$$

Now z has order 6 and $0 = z^6 - 1 = (z^3 - 1)(z + 1)(z^2 - z + 1)$, so that $z^2 - z + 1 = 0$ and $z^2 = z - 1$. Hence $z^3 = -1$, $z^4 = -z$, $z^5 = 1 - z$. From these equations it follows that D_i is simply the coset $t_i \langle z \rangle = \{t_i z^k \mid 0 \le k \le 5\}$ with 0 adjoined.

To show that $(X, \mathcal{T})$ is a Steiner triple system, we need to prove that any two distinct elements x and y of F belong to a unique triple $a + T_i$. Let $f = x - y \in U(F)$. Now f belongs to a unique coset $t_i \langle z \rangle$, and by the observation above $f \in D_i$, so that f is expressible as the difference between two elements in the set T_i, say $f = u_i - v_i$. Writing $a = y - v_i$, we have $x = f + y = (y - v_i) + u_i \in a + T_i$ and $y = (y - v_i) + v_i \in a + T_i$.

Now suppose that x and y belong to another triple $b + T_j$, with $x = b + d_j$ and $y = b + e_j$ where $d_j, e_j \in T_j$. Then $0 \ne f = x - y = d_j - e_j$ and hence $f \in D_j$. Thus $f \in t_j \langle z \rangle$, which means that $j = i$. Also there is clearly only one way to write f as the difference between two elements of T_i. Therefore $d_i = u_i$ and $e_i = v_i$, from which it follows that $a = y - v_i = y - c_i = b$. The proof is now complete. $\qquad \square$

The construction just described produces Steiner triple systems of order 7, 13, 19, 25. Trivially there are Steiner triple systems of orders 1 and 3. In addition there are no Steiner systems of orders 2, 4, 5, 6, 8, 10, 11, 12 by (17.3.7).

In Exercise (17.3.6) below it is indicated how to construct a Steiner triple system of order 9. For a proof of the general case where $n \equiv 3 \pmod 6$ see [2].

Exercises (17.3).

(1) Show $L(1) = 1$, $L(2) = 2$, $L(3) = 12$ where $L(n)$ is the number of Latin squares of order n.

(2) Explain how to construct the following objects: (i) four 5×5 MOLS; (ii) eight 9×9 MOLS. [Hint: for (ii) use the field GF(9) and the polynomial $f = t^2 + 1 \in \mathrm{GF}(3)[t]$. Write the elements of GF(9) in terms of $x = t + (f)$.]

(3) Show that there are at least 48 MOLS of order 6125.

(4) A *quasigroup* is a set Q together with a binary operation $(x, y) \mapsto xy$ such that, given $x, y \in Q$, there is a unique $u \in Q$ such that $ux = y$ and a unique $v \in Q$ such that $xv = y$. Prove that the multiplication table of a finite quasigroup is a latin square.

(5) Conversely, prove that every latin square determines a finite quasigroup.

(6) Construct a Steiner triple system of order 9 by using the following geometric procedure. Start with a 3×3 array of 9 points. Draw all horizontals, verticals and diagonals in the figure. Then draw four curves connecting each corner to the midpoint of a suitable exterior side.

(7) (*Kirkman's[7] schoolgirl problem*) Show that it is possible for nine schoolgirls to walk in three groups of three for four successive days in such a way that each pair of girls walks in the same group on exactly one day.

(8) Let n be a positive integer such that $n \equiv 3 \pmod 6$. Assuming the existence of Steiner triple systems of order n, generalize the preceding problem by showing that it is possible for n schoolgirls to walk in $\frac{n}{3}$ groups of three on $\frac{n-1}{2}$ days without two girls walking together on more than one day.

(9) Use the method of (17.3.8) to construct a Steiner triple system of order 13.

(10) Construct a Steiner triple system of order 25 by starting with the field $\mathbb{Z}_5[t]/(t^2 - t + 1)$. (Note that a root of $t^2 - t + 1$ has order 6.)

17.4 Introduction to error correcting codes

In this age of information technology enormous amounts of data are transmitted electronically over vast distances every second of every day. The data are generally in the form of bit strings, i. e., sequences of 0's and 1's. Inevitably errors occur from time to time during the process of transmission, so that the message received may differ from the message transmitted. An *error correcting code* allows the detection and correction of erroneous messages. The essential idea here is that the possible transmitted codewords should not be too close to one another, i. e., they should not agree in too many entries. This makes it more likely that an error can be detected and the original message recovered. Over the past fifty years an entire mathematical theory of error-correcting codes has evolved.

---

7 Thomas Penyngton Kirkman (1806–1895).

Fundamental concepts

Let Q be a finite non-empty set with q elements; this is called the *alphabet*. A *word w of length n over Q* is an n-tuple of elements of Q, which will be written in the form

$$w = (w_1 w_2 \ldots w_n), \quad w_i \in Q.$$

The set of all words of length n over Q is known as *n-dimensional Hamming space.*[8] Essentially this depends on Q only through $q = |Q|$ and it will be denoted by

$$H_n(q).$$

This is the set of possible messages of length n: notice that $|H_n(q)| = q^n$.

Next suppose that $Q = GF(q)$, the field of order a prime power q; then $H_n(Q)$ can be regarded as an n-dimensional vector space over Q. Words are to be added by adding their entries and multiplied by a scalar by multiplying each entry by the scalar. The *zero word* 0 is the word with every entry equal to 0. In practice Q is usually the field of two elements, in which event Hamming n-space is the set of all strings of 0's and 1's of length n.

It is important to have a measure of how far apart two words in Hamming space are: the natural measure to use is the number of entries in which the words differ. If v and w belong to $H_n(q)$, the *distance* between v and w is defined to be

$$d(v, w) = |\{i \mid v_i \neq w_i\}|,$$

i. e., the number of positions where v and w have different entries. If Q is a field, the *weight* of a word v is defined as its distance from the zero word,

$$wt(v) = d(v, 0),$$

so $wt(v)$ is just the number of non-zero entries of v. Clearly, $d(u, v)$ is the number of errors that have been made if the word u is transmitted and it is received wrongly as v.

The basic properties of the distance function are given in the following result.

(17.4.1). *Let $u, v, w \in H_n(q)$. Then:*
(i) $d(v, w) \geq 0$ *and* $d(v, w) = 0$ *if only if* $v = w$;
(ii) $d(v, w) = d(w, v)$;
(iii) $d(u, w) \leq d(u, v) + d(v, w)$.

These properties assert that the function $d : H_n(q) \times H_n(q) \to \mathbb{N}$ *is a metric on Hamming space $H_n(q)$.*

---

8 Richard Wesley Hamming (1915–1998).

Proof of (17.4.1). Statements (i) and (ii) are obviously true. To prove (iii) note that u can be changed to v by $d(u, v)$ entry changes and v can then be changed to w by $d(v, w)$ changes. Thus u can be changed to w by $d(u, v) + d(v, w)$ entry changes. Therefore $d(u, w) \le d(u, v) + d(v, w)$. □

Codes

A *code of length n* over an alphabet Q with q elements, or briefly a *q-ary code of length n*, is a subset C of $H_n(Q)$ with at least two elements. The elements of C are called *codewords*. Codewords are transmitted in actual messages.

A q-ary code C is said to be *e-error detecting* if $c_1, c_2 \in C$ and $d(c_1, c_2) \le e$ always imply that $c_1 = c_2$. Thus the distance between distinct codewords is always greater than e. Equivalently, a codeword cannot be transmitted and received as a different codeword if e or fewer errors have occurred. In this sense the code C is able to detect up to e errors.

Next a q-ary code of length n is called *e-error correcting* if, for every w in $H_n(q)$, there is at most one codeword c such that $d(w, c) \le e$. This means that if a codeword c is received as a different word w and at most e errors have occurred, it is possible to recover the original codeword by examining all words v in $H_n(q)$ such that $d(w, v) \le e$: exactly one of these is a codeword and it must have been the transmitted codeword c. Clearly a code which is e-error correcting is e-error detecting.

An important parameter of a code is the shortest distance between distinct codewords; this is called the *minimum distance* of the code. The following result is basic.

(17.4.2). *Let C be a code with minimum distance d. Then:*
(i) *C is e-error detecting if and only if $d \ge e + 1$;*
(ii) *C is e-error correcting if and only if $d \ge 2e + 1$.*

Proof. (i) Suppose that $d \ge e + 1$. If c_1, c_2 are distinct codewords, then $d(c_1, c_2) \ge d \ge e + 1$. Hence C is e-error detecting. To prove the converse, assume that $d \le e$. By definition of d there exist $c_1 \ne c_2$ in C such that $d(c_1, c_2) = d \le e$, so that C is not e-error detecting.

(ii) Assume that C is not e-error correcting, so there is a word w and codewords $c_1 \ne c_2$ such that $d(c_1, w) \le e$ and $d(w, c_2) \le e$. Then

$$d \le d(c_1, c_2) \le d(c_1, w) + d(w, c_2) \le 2e$$

by (17.4.1). Hence $d < 2e + 1$.

Conversely, assume that $d < 2e + 1$ and let c_1 and c_2 be codewords at distance d apart. Put $f = [\frac{1}{2}d]$, i. e., the greatest integer $\le \frac{1}{2}d$; thus $f \le \frac{1}{2}d \le e$. We claim that $d - f \le e$. This is true when d is even since $d - f = d - \frac{1}{2}d = \frac{1}{2}d \le e$. If d is odd, $f = \frac{d-1}{2}$ and $d - f = \frac{d+1}{2} < e + 1$; therefore $d - f \le e$. Next we can pass from c_1 to c_2 by changing exactly d entries. Let w be the word obtained from c_1 after the first f entry changes. Then $d(c_1, w) = f \leq e$, while $d(c_2, w) = d - f \leq e$. Therefore C is not e-error correcting. □

Corollary (17.4.3). *If a code has minimum distance d, then its maximum error detection capacity is $d - 1$ and its maximum error correction capacity is $[\frac{d-1}{2}]$.*

Example (17.4.1). Consider the binary code C of length 5 with the three codewords

$$c_1 = (10010), \quad c_2 = (01100), \quad c_3 = (10101).$$

Clearly the minimum distance of C is 3. Hence C is 2-error detecting and 1-error correcting. For example, suppose that c_2 is transmitted and is received as $w = (11000)$, so that two entry errors have occurred. The error can be detected since $w \notin C$. But C is not 2-error correcting since if $v = (11100)$, then $d(c_2, v) = 1$ and $d(c_3, v) = 2$. Thus if v is received and up to two errors occurred, we cannot tell whether c_2 or c_3 was the transmitted codeword.

Bounds for the size of a code
It is evident from (17.4.2) that for a code to have good error correcting capability it must have large minimum distance. But the price to be paid for this is that fewer codewords are available. An interesting question is: what is the maximum size of a q-ary code with length n and minimum distance d. We begin with a lower bound, which guarantees the existence of a code of a certain size.

(17.4.4) (The Varshamov–Gilbert lower bound). *Let n, q, d be positive integers with $d \leq n$. Then there is a q-ary code of length n and minimum distance d in which the number of codewords is at least*

$$\frac{q^n}{\sum_{i=0}^{d-1} \binom{n}{i}(q-1)^i}.$$

Before embarking on the proof we introduce an important concept, the *r-ball with center w,*

$$B_r(w).$$

This is the set of all words in $H_n(q)$ at distance r or less from w. Thus a code C is e-error correcting if and only if the e-balls $B_e(c)$ with c in C are pairwise disjoint.

Proof of (17.4.4). The first step is to establish a formula for the size of an r-ball,

$$|B_r(w)| = \sum_{i=0}^{r} \binom{n}{i}(q-1)^i.$$

To see this observe that in order to construct a word in $B_r(w)$, we must alter at most r entries of w. Choose the i entries to be altered in $\binom{n}{i}$ ways and then replace each one by a different element of the alphabet Q in $(q-1)^i$ ways. This gives a count of $\binom{n}{i}(q-1)^i$ words at distance i from w; the formula now follows at once.

To start the construction choose any q-ary code C_0 of length n with minimum distance d; for example, C_0 might consist of the zero word and a single word of weight d. If the union of the $B_{d-1}(c)$ with $c \in C_0$ is not $H_n(q)$, there is a word w whose distance from every word in C_0 is at least d. Let $C_1 = C_0 \cup \{w\}$; this is a larger code than C_0 which has the same minimum distance d. Repeat the procedure for C_1 and then as often as possible. Eventually a code C with minimum distance d will be obtained which cannot be enlarged; when this occurs, we have $H_n(q) = \bigcup_{c \in C} B_{d-1}(c)$. Therefore

$$q^n = |H_n(q)| = \left| \bigcup_{c \in C} B_{d-1}(c) \right| \leq |C| \cdot |B_{d-1}(c)|$$

for any fixed $c \in C$. Hence $|C| \geq q^n / |B_{d-1}(c)|$, so the bound has been established. □

Next we give an upper bound for the size of an e-error correcting code.

(17.4.5) (The Hamming upper bound). *Let C be a q-ary code of length n which is e-error correcting. Then*

$$|C| \leq \frac{q^n}{\sum_{i=0}^{e} \binom{n}{i}(q-1)^i}.$$

Proof. Since C is e-error correcting, the e-balls $B_e(c)$ for $c \in C$ are pairwise disjoint. Hence

$$\left| \bigcup_{c \in C} B_e(c) \right| = |C| \cdot |B_e(c)| \leq |H_n(q)| = q^n$$

for any fixed $c \in C$. Therefore $|C| \leq q^n / |B_e(c)|$, as required. □

A q-ary code C of length n for which the Hamming upper bound is attained is called a *perfect code*. In this case by (17.4.5)

$$|C| = \frac{q^n}{\sum_{i=0}^{e} \binom{n}{i}(q-1)^i},$$

and clearly this happens precisely when $H_n(q)$ is the union of the disjoint balls $B_e(c)$, $c \in C$, i. e., every word lies at distance $\leq e$ from exactly one codeword. Perfect codes are desirable since they have the largest number of codewords for the given error correcting capacity; however they are also quite rare.

Example (17.4.2) (The binary repetition code). A very simple example of a perfect code is the binary code C of length $2e + 1$ with just two codewords,

$$c_0 = (0, 0, \ldots, 0) \quad \text{and} \quad c_1 = (1, 1, \ldots, 1).$$

Clearly C has minimum distance $d = 2e + 1$ and its maximum error correcting capacity is e by (17.4.3). A word w belongs $B_e(c_0)$ if more of its entries equal 0 than 1; otherwise $w \in B_e(c_1)$. Thus $B_e(c_0) \cap B_e(c_1) = \emptyset$ and $B_e(c_0) \cup B_e(c_1) = H_{2e+1}(2)$.

Linear codes

Let $Q = \mathrm{GF}(q)$, the field of q elements; of course q is now a prime power. The Hamming space $H_n(q)$ is the n-dimensional vector space Q_n of all n-row vectors over Q. A q-ary code C of length n is called *linear* if it is a subspace of $H_n(q)$. Linear codes form an important class of codes; they have the advantage that they can be specified by giving a basis instead of listing all the codewords. Linear codes can also be described by matrices, as will be seen in the sequel.

A computational advantage of linear codes is indicated by the next result.

(17.4.6). *The minimum distance of a linear code equals the minimal weight of a non-zero codeword.*

Proof. Let C be a linear code. If $c_1, c_2 \in C$, then $d(c_1, c_2) = wt(c_1 - c_2)$ and $c_1 - c_2 \in C$. It follows that the minimum distance equals the minimum weight. $\square$

A point to keep in mind here is that to find the minimum distance of a code C one must compute $\binom{|C|}{2}$ distances, whereas to find the minimum weight of C only the distances from the zero word need be found, so that at most $|C| - 1$ computations are necessary.

As with codes in general, it is desirable to have linear codes with large minimum distance and as many codewords as possible. There is a version of the Varshamov–Gilbert lower bound for linear codes.

(17.4.7). *Let d and n be positive integers with $d \leq n$ and let q be a prime power. Then there is a linear q-ary code of length n and minimum distance d in which the number of codewords is at least*

$$\frac{q^n}{\sum_{i=0}^{d-1} \binom{n}{i}(q-1)^i}.$$

Proof. We refer to the proof of (17.4.4). To start the construction choose a linear q-ary code C_0 of length n and minimum distance d; for example, the subspace generated by a single word of weight d will suffice. If $\bigcup_{c \in C_0} B_{d-1}(c) \neq H_n(q)$, choose a word w in $H_n(q)$ which belongs to no $B_{d-1}(c)$ with c in C_0. Thus $w \notin C_0$. Define C_1 to be the subspace generated by C_0 and w. We claim that C_1 still has minimum distance d. To prove this it is sufficient to show that $wt(c') \geq d$ for any c' in $C_1 - C_0$; this is because of (17.4.6). Write $c' = c_0 + aw$ where $c_0 \in C_0$ and $0 \neq a \in Q$. Then

$$wt(c') = wt(c_0 + aw) = wt(-a^{-1}c_0 - w) = d(-a^{-1}c_0, w) \geq d$$

by choice of w, since $-a^{-1}c_0 \in C_0$. Note also that $\dim(C_0) < \dim(C_1)$.

Repeat the argument above for C_1, and then as often as possible. After at most n steps we arrive at a subspace C with minimum distance d such that $\bigcup_{c \in C} B_{d-1}(c) = H_n(q)$. It follows that $|C| \cdot |B_{d-1}(c)| \geq q^n$ for any c in C, which gives the bound. $\square$

Example (17.4.3). Let $q = 2$, $d = 3$ and $n = 31$. According to (17.4.7) there is a linear binary code C of length 31 with minimum distance 3 such that

$$|C| \geq \frac{2^{31}}{1 + 31 + \binom{31}{2}} = 4{,}320{,}892{,}652.$$

In addition C is a subspace of $H_{31}(2)$, so its order is a power of 2. Hence $|C| \geq 2^{23} = 8{,}388{,}608$. In fact there is a larger linear code of this type with 2^{26} codewords, a so-called Hamming code – see Example (17.4.7) below.

The generator matrix and check matrix

Let C be a linear q-ary code of length n and let k be the dimension of C as a subspace of $H_n(q)$. Thus $k \leq n$ and $|C| = q^k$. Choose an ordered basis $\{c_1, c_2, \ldots, c_k\}$ for C and write

$$G = \begin{bmatrix} c_1 \\ c_2 \\ \vdots \\ c_k \end{bmatrix}.$$

This $k \times n$ matrix over $Q = GF(q)$ is called a *generator matrix* for C. If c is any codeword, $c = a_1 c_1 + \cdots + a_k c_k$ for suitable $a_i \in Q$. Thus $c = aG$ where $a = (a_1, \ldots, a_k) \in H_k(q)$. Hence each codeword is uniquely expressible in the form aG with $a \in H_k(q)$. It follows that the code C is *the row space of the matrix G*, i. e., the subspace of $H_n(q)$ generated by all the rows of G. Notice that the rank of G is k since its rows are linearly independent.

Recall from Section 8.1 that the null space N of G consists of all n-column vectors x^T such that $Gx^T = 0$: here of course $x \in H_n(q)$. Choose an ordered basis for N and use the transposes of its elements to form the rows of a matrix H. This is called a *check matrix* for C. Since G has rank k, we can apply (8.3.8) to obtain $\dim(N) = n - k$, so that H is an $(n - k) \times n$ matrix over Q. Since the columns of H^T belong to N, the null space of G, we obtain the important equation

$$GH^T = 0.$$

Keep in mind that the matrices G and H depend on the choice of bases for C and N. At this point the following result about matrices is relevant.

(17.4.8). *Let G and H be $k \times n$ and $(n - k) \times n$ matrices respectively over $Q = GF(q)$, each having linearly independent rows. Then the following statements are equivalent:*
(i) $GH^T = 0$;
(ii) *row space* $(G) = \{x \in H_n(q) \mid xH^T = 0\}$;
(iii) *row space* $(H) = \{x \in H_n(q) \mid xG^T = 0\}$.

Proof. Let $S = \{x \in H_n(q) \mid xH^T = 0\}$; then $x \in S$ if and only if $0 = (xH^T)^T = Hx^T$, i. e., x^T belongs to null space(H). This implies that S is a subspace and $\dim(S) = n - (n - k) = k$. Now assume that $GH^T = 0$. If $x \in$ row space(G), then $x = yG$ for some k-row vector y. Hence $xH^T = yGH^T = 0$ and $x \in S$. Thus row space$(G) \subseteq S$. But $\dim(\text{row space}(G)) = k = \dim(S)$, so that $S =$ row space(G). Thus (i) implies (ii). It is clear that (ii) implies (i), and thus (i) and (ii) are equivalent.

Next observe that $GH^T = 0$ if and only if $HG^T = 0$, by taking the transpose. Thus the roles of G and H are interchangeable, which means that (i) and (iii) are equivalent. $\qquad\square$

Let us now return to the discussion of a linear q-ary code C of length n with generator matrix G and check matrix H. From (17.4.8) we conclude that

$$C = \text{row space}(G) = \{w \in H_n(q) \mid wH^T = 0\}.$$

Thus the check matrix H provides a convenient way to determine if a given word w is a codeword. At this point we have proved half of the next result.

(17.4.9).
(i) *If C is a linear q-ary code with generator matrix G and check matrix H, then $GH^T = 0$ and $C = \{w \in H_n(q) \mid wH^T = 0\}$.*
(ii) *If G and H are $k \times n$ and $(n - k) \times n$ matrices respectively over $GF(q)$ with linearly independent rows and if $GH^T = 0$, then $C = \{w \in H_n(q) \mid wH^T = 0\}$ is a linear q-ary code of length n and dimension k with generator matrix G and check matrix H.*

Proof. To prove (i) note that $C =$ row space(G) and we showed that $GH^T = 0$, so the result follows at once from (17.4.8). Now for (ii): clearly C is a subspace of $H_n(q)$, so it is a linear q-ary code of length n. Also C is the row space of G by (17.4.8). Hence $\dim(C) = k$ and G is a generator matrix for C. Finally, the null space of G consists of all w in $H_n(q)$ such that $Gw^T = 0$, i. e., $wG^T = 0$; this is the row space of H by (17.4.8). Hence G and H are corresponding generator and check matrices for C. $\qquad\square$

On the basis of (17.4.9) we show how to construct a linear q-ary code of length n and dimension $n - \ell$ with check matrix equal to a given $\ell \times n$ matrix H over $GF(q)$ of rank ℓ. Define $C = \{x \in H_n(q) \mid xH^T = 0\}$; this is a linear q-ary code. Pass from H

to its reduced row echelon form $H' = [I_\ell \mid A]$ where A is an $\ell \times (n - \ell)$ matrix: note that interchanges of columns, i. e., of word entries, may be necessary to achieve this. Hence $H' = EHF$ for some non-singular E and permutation matrix F. Writing G' for $[-A^T \mid I_{n-\ell}]$, we have

$$G'H'^T = [-A^T \mid I_{n-\ell}] \begin{bmatrix} I_\ell \\ A^T \end{bmatrix} = 0.$$

Hence $0 = G'H'^T = (G'F^T)H^T E^T$, so $(G'F^T)H^T = 0$ because E^T is non-singular. Put $G = G'F^T$; thus $GH^T = 0$ and by (17.4.9) we have that G is a generator matrix and H a check matrix for C. Also $\dim(C) = \text{rank}(G) = n - \ell$. Note that if no column interchanges are needed to go from H to H', then $F = I$ and $G = G'$.

Example (17.4.4). Consider the matrix

$$H = \begin{bmatrix} 1 & 1 & 1 & 0 & 1 & 0 & 0 \\ 1 & 1 & 0 & 1 & 0 & 1 & 0 \\ 1 & 0 & 1 & 1 & 0 & 0 & 1 \end{bmatrix}$$

over GF(2). Here $q = 2$, $n = 7$ and $\ell = 3$. The rank of H is 3, so it determines a linear binary code C of dimension $7 - 3 = 4$. Put H in reduced row echelon form,

$$H' = \begin{bmatrix} 1 & 0 & 0 & & 0 & 1 & 1 & 1 \\ 0 & 1 & 0 & & 1 & 1 & 0 & 1 \\ 0 & 0 & 1 & & 1 & 1 & 1 & 0 \end{bmatrix} = [I_3 \mid A].$$

No column interchanges were necessary here, so

$$G = G' = [-A^T \mid I_4] = \begin{bmatrix} 0 & 1 & 1 & 1 & 0 & 0 & 0 \\ 1 & 1 & 1 & 0 & 1 & 0 & 0 \\ 1 & 0 & 1 & 0 & 0 & 1 & 0 \\ 1 & 1 & 0 & 0 & 0 & 0 & 1 \end{bmatrix}$$

is a generator matrix for C. The rows of G form a basis for the linear code C.

A useful feature of the check matrix is that from it one can read off the minimum distance of the code.

(17.4.10). *Let H be a check matrix for a linear code C. Then the minimum distance of C equals the largest integer m such that every set of $m - 1$ columns of H is linearly independent.*

Proof. Let d be the minimum distance of C and note that d is the minimum weight of a non-zero codeword, say $d = wt(c)$. Then $cH^T = 0$, which implies that there exist d linearly dependent columns of H. Hence $m - 1 < d$ and $m \le d$. Also, by maximality of m there exist m linearly dependent columns of H, so $wH^T = 0$ where w is a non-zero word with $wt(w) \leq m$. But $w \in C$; therefore $d \leq wt(w) \leq m$ and hence $d = m$. □

Example (17.4.5). Consider the code C in Example (17.4.4). Every pair of columns of the check matrix H is linearly independent, i. e., the columns are all different. On the other hand, columns 1, 4 and 5 are linearly dependent since their sum is zero. Therefore $m = 3$ for this code and the minimum distance is 3 by (17.4.10). Consequently C is a 1-error correcting code.

Using the check matrix to correct errors

Let C be a linear q-ary code with length n and minimum distance d and let H be a check matrix for C. Note that by (17.4.3) C is e-error correcting where $e = [\frac{d-1}{2}]$. Suppose that a codeword c is transmitted, but is received as a word w, and that *at most e* errors in the entries have been made. Here is a procedure that will correct the errors and recover the original codeword c.

Write $w = u + c$ where u is the error; thus $wt(u) \leq e$. Now $|H_n(q) : C| = q^{n-k}$ where $k = \dim(C)$. Choose a transversal to C in $H_n(q)$, say $\{v_1, v_2, \ldots, v_{q^{n-k}}\}$, by requiring that v_i be a word of *shortest length* in its coset $v_i + C$. (There may be more than one choice for v_i.) For any $c_0 \in C$ we have $(v_i + c_0)H^T = v_i H^T$, which depends only on i. Now assume that w belongs to the coset $v_i + C$. Then $wH^T = v_i H^T$, which is called the *syndrome* of w. Writing $w = v_i + c_1$ with $c_1 \in C$, we have $u = w - c \in v_i + C$, so that $wt(v_i) \leq wt(u) \leq e$ by choice of v_i. Hence $w = u + c = v_i + c_1$ belongs to $B_e(c) \cap B_e(c_1)$. But this implies that $c = c_1$ since C is e-error correcting. Therefore $c = w - v_i$ and the transmitted codeword has been identified.

In summary here is the procedure to identify the transmitted codeword c. It is assumed that the transversal $\{v_1, v_2, \ldots, v_{q^{n-k}}\}$ has been chosen as described above, with each v_i of shortest length in its coset.

(i) Suppose that w is the word received with at most e errors; first compute the syndrome wH^T.

(ii) By comparing wH^T with the syndromes $v_i H^T$, find the unique i such that $wH^T = v_i H^T$.

(iii) Then the transmitted codeword was $c = w - v_i$.

Example (17.4.6). The matrix

$$H = \begin{bmatrix} 1 & 0 & 1 & 1 & 0 \\ 0 & 0 & 1 & 1 & 1 \\ 1 & 1 & 0 & 1 & 1 \end{bmatrix}$$

determines a linear binary code C with length 5 and dimension $5 - 3 = 2$; thus H is a check matrix for C. Clearly C has minimum distance 3, so it is 1-error correcting. Also $|C| = 2^2 = 4$ and $|H_5(2) : C| = 2^5/4 = 8$. By reducing H to reduced row echelon form as in Example (17.4.4), we find a generator matrix for C to be

$$G = \begin{bmatrix} 0 & 1 & 1 & 1 & 0 \\ 1 & 0 & 1 & 0 & 1 \end{bmatrix}.$$

Thus C is generated by (01110) and (10101), so in fact C consists of (00000), (01110), (10101) and (11011).

Next enumerate the eight cosets of C in $H_5(2)$ with $C_1 = C$ and choose a word of minimum weight from each coset; these are shown in bold face.

$$C_1 = \{(\mathbf{00000}), (01110), (10101), (11011)\}$$
$$C_2 = \{(\mathbf{10000}), (11110), (00101), (01011)\}$$
$$C_3 = \{(\mathbf{01000}), (00110), (11101), (10011)\}$$
$$C_4 = \{(\mathbf{00100}), (01010), (10001), (11111)\}$$
$$C_5 = \{(\mathbf{11000}), (10110), (01101), (00011)\}$$
$$C_6 = \{(01100), (\mathbf{00010}), (11001), (10111)\}$$
$$C_7 = \{(10100), (11010), (\mathbf{00001}), (01111)\}$$
$$C_8 = \{(11100), (\mathbf{10010}), (01001), (00111)\}$$

The coset syndromes are computed as (000), (101), (001), (110), (100), (111), (011), (010).

Now suppose that the word $w = (11111)$ is received with at most one error in its entries: note that $w \notin C$, so w is not a codeword. The syndrome of w is $wH^T = (110)$, which is the syndrome of elements in the coset C_4, with coset representative $v_4 = (00100)$. Hence the transmitted codeword was $c = w - v_4 = (11011)$.

Hamming codes

Let C be a linear q-ary code of length n and dimension k. Assume that the minimum distance of C is at least 3, so that C is 1-error correcting. A check matrix H for C has size $\ell \times n$ where $\ell = n - k$, and by (17.4.10) no column of H can be a multiple of another column.

Now consider the problem of constructing such a linear code which is as large as possible for given q and $\ell > 1$. Then H should have as many columns as possible, subject to no column being a multiple of another one. Now there are $q^\ell - 1$ non-zero ℓ-column vectors over $GF(q)$, but each of these is a multiple of $q - 1$ other columns. So the maximum possible number of columns for H is $n = \frac{q^\ell - 1}{q - 1}$. Note that the columns of the identity $\ell \times \ell$ matrix can be included among those of H, so that H has rank ℓ. It follows that the matrix H determines a linear q-ary code C of length

$$n = \frac{q^\ell - 1}{q - 1}.$$

The minimum distance of H is at least 3 by construction, and in fact it is exactly 3 since we can include among the columns of H three linearly dependent ones, $(10\ldots0)^T$, $(110\ldots0)^T$, $(010\ldots0)^T$. Thus C is 1-error correcting: its dimension is $k = n - \ell$ and its order is q^n. Such a code is known as a *Hamming code*. It is not surprising that Hamming codes have optimal properties, as the next result shows.

(17.4.11). *Hamming codes are perfect.*

Proof. Let C be a q-ary Hamming code of length n constructed from a check matrix with ℓ rows. Then

$$|C| = q^{n-\ell} = \frac{q^n}{q^\ell} = \frac{q^n}{1 + n(q-1)}$$

since $n = \frac{q^\ell - 1}{q-1}$. Thus C attains the Hamming upper bound of (17.4.5), so it is a perfect code. $\qquad\square$

Example (17.4.7). Let $q = 2$ and $\ell = 4$. A Hamming code C of length $n = \frac{2^4-1}{2-1} = 15$ and dimension $n - \ell = 11$ can be constructed from the 4×15 check matrix

$$H = \begin{bmatrix} 1 & 0 & 0 & 0 & 1 & 1 & 1 & 0 & 0 & 0 & 0 & 1 & 1 & 1 & 1 \\ 0 & 1 & 0 & 0 & 1 & 0 & 0 & 1 & 1 & 0 & 1 & 0 & 1 & 1 & 1 \\ 0 & 0 & 1 & 0 & 0 & 1 & 0 & 1 & 0 & 1 & 1 & 1 & 0 & 1 & 1 \\ 0 & 0 & 0 & 1 & 0 & 0 & 1 & 0 & 1 & 1 & 1 & 1 & 1 & 0 & 1 \end{bmatrix}.$$

Here $|C| = 2^{n-\ell} = 2^{11} = 2048$. Similarly, by taking $q = 2$ and $\ell = 5$ we can construct a perfect linear binary code of length 31 and dimension 26.

Perfect codes

We conclude with an analysis of perfect codes which will establish the unique position of the Hamming codes.

(17.4.12). *Let C be a perfect q-ary code where $q = p^a$ and p is a prime. Assume that C is 1-error correcting. Then:*
(i) *C has length $\frac{q^s-1}{q-1}$ for some $s \geq 1$;*
(ii) *if C is linear, it is a Hamming code.*

Proof. (i) Let C have length n. Then $|C| = \frac{q^n}{1+n(q-1)}$ since C is perfect and 1-error correcting. Hence $1 + n(q-1)$ divides q^n, so it must be a power of p, say $1 + n(q-1) = p^r$. By the Division Algorithm we can write $r = sa + t$ where $s, t \in \mathbb{Z}$ and $0 \leq t < a$. Then

$$1 + n(q-1) = p^r = (p^a)^s p^t = q^s p^t = (q^s - 1)p^t + p^t.$$

Therefore $q - 1$ divides $p^t - 1$. However, $p^t - 1 < p^a - 1 = q - 1$, which shows that $p^t = 1$ and $t = 0$. Hence $1 + n(q-1) = p^{as} = q^s$ and it follows that $n = \frac{q^s-1}{q-1}$.

(ii) Now assume that C is linear. Since $|C| = \frac{q^n}{1+n(q-1)}$ and we have shown in (i) that $1 + n(q-1) = q^s$, it follows that $|C| = q^n/q^s = q^{n-s}$. Hence $\dim(C) = n - s$ and a check matrix H for C has size $s \times n$. The number of columns of H is $n = \frac{q^s-1}{q-1}$, which is the maximum number possible, and no column is a multiple of another one since C is 1-error correcting and thus has minimum distance ≥ 3. Therefore C is a Hamming code. $\qquad\square$

Almost nothing is known about perfect q-ary codes when q is not a prime power. Also there are very few perfect linear q-ary codes which are e-error correcting with $e > 1$. Apart from binary repetition codes of odd length – see Exercise (17.4.3) below – there are just two examples, a binary code of length 23 and a ternary code of length 11. These remarkable examples, known as the *Golay codes*, are of great importance in algebra: see [22] for details.

Exercises (17.4).
(1) Give an example of a code for which the minimum distance is different from the minimum weight of a non-zero codeword.
(2) How many words in $H_n(q)$ have weights in the range i to $i + k$.
(3) Let C be the set of all q-ary words of the form $(aa \ldots a)$ of length n where $a \in \mathrm{GF}(q)$.
   (i)   Show that C is a linear q-ary code of dimension 1.
   (ii) Find the minimum distance and error correcting capacity of C.
   (iii) Write down a generator matrix and a check matrix for C.
   (iv) Show that when $q = 2$, the code C is perfect if and only if n is odd.
(4) Let C be a q-ary code of length n and minimum distance d. Establish the *Singleton upper bound*, i. e., $|C| \leq q^{n-d+1}$. [Hint: two codewords with the same first $n - d + 1$ entries are equal.]
(5) If C is a linear q-ary code of length n and dimension k, prove that the minimum distance of C is at most $n - k + 1$.
(6) Let C be a linear q-ary code of length n and dimension k. Suppose that G is a generator matrix for C and that $G' = [I_k \mid A]$ is the reduced row echelon form of G. Prove that there is a check matrix for C of the form $[-A^T \mid I_{n-k}]$ up to a permutation of columns. [Hint: write $G' = EGF$ where E is non-singular and F is a permutation matrix. Let $H' = [-A^T \mid I_{n-k}]$ and prove that $H = H'F$ is a check matrix associated with G for the code C.]
(7) A linear binary code C has basis $\{(101110), (011010), (001101)\}$. Find a check matrix for C and use it to determine the error-correcting capacity of C.
(8) A check matrix for a linear binary code C is

$$\begin{bmatrix} 1 & 1 & 0 & 1 & 1 \\ 0 & 1 & 0 & 0 & 1 \\ 1 & 1 & 1 & 0 & 0 \end{bmatrix}.$$

   (i) Find a basis for C.
   (ii) Find the minimum distance and error correcting capacity of C.

(iii) If a word (01111) is received and at most one entry is erroneous, use the syndrome method to find the transmitted codeword.

(9) (*An alternative decoding procedure for small n*). Let C be a linear q-ary code of length n with error correcting capacity e. Let H be a check matrix for C. Suppose that a word w is received with at most e errors. Show that the following procedure will find the transmitted codeword.
  (i)   Enumerate all words u in $H_n(q)$ of weight $\leq e$; these are the possible errors.
  (ii) Find the syndrome uH^T of each word u from (i).
  (iii) Compute the syndrome wH^T and compare it with each uH^T: prove that there is a unique word u in $H_n(q)$ of with weight at most e such that $uH^T = wH^T$.
  (iv) Conclude that the transmitted codeword was $w - u$.

(10) Use the method of Exercise (17.4.9) to find the transmitted codeword in Exercise (17.4.8).

(11) (*Dual codes*). Let C be a linear q-ary code of length n and dimension k. Define the *dot product* of two words v, w in $H_n(q)$ by $v \cdot w = \sum_{i=1}^{n} v_i w_i$. Then define $C^{\perp} = \{w \in H_n(q) \mid w \cdot c = 0, \forall c \in C\}$.
  (i)   Show that $C^{\perp}$ is a linear q-ary code of length n: this is called the *dual code* of C.
  (ii) Let G be a generator matrix and H an associated check matrix for C. Prove that G is a check matrix and H a generator matrix for $C^{\perp}$.
  (iii) Prove that $\dim(C^{\perp}) = n - k$ and $|C^{\perp}| = q^{n-k}$.

(12) Let C be a binary Hamming code of length 7. Find a check matrix for the dual code $C^{\perp}$ and show that its minimum distance is 4.

17.5 Algebraic models for accounting systems

This final section will see a complete change of topic. The aim is to create an algebraic structure that is able to simulate, as realistically as possible, the everyday operations of the accounting system of a privately or publicly owned company. The first step is to identify the essential components of such a system and then to decide which algebraic structures will be most useful.

Accounts

An accounting system comprises first and foremost a finite set of *accounts*. The accounts of a company generally fall into three categories:
(i)   *Asset accounts*, which represent anything owned by the company.
(ii) *Liability accounts*, which record what is owed by the company to external entities.
(iii) *Equity accounts*; these accounts show the net worth of the company.

It should be possible to assign a "value" to each account at any instant. In practice these are most likely to be sums of money, but they might be any items that can be bought or sold. The first step is to identify an algebraic structure which can accommodate the account values. The structure chosen must be sufficient to allow the standard accounting operations, which certainly include addition and subtraction. While multiplication has a less natural role in accounting computations, its inclusion provides a richer mathematical environment within which to work.

What comes to mind first as a candidate for account values is probably an integral domain: clearly zero divisors are to be avoided. However, there is a further necessary property that account values should have, namely positivity and negativity, concepts that do not exist in integral domains in general.

Let R be an integral domain and assume that it has a *positive subset*: that is a non-empty subset P, not containing 0_R, with the following properties.
(i)   For each $a \in R$ exactly one of the following statements holds: $a \in P$, $a = 0$, $-a \in P$;
(ii) if $a, b \in P$, then $a + b \in P$ and $ab \in P$.

The notion of positivity allows the introduction of an order relation on R as follows: $a \leq b$ means that either $a = b$ or $b - a \in P$. On the basis of this definition it is straightforward to show that $\leq$ is a linear order on R. It is also easy to establish the following facts for $a, b, c \in R$:
(iii) if $a \leq b$, then $a + c \leq b + c$.
(iv) if $a \leq b$ and $0 \leq c$, then $ac \leq bc$.

An integral domain with a linear order satisfying (iii) and (iv) is called a *linearly ordered domain*. Thus a domain which has a positive subset is linearly ordered. Conversely, if R is a linearly ordered domain, then it has positive subsets, for example $\{r \in R \mid r \neq 0, \; 0 \leq r\}$.

From now on it will be assumed that all account values in an accounting system belong to some ordered integral domain R. Of course, in practice R is likely to be one of the domains

$$\mathbb{Z}, \quad \mathbb{Q}, \quad \mathbb{R}.$$

Each non-zero account value is either positive or negative. The standard convention in accounting is that the value of an asset account should normally be positive and the value of a liability account negative. Accounts such as an equity account will usually have a negative balance, but it might be positive if the company is losing money. While this last convention might seem counterintuitive, the explanation is that the equity, i. e., net assets of the company, belongs to the shareholders, so it is owed to them by the company.

Balance vectors

Let the accounts of an accounting system be written in a fixed order as $a_1, a_2, \ldots, a_n$. At any instant each account a_i has a *value* v_i in a fixed linearly ordered domain R. The state of the accounting system at that instant is described by the list of all account values in order, which will be written as a column vector

$$\mathbf{v} = \begin{bmatrix} v_1 \\ v_2 \\ \vdots \\ v_n \end{bmatrix}$$

Occasionally for typographical convenience $\mathbf{v}$ may be written as the transposed row vector $[v_1 \ v_2 \ \ldots \ v_n]^T$. The set of all n-column vectors over R is denoted by R^n. This is a free R-module of rank n.

There is a critical property that the vector of account values $\mathbf{v}$ must have at all times. Since the accounting system must always be in balance, the sum of all the entries of $\mathbf{v}$ should equal zero. For this reason a vector $\mathbf{v}$ in R^n is called a *balance vector* if

$$\sum_{i=1}^{n} v_i = 0.$$

The set of all balance vectors in R^n is termed the *balance space* and is denoted by

$$\mathrm{Bal}_n(R).$$

If $n = 1$, the only balance vector is zero. Thus we can assume that $n > 1$.

The balance space is a submodule of the free module R^n, and in fact it is also a free R-module. To establish this, we will need to produce an R-basis. Let

$$\mathbf{e}(i, j), \quad i \neq j,$$

denote the n-column vector whose ith entry is 1, jth entry -1 and all entries are 0. This is obviously a balance vector. The $\mathbf{e}(i, j)$ are called the *elementary balance vectors*. More generally, a balance vector with exactly two non-zero entries is said to be a *simple* balance vector.

(17.5.1). *Let R be an ordered domain and let $n > 1$ be an integer. Then the elementary balance vectors $\mathbf{e}(1, 2), \mathbf{e}(2, 3), \ldots, \mathbf{e}(n-1, n)$ form an R-basis of* $\mathrm{Bal}_n(R)$*, so that* $\mathrm{Bal}_n(R)$ *is a free module of rank $n - 1$.*

Proof. In the first place $\mathbf{e}(1, 2), \ldots, \mathbf{e}(n - 1, n)$ are linearly independent over R. To see this, let $r_1, \ldots, r_{n-1} \in R$; then

$$r_1\,\mathbf{e}(1, 2) + r_2\,\mathbf{e}(2, 3) + \cdots + r_{n-1}\,\mathbf{e}(n - 1, n) = \begin{bmatrix} r_1 \\ r_2 - r_1 \\ r_3 - r_2 \\ \vdots \\ r_{n-1} - r_{n-2} \\ -r_{n-1} \end{bmatrix}.$$

This column vector can only equal $\mathbf{0}$ if $r_1 = r_2 = \cdots = r_{n-1} = 0$, which establishes the linear independence.

It remains to prove that an arbitrary balance vector $\mathbf{b}$ is expressible as a linear combination of $\mathbf{e}(1, 2), \ldots, \mathbf{e}(n - 1, n)$. Since $\mathbf{b}$ is a balance vector, it can be written in the form

$$\mathbf{b} = [b_1\ b_2\ \ldots\ b_{n-1}\ -b_1 - b_2 - \cdots - b_{n-1}]^T.$$

Let $v_i = b_1 + b_2 + \cdots + b_i$, where $1 \le i \le n - 1$. Then by a simple computation

$$v_1\,\mathbf{e}(1, 2) + v_2\,\mathbf{e}(2, 3) + \cdots + v_{n-1}\,\mathbf{e}(n - 1, n) = \mathbf{b},$$

which completes the proof. □

The type of a balance vector

Let n be a positive integer and R an ordered domain. The *type* of a balance vector $\mathbf{v} \in \mathrm{Bal}_n(R)$,

$$\mathrm{type}(\mathbf{v}),$$

is defined to be the n-column vector whose ith entry is 0, $+$ or $-$ according as $v_i = 0$, $v_i > 0$ or $v_i < 0$ respectively. For example, if $\mathbf{v}$ is the balance vector $[-300\ 400\ -100\ 0]^T$ in $\mathrm{Bal}_4(\mathbb{Z})$, the type of $\mathbf{v}$ is $[-\ +\ -\ 0]^T$.

The zero vector has type $\mathbf{0}$, while the type of a simple balance vector contains a single $+$ and a single $-$, with other entries 0. Notice that any non-zero type must have at least one $+$ and at least one $-$.

Let $\mathbf{s}$ and $\mathbf{t}$ be types of balance vectors in $\mathrm{Bal}_n(R)$: thus the entries of the vectors $\mathbf{s}, \mathbf{t}$ are 0, $+$ or $-$. A binary relation $\le$ on the set of types of vectors in $\mathrm{Bal}_n(R)$ is defined as follows: $\mathbf{s} \le \mathbf{t}$ is to mean that $s_i = t_i$ or $s_i = 0$ for $i = 1, 2, \ldots, n$. Thus $\mathbf{s}$ and $\mathbf{t}$ have the same configuration of $+$ and $-$ signs except that $\mathbf{s}$ may have more zeros than $\mathbf{t}$. This relation is reflexive, transitive and antisymmetric, so it is a *partial order* on the set of types. It is not a linear order: for example, the types $[+ - 0]^T$ and $[- + 0]^T$ are incomparable.

As usual with partial orders, the set of types can be visualized by means of a Hasse diagram: the less complex types occur lower down in the diagram. At the lowest point will be the type of the zero vector **0**, while type(**v**) sits directly below type(**w**) if the entries of type(**v**) and type(**w**) are the same except that type(**v**) has one more zero entry.

A natural measure of the complexity of a balance vector is the total number of non-zero entries. For any **v** in $\mathrm{Bal}_n(R)$ define the *level* of **v** to be the number of non-zero entries, with the convention that the zero vector is at level 1. Thus the level of a balance vector is a positive integer. Clearly, if k is any integer satisfying $1 \le k \le n$, then $\mathrm{Bal}_n(R)$ has vectors of level k. One can think of the types of balance vectors as being classified in a hierarchy of levels. At level 1 is the zero vector, at level 2 the simple balance vectors, and thereafter balance vectors of increasing complexity.

Next we derive formulas for the number of types at any given level: apart from their intrinsic interest, these formulas provide insight into the distribution of balance vector types over the various levels.

(17.5.2). *Let n be an integer greater than 1.*
(i)   *The number of types of n-balance vectors at level r is $\binom{n}{r}(2^r - 2)$, where $1 < r \le n$.*
(ii) *The total number of types of n-balance vectors is $3^n - 2^{n+1} + 2$.*

Proof. (i) In order to construct an n-balance type at level r one must first pick the r "slots" in which a + or − is to be placed; this may be done in $\binom{n}{r}$ ways. Then one has to count the number of ways of placing a + or − in each of the r chosen slots, taking care not to have a + in every slot or a − in every slot. This can be done in $2^r - 2$ ways. The remaining $n - r$ slots get 0's, so the type has been determined. Hence the number of types at level r is $\binom{n}{r}(2^r - 2)$.

(ii) The total number of n-balance types is the sum of the numbers of types at levels 1 through n. Since there is just one type of level 1, this is

$$1 + \sum_{r=2}^{n} \binom{n}{r}(2^r - 2) = 1 + \sum_{r=2}^{n} \binom{n}{r}2^r - 2\sum_{r=2}^{n} \binom{n}{r}.$$

By the Binomial Theorem

$$\sum_{r=0}^{n} \binom{n}{r}2^r = (1 + 2)^n = 3^n \quad \text{and} \quad \sum_{r=0}^{n} \binom{n}{r} = (1 + 1)^n = 2^n.$$

Hence the total number of types is $1 + (3^n - 1 - 2n) - 2(2^n - 1 - n) = 3^n - 2^{n+1} + 2$.   □

Example (17.5.1). For a system with three accounts, there are 13 types of balance vectors. They are listed below in descending order of levels:

$$
\text{level 3}: \quad
\begin{bmatrix} + \\ + \\ - \end{bmatrix}, \quad
\begin{bmatrix} - \\ + \\ + \end{bmatrix}, \quad
\begin{bmatrix} + \\ - \\ + \end{bmatrix}, \quad
\begin{bmatrix} - \\ - \\ + \end{bmatrix}, \quad
\begin{bmatrix} + \\ - \\ - \end{bmatrix}, \quad
\begin{bmatrix} - \\ + \\ - \end{bmatrix}
$$

$$
\text{level 2}: \quad
\begin{bmatrix} + \\ - \\ 0 \end{bmatrix}, \quad
\begin{bmatrix} 0 \\ + \\ - \end{bmatrix}, \quad
\begin{bmatrix} - \\ 0 \\ + \end{bmatrix}, \quad
\begin{bmatrix} - \\ + \\ 0 \end{bmatrix}, \quad
\begin{bmatrix} 0 \\ - \\ + \end{bmatrix}, \quad
\begin{bmatrix} + \\ 0 \\ - \end{bmatrix}
$$

$$
\text{level 1}: \quad
\begin{bmatrix} 0 \\ 0 \\ 0 \end{bmatrix}
$$

On the basis of (17.5.2) it can be shown that the maximum number of types of n-balance vectors occurs at level

$$
\left[\frac{2n+2}{3} \right],
$$

or roughly two thirds of the way up the hierarchy of levels. For a proof of this fact see [3, Section (3.2)].

Transactions

The next question to be addressed is how changes are made to the account values in an accounting system. Such changes are called *transactions* and when one is applied, a flow of value occurs between accounts of the system. Some account balances will increase, others decrease and many will likely be unaffected by the transaction. After the transaction has been applied, the system must still be in balance, which shows that a transaction is actually a function defined on the balance space. What is more, the effect of the transaction is to add a balance vector to the original balance vector.

After these preliminary observations we are ready to give the formal definition. Let $\mathbf{v} \in \mathrm{Bal}_n(R)$ be a fixed balance vector over a linearly ordered domain R. A *transaction* on the balance space $\mathrm{Bal}_n(R)$ is a function

$$
\tau_\mathbf{v} : \mathrm{Bal}_n(R) \rightarrow \mathrm{Bal}_n(R),
$$

which is defined by the rule

$$
\tau_\mathbf{v}(\mathbf{x}) = \mathbf{x} + \mathbf{v}, \ \mathbf{x} \in \mathrm{Bal}_n(R).
$$

Notice that $\mathbf{x} + \mathbf{v}$ is a balance vector. Thus, when the transaction $\tau_\mathbf{v}$ is applied to a balance vector $\mathbf{x}$, it produces the new balance vector $\mathbf{x}' = \mathbf{x} + \mathbf{v}$. The vector $\mathbf{v} = \mathbf{x}' - \mathbf{x}$ is called the *transaction vector*. We may also refer to the vector **v** as the transaction. The set of all transactions is denoted by

$$\text{Trans}_n(R) = \{\tau_{\mathbf{v}} \mid \mathbf{v} \in \text{Bal}_n(R)\}.$$

Notice that $\tau_{\mathbf{0}}$ is the *identity transaction*. Of course $\tau_{\mathbf{v}}$ is called an elementary or simple transaction according as **v** is an elementary or simple balance vector.

For transactions $\tau_{\mathbf{v}}$ and $\tau_{\mathbf{w}}$, the composite $\tau_{\mathbf{v}} \circ \tau_{\mathbf{w}}$ sends $\mathbf{x} \in \text{Bal}_n(R)$ to $\mathbf{x} + \mathbf{w} + \mathbf{v}$, as does $\tau_{\mathbf{w}} \circ \tau_{\mathbf{v}}$. Therefore

$$\tau_{\mathbf{v}} \circ \tau_{\mathbf{w}} = \tau_{\mathbf{v}+\mathbf{w}} = \tau_{\mathbf{w}} \circ \tau_{\mathbf{v}}.$$

Also $\tau_{\mathbf{v}} \circ \tau_{-\mathbf{v}} = \tau_{\mathbf{0}}$ and $\tau_{-\mathbf{v}}$ is the inverse of the transaction $\tau_{\mathbf{v}}$. These equations show that $\text{Trans}_n(R)$ is an abelian group with $\tau_{\mathbf{0}}$ playing the role of the identity element. In addition $\text{Trans}_n(R)$ has an R-module structure given by $r\tau_{\mathbf{v}} = \tau_{r\mathbf{v}}$. Thus $(r\tau_{\mathbf{v}})(\mathbf{x}) = \mathbf{x} + r\mathbf{v}$. It is easy to check the validity of module axioms.

It is apparent that the modules $\text{Trans}_n(R)$ and $\text{Bal}_n(R)$ are very similar. In fact by a simple calculation we have:

(17.5.3). *The assignment* $\mathbf{v} \mapsto \tau_{\mathbf{v}}$ *determines an isomorphism of R-modules* $\tau : \text{Bal}_n(R) \to \text{Trans}_n(R)$.

Example (17.5.2). Here is a simple example of a transaction. A company finishes off pieces of machinery that it has purchased and then sells them. The cost per unit is $100 and the sales price is $150. The sale of a single unit gives rise to a transaction which involves the cash, inventory and equity accounts. Thus $150 is deposited in the cash account and $100 is deducted from inventory. The profit of $50 goes into the equity account, but note that this amount should be subtracted from the equity balance since this is owed to the owners of the company. Therefore the transaction vector has cash entry 150, inventory −100, equity −50: all other accounts have zero values since they are not impacted by the transaction.

Abstract accounting systems

Thus far it has been shown how we can represent the state of an accounting system by a balance vector and a change in the state of the system by a transaction, which in turn may be identified with a balance vector. Now in any real life accounting system there will be certain transactions that would be regarded as improper, for example, if they are contrary to sound business practice or if they violate government regulations. To exclude such undesirable operations, an accounting system should come equipped with a list of transactions that are regarded as valid operations for the system. These will be called *allowable transactions*.

Another feature of an accounting system is that, even if a transaction is allowable, it might still be rejected if it caused an unacceptable balance to appear in some account. Thus we recognize the possible need to have a list of *allowable balances* as well.

The discussion suggests that a basic model of an accounting system should include three elements: (i) a set of accounts in a specified order; (ii) a set of allowable transactions; (iii) a set of allowable balance vectors.

The formal definition follows. An *abstract accounting system* over an ordered domain R is a triple

$$\mathcal{A} = (A \mid T \mid B)$$

where A is an ordered set of n accounts, and T and B are subsets of $\mathrm{Bal}_n(R)$ called the sets of *allowable transactions* and *allowable balances* respectively. Here B is required to be non-empty. If there are no limits on balances and $B = \mathrm{Bal}_n(R)$, then $\mathcal{A}$ is an *unbounded* system.

The accounting system as an automaton

An accounting system operates in a natural way as a state output automaton in the sense of Section 1.3 – see Exercise (1.3.8). The state of the system is the balance vector representing the current position and the input symbols are the transaction vectors. Suppose that the current state is described by the balance vector **b** and that a transaction represented by a balance vector **v** is applied. The automaton scans the input **v**. If **v** is an allowable transaction, the transaction is applied to the system. The potential new balance is $\mathbf{b}' = \mathbf{b} + \mathbf{v}$. The automaton reads $\mathbf{b}'$. If this is an allowable balance, it becomes the next state, i. e., the new balance vector for the system. Otherwise the transaction is rejected, an error message is printed and the balance remains **b**.

In practice the allowable transactions will be of two sorts. There may be specific allowable transactions with fixed entries, for example, rent or mortgage payments. Then there may be entire types of transactions that are allowable: a transaction in a retail firm which debits cash and credits inventory and profit/loss would be of this type. It is therefore reasonable to replace the set T by two subsets

$$T_0 \quad \text{and} \quad T_1$$

and write

$$\mathcal{A} = (A \mid T_0, T_1 \mid B)$$

where T_0 is the list of allowable transaction types and T_1 is the list of specific allowable transactions. The identity transaction is allowable in any system since it has no effect on balances.

Accounting systems with one account are uninteresting since the balance is always zero. Systems with two accounts are scarcely more interesting: the two accounts have balance vectors whose entries are negatives of each other. The simplest interesting accounting system has three accounts. Such a system could represent the uncomplicated financial position of a small firm or an individual with few assets and liabilities, with one account representing total assets, one total liabilities and a third account specifying the net equity.

The objective in the rest of the section is to determine to what extent algebraic concepts such as subsystem, quotient system and homomorphism can be applied to abstract accounting systems in a way that reflects the workings of a real accounting system.

Subaccounting Systems

As a first step the concept of a subaccounting system will be introduced. In order to come up with a useful definition, one needs to look at a real life accounting system. In the case of a large firm there are likely to be subdivisions with a degree of autonomy. Such a subdivision might have a set of accounts under its control and be able to execute certain transactions on these accounts, although such transactions would still need approval at the company level. The subdivision's allowable transactions would not affect accounts which are outside its control. Of course, allowable balances for the unit would always have to be compatible with those for the entire system.

These observations suggest how a subaccounting system should be defined. Consider a system with n accounts over an ordered domain R

$$\mathcal{A} = (A\,|\,T\,|\,B),$$

with the usual notation and conventions. An accounting system $\mathcal{A}' = (A'\,|\,T'\,|\,B')$ over R is said to be a *subaccounting system* of $\mathcal{A}$ if the following conditions are satisfied:
(i) $A' \subseteq A$;
(ii) $T' = \{\mathbf{v}|_{A'} \mid \mathbf{v} \in T,\ \mathrm{sppt}(\mathbf{v}) \subseteq A'\}$, so that $\mathbf{v}|_{A'}$ is a balance vector;
(iii) if $\mathbf{b} \in B$, then $\mathbf{b}|_{A'}$ is a balance vector and $B' = \{\mathbf{b}|_{A'} \mid \mathbf{b} \in B\}$.

Here the vector $\mathbf{v}|_{A'}$ is obtained from $\mathbf{v}$ by omitting entries for accounts that are not in A'. Also, $\mathrm{sppt}(\mathbf{v})$, the *support* of $\mathbf{v}$, is the set of accounts for which $\mathbf{v}$ has a non-zero entry. By (i) each account of $\mathcal{A}'$ is an account of $\mathcal{A}$. The effect of (ii) is that the allowable transactions of $\mathcal{A}'$ are the restrictions to A' of allowable transactions of $\mathcal{A}$ that do not affect accounts outside A'. Finally, (iii) asserts that the allowable balance vectors of $\mathcal{A}'$ are precisely the restrictions to A' of allowable balance vectors of $\mathcal{A}$.

It is obvious that every accounting system is a subsystem of itself. A subsystem of a system $\mathcal{A}$ with fewer accounts than $\mathcal{A}$ is called a *proper subsystem* of $\mathcal{A}$. It is possible that an accounting system has no proper subsystems. Here is a criterion for the existence of proper subsystems.

(17.5.4). *An accounting system* $\mathcal{A} = (A| T| B)$ *has a proper subsystem if and only if there is a proper non-empty subset A' of A such that* $\mathbf{b}|_{A'}$ *is a balance vector whenever* $\mathbf{b} \in B$.

Proof. Let A' be a subset of A satisfying the given condition and define

$$T' = \{\mathbf{v}|_{A'} \mid \mathbf{v} \in T, \text{ sppt}(\mathbf{v}) \subseteq A'\} \quad \text{and} \quad B' = \{\mathbf{b}|_{A'} \mid \mathbf{b} \in B\}.$$

By hypothesis B' is non-empty and consists of balance vectors. If $\mathbf{v} \in T$ and $\text{sppt}(\mathbf{v}) \subseteq A'$, it follows that $\mathbf{v}|_{A'}$ is a balance vector. Thus T' consists of balance vectors. Define $\mathcal{A}' = (A'| T'| B')$; this is an accounting system. Moreover, $\mathcal{A}'$ is a proper subsystem of $\mathcal{A}$ since $A' \neq A$.

Conversely, if $\mathcal{A}$ has a proper subsystem, the condition on A' is satisfied. $\qquad\square$

Quotients of Accounting Systems

Next it will be shown how to define a quotient structure on an accounting system. In forming a quotient the balances of certain sets of accounts are combined. A practical use for this is to model the standard accounting operation of generating a report or summary.

Consider an accounting system $\mathcal{A} = (A| T| B)$ with n accounts over an ordered domain R. To construct a quotient of $\mathcal{A}$ we introduce an equivalence relation E on the set of accounts A. Thus A is partitioned by E into, say, $\bar{n}$ distinct equivalence classes $[a]_E$ where $\bar{n} \leq n$, $a \in A$. The account set of the quotient system is to be the set of equivalence classes

$$\overline{A}_E = \{[a]_E \mid a \in A\}.$$

The set $\overline{A}_E$ can be ordered by the smallest account subscript in each equivalence class.

Next we need to specify the allowable transactions and balance vectors for the quotient system. Let $\mathbf{v} \in \text{Bal}_n(R)$ and define a vector $\overline{\mathbf{v}} = \overline{\mathbf{v}}_E$ in $\text{Bal}_{\bar{n}}(R)$ by the rule

$$\overline{v}_i = \sum_{a_j E a_i} v_j,$$

the summation being over all j for which $a_j \, E \, a_i$. What this means is that the entries of $\mathbf{v}$ are totaled for accounts belonging to the same E-equivalence class. Observe that

$$\sum_{i=1}^{\bar{n}} \overline{v}_i = \sum_{i=1}^{n} v_i = 0,$$

so that $\overline{\mathbf{v}} \in \text{Bal}_{\bar{n}}(R)$.

From this definition we quickly derive the rules

$$\overline{\mathbf{u} + \mathbf{v}} = \overline{\mathbf{u}} + \overline{\mathbf{v}} \quad \text{and} \quad \overline{r\mathbf{v}} = r\overline{\mathbf{v}},$$

where $\mathbf{u}, \mathbf{v} \in \mathrm{Bal}_n(R)$ and $r \in R$. Hence the assignment $\mathbf{v} \mapsto \overline{\mathbf{v}}$ determines a homomorphism of R-modules from $\mathrm{Bal}_n(R)$ to $\mathrm{Bal}_{\overline{n}}(R)$. It is clearly surjective.

Example (17.5.3). Let $A = \{a_1, a_2, a_3, a_4, a_5\}$ and let E be the equivalence relation on A with associated partition

$$A = \{a_1\} \cup \{a_2, a_4, a_5\} \cup \{a_3\}.$$

The elements of $\overline{A}_E$ in order are

$$[a_1] = \{a_1\}, \quad [a_2] = \{a_2, a_4, a_5\}, \quad [a_3] = \{a_3\}.$$

If we take a typical vector $[v_1 \, v_2 \, v_3 \, v_4 \, -v_1 \, -v_2 \, -v_3 \, -v_4]^T$ in $\mathrm{Bal}_5(R)$, then, according to the definition,

$$\overline{\mathbf{v}}_E = \begin{bmatrix} v_1 \\ -v_1 - v_3 \\ v_3 \end{bmatrix} \in \mathrm{Bal}_3(R).$$

Returning to the general situation, we are ready to formulate the definition of the quotient of the accounting system $A = (A|\,T|\,B)$ determined by the equivalence relation E on A. The account set of the quotient is to be

$$\overline{A}_E = \{[a_{i_j}]_E \mid j = 1, 2, \ldots, \overline{n}\},$$

that is, the set of distinct E-equivalence classes. Next define the allowable sets of transactions and balances to be respectively

$$\overline{T}_E = \{\overline{\mathbf{v}}_E \mid \mathbf{v} \in T\} \quad \text{and} \quad \overline{B}_E = \{\overline{\mathbf{v}}_E \mid \mathbf{v} \in B\}.$$

These sets are, of course, the images of T and B under the mapping in which $\mathbf{v} \mapsto \mathbf{v}_E$. Finally, the *quotient system* of A by E is defined to be

$$A/E = (\overline{A}_E|\,\overline{T}_E|\,\overline{B}_E).$$

Thus the accounts of A/E are the E-equivalence classes of accounts of A, while the allowable transactions and balances of A/E arise from the corresponding entities of A by application of the function $\mathbf{v} \mapsto \overline{\mathbf{v}}_E$, i. e., by summing vector entries in each equivalence class.

Reports

An example of how a quotient system occurs in accounting practice is the creation of a report on an accounting system. This occurs when the account set is divided up into a number of control groups and the resulting partition yields an equivalence relation on the account set. This produces a quotient system that generates a report on the control groups. The simplest case arises from the partition

$$A = A_a \cup A_\ell \cup A_e$$

where A_a, A_ℓ, A_e are the respective sets of asset accounts, liability accounts, equity accounts. The resulting quotient system $\mathcal{A}/E$ has three accounts, namely total assets, total liabilities, net equity. This quotient system produces the most basic type of report on the system.

Homomorphisms of accounting systems

Next we explain the concept of a homomorphism between accounting systems. This provides a means of relating the operations of the two systems. The challenge is to come up with the "right" definition of a homomorphism, that is one that will accord with practice.

Consider two accounting systems

$$\mathcal{A} = (A|\,T|\,B) \quad \text{and} \quad \mathcal{A}' = (A'|\,T'|\,B')$$

over the same ordered domain R, with account sets $A = \{a_1, a_2, \ldots, a_n\}$ and $A' = \{a_1', a_2', \ldots, a_{n'}'\}$. To define a homomorphism from $\mathcal{A}$ to $\mathcal{A}'$ we start with a function between the account sets

$$\theta : A \to A'.$$

This is used to generate a function between balance modules

$$\theta^* : \mathrm{Bal}_n(R) \to \mathrm{Bal}_{n'}(R),$$

where $\theta^*(\mathbf{v})$ is defined by the following rule. If $a_i' \in \theta(\mathcal{A}) = \mathrm{Im}(\theta)$, then for $i = 1, 2, \ldots, n'$

$$\left(\theta^*(\mathbf{v})\right)_i = \sum_{\theta(a_j) = a_i'} v_j,$$

the sum being formed over all j for which $\theta(a_j) = a_i'$. On the other hand, if $a_i' \notin \mathrm{Im}(\theta)$, then by definition

$$\left(\theta^*(\mathbf{v})\right)_i = 0.$$

Notice that $\theta^*(\mathbf{v})$ is a balance vector since its components outside $\mathrm{Im}(\theta)$ are all zero.

Thus the function θ^* sums all entries of $\mathbf{v}$ that correspond to accounts mapped by θ to the same account in A' and it assigns a zero entry to any account of A' which is not in $\text{Im}(\theta)$. It is a simple matter to deduce from the definition that

$$\theta^*(\mathbf{u} + \mathbf{v}) = \theta^*(\mathbf{u}) + \theta^*(\mathbf{v}) \quad \text{and} \quad \theta^*(r\mathbf{v}) = r\theta^*(\mathbf{v})$$

for $\mathbf{u}, \mathbf{v} \in \text{Bal}_n(R)$ and $r \in R$. These equations show that θ^* is a homomorphism of R-modules.

Example (17.5.4). As an example take the sets $A = \{a_1, a_2, a_3, a_4, a_5\}$ and $A' = \{a_1', a_2', a_3', a_4'\}$ and consider the function $\theta : A \rightarrow A'$ which is defined by the rules

$$\theta(a_1) = a_1', \quad \theta(a_2) = a_3', \quad \theta(a_3) = a_2', \quad \theta(a_4) = a_2', \quad \theta(a_5) = a_2'.$$

Let $\mathbf{v} = [v_1 \; v_2 \; v_3 \; v_4 \; -v_1 - v_2 - v_3 - v_4]^T \in \text{Bal}_5(\mathbb{Z})$: then $\mathbf{v}$ is sent by $\theta^* : \text{Bal}_5(\mathbb{Z}) \rightarrow \text{Bal}_4(\mathbb{Z})$ to

$$\theta^*(\mathbf{v}) = [v_1 \quad -v_1 - v_2 \quad v_2 \quad 0]^T.$$

To see where this comes from, note that only $\theta(a_1)$ equals a_1', so $(\theta^*(\mathbf{v}))_1 = v_1$: similarly $\theta(a_2) = a_3'$ and $(\theta^*(\mathbf{v}))_3 = v_2$. Next $\theta(a_3) = \theta(a_4) = \theta(a_5) = a_2'$; therefore $(\theta^*(\mathbf{v}))_2 = v_3 + v_4 + (-v_1 - v_2 - v_3 - v_4) = -v_1 - v_2$. Finally, $(\theta^*(\mathbf{v}))_4 = 0$, since $a_4' \notin \text{Im}(\theta) = \{a_1', a_2', a_3'\}$.

We return to the general situation with two accounting systems $\mathcal{A} = (A | T | B)$ and $\mathcal{A}' = (A' | T' | B')$, together with a function $\theta : A \rightarrow A'$. Then θ is said to determine a *homomorphism of accounting systems*, (also denoted by θ),

$$\theta : \mathcal{A} \rightarrow \mathcal{A}',$$

provided that the following conditions are met:
(i) if $\mathbf{v} \in T$, then $\theta^*(\mathbf{v}) \in T'$, i.e., $\theta^*(T) \subseteq T'$;
(ii) if $\mathbf{b} \in B$, there exists a $\mathbf{b}' \in B'$ such that $\theta^*(\mathbf{b})|_{\text{Im}(\theta)} = \mathbf{b}'|_{\text{Im}(\theta)}$.

What this means is that θ^* sends allowable transactions of $\mathcal{A}$ to allowable transactions of $\mathcal{A}'$ which affect only accounts in $\text{Im}(\theta)$. On the other hand, θ^* sends an allowable balance vector of $\mathcal{A}$ to a balance vector that agrees in its $\text{Im}(\theta)$-entries with some allowable balance vector $\mathbf{b}'$ of $\mathcal{A}'$.

Notice that the vector $\mathbf{b}'$ in condition (ii) may not be unique; thus the homomorphism depends on the assignment $\mathbf{b} \mapsto \mathbf{b}'$, as well as the function θ. The vectors $\theta^*(\mathbf{b})$ and $\mathbf{b}'$ in condition (ii) may be different: indeed $\mathbf{b}'$, unlike $\theta^*(\mathbf{b})$, could have non-zero entries for accounts in A' that are not in $\text{Im}(\theta)$, and $\theta^*(\mathbf{b})$ might not belong to B'. There is reason to permit this phenomenon: in practice it might be unacceptable for certain accounts in $A' - \text{Im}(\theta)$ to have zero values. Thus it would be unreasonable to insist that the entries for such accounts in allowable balance vectors equal 0. This feature of the homomorphism concept limits its utility in the general case. For the reasons just set out, the homomorphism concept is most useful in the context of unbounded accounting systems, where balance restrictions are not imposed. For then two homomorphisms can be composed in the natural way.

To see this consider two homomorphisms of unbounded accounting systems $\theta : \mathcal{A} \rightarrow \mathcal{A}'$ and $\phi : \mathcal{A}' \rightarrow \mathcal{A}''$. It can be shown that for the corresponding module homomorphisms the equation $(\phi\theta)^* = \phi^*\theta^*$ holds. From this it follows that $\phi\theta : \mathcal{A} \rightarrow \mathcal{A}''$ is a homomorphism. What is more, the composition is associative. A consequence of these considerations is that *unbounded accounting systems constitute a category*, the morphisms being the homomorphisms between systems: for a more on this topic see Exercise (17.5.6) below.

Injective and surjective homomorphisms

There are three special types of homomorphisms that are occur in practice. Suppose that $\theta : \mathcal{A} \rightarrow \mathcal{A}'$ is a homomorphism between accounting systems $\mathcal{A} = (A, T, B)$ and $\mathcal{A}' = (A', T', B')$. Put $n = |A|$ and $n' = |A'|$. Let θ^* be the induced module homomorphism defined above: thus $(\theta^*(\mathbf{v}))_i = \sum_{\theta(a_j)=a'_i} v_j$ if $a_i \in \text{Im}(\theta)$ and $(\theta^*(\mathbf{v}))_i = 0$ otherwise. First of all we observe:

(17.5.5).
(i) *If θ is an injective function, then θ^* is injective.*
(ii) *If θ is a surjective function, then θ^* is surjective.*

Proof.
(i) Let θ be injective and suppose that $\theta^*(\mathbf{v}) = \mathbf{0}$ for some $\mathbf{v} \neq \mathbf{0}$. Then $v_i \neq 0$ for some i. Now $\theta(a_i) = a'_j$ for some j and i is uniquely determined by j. Therefore $(\theta^*(\mathbf{v}))_j = v_i$ and hence $\theta^*(\mathbf{v}) \neq \mathbf{0}$, a contradiction that shows θ^* to be injective.
(ii) Assume that θ is surjective and let $\mathbf{u} \in \text{Bal}_{n'}(\mathbb{Z})$. If $1 \leq i \leq n'$, there exists $j(i) \leq n$ such that $\theta(a_{j(i)}) = a'_i$. Define $\mathbf{v} \in \text{Bal}_n(R)$ by $v_{j(i)} = u_i$ and $v_k = 0$ if $k \neq j(i)$ for all i. Then

$$\left(\theta^*(\mathbf{v})\right)_i = \sum_{\theta(a_k)=a'_i} v_k = v_{j(i)} = u_i.$$

Since $v_k = 0$ if $k \neq j(i)$ for all i, it follows that $\theta^*(\mathbf{v}) = \mathbf{u}$. Finally, $\mathbf{v}$ is a balance vector since $\sum_{k=1}^{n} v_k = \sum_{i=1}^{n'} v_{j(i)} = \sum_{i=1}^{n'} u_i = 0$. Therefore θ^* is surjective. $\square$

Assume that $\theta : \mathcal{A} \rightarrow \mathcal{A}'$ is a homomorphism of accounting systems. Consider first the case when the function $\theta : A \rightarrow A'$ is injective and $\theta : \mathcal{A} \rightarrow \mathcal{A}'$ is called an *injective homomorphism of accounting systems*. Then θ^* is injective by (17.5.5). An allowable transaction of $\mathcal{A}$ is mapped by θ^* to an allowable transaction of $\mathcal{A}'$ with zero entries for accounts in $A' - \text{Im}(\theta)$. In addition, each allowable balance vector of $\mathcal{A}$ is mapped by θ^* to the restriction to $\text{Im}(\theta)$ of some allowable balance vector of $\mathcal{A}'$. This type of homomorphism is realized when new accounts are added to an existing accounting system or when an accounting system is inserted into a larger system.

The case when the function $\theta : A \to A'$ is surjective is also of interest. Then the conditions in the definition of a homomorphism guarantee that $\theta^*(T) \subseteq T'$ and $\theta^*(B) \subseteq B'$ since $\text{Im}(\theta) = A'$. Also θ^* is surjective by (17.5.5). If the stronger conditions

$$\theta^*(T) = T' \quad \text{and} \quad \theta^*(B) = B',$$

are valid, then $\theta : \mathcal{A} \to \mathcal{A}'$ is called a *surjective homomorphism of accounting systems*.

Finally, a homomorphism $\theta : \mathcal{A} \to \mathcal{A}'$ is an *isomorphism* if it is both injective and surjective. Under these circumstances $\theta : A \to A'$ is a bijection and it sets up a one-one correspondence between the accounts, allowable transactions and allowable balances of $\mathcal{A}$ and the corresponding entities of $\mathcal{A}'$. If there is an isomorphism from $\mathcal{A}$ to $\mathcal{A}'$, then $\mathcal{A}$ and $\mathcal{A}'$ are said to be *isomorphic* systems and the notation

$$\mathcal{A} \simeq \mathcal{A}'$$

is used. Thus in essence isomorphic accounting systems operate by the same set of rules, although their account sets may be different. If $\theta : \mathcal{A} \to \mathcal{A}'$ is an isomorphism of accounting systems, there is an inverse, namely the isomorphism of accounting systems which is induced from the set bijection θ^{-1}: see Exercise (17.5.5) below.

The canonical homomorphism

A natural example of a surjective homomorphism of accounting systems arises on forming a quotient of an accounting system. Suppose that $\mathcal{A} = (A|\, T|\, B)$ is an accounting system and E is an equivalence relation on the account set A. Then, as we have seen, there is a corresponding quotient system

$$\mathcal{A}/E = (\overline{A}|\, \overline{T}|\, \overline{B})$$

where $\overline{A}$ is the set of E-equivalence classes, $\overline{T} = \{\overline{\mathbf{v}}_E \mid \mathbf{v} \in T\}$, and $\overline{B} = \{\overline{\mathbf{b}}_E \mid \mathbf{b} \in B\}$, Recall that $\overline{\mathbf{v}}_E$ is defined by the rule $(\overline{\mathbf{v}}_E)_i = \overline{v}_i = \sum_{a_j E a_i} v_j$. There is a natural surjective function

$$\sigma_E : A \to \overline{A}$$

defined by sending each $a \in A$ to its E-equivalence class; thus $\sigma_E(a_i) = [a_i]_E$. From (17.5.5) we see that σ_E induces a surjective R-module homomorphism

$$\sigma_E^* : \text{Bal}_n(R) \to \text{Bal}_{\overline{n}}(R)$$

where $\bar{n} = |\bar{A}| =$ the number of E-equivalence classes. The definition of σ_E^* shows that

$$(\sigma_E^*(\mathbf{v}))_i = \sum_{\sigma_E(a_j) = \sigma_E(a_i)} v_j = \sum_{a_j E a_i} v_j = \bar{v}_i,$$

and therefore $\sigma_E^*(\mathbf{v}) = \bar{\mathbf{v}}_E$. Consequently, $\sigma_E^*(T) = \bar{T}$ and $\sigma_E^*(B) = \bar{B}$, equations which show that

$$\sigma_E : \mathcal{A} \to \mathcal{A}/E$$

is a surjective homomorphism of accounting systems. This will be called the *canonical homomorphism* from $\mathcal{A}$ to $\mathcal{A}/E$. It is easy to remember since what it does is to combine all accounts belonging to the same E-equivalence class. This is analogous to how the group elements in a coset are combined in a quotient group.

These conclusions are summarized in the following result.

(17.5.6). *Let $\mathcal{A} = (A|\, T\,|\, B)$ be an accounting system and let E be an equivalence relation on the account set A. Then the assignment $a \mapsto [a]_E$ determines a surjective homomorphism $\sigma_E : \mathcal{A} \to \mathcal{A}/E$.*

The image of a homomorphism

A natural feature of a homomorphism of accounting systems $\theta : \mathcal{A} \to \mathcal{A}'$ is its *image*

$$\mathrm{Im}(\theta).$$

This is an accounting system contained within $\mathcal{A}'$, but not necessarily as a subsystem. The account set of the system $\mathrm{Im}(\theta)$ is the image of the set function θ, which is also written $\mathrm{Im}(\theta)$ or $\theta(A)$. Let

$$\theta_0 : A \to \theta(A)$$

be the surjective function sending a_i to $\theta(a_i)$. Then θ_0 induces a surjective homomorphism of modules

$$\theta_0^* : \mathrm{Bal}_n(R) \to \mathrm{Bal}_{\bar{n}}(R)$$

where $n = |A|$, $\bar{n} = |\theta(A)|$ and θ_0^* is defined by the usual rule

$$(\theta_0^*(\mathbf{v}))_i = \sum_{\theta(a_j) = \theta(a_i)} v_j.$$

Here, as before, the sum is formed over all accounts with the same θ-value as a_i. Thus $\theta_0^*(T)$ and $\theta_0^*(B)$ are subsets of $\mathrm{Bal}_{\bar{n}}(R)$. The *image* of θ is defined to be the accounting system

$$\mathrm{Im}(\theta) = \big(\theta(A) \mid \theta_0^*(T) \mid \theta_0^*(B)\big).$$

Suppose that $\mathbf{v} \in T \cup B$; thus $\theta_0^*(\mathbf{v})$ is a typical allowable vector for the accounting system $\mathrm{Im}(\theta)$. If $\mathbf{v} \in T$, then $\theta^*(\mathbf{v}) \in T'$ differs from $\theta_0^*(\mathbf{v})$ only through its $A' - \theta(A)$-entries, all of which are zero. If $\mathbf{b} \in B$, then by definition of a homomorphism $\theta^*(\mathbf{b})|_{\mathrm{Im}(\theta)} = \mathbf{b}'|_{\mathrm{Im}(\theta)}$ for some $\mathbf{b}' \in B'$. Again $\theta^*(\mathbf{b}) \in B'$ differs from $\theta_0^*(\mathbf{b})$ only in its $A' - \theta(A)$-entries, but these need not be zero.

One might expect $\mathrm{Im}(\theta)$ to be a subaccounting system of $\mathcal{A}'$, but this is only true with additional conditions.

(17.5.7). *Let $\theta : \mathcal{A} \to \mathcal{A}'$ be a homomorphism of accounting systems arising from $\theta : A \to A'$. Define $\theta_0 : A \to \theta(A)$ as above. Then $\mathrm{Im}(\theta)$ is a subaccounting system of $\mathcal{A}'$ if and only if the following conditions are satisfied.*
(i) If $\mathbf{v}' \in T' \cup B'$, then $\mathbf{v}'|_{\theta(A)}$ is a balance vector.
(ii) If $\mathbf{v}' \in T'$ and $\mathrm{sppt}(\mathbf{v}') \subseteq \theta(A)$, then $\mathbf{v}'|_{\theta(A)} \in \theta_0{}^*(T)$.
(iii) If $\mathbf{b}' \in B'$, then $\mathbf{b}'|_{\theta(A)} \in \theta_0{}^*(B)$.

This is true because the conditions in (17.5.7) are exactly what is required for the image to be a subsystem.

Isomorphism theorems

There is an interplay between homomorphisms and quotients in many branches of algebra, often culminating in so-called isomorphism theorems: for example for groups, rings and modules. We will record two isomorphism theorems for abstract accounting systems.

Let $\mathcal{A} = (A \mid T \mid B)$ and $\mathcal{A}' = (A' \mid T' \mid B')$ be accounting systems. A homomorphism $\theta : \mathcal{A} \to \mathcal{A}'$ is defined by a set map from A to A', which is also denoted by θ. There is a corresponding equivalence relation on A

$$E_\theta$$

that is determined by the function θ: thus $a_i E_\theta a_j$ if and only if $\theta(a_i) = \theta(a_j)$: for this see Exercise (1.3.9).

With this notation we can state a theorem which effectively identifies the image of a homomorphism of accounting systems with a quotient system.

(17.5.8). *The assignment $[a]_{E_\theta} \mapsto \theta(a)$ determines an isomorphism of accounting systems $\psi : \mathcal{A}/E_\theta \to \mathrm{Im}(\theta)$.*

Proof. Define $\theta_0 : A \to \theta(A)$ by $\theta_0(a_i) = \theta(a_i)$. Then

$$\mathrm{Im}(\theta) = \big(\theta(A) \mid \theta_0^*(T) \mid \theta_0^*(B)\big).$$

Consider the quotient system $\mathcal{A}/E_\theta = (\overline{A}|\,\overline{T}|\,\overline{B})$ where $\overline{A}$ is the set of E_θ-equivalence classes $[a_i]_{E_\theta}$. A function between account sets

$$\psi : \overline{A} \to \theta(A)$$

is defined by $\psi([a_i]_{E_\theta}) = \theta(a_i)$. The aim is to prove that ψ induces an isomorphism of accounting systems from $\mathcal{A}/E_\theta$ to $\mathrm{Im}(\theta)$. The first thing to note is that $\theta(a_i)$ depends only on the equivalence class $[a_i]_{E_\theta}$, not on a_i, so that ψ is a well-defined function. Next, if $\psi([a_i]_{E_\theta}) = \psi([a_j]_{E_\theta})$, then $\theta(a_i) = \theta(a_j)$ and hence $[a_i]_{E_\theta} = [a_j]_{E_\theta}$. Therefore ψ is injective: it is obviously surjective, so ψ is a bijection.

It remains to show that ψ induces a homomorphism of accounting systems $\psi : \mathcal{A}/E_\theta \to \mathrm{Im}(\theta)$, for which purpose it suffices to prove that $\psi^*(\overline{T}) = \theta_0^*(T)$ and $\psi^*(\overline{B}) = \theta_0^*(B)$. Let $\overline{\mathbf{v}} \in \overline{T} \cup \overline{B}$; then by definition of the quotient system $\mathcal{A}/E_\theta$, there exists $\mathbf{v} \in T \cup B$ such that $\overline{\mathbf{v}} = \sigma_{E_\theta}^*(\mathbf{v})$, where $\sigma_{E_\theta} : A \to \overline{A}$ is the canonical homomorphism in which $a_i \mapsto [a_i]_{E_\theta}$. Then we have

$$\overline{v}_i = \left(\sigma^*_{E_\theta}(\mathbf{v})\right)_i = \sum_{\sigma_{E_\theta}(a_j)=\sigma_{E_\theta}(a_i)} v_j = \sum_{\theta_0(a_j)=\theta_0(a_i)} v_j = \left(\theta_0^*(\mathbf{v})\right)_i.$$

In addition

$$\left(\psi^*(\overline{\mathbf{v}})\right)_i = \sum_{\psi([a_j])=\psi([a_i])} \overline{v}_j = \overline{v}_i$$

since ψ is injective. Therefore $\left(\psi^*(\overline{\mathbf{v}})\right)_i = \left(\theta_0^*(\mathbf{v})\right)_i$ for all i and hence

$$\psi^*(\overline{\mathbf{v}}) = \theta_0^*(\mathbf{v}).$$

It follows that $\psi^*(\overline{T}) \subseteq \theta_0^*(T)$ and $\psi^*(\overline{B}) \subseteq \theta_0^*(B)$. The equation $\psi^*(\overline{\mathbf{v}}) = \theta_0^*(\mathbf{v})$ holds for any $\mathbf{v} \in T \cup B$ with $\overline{\mathbf{v}} = \sigma_{E_\theta}^*(\mathbf{v})$, so we conclude that $\psi^*(\overline{T}) = \theta_0^*(T)$ and $\psi^*(\overline{B}) = \theta_0^*(B)$, which completes the proof. $\qquad\square$

Quotients of quotients

A further isomorphism theorem arises when one considers quotients of a quotient accounting system. Let $\mathcal{A} = (A|\,T|\,B)$ be an accounting system and let E be an equivalence relation on the account set A. Then $\mathcal{A}/E = (\overline{A}|\,\overline{T}|\,\overline{B})$ has as its account set $\overline{A}$, the set of all E-equivalence classes $[a_i]_E$. Now suppose that F is an equivalence relation on $\overline{A}$, so that it is possible to form the quotient of the quotient system $\mathcal{A}/E$ by F, i.e.,

$$(\mathcal{A}/E)/F.$$

The key to understanding this more complex object is the observation that E and F determine a new equivalence relation on A denoted by

$$E \# F,$$

where by definition $a_i\, (E \# F)\, a_j \;\leftrightarrow\; [a_i]_E\, F\, [a_j]_E$.

In terms of partitions, $E \# F$ arises by taking the partition of A determined by E and forming the union of all subsets in this partition that belong to same subset in the partition corresponding to F. This procedure leads to a partition with larger subsets than E which determines the equivalence relation $E \# F$. We can therefore form the quotient system

$$\mathcal{A}/(E \# F).$$

The connection with quotients of quotient systems is shown by the final theorem. This should be compared with the Third Isomorphism Theorems for groups and rings.

(17.5.9). *Let $\mathcal{A} = (A|\, T|\, B)$ be an accounting system. Suppose that $\mathcal{A}/E$ is a quotient of $\mathcal{A}$ and $(\mathcal{A}/E)/F$ a quotient of $\mathcal{A}/E$. Then*

$$(\mathcal{A}/E)/F \simeq \mathcal{A}/(E \# F).$$

This result is intuitively reasonable, but it will not be proved here. For a detailed proof together with an example see [3].

The basic model of an accounting system described here can be extended to a 10-tuple by introducing further algebraic concepts. The extended model incorporates additional features of real life systems such as authorization of transactions, control of accounts, ability to generate reports on units of the company. For details the reader is referred to [3].

Exercises (17.5).

(1) Let R be an integral domain. Prove that R can be linearly ordered if and only if it has a positive subset.

(2) Prove that a linearly ordered domain has characteristic zero.

(3) Prove that every transaction on an accounting system with n accounts can be expressed as a composite of $n-1$ simple transactions. [Hint: use (17.5.1) and (17.5.3).]

(4) Let $\mathcal{A} = (A|T|B)$ be an accounting system over an ordered domain R. Let E be an equivalence relation on A. Recall that $\bar{\mathbf{v}}_E$ is defined by $(\bar{\mathbf{v}}_E)_i = \sum_{a_j E a_i} v_j$. Prove that $\overline{\mathbf{u}_E + \mathbf{v}_E} = \bar{\mathbf{u}}_E + \bar{\mathbf{v}}_E$ and $\overline{r\mathbf{v}_E} = r\bar{\mathbf{v}}_E$, where $\mathbf{u}$ and $\mathbf{v}$ are balance vectors over R and $r \in R$.

(5) Let $\theta : \mathcal{A} \to \mathcal{A}'$ be an isomorphism of accounting systems with account sets $A = \{a_1, \ldots, a_n\}$ and $A' = \{a_1', \ldots, a_n'\}$ respectively. Prove that there is a permutation $\pi \in S_n$ such that $\theta(a_i) = a_{\pi(i)}'$ for $1 \le i \le n$. Then deduce that there is an inverse isomorphism from $\mathcal{A}'$ to $\mathcal{A}$ induced by θ^{-1}.

(6) Let $\theta : \mathcal{A} \to \mathcal{A}'$ and $\phi : \mathcal{A}' \to \mathcal{A}''$ be homomorphisms of unbounded accounting systems. (i) Prove that $(\phi\theta)^* = \phi^*\theta^*$. (ii) Deduce that the composite $\phi\theta : \mathcal{A} \to \mathcal{A}''$ is also a homomorphism. (iii) Conclude that the unbounded accounting systems form a category with homomorphisms as morphisms. [Hint: to prove (i) use the definition of θ^* and ϕ^* to compute $\phi^*(\theta^*(\mathbf{v}))$.]

Bibliography

[1]   I. T. Adamson. Introduction to Field Theory. Cambridge University Press, Cambridge, 2nd ed. 1982.

[2]   P. J. Cameron. Combinatorics. Cambridge University Press, Cambridge, 1994.

[3]   S. Cruz Rambaud, J. Garcia Perez, R. A. Nehmer, D. J. S. Robinson. Algebraic models for accounting systems. World Scientific, Singapore 2010.

[4]   K. Doerk and T. O. Hawkes. Finite Soluble Groups. De Gruyter, Berlin, 1992.

[5]   L. Fuchs. Infinite Abelian Groups. 2 vols. Academic Press, New York, 1970-3.

[6]   D. Goldrei. Classic Set Theory. Chapman and Hall, London, 1996.

[7]   P. R. Halmos. Naive Set Theory. Undergrad. Texts in Mathematics, Springer-Verlag, New York, 1974.

[8]   T. W. Hungerford. Algebra. Vol. 73, Graduate Texts in Mathematics, Springer-Verlag, New York, 2003.

[9]   I. M. Isaacs. Character Theory of Finite Groups. Dover, New York, 1994.

[10]   G. J. Janusz. Algebraic Number Fields. Vol. 7, Graduate Studies in Mathematics, Amer. Math. Soc., Providence, RI, 2nd ed., 1996.

[11]   R. C. Lyndon and P. E. Schupp. Combinatorial group theory. Springer-Verlag, Berlin, 1977.

[12]   S. Mac Lane. Categories for the working mathematician. Vol. 5, Graduate Texts in Mathematics, Springer-Verlag, New York, 1971.

[13]   I. Niven. Irrational Numbers. Vol. 11, Carus Math. Monographs, Wiley, New York, 1956.

[14]   B. C. Pierce. Basic category theory for computer scientists. Foundations of computing series. The MIT Press, Cambridge, MA, 1991.

[15]   D. J. S. Robinson. A Course in the Theory of Groups. Vol. 80, Graduate Texts in Mathematics, Springer-Verlag, New York, 2nd ed., 1996.

[16]   D. J. S. Robinson. A Course in Linear Algebra with Applications. World Scientific, Singapore, 2nd ed., 2006.

[17]   J. J. Rotman. An Introduction to the Theory of Groups. Vol. 148, Graduate Texts in Mathematics, Springer-Verlag, New York, 4th ed., 1995.

[18]   J. J. Rotman. Galois Theory. Springer-Verlag, New York, 2nd ed., 1998.

[19]   J. J. Rotman. Advanced Modern Algebra. Vol. 114, Graduate Studies in Mathematics, Amer. Math. Soc., Providence, RI, 2nd ed., 2010.

[20]   J.-P. Serre. Linear representations of finite groups. Springer-Verlag, New York, 1977.

[21]   B. van der Waerden. A History of Algebra. Springer-Verlag, Berlin, 1985.

[22]   J. H. van Lint. Introduction to Coding Theory. Vol. 86, Graduate Texts in Mathematics, Springer-Verlag, Berlin, 3rd ed., 1999.

https://doi.org/10.1515/9783110691160-018

Index